Modern Chromatographic Analysis of the Vitamins

CHROMATOGRAPHIC SCIENCE

A Series of Monographs

Editor: JACK CAZES
Fairfield, Connecticut

Volume 1: Dynamics of Chromatography (out of print)
J. Calvin Giddings

Volume 2: Gas Chromatographic Analysis of Drugs and Pesticides
Benjamin J. Gudzinowicz

Volume 3: Principles of Adsorption Chromatography: The Separation of Nonionic Organic Compounds (out of print)
Lloyd R. Snyder

Volume 4: Multicomponent Chromatography: Theory of Interference (out of print)
Friedrich Helfferich and Gerhard Klein

Volume 5: Quantitative Analysis by Gas Chromatography
Joseph Novák

Volume 6: High-Speed Liquid Chromatography
Peter M. Rajcsanyi and Elisabeth Rajcsanyi

Volume 7: Fundamentals of Integrated GC-MS (in three parts)
Benjamin J. Gudzinowicz, Michael J. Gudzinowicz, and Horace F. Martin

Volume 8: Liquid Chromatography of Polymers and Related Materials
Jack Cazes

Volume 9: GLC and HPLC Determination of Therapeutic Agents (in three parts)
Part 1 edited by Kiyoshi Tsuji and Walter Morozowich
Part 2 and 3 edited by Kiyoshi Tsuji

Volume 10: Biological/Biomedical Applications of Liquid Chromatography
Edited by Gerald L. Hawk

Volume 11: Chromatography in Petroleum Analysis
Edited by Klaus H. Altgelt and T. H. Gouw

Volume 12: Biological/Biomedical Applications of Liquid Chromatography II
Edited by Gerald L. Hawk

Volume 13: Liquid Chromatography of Polymers and Related Materials II
Edited by Jack Cazes and Xavier Delamare

Volume 14: Introduction to Analytical Gas Chromatography: History, Principles, and Practice
John A. Perry

Volume 15: Applications of Glass Capillary Gas Chromatography
Edited by Walter G. Jennings

Volume 16: Steroid Analysis by HPLC: Recent Applications
Edited by Marie P. Kautsky

Volume 17: Thin-Layer Chromatography: Techniques and Applications
Bernard Fried and Joseph Sherma

Volume 18: Biological/Biomedical Applications of Liquid Chromatography III
Edited by Gerald L. Hawk

Volume 19: Liquid Chromatography of Polymers and Related Materials III
Edited by Jack Cazes

Volume 20: Biological/Biomedical Applications of Liquid Chromatography IV
Edited by Gerald L. Hawk

Volume 21: Chromatographic Separation and Extraction with Foamed Plastics and Rubbers
G. J. Moody and J. D. R. Thomas

Volume 22: Analytical Pyrolysis: A Comprehensive Guide
William J. Irwin

Volume 23: Liquid Chromatography Detectors
Edited by Thomas M. Vickrey

Volume 24: High-Performance Liquid Chromatography in Forensic Chemistry
Edited by Ira S. Lurie and John D. Wittwer, Jr.

Volume 25: Steric Exclusion Liquid Chromatography of Polymers
Edited by Josef Janča

Volume 26: HPLC Analysis of Biological Compounds: A Laboratory Guide
William S. Hancock and James T. Sparrow

Volume 27: Affinity Chromatography: Template Chromatography of Nucleic Acids and Proteins
Herbert Schott

Volume 28: HPLC in Nucleic Acid Research: Methods and Applications
Edited by Phyllis R. Brown

Volume 29: Pyrolysis and GC in Polymer Analysis
Edited by S. A. Liebman and E. J. Levy

Volume 30: Modern Chromatographic Analysis of the Vitamins
Edited by André P. De Leenheer, Willy E. Lambert, and Marcel G. M. De Ruyter

Volume 31: Ion-Pair Chromatography: Theory and Biological and Pharmaceutical Applications
Edited by Milton T. W. Hearn

Volume 32: Therapeutic Drug Monitoring and Toxicology by Liquid Chromatography
Edited by Steven H. Y. Wong

Other Volumes in Preparation

PREFACE

The expansion of information about vitamins and their analogs, and the advent of newer techniques for the determination of the vitamins in different matrices, are reflected in this work.

The first part of the book covers the fat-soluble vitamins A, D, E, and K; the second part covers the water-soluble vitamins and coenzyme forms of ascorbate, folate, nicotinate, thiamine, flavins, pyridoxin, cyanocobalamin, and biotin. For each vitamin or vitamin group, a detailed overview is given of the analytical techniques for the quantitative analysis of the vitamin in different matrices. Special attention is given to the chromatographic methods that allow the separation and the selective detection of the different forms (or metabolites) of a vitamin. In addition, information of metabolism and biochemical functions of the vitamins is also included.

The book is designed primarily for those who have acquaintance with the different chromatographic techniques, such as GC, GC-MS, HPLC, and TLC. However, useful information for the beginner in this field is also presented in the set-up of a new analytical system for the quantitative determination of a vitamin.

This work could not have been presented without the contributions and cooperation of the numerous authors and we are very grateful for their efforts.

André P. De Leenheer
Willy E. Lambert
Marcel G. M. De Ruyter

CONTENTS

CONTRIBUTORS

Johannes Abels Institute of Hematology, Erasmus University, Rotterdam, The Netherlands

Stella G. Brownlee Nutrient Composition Laboratory, Beltsville Human Nutrition Research Center, U.S. Department of Agriculture, Beltsville, Maryland

Mai H. Bui-Nguyén Swiss Federal Dairy Research Institute, Swiss Institute of Vitamins, Lausanne, Switzerland

Albert E. Claeys Laboratorium voor Analytische Chemie, Rijksuniversiteit Gent, Gent, Belgium

Mariann E. Cortissoz Nutrient Composition Laboratory, Beltsville Human Nutrition Research Center, U.S. Department of Agriculture, Beltsville, Maryland

Veerle O. R. C. De Bevere Laboratoria voor Medische Biochemie en vor Klinische Analyse, Rijksuniversiteit Gent, Gent, Belgium

André P. De Leenheer Laboratoria voor Medische Biochemie en voor Klinische Analyse, Rijksuniversiteit Gent, Gent, Belgium

Marcel G. M. De Ruyter * Laboratoria voor Medische Biochemie en voor Klinische Analyse, Rijksuniversiteit Gent, Gent, Belgium

Present affiliation: Psychiatric Center St. Jozef, Munsterbilzen, Belgium

Jan X. de Vries Medical Clinic, University of Heidelberg, Federal Republic of Germany

François Frappier Laboratorie de Chemie Organique Biologique, Pierre and Marie Curie University, Paris, France

Michel Gaudry Laboratorie de Chemie Organique Biologique, Pierre and Marie Curie University, Paris, France

Nikolaus Hengen Human Pharmacology Institute, Ciba-Geigy GmbH, Tübingen, Federal Republic of Germany

Glenville Jones* Research Institute, The Hospital for Sick Children, Toronto, Ontario, Canada

Takashi Kawasaki Department of Biochemistry, Hiroshima University School of Medicine, Hiroshima, Japan

Willy E. Lambert Laboratoria voor Medische Biochemie en voor Klinische Analyse, Rijksuniversiteit Gent, Gent, Belgium

Marc F. L. Lefevere Laboratorium voor Analytische Chemie, Rijksuniversiteit Gent, Gent, Belgium

David R. Light† Department of Chemistry, Massachusetts Institute of Technology, Cambridge, Massachusetts

Jan Lindemans Institute of Hematology, Erasmus University, Rotterdam, The Netherlands

John J. McCormack Vermont Regional Cancer Center and Department of Pharmacology, College of Medicine, University of Vermont, Burlington, Vermont

Catherine E. Maire Nutrient Composition Laboratory, Beltsville Human Nutrition Research Center, U.S. Department of Agriculture, Beltsville, Maryland

H. L. J. Makin Department of Chemical Pathology, The London Hospital Medical College, London, England

Michael A. Marletta Department of Nutrition and Food Science, Massachusetts Institute of Technology, Cambridge, Massachusetts

Hans J. Nelis Laboratoria voor Medische Biochemie en voor Klinische Analyse, Rijksuniversiteit Gent, Gent, Belgium

Robert A. Newman Vermont Regional Cancer Center and Department of Pharmacology, College of Medicine, University of Vermont, Burlington, Vermont

---

Present affiliation:
*Departments of Biochemistry and Medicine, Queen's University, Kingston, Ontario, Canada
†Biocatalysis Department, Genentech Inc., San Francisco, California

Hiroshi Sanemori Department of Biochemistry, Hiroshima University School of Medicine, Hiroshima, Japan

D. A. Seamark Department of Chemical Pathology, The London Hospital Medical College, London, England

D. J. H. Trafford Department of Chemical Pathology, The London Hospital Medical College, London, England

Joseph T. Vanderslice Nutrient Composition Laboratory, Beltsville Human Nutrition Research Center, U.S. Department of Agriculture, Beltsville, Maryland

Modern Chromatographic Analysis of the Vitamins

1

VITAMIN A: RETINOL, CAROTENOIDS, AND RELATED COMPOUNDS

WILLY E. LAMBERT, HANS J. NELIS, MARCEL G. M. DE RUYTER, and ANDRÉ P. DE LEENHEER*

Laboratoria voor Medische Biochemie en voor Klinische Analyse, Rijksuniversiteit Gent, Gent, Belgium

**Present affiliation:* Psychiatric Center St. Jozef, Munsterbilzen, Belgium

PART A: RETINOL AND RELATED COMPOUNDS

1. INTRODUCTION

1.1 Chemical Properties

The most recent nomenclature changes recommended by the International Union of Pure and Applied Chemistry [1] prescribe the term *retinol* and variations of it for various vitamin A derivatives. On this basis all derivatives and isomers of vitamin A have been renamed; e.g., retinene or vitamin A aldehyde becomes retinal, vitamin A acid becomes retinoic acid, and vitamin A_2 aldehyde becomes 3-dehydroretinal. This new nomenclature will be followed further in this chapter. The term vitamin A is still widely used and includes all products structurally or biologically related to retinol.

The basic structure is retinol, or 9,13-dimethyl-7-(1,1,5-trimethyl-6-cyclohexene-5-yl)-7,9,11,13-nonatetraene-15-ol.

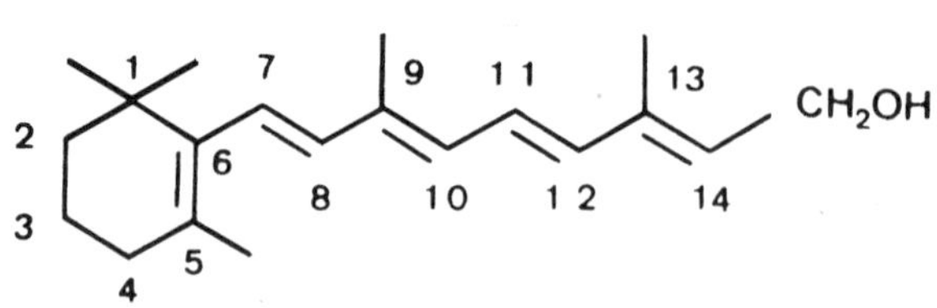

The most common parent compounds, together with some chemical properties, are given in Table 1.1. In the latter only data useful in the detection of compounds after a chromatographic separation step are summarized.

To specify the exact structure of a vitamin A compound unambiguously, the isomeric form must also be defined. The four double bonds in the side chain theoretically permit 16 geometric isomers ($N = 2^n$, N = number of possible isomers, n = number of exocyclic double bonds). Only three isomers are "unhindered," i.e., 9-*cis*, 13-*cis*, and 9,13-di-*cis*, whereas the other isomeric forms are "hindered," e.g., 11-*cis* and 11,13-di-*cis*.

The formation of isomers from all-trans is catalyzed by light, iodine, and higher temperatures; the relative amount of the different isomers is dependent on the wavelength and on the solvent used. Consequently it is imperative that special precautions be taken when working with vitamin A compounds, e.g., subdued light and absence of drastic reagents. These problems will be encountered during each step of vitamin A analysis and make research in the vitamin A field a real analytical challenge. Since it is not the scope of this chapter to provide complete information on synthetic chemistry and on chemical properties of vitamin A we therefore refer to some excellent reviews by Kofler and Rubin [6], Olson [7], Isler et al. [8,9], and Schwieter and Isler [10].

1.2 Biochemistry

In the beginning of the twentieth century a large number of reports mention the existence of a lipid-soluble compound in certain foods (e.g., egg yolk and cod-liver oil). This compound was essential for life, restored and maintained growth, and prevented xerophthalmia and night blindness in rats kept on a diet without this compound [11-14]. This compound was demonstrated to be vitamin A. During the last decades knowledge of the intestinal absorption, storage, plasma transport, and metabolism of vitamin A has expanded greatly. However, little is still known about its general mode of action, except for its role in the visual process.

1.2.1 Absorption and Storage

Although vitamin A is found especially in liver, milk, butter, and egg yolk, carrots are the main source of β-carotene (provitamin A). The dietary intake mainly consists of vegetable carotenoids (β-carotene) and long-chain retinyl esters from animal sources.

In the intestinal mucosa β-carotene is first cleaved to two molecules of retinal by β-carotene-15,15'-dioxygenase [15] and consecutively the retinal is reduced to retinol by retinaldehyde reductase [16,17]. In the intestine, the dietary esters (mainly retinyl palmitate) are hydrolyzed and the resulting retinol is absorbed in the mucosa cells [18].

TABLE 1.1 UV[a] and Fluorescence[b] Properties of Vitamin A-Related Compounds

Compound	Solvent	λ_{max} (nm)	$\varepsilon \times 10^{-3}$ (liters/mol. cm)	Solvent	λ_{ex} (nm)	λ_{em} (nm)	Relative fluorescence, intensity at 25°C
Retinol (A_1)							
All-*trans*	Ethanol	324	51.6				
	Petroleum ether	324	52.4	Cyclohexane	330	478	0.71
				Dioxane	334	476	1.00
13-*cis*	Ethanol	328	48.3	Ethanol	325	480	0.36
11-*cis*	Ethanol	319	39.4				
11,13-di-*cis*	Ethanol	311	26.0				
9-*cis*	Ethanol	323	42.3				
9,13-di-*cis*	Ethanol	324	39.5				
Retinal (A_1)							
All-*trans*	Ethanol	381	43.4	Ether-isopentane-ethanol (5:5:2)	420	500	—
13-*cis*	Ethanol	375	35.7				
		257	9.5				
11-*cis*	Ethanol	377	24.9				
		290	9.2				
		254	17.4				
11,13-di-*cis*	Ethanol	377	19.9				
9-*cis*	Ethanol	373	36.1				
9,13-di-*cis*	Ethanol	368	32.4				
Retinoic acid (A_1)							
All-*trans*	Ethanol	352	44.4	Ethanol	334	484	0.03
				Dioxane	348	472	0.008
13-*cis*	Ethanol	359	38.3				
11,13-di-*cis*	Ethanol	346	—				
9-*cis*	Ethanol	348	35.5				
9-13-di-*cis*	Ethanol	352	33.9				

[a]Values are by Cama et al. [2] and by Boldingh et al. [3].
[b]By Olson and Pungapongse [4] and Balke and Becker [5].

Retinol obtained from the esters or synthesized from β-carotene is reesterified in the mucosa cells (mainly with palmitic acid), and in association with lymph chylomicrons the esters enter the bloodstream via the thoracic duct and the vena cava cranialis [18]. Only a small fraction of the esters (10%) is transported directly through the vena porta to the liver [19]. In the liver the chylomicrons containing retinyl esters are hydrolyzed and reesterified, mainly with palmitic acid, and stored associated with lipid droplets [20]. Free retinol, mobilized from these stores after hydrolysis, is transported in plasma associated with a specific transport protein (RBP, retinol binding protein) and with prealbumin. Excellent reviews on this protein complex were given by Goodman in 1976 and 1979 [21,22].

1.2.2 Metabolism

In the liver retinol is oxidized to retinal and to retinoic acid. Whereas retinol and retinal are metabolically interconvertible in the body, the oxidation to retinoic acid is irreversible, and retinoic acid cannot be stored in the body. It is further metabolized by conjugation with glucuronic acid and excreted in the urine and in the bile [23-25].

In tissues (e.g., testes, mucosa, and liver) the same oxidative metabolism takes place and retinoic acid is converted to more polar compounds, such as 5,6-epoxyretinoic acid [26,27], 4-ketoretinoic acid, and 4-hydroxyretinoic acid [28-30]. These derivatives also are glucuronoconjugated and appear in the bile [31]. All these experiments were performed in rats or hamsters, after administration of a vitamin A compound. Until now, only a minimal insight has been achieved into the pathways involved in the formation of these metabolites, their interrelationships and their biological activity. However, recently Zile et al. [32] and Roberts et al. [29] postulated that the further oxidation and conjugation of retinoic acid is a pathway in the deactivation of retinoic acid.

1.2.3 Biological Activity

The term vitamin A is used for compounds that exhibit the biological activity of retinol. Generally, vitamin A is necessary for growth support and health in higher animals. In its absence growth of these animals ceases, and after a time they die. Retinol is essential for vision and is capable of preventing xerophthalmia and blindness [33]. Furthermore, it has a role in spermatogenesis and in the maintenance of differentiated epithelia. Excellent reviews on this subject are given by Moore [34], McLaren [35], and Harris [36]. Retinoic acid cannot be reduced to retinal and retinol; consequently, rats maintained on retinoic acid instead of retinol grow normally but become blind. However, retinoic acid can maintain the other functions of retinol. This leads to two possibilities: first, retinoic acid is the active metabolite of retinol;

second, both retinol and retinoic acid are metabolized to a common active form. More recently, interest in retinoic acid increased as it was demonstrated to inhibit both the incidence and the severity of chemically induced tumors [37,38], without the toxic side effects due to storage in the liver. With these results in mind, a number of retinoic acid analogs, retinoids, have been tested for their antineoplastic activity [39-41] and in dermatology [42,43].

1.3 Vitamin A Assays

1.3.1 Food and Drug Analysis

Vitamin A and β-carotene are often added to food (e.g., butter and cheese) to ensure adequate dietary intake. During preparation and storage of foods, the vitamin A is extremely susceptible to oxidation and isomerization; therefore, control of the vitamin A level is necessary.

1.3.2 Chromatography

In order to obtain better specificity and higher sensitivity in the analysis of biological fluids, food, and drugs, a chromatographic separation step is often necessary prior to the end measurement. Different techniques are available. One of them is thin-layer chromatography (TLC), which today can give acceptable precision as automatic application and scanning in situ have become available. Gas chromatography is the most traditional separation technique; however, incomplete derivatization and degradation at higher temperatures are drawbacks encountered in the work with vitamin A compounds. On the other hand, liquid chromatography (LC) allows separation at room temperature and in most cases does not need a derivatization before the detection. Therefore, the list of applications of LC has become increasingly impressive during the last few years, and clearly demonstrates the undeniable value of chromatography in regard to the more classic colorimetric determinations.

1.3.3 Sample Preparation

Sample preparation prior to the final chromatographic analysis is highly dependent upon the nature of the sample matrix. A minimal preparation is necessary for the analysis of irradiated concentrated standard solutions. As a biological matrix is more complex, often sophisticated sample preparations are necessary. In addition, to avoid isomerization and oxidation, all manipulations must be carried out in subdued light in amberized glassware and in the absence of drastic reagents.

1.3.3.1 Retinyl Esters and Retinol. Frequently, gel permeation chromatography (Styragel) [44,45] and column chromatography (Na_2SO_4, NaCl) [46] are applied to the quantitative determination of retinyl palmitate in margarines and oils. In food and cereals, total vitamin A content is measured after hydrolysis of the retinyl esters to retinol. The sample is then treated with ethanolic KOH at 60-80°C for 20-45 min.

Pyrogallol, hydroquinone [47], and butylated hydroxytoluene [46] are the most common antioxidants used during this manipulation. After saponification the free retinal is extracted into *n*-hexane [48].

In the analysis of serum, retinol is liberated from its binding protein by denaturation with acetonitrile [49], ethanol, or methanol [50]. Finally, extraction into one of the common solvents for fat-soluble compounds (e.g., *n*-hexane, or a mixture of petroleum ether, dichloromethane, and isopropanol) [51] is followed by evaporation and a chromatographic separation.

1.3.3.2 Retinal. Retinal isomers play an important role in vision. To have an exact idea about the different isomeric forms of retinal in the eye no isomerization is allowed during sample manipulation.

The use of a specific detergent (Ammonyx LO) [52] is described to keep the retinal isomers in their initial isomeric form; however, very poor extraction recoveries into dichloromethane were obtained by this procedure. Another approach is the use of an excess of hydroxylamine under denaturating conditions; the chromophore of retinal bearing natural products is converted into the corresponding retinaloxime with complete retention of the original cis and trans configurations. These retinaloximes can be quantitatively extracted with dichloromethane and analyzed by high-pressure liquid chromatography (HPLC) [53].

1.3.3.3 Retinoic Acid. In the search for retinoic acid in biological fluids, different extraction procedures have been developed. As retinoic acid contains a carboxylic function, a simple extraction of the acidified sample with diethyl ether [54], or a *n*-hexane-methylene chloride-isopropanol mixture [55] seems sufficient. However, preextraction and removal of possibly interfering compounds is recommended, as repeated injections of dirty samples are disastrous for an HPLC column [56].

2. THIN-LAYER CHROMATOGRAPHY

2.1 Sample Handling

Oxidative degradation and isomerization of the vitamin A compounds always hamper analysis in this field. Therefore, especially in thin-layer chromatography, a number of precautions must be considered. Indeed, in addition to the classic rules described by Stahl in 1969 [57] (e.g., constant temperature, saturated chambers, freshly prepared eluents, and plates without drafts), the sample application and chromatography must be performed under an N_2 or argon atmosphere and in subdued light to prevent isomerization and oxidation.

2.2 Stationary Phases

Most of the TLC work in vitamin A research is performed on silica gel plates. The silica can be activated by heating to a temperature of 120°C for a period of time (30 min to 1 hr). Pretreatment of the plates by spraying with a solution of an antioxidant has been tried for prevention of degradation on the plates. Also, reversed-phase TLC was tested and could not prevent formation of decomposition products. Maybe shorter developing times, as obtained in HPTLC, afford a satisfactory solution to this problem.

2.3 Mobile Phases

Nearly all the solvent systems used contain a high percentage of an apolar solvent (petroleum ether, cyclohexane, or *n*-hexane) and a small amount of a more polar solvent (diethyl ether, acetone, or methylheptenone). In the reversed-phase mode, a polar solvent, based on methanol, is used for development.

2.4 Detection

After separation, compounds can be detected based upon their chemical properties: characteristic fluorescence under ultraviolet (UV) light or intense color of the spot on the plate, typical color development when sprayed with a reagent, Carr-Price, or a modified version with trifluoroacetic acid or trichloroacetic acid [58]. Finally, after visualization, compounds can be eluted or scraped off followed by spectrophotometric measurement. Detection in thin-layer radiochromatography is performed by an x-ray film by scanning or by counting the radioactivity after the spot has been scraped off into a counting vial.

2.5 Applications

2.5.1 Isomers and Degradation Products

Partial separation of retinol isomers has been reported by Planta et al. [59] using petroleum ether and methylheptenone (11:2). However, no clear resolution of 9-*cis* and all-trans and of 13-*cis* and 9,13-di-*cis* isomers was obtained. In the same system, 3-dehydroisomers (A_2) always move slower than the corresponding isomers of retinol. Furthermore, all-trans forms always move slowest. The same observations were made by Herting (personal communication), who separated 9-*cis* from all-*trans*-retinal using *n*-hexane-diethyl ether-acetic acid (70:30:1); however, again, 13-*cis* and 9,13-di-*cis* retinal remained unresolved.

In the analysis of retinal isomers, the methoxime derivatization produces syn and anti derivatives [60] of each isomer and facilitates their identification [61,62]. Syn and anti retinal methoximes are separated by a mixture of petroleum ether and acetone (95:5) and the corresponding isomers are separated with cyclohexane-toluene-ethyl acetate (50:50:20).

Preparative TLC was applied to the separation of retinyl acetate degradation products using a mobile phase of hexane-diethyl ether in ratios of 95:5-10:90. Under these circumstances, Parizkova and Blattna [63] separated 14 substances; however, no identification is given by these authors.

Thin-layer radiochromatography is still the method of choice for a check of radiochemical purity, since in other methods, such as LC, some impurities can remain undetectable on top of the column. Solvent systems used were heptane-chloroform-methanol (8:1:1) and benzene-chloroform-methanol (4:1:1). In a typical experiment detection was performed by exposing the TLC plate to an x-ray film; thus the radiation produces an image proportional in intensity to the amount of radioactivity present [64].

In 1964, Varma et al. [65] described the separation of retinol, retinal, retinoic acid, and retinyl acetate standards. However, different mixtures of cyclohexane and increasing amounts of either methanol or ethanol were necessary. On the other hand Fung et al. [66] clearly separated all those compounds in a single run with acetone-light petroleum ether (18:82).

2.5.2 Retinoids in Biological Samples

TLC fractionation and quantitation of β-carotene, retinol, retinal, retinyl esters, and retinoic acid in human serum was done by Targan et al. [67] with benzene-ethyl ether (90:10). TLC was also applied for the search of possible vitamin A metabolites [68,69]. This work was mainly done by thin-layer radiochromatography and contributed a great deal to the identification of the β-glucuronides of retinol and retinoic acid, since a glucuronidase test on these "unknowns" resulted in products that cochromatographed with retinol and retinoic acid standards. For this purpose, solvent systems were used containing a high percentage of more polar solvents, such as methanol and diethyl ether [70,71].

2.5.3 Vitamin A and Other Vitamins

TLC separation of vitamin A (palmitate) and vitamin D_3 and E in pharmaceutical preparations is described by Baczyk et al. [72]. Pure chloroform was used as solvent, whereas after development the spots were located under UV light, scraped off, and determined spectrophotometrically.

2.5.4 Conclusions

Although thin-layer chromatography remains a convenient technique in terms of its low costs and simplicity, several points must be considered. First of all, vitamin A compounds are readily oxidized when exposed to air and are very light sensitive. In TLC the contact to air can be extremely high and produces artefacts on the plate and obviously inhibits quantitative work. Furthermore, some of the vitamin A compounds are present in serum only at an extremely low level (e.g., retinoic acid); therefore, product loss really is not permitted. The more modern TLC techniques, such as HPTLC and circular TLC on HPTLC plates in hermetically sealed chambers, where conditions can be more easily controlled, could be better adapted than classic TLC for vitamin A analysis.

3. GAS-LIQUID CHROMATOGRAPHY

3.1 General Considerations

Due to their thermal lability, vitamin A compounds are not very well suited for gas-liquid chromatographic (GLC) analysis. Derivatization is imperative to prevent the formation of anhydroretinol, which is enhanced by temperatures above 150°C and by exposure to metal columns [73]. Several attempts have been made to minimize the degradation of the compounds, such as carotene-treated supports [74], low temperatures, and high carrier gas flow rates [75]. Catalytic perhydrogenation was examined to form the stable perhydrovitamin A [76]. However, under these hydrogenation conditions two products are formed. One of the products is the same whether the starting compound is retinol or retinyl acetate. The second product is the saturated alcohol or the saturated ester, depending upon when the starting compound was in the alcohol or acetate form. To the derivatization and the gas chromatographic conditions special attention is paid in the following section.

3.2 Derivatizations

Some essential characteristics of derivatives and derivatization procedures are listed below.

1. The derivatives must be easily formed in order to fit a routine analytical procedure.
2. The derivatization must be reproducible and linear over a certain range, even in the presence of endogenous material.
3. The derivatives should be stable over time, temperature, and the GLC conditions, and nonreactive with column tubing or support material.

4. Not imperative, but preferable are high yields and the formation of only one derivative.

Without derivatization, GLC of retinol results in broad, tailing peaks of various degradation products [77,78]. According to the various functional groups of retinol, retinal, and retinoic acid, different types of derivatization have been proposed. Table 1.2 lists these methods for derivatization.

3.2.1 Retinol

The primary alcohol group of retinol can be transformed to an ether either by silylation or alkylation. Derivatization to esters is not possible as these tend to degrade under GC conditions to anhydroretinol [74].

Vecchi et al. [79,80,80a] described the formation of trimethylsilyl derivatives. First they tried the silylating mixture as described by Sweeley et al. [81], however, the use of *N*,*O*-bis(trimethylsilyl)acetamide (BSA) in pyridine resulted in a more quantitative reaction. These trimethylsilyl (TMS) derivatives are chromatographed at 170°C on aged 7% silicone DC-QF-1 columns. Freshly prepared columns are preconditioned during several days by repeated injections of BSA. In this manner, all-*trans* and 13-*cis*-retinol can easily be separated. However, for 9-*cis* and 13-*cis* isomers and 11-*cis* and 11,13-di-*cis* isomers, the method is not successful.

In terms of short analysis time, peak asymmetry, and plate height, QF-1 and SP 2401 stationary phases are best suited for chromatography of the TMS derivatives of retinol. Although the chromatography was acceptable, this method can hardly be used for routine analysis, as highly purified samples are necessary, which would result in a time-consuming sample treatment (unpublished results by the authors).

Formation of the methyl ether of retinol, either obtained by treating the lithium derivative with dimethylsulfate [75] or by reaction with CH_3I and DMSO [82], resulted in several artefacts on chromatography on a "fresh" 1% SE-30 column at 150°C using argon flow rate of 150 ml/min. However, on an aged column, methyl retinyl ether can also be chromatographed without appreciable destruction [74].

3.2.2 Retinal

For retinal, methoxime derivatives were obtained by treatment with a saturated solution of methoxylamine hydrochloride in pyridine for 15 min at 60°C under N_2 atmosphere. Gas chromatography was performed on a 1.80 m × 0.2 cm column packed with 3% QF-1 on acid-washed and hexamethyldichlorosilane-treated Chromosorb W. N_2 was used as carrier gas at a linear velocity of 10 cm/sec; oven temperature was 175°C and injector and detector temperatures were 190°C. The chromatogram

TABLE 1.2 Derivatization of Vitamin A Compounds for Gas Chromatographic Analysis

Compound	Method	Reaction product	Reference
Retinol	Catalytic hydrogenation	Perhydrovitamin A	76
Retinol	*N*,*O*-bis(trimethylsilyl)acetamide in pyridine	Trimethylsilyl ether	80
Retinol	Dimethylsulfate treatment of Li derivative of retinol	Methyl retinyl ether	75
Retinol	Methyliodide and DMSO and K-*t*-butoxide	Methyl retinyl ether	82
Retinal	Methoxylamine HCl in pyridine for 15 min at 60°C	Syn- and anti-retinal methoxime	60
Retinoic acid	Diazomethane in ether	Methylretinoate	56,75
Retinoic acid	Dimethylformamide dimethylacetal	Methylretinoate	85
Retinoic acid	Diazomethane in ether-hexamethyl disilazane + trimethylchlorosilane in pyridine	Methylretinoate-TMS ether	84

revealed two components identified as syn and anti derivatives. Although the method can be useful to identify retinal, it can hardly be used for quantitative purposes [60].

3.2.3 Retinoic Acid

The most popular derivatives of organic acids are obviously esters. Both TMS and methyl esters of retinoic acid can be formed. In our laboratory we could not perform TMS derivatization with acceptable reproducibility. Also, the linearity was not so good. Different reagents can be proposed for methylation, such as diazomethane, anhydrous HCl-methanol, anhydrous methanol-BF_3, and dimethyl formamide and dimethylacetal. In view of the lability of vitamin A compounds the mildest procedure should be used. Retinoic acid can be treated [56,75] with an ethereal solution of diazomethane. As the reaction is instantaneous, after a short time the ether can be evaporated. The methylretinoate obtained is the most stable derivative of all vitamin A compounds and has excellent gas chromatographic properties. Retinoic acid is completely methylated by this procedure with no apparent cis/trans isomerization or other chemical transformation. In our study the QF-1 phase gives the best results for GC of methylretinoate regarding both peak symmetry and plate height.

3.3 Without Derivatization

By GLC on a column containing 1% SE-30 coated on Gas Chrom P at 150°C and a flow rate of 150 ml/min, retinol and retinyl acetate are completely converted to anhydroretinol. However, pretreatment of the column with β-carotene reduced dehydration and deacetylation to 30-50%. By increasing the flow rate to 880 ml/min only 6% by-product was formed. The protective effect of β-carotene lasts for 2-3 days and is based on its reaction with the active sites of the column [75]. Also, treatment with hydroquinone significantly reduces artefact formation; unfortunately, the protection lasts only for a few hours due to the sublimation of this product at high temperatures.

3.4 Application to Biological Samples

Gas chromatography-mass spectrometry (GC-MS) was used by Hughes et al. [83] for an estimation of the vitamin A status in rats. To determine the vitamin A body pool these authors administered [11,12-$^2H_2$]-retinyl acetate to rats. Ratios of unlabeled and labeled retinol were determined in plasma and saponified liver as their TMS derivatives. The effluent of the GC was passed through an MS apparatus programmed in the selected ion-monitoring (SIM) mode to monitor the prominent ion fragments at m/z 255 and 257, respectively, resulting from loss of the terminal CH_2O-TMS group. As mentioned previously, methylretinoate has excellent chromatographic properties, and it is not surprising that most of the research work with GC was done on this component. Different detection systems have been studied in our laboratory (flame ionization detector, FID; multiple ion detection, MID; and electron capture detector, ECD). Since methylretinoate contains a conjugated polyene system and a keto function it is electron-capture active (Fig. 1.1); however, MID provides higher specificity and sensitivity.

GC was used by Dunagin et al. [84] for the identification of metabolites of retinal and retinoic acid in rat bile. Methylretinoate was chromatographed on Gas Chrom P coated with 3% SE-30 at 180°C and using an argon flow rate of 50 ml/min. For radioactive components detection was performed by condensing the compound followed by elution and spectrophotometric or radioactivity measurement, or by condensing directly in a liquid scintillation mixture followed by counting. We successfully applied GC-MS with selected peak monitoring on a magnetic instrument for the demonstration of endogenous retinoic acid in human serum [56]. Chromatography of the purified and diazomethane-treated extracts was performed under the following conditions.

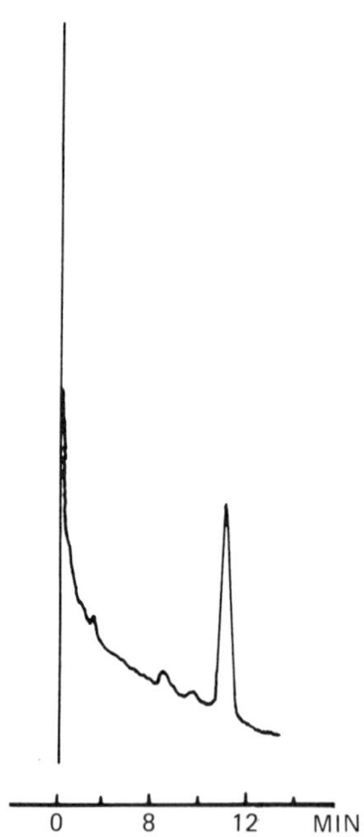

Figure 1.1 Injection of 4 ng methylretinoate on a 1.8 m 2% FFAP column at 180°C with ECD at 300°C.

The column was a 1.80 m × 2 mm inner diameter (ID) glass silanized column, filled with 1% free fatty acid phase (FFAP) coated on Gas-Chrom Q (100-120 mesh). Temperatures were 210°C (injected port) and 190°C for the column. Electron energy was 20 eV. The MID device was focused on ions at *m/z* 314 ($M^{+\cdot}$, methylretinoate), and *m/z* 317 ($M^{+\cdot}$, methyl-d_3-retinoate). Levels down to 3 ng/ml could be demonstrated using this technique. Chiang [85] also used GC-MS for low-level quantitation of retinoic acid. Methylretinoate was formed by reaction with dimethylformamide dimethylacetal. Separation was done on a glass GC column (6 ft × 2 mm ID) packed with 3% SE-30 on 80-100 mesh Chromosorb W HP at a column temperature of 230 and 250°C for injector. Ions at *m/z* 314 and 316 ([^{14}C]retinoic acid methyl ester) were monitored; however, when analyzing more than 5 ml of nonsupplemented plasma, no retinoic acid was detected. The high column temperature could probably destroy a considerable amount of the product, thus increasing severely the detection limit. An overview of the gas chromatographic conditions for different vitamin A compounds is given in Table 1.3. Both QF-1 (trifluoropropylsilicone gum) and SE-30 (methylsilicone gum) are undoubtedly the most popular liquid phases.

In conclusion we can say that GC in general is not very well suited for vitamin A analysis due to the production of artefacts at the higher temperatures required. Therefore, in the last few years, most research groups in the vitamin A field focused on milder chromatographic separation procedures, such as modern LC.

4. HIGH-PERFORMANCE LIQUID CHROMATOGRAPHY

4.1 Introduction

Modern LC does not need derivatization and enables a rapid and nondestructive analysis at room temperature. Therefore, LC has become an extremely valuable method for separation, identification, and quantitation of the different vitamin A compounds.

A wide variety of column packing materials and eluent compositions has been used for vitamin A analysis. The most useful packings are silica packings for normal phase; in the reversed-phase mode most separations are performed on octadecyl columns. Eluents for normal phase usually contain *n*-hexane or petroleum ether with a low percentage of a more polar solvent. In the reversed-phase mode, mixtures of water or buffers with alcohols or acetonitrile are the most common eluents.

Detection is performed either by UV at the λ_{max} of the compound of interest or by fluorescence detection (Table 1.1). The detection limits of most LC procedures are in the low nanogram or even in the picogram range for injected standards. The rapid development of the liquid chromatographic technique in the last few years really offers the

TABLE 1.3 Gas Chromatographic Conditions for Vitamin A Compounds on Various Supports

Compound	Column	Temperature (°C)			Carrier gas flow rate	Detector	Reference[a]
		Injector	Oven	Detector			
Retinol							
TMS-retinol	Gas-chrom Q, 80-100 mesh, 5% silicone QF-1, 2 m × 2.3 mm	245	160	–	He 70 ml/min	FID	79
TMS-retinol	Chromosorb W-HP, 7% silicone DC-QF-1, 6 ft	–	170	–		FID	80
TMS-retinal	Gas-chrom Q, 80-100 mesh, 1-3% QF-1, 1.8 m × 2 mm	195	175	195	N_2 7 cm/min	FID	UR
Methyl ether, retinal, methyl-retinoate, anhydroretinol	Gas-chrom P, 60-80 mesh, 1% SE-30, 35 cm	–	150	–	Ar 150 ml/min		74
Retinal							
Retinal methoxime	Chromosorb W, 80-100 mesh, 3% QF-1, 1.8 m × 2 mm	190	175	190	N_2 10 cm/min	FID	60
Retinoic acid							
TMS-retinoic acid	Gas-chrom Q, 80-100 mesh, 3% QF-1, 1.8 m × 2 mm	190	176	190	N_2 7 cm/min	FID	UR
Methyl ester	Gas-chrom Q, 80-100 mesh, 2% FFAP, 1.8 m × 2 mm	190	180	300		ECD	UR
Methyl ester	Gas-chrom P, 100-120 mesh, 3% SE-30, 4.5 ft	–	180	–	Ar 50 ml/min	FID	84
Methyl ester	Gas-chrom Q, 100-120 mesh, 1% FFAP, 1.8 m × 2 mm	210	190	265-270	He 30 ml/min	MID	56
Methyl ester	Chromosorb W HP, 80-100 mesh, 3% SE-30, 6 ft × 2 mm	250	230		He 50 ml/min	MID	85

[a]UR, unpublished results by De Ruyter et al.

greatest opportunities for research work on vitamin A metabolism and isomerization.

In this section we will review applications of these techniques (normal phase and reversed phase) to the vitamin A field. For readers interested in a detailed theoretical description of the different LC approaches we refer to basic LC books [86-88].

4.2 Normal-Phase Chromatography

In this technique a polar column packing and a nonpolar eluent are used. The compounds to be separated are eluted in the order of least to most polar.

4.2.1 Chromatography on Underivatized Silica

Adsorption chromatography is well suited for separation of isomers and for functional group differentiation. As different batches of stationary phases have active sites that widely differ in sorption energy, adequate regulation of activity must be ensured. Adding a small percentage of water or alcohol to the nonpolar eluent provides this regulation and also reduces peak tailing [89]. These modulators bind with the most active places and increase the surface homogeneity of the column packing.

Silica is the most popular adsorbent used in LC and is commercially available as Lichrosorb, Lichrospher, Spherisorb, Nucleosil, Zorbax, etc. The active sites on the silica surface are supposed to be the "silanol" groups; these are the hydroxyl functions on the silicon atoms. The surface area is a function of the average pore diameter, which normally ranges from 60 to 800 Å. With pore sizes smaller than 60 Å exclusion effects can be observed. Silica reacts weakly acidic, and therefore basic substances are strongly retained or even irreversibly adsorbed. Addition of bases, e.g., triethylamine, reduces tailing of basic substances; in a similar way addition of an acid (e.g., acetic acid) reduces tailing of acidic compounds. However, care should be taken not to exceed a pH of 9, as the silica starts to dissolve and repacking of the column is necessary.

4.2.1.1 Retinol. Various kinds of silica supports have been tried for the analysis of retinol as well as for retinol isomers. Table 1.4 presents the relevant data from some research groups in this field.

n-Hexane is used in almost all eluents, together with a low percentage of an organic modifier. In most of the isomer separations, a ternary solvent is used, except by Paanakker [94] who uses 5% dioxane in *n*-hexane to effect adequate separation of the isomers.

For determination of retinol in food substances, saponification of the esters is performed using ethanolic KOH; hydroquinone [47] and ascorbic acid [91] were used as antioxidants during this manipulation.

TABLE 1.4 Liquid Chromatographic Conditions for Separation of Retinol and Retinol Isomers

Compound[a]	Packing material	Eluent composition	Remarks, sample origin	Not separated isomers[b]	Detection limit	Reference
Retinol	Lichrosorb Si 60	*n*-Hexane-ethanol (97:3)	Food	-	15 ng[c]	47
Retinol	Micropak Si 10, 10 μm	Petroleum ether-dichloromethane-isopropanol (80:19.3:0.7 v/v)	Serum (100 μl)	-	50 ng/ml	51
11-*cis*, 13-*cis*, 9-*cis*, all-*trans*-retinol	Zorbax SIL	*n*-Hexane-ethyl acetate (88:12)	Standards	11-*cis*, 13-*cis*	—	90
	Zorbax SIL	*n*-Hexane-diethyl ether (80:20)	Standards	11-*cis*, 13-*cis*	—	90
	Zorbax SIL	*n*-Hexane-ethyl acetate-dichloromethane (83.2:7.5:7.7)	Standards	+	—	90
	Zorbax SIL	*n*-Hexane-ethyl acetate-dichloromethane (86.1:6.2:7.7)	Standards	+	—	90
13-*cis*, all-*trans*-retinol	Zorbax SIL, 5 μm	*n*-Hexane-dichloromethane-isopropanol (300:200:1.5)	Food, oil (3 g)	+	1 μg/g	91
Retinol	μPorasil	*n*-Hexane-chloroform-ethanol (50:50:1)	Food	-	10 ng[c]	92
Retinol	Lichrosorb Si 60, 7 μm	*n*-Hexane-dichloromethane-isopropanol (90:9:1)	Serum	-	15 ng[c]	93
11-*cis*, 13-*cis*, 9-*cis*, all-*trans*-retinol	Si 60, 5 μm	*n*-Hexane-dioxane (95:5)	Standards	+	0.5 ng[c]	94
Retinol	RSIL Silica, 5 μm	*n*-Hexane-isopropanol (99.75:0.25)	Corpus luteum	-	—	95
	RSIL Silica, 5 μm	*n*-Hexane-dichloromethane-isopropanol (200:50:1.75)	Corpus luteum	-	—	95

Dennison and Kirk [92], however, do not mention the use of one of them.

For quantitative work, use of an internal standard (IS) is highly recommended in order to compensate for possible losses during the extraction procedure. The choice of the IS has to be based on structural and chromatographic similarity with the compound of interest. These similarities ensure an identical behavior during the extraction procedure and the chromatographic separation as well. Furthermore, the IS has to be added to the sample before any treatment [51] as the extraction obviously is one of the most critical steps in the analysis. For quantitative work with a high degree of precision internal standardization is highly recommended [47,92,93].

In 1968, Zile and DeLuca [96] described a chromatographic separation of retinol isomers on a large gravity-fed silicic acid column eluted with petroleum ether and diethyl ether. The column was 60 cm long and consisted of four 15 cm sections decreasing in diameter. The uppermost section had an ID of 13, the lowermost of 4.5 mm. 11,13-Di-*cis* retinol eluted before 13-*cis* and 9.13-di-*cis* isomers, which were not separated; also, all-*trans*-, and 9-*cis*-retinol were not separated. The addition of *dl*-tocopherol as an antioxidant was necessary to obtain complete recovery and to avoid the destruction of the vitamin A compounds. In the last few years, smaller particles and better packing techniques have allowed the use of smaller columns and resulted in faster analyses. In this way on-column destruction has been diminished.

4.2.1.2 Retinal. Table 1.5 lists different eluent compositions and column packings used for the separation of retinal isomers, as reported in the literature. The most popular solvent mixture is *n*-hexane or petroleum ether with a low percentage of a more polar moderator, diethyl ether. Only Paanakker and Groenendijk [94] used dioxane instead of diethyl ether as it has lower volatility and covers a larger polarity range than diethyl ether. A typical separation of retinal isomers is shown in Fig. 1.2. Following the classification system proposed by Snyder [97], diethyl ether and dioxane belong to different selectivity groups; however, the results obtained with those two moderators were very similar. A striking difference with the chromatography of retinol is the difference in elution pattern. In the retinol series the 11-*cis* isomer is eluted first, whereas in the retinal series the 13-*cis* isomer has a shorter retention time. However, 13-demethyl retinal isomers again follow the retinol sequence of elution pattern [98]. 3-Dehydroretinal, as described by Tsukida et al. in 1980 [99], is extremely unstable and deteriorates during its contact with silica. Therefore, the amount of silica, and consequently the internal diameter of the LC columns, has to be kept to a minimum. To overcome this problem of deterioration repeated injections of a mixture of butylated hydroxyanisole (BHA) and butylated hydroxytoluene (BHT) prior to sample

TABLE 1.5 Liquid Chromatographic Conditions for Separations of Retinal-Related Compounds and Retinal Isomers

			Remarks			
Compound[a]	Packing material	Eluent composition	Sample origin	Not separated isomers[b]	Detection limit	Reference
Retinal						
13-*cis*, 11-*cis*, 9-*cis*, all-*trans*-retinal	μPorasil	Petroleum ether-diethyl ether (98:2)	Standards	+	20 ng[c]	101
9,13-di-*cis*, 13-*cis*, 11-*cis*, 9-*cis*, all-*trans*-retinal	μPorasil	*n*-Hexane-diethyl ether (98:2)	Standards	+	-	102
13-*cis*, 11-*cis*, 9-*cis*, 7-*cis*, all-*trans*-retinal	μPorasil	*n*-Hexane-diethyl ether (98:2)	Standards	+	-	103
13-*cis*, 9,13-di-*cis*, 11-*cis*, 9-*cis*, all-*trans*-retinal	Spherisorb 5 μm	Petroleum ether-diethyl ether (90:10)	Standards	9,13-di-*cis* and 13-*cis*	-	104
13-*cis*, 11-*cis*, 9-*cis*, all-*trans*-retinal	Zorbax SIL	*n*-Hexane-diethyl ether (88:12)	Standards	+	-	105
13-*cis*, 11-*cis*, 9-*cis*, 7-*cis*, all-*trans*-retinal	Jascopack silica	Petroleum ether-diethyl ether (92:8)	Standards	+	-	106
13-*cis*, 11-*cis*, 9-*cis*, 7-*cis*, all-*trans*-retinal	Shimadzu silica	*n*-Hexane-diethyl ether (88:12)	Standards	+	-	107
13-*cis*, 11-*cis*, 9-*cis*, all-*trans*-retinal	Zorbax SIL	*n*-Hexane-diethyl ether (88:12)	Standards	+	-	108

13-*cis*, 11-*cis*, 9-*cis*, 7-*cis*, all-*trans*-retinal	Zorbax SIL	Petroleum ether-diethyl ether (88:12)	Standards	+	-	109
13-*cis*, 11-*cis*, 9-*cis*, all-*trans*-retinal	Si 60 5 μm	*n*-Hexane-dioxane (97.5:2.5)	Rod out. segment	+	0.5 ng[c]	94
Retinal-related compounds						
13-*cis*, 9,13-di-*cis*, 11-*cis*, 9-*cis*, all-*trans*-14-methylretinal	Corasil II	*n*-Hexane-diethyl ether (98.5:1.5)	Standards	+	-	110
13-*cis*, 9,13-di-*cis*, 11-*cis*, 9-*cis*, 7-*cis*, all-*trans*-retinal	μPorasil	*n*-Hexane-anhydrous ether (96:4)	Standards	9,13-di-*cis* and 11-*cis*	-	111
11-*cis*, 9-*cis*, 7-*cis*, all-*trans*-13-demethylretinal	μPorasil	*n*-Hexane-anhydrous ether (96:4)	Standards	+	-	111
13-*cis*, 11-*cis*, 9,13-di-*cis*, 9-*cis*, 7-*cis*, all-*trans*-14-methylretinal	μPorasil	*n*-Hexane-anhydrous ether (96:4)	Standards	+	-	111
11-*cis*, 13-*cis*, 9-*cis*, 7-*cis*, all-*trans*-13-demethylretinal	Lichrosorb Si 60	Petroleum ether-diethyl ether (90:10)	Standards	+ Other sequence	-	98
13-*cis*, 11-*cis*, 9-*cis*, 7-*cis*, all-*trans* dehydroretinal and retinal	μPorasil	*n*-Hexane-diethyl ether (94:6)	Standards	+	-	99

[a]The isomers elute in the sequence given.
[b]-, not mentioned; +, all isomers mentioned are separated.
[c]Absolute amount.

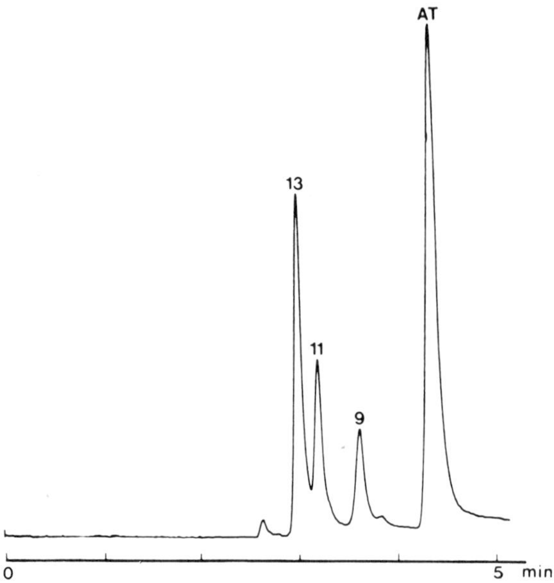

Figure 1.2 Chromatogram of a test mixture of four cis/trans isomers of retinal. Mobile phase, *n*-hexane-dioxane (2.5%); pressure, 3000 psi; flow rate, 17.1 μl/sec; wavelength, 360 nm. 11 = 11-*cis*-retinal; 13 = 13-*cis*-retinal; 9 = 9-*cis*-retinal; AT = all-*trans*-retinal. Column length, 25 cm, ID, 3.0 mm. Stationary phase, Si 60 (5 μm). (From Ref. 94.)

application were proposed by Vecchi and Vesely in 1973 [80a]. Some authors [53,100] describe the chromatography of retinal oximes on MicroPak Si-5 with *n*-hexane-dioxane (25:1) (Fig. 1.3) and on a Lichrosorb 10 μm column with *n*-hexane-dioxane (90:10). The formation of retinal oxime allows extraction with nearly complete retention of the original isomeric form. However, of each retinal isomer a mixture of syn and anti retinal oximes is formed during the derivatization, which can be a serious drawback for quantitative work.

The separations as described in Table 1.5 were part of isomerization studies. A great solvent dependence on quantum yield of isomerization was observed. Aprotic solvent effects were extensively described by several research groups [102,103,105].

The formation of hindered isomers (11-*cis*-retinal) appears to be favored in solvents of high dielectric constants, such as acetone and acetonitrile.

4.2.1.3 Retinyl Esters. The chromatographic conditions for separation of various esters are brought together in Table 1.6.

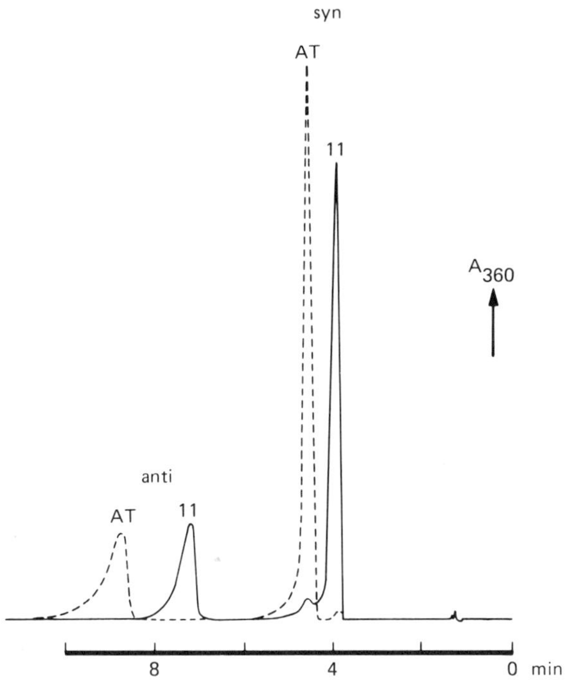

Figure 1.3 Representative high-performance liquid chromatography chromatogram of extracted rhodopsin before (——) and after illumination (- - -). Column MicroPak Si-5 25 cm × 2 mm; eluent, hexane-dioxane (25:1); flow rate, 90 ml/hr. Abbreviations, AT: all-*trans*; 11: 11-*cis*-retinaloxime. (From Ref. 53.)

Chromatography of retinyl esters on gravity-fed silicic acid columns as described by Zile and DeLuca [96] is seriously hampered by degradation of the compounds. It is likely that acid-catalyzed dehydration reactions occurred as chromatography of retinyl esters has been successfully performed on alkali-treated silicic acid columns [114]. For baseline separation of the different isomers on silica or alumina columns, addition of dioxane (0.05%) or diethyl ether (1%) to the *n*-hexane is necessary [80a,94]. Liquid-liquid chromatography with *n*-hexane as eluent and 3,3'-oxydipropionitrile as the stationary phase on a Mercko-sorb SI 60 support enabled the separation of several retinyl acetate isomers [80a]. However, for quantitative work, these authors prefer to apply an adsorption system on Corasil II with *n*-hexane containing 0.1% dioxane as the eluting solvent. The use of either naphthalene or ethylbenzoate as IS for quantitative work clearly was not the best choice, as no structural relationship can be observed between these IS

TABLE 1.6 Liquid Chromatographic Conditions for Separation of Retinyl Esters

			Remarks			
Compound[a]	Packing material	Eluent composition	Sample origin	Not separated isomers[b]	Detection limit	Reference
Retinyl acetate, vitamin D_2	Zorbax SIL	50% Water saturated heptane-ethyl acetate-dichloromethane (84:4:12)	Tablet	-	-	112
Retinyl palmitate	Lichrosorb SI 60 5 μm	Heptane-diisopropyl ether (95:5)	Margarine	-	-	46
13-*cis*, 11-*cis*, 9-*cis*, all-*trans*-retinyl palmitate	SI 60 5 μm	*n*-Hexane-dioxane (100:0)	Standards	13-*cis* and 11-*cis*	-	94
13-*cis*, 11-*cis*, 9-*cis*, all-*trans*-retinyl palmitate	SI 60 5 μm	*n*-Hexane-dioxane (99.95:0.05)	Standards	+	-	94
13-*cis*, 11-*cis*, 9-*cis*, all-*trans*-retinyl palmitate	SI 60 5 μm	*n*-Hexane-dioxane (99.9:0.1)	Standards	+	-	94
11,13-di-*cis*, 11-*cis*, 13-*cis*, 9,13-di-*cis*, 9-*cis*, all-*trans*-retinyl palmitate	Lichrosorb AloxT 5 μm	Isooctane-diisopropyl ether (98:2)	Standards	+	-	113

11,13-di-*cis*, 11-*cis*, 13-*cis*, 9,13-di-*cis*, all-*trans*-retinyl acetate	Lichrosorb AloxT 5 μm	Isooctane-isopropyl ether (98:2)	Standards	+	-	113
All-*trans*-retinyl palmitate, oleate, acetate	Lichrosorb AloxT 5 μm	Isooctane-isopropyl ether (98:2)	Standards	+	-	113
13-*cis*, 9-*cis*, all-*trans*-retinyl acetate	Al_2O_3 10 μm	*n*-Hexane-diethyl ether (99:1)	Standards	+	-	80
13-*cis*, all-*trans*-retinyl acetate	Corasil II 10 μm	*n*-Hexane-dioxane (99.9:0.1)	Food	+	-	80
11,13-di-*cis*, 11-*cis*, 13-*cis*, 9-*cis*, all-*trans*-retinyl acetate	Merckosorb SI 60 30 μm 3,3'-oxydi-propionitrile	*n*-Hexane	Standards	+	-	80
Retinyl acetate, retinyl palmitate	Silicic acid	Petroleum ether-diethyl ether (100:0 → 70:30)	Standards	+	-	96

[a]The isomers elute in the sequence given.
[b]-, Not mentioned; +, all isomers mentioned are separated.

and the retinyl esters to be determined. Reproducibility data of the liquid liquid system are not given, but for the quantitation of retinol isomers in the presence of the corresponding ester, excellent relative standard deviations of 2.5 and 0.7% were obtained.

Addition of 1% methyl *t*-butyl ether to hexane also allowed the separation of the four predominant sterically unhindered isomers (all-*trans*, 13-*cis*, 9-*cis*, and 9,13-di-*cis*) of retinyl palmitate on a silica column. However, again, the internal standard (fluoranthene) showed no structural relationship with the compounds of interest [80b]. The separation of retinyl palmitate, oleate, and acetate on an alumina support was described by Steuerle in 1981 [113]; however, most of the separations of retinyl esters are performed on a reversed-phase system and are discussed later.

4.2.1.4 Retinoic Acid. The chromatographic conditions for retinoic acid separations are given in Table 1.7.

In 1968, Zile and DeLuca [96] described the influence of lipid impurities on the chromatographic behavior of vitamin A compounds. In the presence of lipid material, chromatography on gravity-fed silicic acid columns is seriously affected, resulting in a decrease of k' values and in a drastic decrease of plate count (N). Therefore, extracts of biological material should be extremely pure if reproducible retention volumes are to be obtained.

In order to prevent the acid-catalyzed destruction of vitamin A compounds, Grain et al. in 1967 [114] and also, in 1971, Kleiner-Bossaler and DeLuca [117] treated the silicic acid for 2 hr with 0.1 N NaOH. After treatment the total recovery for retinol, retinal, and retinyl esters increased to almost 90%; however, for retinoic acid only 60% was obtained. Adsorption of 10-50 mg α-tocopherol on 20 g basic silicic acid [117] increased the recovery of retinoic acid to 80-90%. Chromatography on silica systems is extremely sensitive to water, whereas the use of gradient systems often results in serious problems of reproducibility. However, isocratic separations of closely related compounds can easily be performed on a silica system.

Straight-phase adsorption chromatography has been used in several studies on the oxidative metabolism of retinol. Administration of retinol or retinyl esters resulted in the appearance of retinoic acid in bile, blood, and urine of rats [54,115,118,119]. In all those experiments, massive doses of precursor were administered before retinoic acid could be demonstrated. The use of smaller columns (15×0.32 cm ID) and selective detection at 350 nm at the highest sensitivity setting allowed us to demonstrate the presence of endogenous retinoic acid in human serum under normal physiological conditions without administration of any precursor [56]. The use of a suitable internal standard (all-*trans*-13-demethylretinoic acid) closely related to all-*trans*-retinoic acid also allowed quantitative work [116a,b].

TABLE 1.7 Liquid Chromatographic Conditions for Separation of Retinoic Acid and Retinoic Acid Isomers

			Remarks			
Compound[a]	Packing material	Eluent composition	Sample	Not separated isomers[b]	Detection limit	Reference
Retinoic acid	Silicic acid, Bio-Sil HA	Petroleum ether-diethyl ether (100:0 → 70:30)	Standards	-	-	96
13-*cis*, all-*trans*-retinoic acid	Partisil 10 μm	Dichloromethane-glacial acetic acid (99.5:0.5)	Serum	+	10-20 ng/ml	54
All-*trans*-methylretinoate, 4-oxo-all-*trans*-methylretinoate, 4-hydroxy-all-*trans*-methylretinoate	μPorasil	Tetrahydrofuran-hexane (7.5:92.5)	Standards	-	-	115
13-*cis*, 9-*cis*, all-*trans*-retinoic acid	RSIL 5 μm	Petroleum ether-acetic acid-acetonitrile (99.5:0.3:0.2)	Serum	-	1 ng/ml	56,116
Retinoic acid	Basic silicic acid, 100 mesh	Methanol-diethyl ether (90:10 → 100:0)	Kidney tissue	-	-	117
Retinoic acid	Basic silicic acid, 325 mesh	Methanol (100)	Intestine	-	-	114
13-*cis*, all-*trans*-methylretinoate	Partisil PXS 10 μ	*n*-Hexane-isopropanol (99.95:0.05)	Trachea cultures	+	-	119

[a]The isomers elute in the sequence given.
[b]-, Not mentioned; +, all isomers mentioned are separated.

This method allows detection of the retinoic acid down to 1 ng/ml with a linear detector response up to 10 ng/ml. The within-run precision (CV) was 6.9% at the 2.7 ng/ml level (n = 9). The analysis was applied to serum samples of 37 volunteers, and a mean physiological concentration of 3.5 ng/ml was found. A typical chromatogram of a serum extract is shown in Fig. 1.4. The addition of acetic acid to the eluent was necessary to prevent broad tailing peaks for retinoic acid and the internal standard as well.

The influence of the amount of acetic acid added on the k' values of different isomers of retinoic acid is given in Fig. 1.5.

The same LC system also allowed separation of several isomers of both retinoic acid and methylretinoate, as shown in Fig. 1.6. The chromatogram clearly demonstrates the high separating power of the silica system we used; however, for separations of retinoic acid from more polar metabolites, such as epoxy, keto, and hydroxy derivatives,

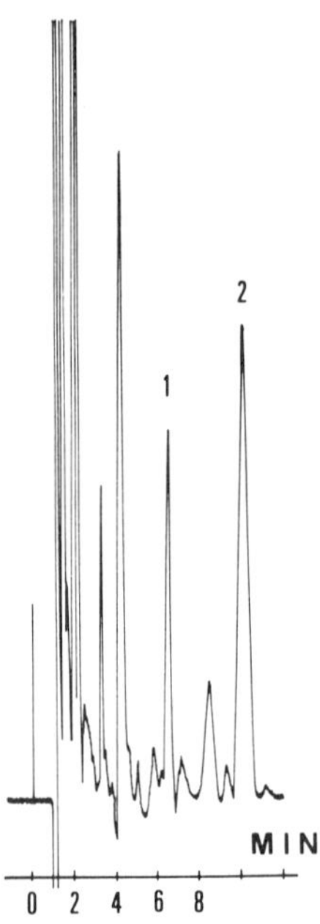

Figure 1.4 Representative chromatogram of a serum extract for retinoic acid analysis. Peak identification is as follows: (1) all-*trans*-retinoic acid, (2) internal standard all-*trans*-13-demethylretinoic acid. Eluent composition: petroleum ether-acetonitrile-acetic acid (99.5:0.2:0.3). Flow rate, 0.75 ml/min. Detection, 350 nm. Column: 15 × 0.32 cm 5 μm RSIL; the temperature was ambient. (From Ref. 56.)

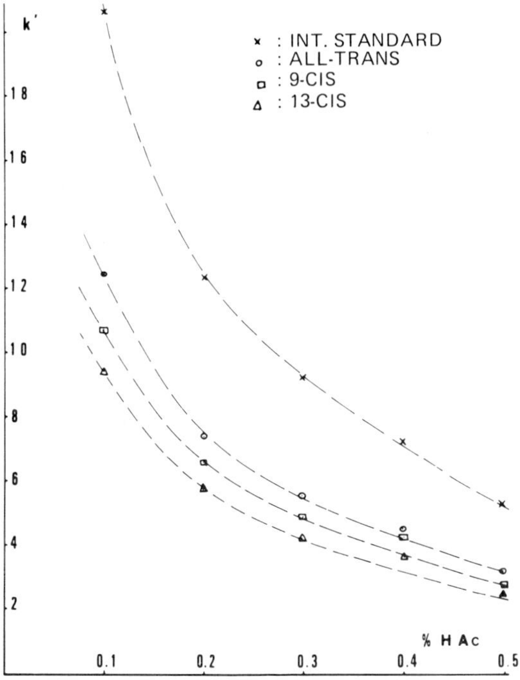

Figure 1.5 Relation of capacity ratio (k') of retinoic acid isomers and of the internal standard versus the acetic acid concentration in the eluent.

reversed-phase chromatography is a better choice as the more polar compounds elute before retinoic acid and are not retained on top of the column.

The combination of reversed-phase and normal-phase chromatography of retinoic acid and the diazomethane-treated sample often is used in metabolic studies for the identification of retinoic acid [56,119]. Cochromatography of the metabolite with the synthetic standards (all-*trans*-retinoic acid or all-*trans*-methylretinoate) in two completely different LC systems provides strong evidence for the identity of the compound of interest.

4.2.2 Chromatography on Chemically Bonded Phases

In 1977 Pilkiewicz et al. [52] reported a nonisomerizing isolation procedure for the identification of the isomeric form of protein-linked retinals. However, this method requires lipids or detergents (Ammonyx-LO)

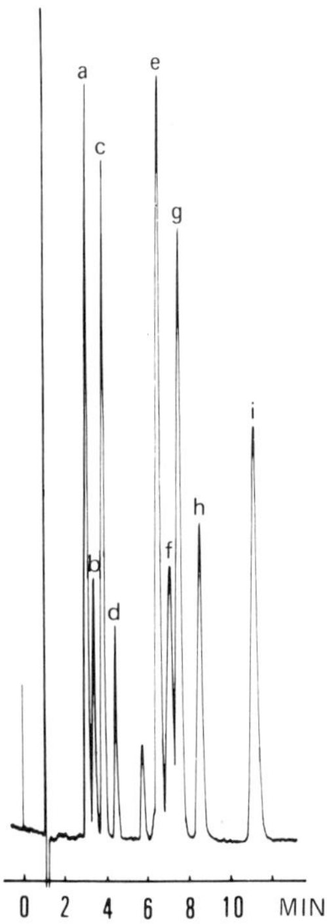

Figure 1.6 Separation of different retinoids on the silica system. Conditions were as in Fig. 1.4. (a) 13-*cis*-methylretinoate, (b) 9-*cis*-methylretinoate, (c) all-*trans*-methylretinoate, (d) 11,13-di-*cis*-methylretinoate, (e) 13-*cis*-retinoic acid, (f) 9-*cis*-retinoic acid, (g) all-*trans*-retinoic acid, (h) 11,13-di-*cis*-retinoic acid, (i) internal standard, all-*trans*-13-demethylretinoic acid.

for solubilization; after a few injections these lipids rapidly deactivate a silica column. The retention time decreases, the resolution is lost, and regeneration of the system is necessary. To prevent this time-consuming process of regeneration (especially for silica systems) the authors used a bonded-phase column packing with nitrile groups. These CN packings are compatible with both reversed- and normal-phase chromatography and are less sensitive to the interfering lipids. Chromatography of the retinal isomers is performed in the normal phase mode with 1% diethyl ether in *n*-hexane as eluent. A typical chromatogram is shown in Fig. 1.7.

Although the system is used in the normal phase mode the elution pattern of the retinal isomers on the CN-column differs from that on the previously described silica systems. In the CN system the elution sequence

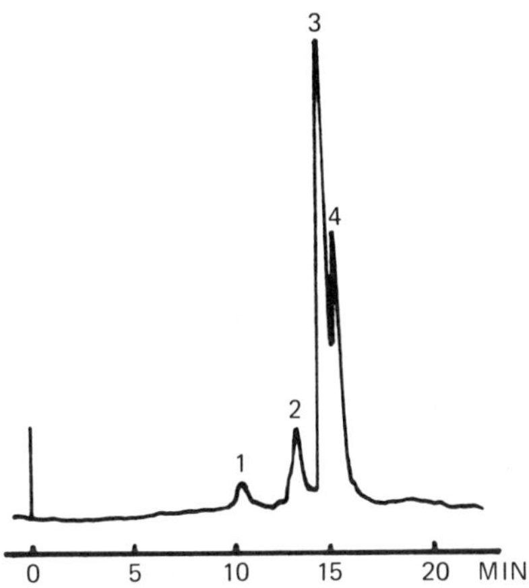

Figure 1.7 Elution of 11-*cis* (1), 13-*cis* (2), 9-*cis* (3), and all-*trans* (4) retinals on μBondapak CN column, 4 mm ID × 30 cm; flow rate, 1.0 ml/min; pressure, 600 psi; solvent, 1% ether in hexane; Waters Model 6000 pump, Schoeffel variable UV detector, λ = 350 nm. Samples were injected under dim red light. [Reproduced with permission from Ref. 52. Copyright Academic Press, Inc. (London), Ltd.]

of 11-*cis* and 13-*cis* retinal is reversed, with the 11-*cis* isomer eluting first. The authors report a broadening of the peaks by injecting a 30 microliter aliquot of methanol on the column. Injection of even a small quantity of a stronger solvent can indeed cause problems of overloading or nonequilibration. It also changes the selectivity and the reproducibility of the system, as described by Rabel in 1980 [120].

For their quantitation of retinol in serum, Biesalski et al. [49] also used a Zorbax CN column eluted with *n*-hexane-isopropanol (95:5) coupled with a fluorescence detector. The authors claim a detection limit of 50 pg for an injection of a standard solution, and a day-to-day coefficient of variation of 6.3% is obtained at the 448 ng/ml level. This is a remarkable result given the fact that no internal standard is used; also, a flow rate of 3 ml/min is rather high for a 15 × 0.46 cm column and negatively affects the efficiency of separation as demonstrated by the van Deemter plot [86].

Liquid-liquid chromatography using 3,3'-oxydipropionitrile as the stationary phase and *n*-hexane as eluent [80] caused serious problems in quantitation of the vitamin A acetate isomers. Such disadvantages as bleeding of the stationary phase and inability to solvent programming resulted in the search for chemically bonded organic stationary phases. The two papers above [49,52] reported the use of a CN-bonded phase that retains the nitrile function of the 3,3'-oxydipropionitrile. Applications of polar bonded phases in the reversed-phase mode is discussed in the next section.

4.3 Reversed-Phase Chromatography

In the retention behavior of a solute in a reversed-phase system only the role of the hydrophobic interactions is clearly understood. Retention is not due to favorable interaction with the stationary phase but to the effect of the mobile phase forcing the compound to the hydrocarbon stationary phase. However, as it is not the aim of this chapter to give a theoretical review on chromatography, we refer to some excellent papers on retention mechanisms on chemically bonded phases by Colin and Guischon and McCormick and Karger in 1977 and 1980 [121,122].

4.3.1 Retinol

Table 1.8 lists some applications of reversed-phase chromatographic separations of retinol.

Food products, such as milk, margarine, and cheese, are often fortified with specified amounts of vitamin A mainly in the form of retinyl esters. The levels of these additives must be monitored according to government rules. Quantitative determinations of vitamin A in margarine, infant formulas, and pharmaceuticals all include a hydrolysis step either to liberate retinol from the entrapping matrix or to hydrolyze the retinyl esters to free retinol. However, no use of an appropriate internal standard is reported [48,91,123,125] although the authors report accidental evaporation as a possible source of error [123], a problem that can easily be overcome by use of an internal standard.

For their measurement of retinol in milk and cheese Bui-Nguyen and Blanc [125] obtained a CV of 6% analyzing the same sample over a period of 1 month. A typical chromatogram of this food analysis is shown in Fig. 1.8.

In 1977, Abe et al. [128] reported a separate determination of retinol and retinyl esters in serum and liver. For quantitative work benzpyrene served as internal standard, although this compound has no structural similarity to retinol. Fluorescence detection was used to avoid coextracted lipids (e.g., β-carotene and vitamins D, E, and K). In our simultaneous determination of serum retinol and α-tocopherol [124] we used *dl*-tocol as IS due to its great structural relationship to α-tocopherol. In regard to retinol, both compounds are liposoluble and both have an alcoholic function; therefore, a similar retention behavior on the LC system was obtained. A typical chromatogram is shown in Fig. 1.9. A within-day and day-to-day precision of 2.5% ($\bar{x}$ = 674 ng/ml, n = 10) and 4.9% ($\bar{x}$ = 635 ng/ml, 20 days), respectively, were

TABLE 1.8 Chromatography of Retinol on Reversed-Phase Systems

Compound[a]	Packing material	Eluent composition	Remarks: Sample origin	Remarks: Not separated isomers[b]	Remarks: Detection limit	Reference
Retinol	LiChrosorb phase 10 µm	Methanol-water (90:10)	Food	-	-	123
13-*cis*, all-*trans*-retinol	Vydac ODS 10 µm	Acetonitrile-water (65:35)	Food (3 g)	+	1 µg/g	91
Retinol, α-tocopherol	RSIL C_{18} HL 10 µm	Methanol	Serum (100 µl)	-	60 ng/ml	124
Retinol	Kieselgel ODS C_{18} 10 µm	Methanol-1% ammonium carbonate solution (95:5)	Pharmaceuticals	-	-	48
Retinol	Nucleosil C_{18} 10 µm	Acetonitrile-water (95:5)	Food	-	20 ng[c]	125
Retinol, retinyl palmitate	Permaphase ODS	Isopropanol-ethanol-water (56:16:28)	Serum	-	-	126
Retinyl acetate, retinol, vitamin E acetate, vitamin D_2, vitamin D_3, β-carotene, retinyl palmitate	Zorbax ODS 6 µm	Acetonitrile-dichloromethane (0.001% triethylamine)-methanol (700:300:X)	Standards	-	-	127

[a]The isomers elute in the sequence given.
[b]-, Not mentioned; +, all isomers mentioned are separated.
[c]Absolute amount.

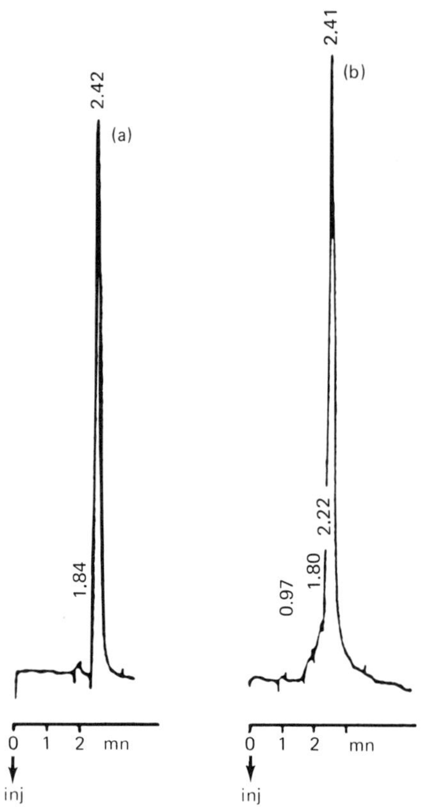

Figure 1.8 (a) Vitamin A retinol standard chromatogram. (b) Vitamin A chromatogram from cheese extract. (From Ref. 125.)

obtained. These results again clearly demonstrate the need and the usefulness of internal standardization.

Recently, Catignani and Bieri [124b] also described a simultaneous determination of retinol and α-tocopherol on a C_{18} column eluted with 5% water in methanol. In this method retinyl acetate and α-tocopheryl acetate were used as internal standards. A serious drawback of this procedure is the injection of the samples in 25% diethyl ether in methanol. This stronger solvent negatively influences column efficiency, especially for the early eluting compounds.

Another approach for reversed-phase chromatography of low-polarity samples was reported by Parris in 1978[126]. Many hydrophobic substances do not completely elute from a reversed-phase liquid chromatographic column when using semiaqueous mobile phases containing methanol or acetonitrile, due to their low solubility in water and methanol.

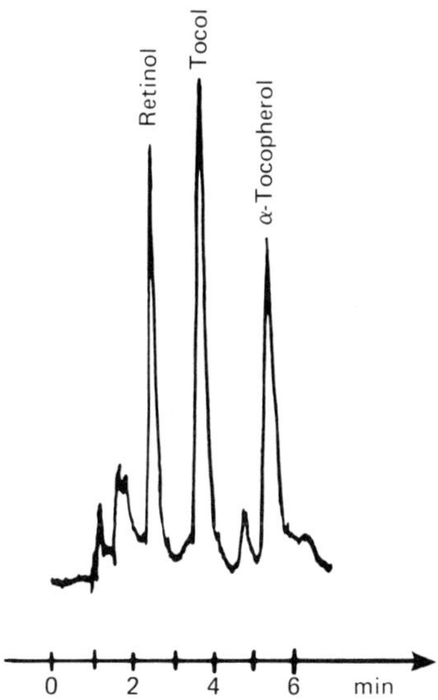

Figure 1.9 Chromatogram of a plasma extract of a normal subject. Peak identity: (1) retinol, (2) internal standard, *dl*-tocol, (3) α-tocopherol. UV detection at 292 nm at 32.10^{-4} AU/cm. (From Ref. 124.)

It is evident that incomplete elution of sample components significantly contributes to the steady deterioration of the column. Therefore, chromatographic eluents are prepared from solvents other than water, i.e., solvents in which low-polarity compounds have an appreciable solubility (e.g., methylene chloride and tetrahydrofuran); this technique offers several advantages. First, sample solubility is no more a problem and column life is extended, and second, due to the higher solubility, a higher sample load and, thus, an improvement of the detection limit can be seen. Recently, Landen [127] reported an application of this nonaqueous reversed-phase chromatography in the vitamin field. The mobile phase consisted of acetonitrile and methylene chloride (0.001% triethylamine), 700:300. The addition of methanol (0.05-4.0%) improved the separation of vitamins D_2 and D_3 but retinol and retinyl acetate cochromatographed on addition of 0.05% methanol. The retention time of retinyl palmitate and β-carotene remained constant up to 5 ml methanol per liter mobile phase. Beyond this level these compounds had higher capacity values. Using the eluent containing 4 ml methanol per liter, the k' value for retinyl acetate remained unchanged but retinol migrated ahead of the retinyl acetate. These separations are shown in Fig. 1.10. The shift of some vitamin A compounds to

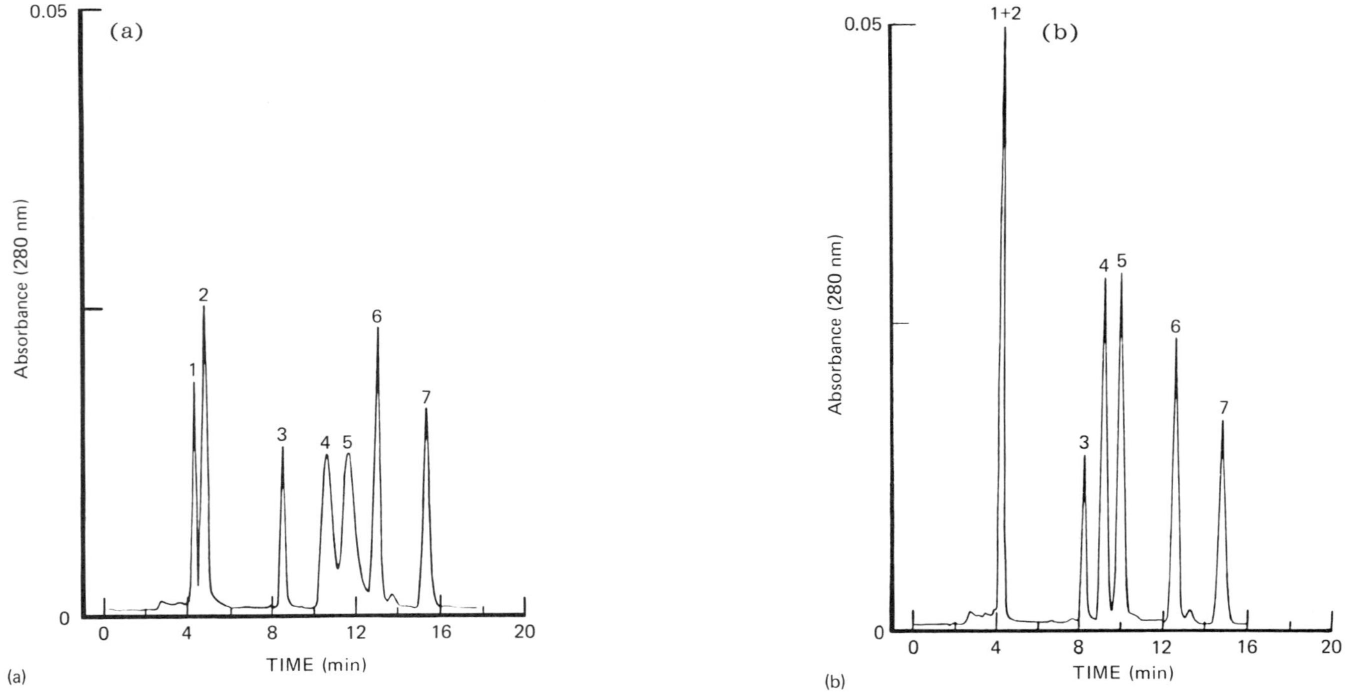

Figure 1.10 (a) Chromatogram of vitamin mixtures. (1) retinyl acetate, (2) retinol, (3) α-tocopheryl acetate, (4) vitamin D_2, (5) vitamin D_3, (6) β-carotene, (7) retinyl palmitate. Detector at 280 nm, 0.05 a.u.f.s. (absorption units full scale), mobile phase, acetonitrile-methylene chloride (700:300), flow rate, 1 ml/min; chart speed, 0.5 cm/min; column, 250 × 4.6 mm ID, Zorbax ODS, injection volume, 50 μl, temperature, 25°C. (b) chromatogram of same vitamin mixture as in a. Experimental conditions were the same as in a, except 2.0 ml methanol was added per liter of mobile phase. (From Ref. 127.)

longer retention times with increasing methanol concentration is a rather strange effect, as in general methanol is recognized to be a strong eluent, shortening the retention of a solute. Bhat et al. [129] reported the effect of 2-alkanols as mobile phase modifiers on the resolution of 11-*cis*, 13-*cis*, 9-*cis*, and all-*trans*-retinol in two nonaqueous high-performance liquid chromatographic systems. They were using a Partisil ODS and a Zorbax CN column in series.

4.3.2 Retinal

Few reports are found concerning the separation of retinal isomers on reversed-phase columns. Separations of several naturally occurring vitamin A compounds (retinoic acid, retinol, retinal, retinyl acetate, and retinyl palmitate) were described by Frolik et al. [130] and McCormick et al. [131] in 1978 and 1980. Mixtures of acetonitrile and 1% ammonium acetate (80:20 or 60:40) were used as eluents; a chromatogram is depicted in Fig. 1.11.

The ammonium acetate was necessary to prevent broadening of the retinoic acid peak; addition of as little as 0.2% was sufficient for a Spherisorb ODS column type; however, for chromatography on a Partisil ODS-2 column, 1% ammonium acetate was necessary to obtain the same symmetrical peaks. Again, most retinoids are only partially soluble in the methanol-water mixtures and the samples must be injected in a *small* volume of a stronger solvent, such as methanol.

4.3.3 Retinyl Esters

Several reports on the simultaneous determination of retinol, retinyl acetate, and other fat-soluble vitamins (e.g., vitamin D_2, vitamin E succinate and acetate) have been published, as demonstrated in Table 1.9. Quality control laboratories in particular routinely perform these analyses. The vitamins A, D, and E have absorption maxima at 325, 294, and 264 nm; thus, detection at 254 or 280 nm is a good compromise. Only for the vitamin A derivatives can a small loss in sensitivity be expected. The analysis does not require the inconvenient and time-consuming saponification step, although the possibility of oxidation during saponification is excluded; again, internal standardization should not be omitted.

In 1978, Roberts et al. [136] described a separation of four retinyl esters on a Spherisorb ODS column with an acetonitrile-water mixture (98:2). Only the retinyl palmitate ($C_{16:0}$) peak was positively identified by cochromatography with a standard. Comparison of the elution order of the ester peaks with known elution patterns from reversed-phase systems allowed tentative identification of retinyl linoleate ($C_{18:2}$), oleate ($C_{18:1}$), and stearate ($C_{18:0}$), respectively. However, under these conditions the "critical" pair of the oleate ($C_{18:1}$) and the palmitate ($C_{16:0}$) could not be separated completely. The same problem

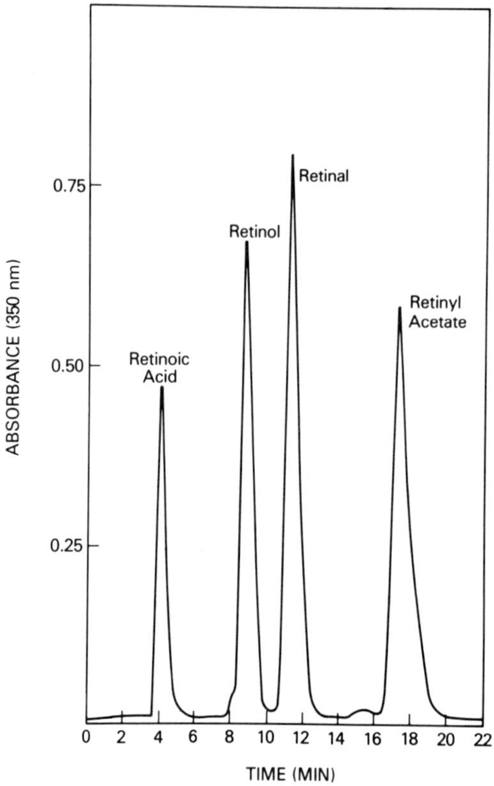

Figure 1.11 High-pressure liquid chromatography of four retinoids. A mixture of retinoic acid, retinol, retinal, and retinyl acetate was injected in 10 µl of 100% methanol onto a Spherisorb ODS column (25 × 0.3 cm). Acetonitrile-1% ammonium acetate, 60:40, was used as the developing solvent. At a pressure of 2620 psi, a flow rate of 1.1 ml/min was achieved. (From Ref. 130.)

was encountered by De Ruyter and De Leenheer [50] on a RSIL C_{18} HL column eluted with pure methanol, as depicted in Fig. 1.12. The use of this heavily loaded silica (18% organic material bonded) was imperative for a simultaneous determination of retinol and retinyl esters, as it allowed the use of pure methanol as mobile phase. If a lower loading was used water addition to the eluent was necessary for retention of retinol, however, precipitation of retinyl esters could occur. In 1979 we developed a separation of the retinyl esters on the basis of chain length and unsaturation. The same C_{18} HL column was used with a methanolic eluent containing silver ions. The addition of silver ions to the mobile

TABLE 1.9 Chromatography of Multivitamin Preparations

Compounds[a]	Packing material	Eluent composition	Sample origin	Detection: Wavelength (nm)	Detection: Detection limit	Reference
Vitamin D_2, vitamin E acetate, vitamin A	Permaphase ODS	Gradient from methanol-water (50:50) to methanol	Standards	254	-	132
Vitamin A acetate, vitamin D_2, vitamin E succinate	Zipax HCP	Methanol-water-phosphoric acid (77:23:0.1)	Tablets	254	-	132
Vitamin A acetate, vitamin D_2, vitamin E succinate	LDC RP8 5 µm	Methanol-water (92:8) (0.1% phosphoric acid)	Tablets	254	-	133
Vitamin A acetate, vitamin D_2, vitamin E acetate	MicroPak CH-10	Methanol-water (98:2)	Tablets	265	-	134
Vitamin A, vitamin A acetate, vitamin D_3, vitamin E alcohol, vitamin E acetate	Bondapak C_{18}	Methanol-water (95:5)	Food	280	250 IU/kg for vitamin A	135

[a]The compounds elute in the sequence given.

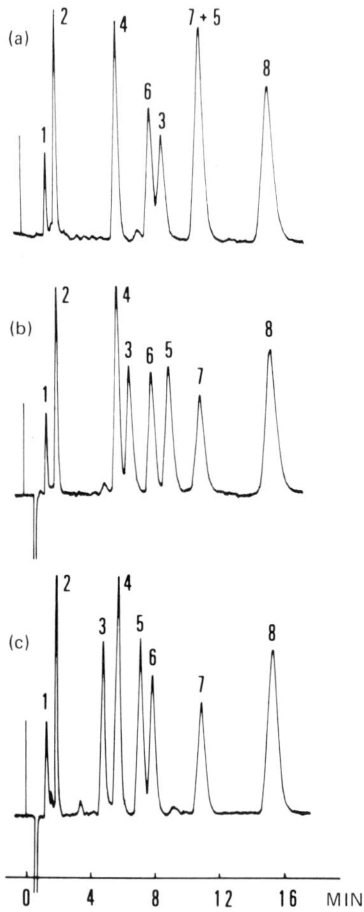

Figure 1.12 LC record of a standard mixture of retinol and retinyl esters. Conditions: column, 15 × 0.32 cm; 10 μm RSIL C_{18} HL; flow rate, 1 ml/min; mobile phase, (a) methanol; (b) methanol with 23.5 × 10^{-3} M Ag^+; (c) methanol with 58.9 × 10^{-3} M Ag^+. (1) retinol, (2) retinyl propionate (C_3), (3) retinyl linoleate ($C_{18:2}$), (4) retinyl laurate (C_{12}), (5) retinyl oleate ($C_{18:1}$), (6) retinyl myristate (C_{14}), (7) retinyl palmitate (C_{16}), (8) retinyl stearate (C_{18}). (Reproduced with permission from Ref. 138. Copyright 1979, American Chemical Society.)

phase decreased the retention of retinyl esters with an isolated unsaturation in the fatty acid chain (retinyl linoleate $C_{18:2}$, and retinyl oleate $C_{18:1}$) (Fig. 1.12a and b). The silver ions did not alter the retention of esters with a saturated fatty acid chain in spite of the unsaturation in the retinyl moiety of the molecule. For the oleate and the linoleate ester a linear relation was found between log k' and the concentration of silver ions in the mobile phase. The decrease of capacity ratios when silver ions are added to the mobile phase is due to the complexation of isolated double bonds with silver, thus favoring the partition to the mobile phase. The lack of effect of silver ions with the unsaturation in the retinyl moiety can be explained by the substitution and conjugation of the double bounds in this polyene chain. Indeed, substitution decreases complexation due to steric hindrance, and conjugation also results in a lower degree of complexation. Recently, Ross [137] separated the various retinyl esters, including the palmitate and the oleate, on an octyl or a phenyl reversed phase eluted with acetonitrile-water mixtures in a two-step discontinuous gradient system (from 88:12 to 98:2). In this system a detection limit of 0.5 ng for retinol and 40 ng for retinyl palmitate was obtained. The larger value for the palmitate ester is due to its higher retention volume and, hence, greater dilution. For quantitative analyses of retinol and retinyl esters the choice of an internal standard should be limited to compounds not found in biological samples and with an intermediate polarity between retinol and the esters.

In our measurements [50,138] of retinol and retinyl esters we used retinyl acetate. Roberts et al. [136] proposed axerophtene, a hydrocarbon analog of retinol, which remains unaltered during sample analysis. Retinyl heptanoate as used by Ross [137] is really an ideal internal standard for the analysis of retinyl esters in biological samples. Two applications of nonaqueous reversed-phase chromatography have been reported by Landen for determinations of retinyl palmitate in oils [44] and fortified cereals [45] (Fig. 1.13). Quantitation was achieved on μBondapak C_{18} 10 μm columns using dichloromethane-acetonitrile (30:70). High-pressure gel permeation chromatography on μStyragel 100 Å with dichloromethane containing 0.001% triethylamine served as a cleanup procedure before the LC quantitation. The necessity of this cleanup step, however, is not very clear, as a lot of reports on retinyl ester quantitation omitted this step without excessive loss of accuracy and reproducibility.

4.3.4 Retinoic Acid

The various chromatographic conditions and separations of retinoic acid are presented in Table 1.10. In reversed-phase chromatography more polar products elute earlier than the apolar compounds. Therefore this technique is of extreme value for detection and separation of

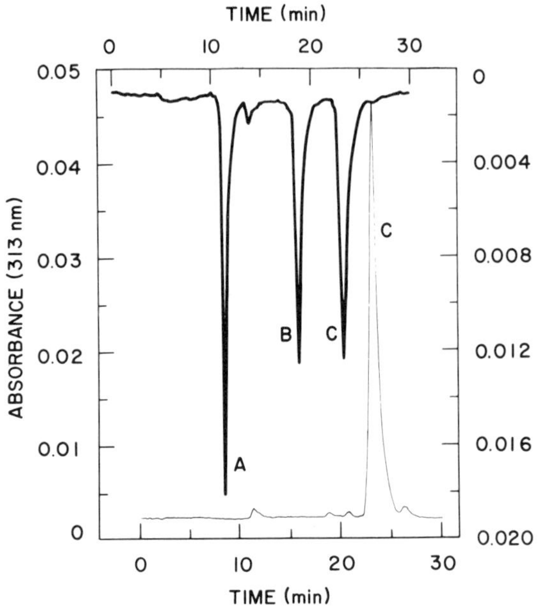

Figure 1.13 Chromatogram of vitamin fraction from wheat-based cereal fortified with vitamins A and E after one pass through μStyragel on C_{18} reverse-phase column. Peak A, BHA; B, α-tocopheryl acetate; C, retinyl palmitate. (From Ref. 45.)

either degradation products [139] or more polar metabolites [115]. For their chromatography McKenzie et al. used methanol-water mixtures without addition of ammonium acetate or acetic acid [139-141]. This resulted in broad tailing peaks as the ionization of retinoic acid was not suppressed.

In 1978, Frolik et al. [119a] reported a quantitative determination of 13-*cis*- and all-*trans*-retinoic acid in human plasma after an oral dosage of 100 mg 13-*cis*-retinoic acid. At the zero time point no 13-*cis*- or all-*trans*-retinoic acid was found on analyzing 0.5 ml of plasma. On the other hand, we extracted 3.5 ml plasma and demonstrated the presence of all-*trans*-retinoic acid on a silica system [56]; identification of the peak was performed on a reversed-phase system. After evaporation of a fraction eluting from the silica system, presumed to contain the retinoic acid, we injected an aliquot on a Lichrospher RP18 10 μm column; an equal k' value was obtained as for a retinoic acid standard. No complete separation of the different isomers of retinoic acid was obtained in this reversed-phase system, as shown in Fig. 1.14, because retention on a reversed-phase column is governed mainly by the number

TABLE 1.10 Chromatographic Conditions for Reversed-Phase Separations of Retinoic Acid and Retinoic Acid Isomers

Compound[a]	Packing material	Eluent composition	Remarks: Sample origin	Remarks: Not separated isomers[b]	Remarks: Detection limit	Reference
Retinoic acid	Partisil 10 ODS	Methanol-water (71:29)	Standard	-	-	139
13-*cis*, all-*trans*-retinoic acid	μBondapak C_{18}	Methanol-0.01 M ammonium acetate (75:25)	Standard	+	1 ng[c]	115
4-Ketoretinoic acid, 4-hydroxyretinoic acid, all-*trans*-retinoic acid	μBondapak C_{18}	Methanol-0.07 M ammonium acetate (65:35)	Standards	-	1 ng[c]	115
13-*cis*, all-*trans*-retinoic acid	Partisil 10 ODS-2 10 μm	Acetonitrile-1% ammonium acetate (65:35)	Serum	+	50 ng/ml	140
Retinoic acid, retinol, retinal, retinyl acetate	Spherisorb ODS	Acetonitrile-1% ammonium acetate (60:40)	Standards	-	-	130
13-*cis*, all-*trans*-retinoic acid	Partisil ODS-2	Acetonitrile-1% ammonium acetate (60:40)	(Standards)	+	-	130
9,11,13-tri-*cis*, 11,13-di-*cis*, 13-*cis*, 9,13-di-*cis*, 11-*cis*, 9-*cis*, all-*trans*-retinoic acid	Partisil 10 ODS	Methanol-water (70:30)	Standards	No complete separation	-	141
13-*cis*, all-*trans*-retinoic acid, retinol, retinyl acetate	Lichrosorb C_{18} 10 μm	Acetonitrile-1% ammonium acetate (80:20)	Serum	+	25 ng/ml	55
13-*cis*, all-*trans*-retinoic acid	Partisil ODS-2	Acetonitrile-1% ammonium acetate (57:42)	Organs	+	-	142
13-*cis*, 9-*cis*, all-*trans*-retinoic acid	Lichrospher RP8 10 μm	Methanol-water-acetic acid (89.7:10:0.2)	Standards	+	-	56

[a]The isomers elute in the sequence given.
[b]-, Not mentioned; +, all isomers mentioned are separated.
[c]Absolute amount.

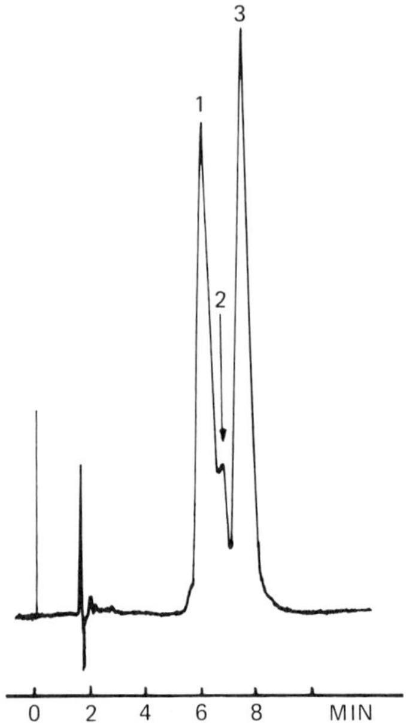

Figure 1.14 Chromatogram of retinoic acid isomers on a C_{18} HL column (15 × 0.32 cm), eluted with methanol-water-acetonitrile (89.7:10:0.3) at a flow rate of 0.75 ml/min with UV detection at 350 nm. Peak identification: (1) 13-*cis*-retinoic acid, (2) 9-*cis*-retinoic acid, (3) all-*trans*-retinoic acid.

of carbon atoms in the molecule. Very recently, Sundaresan and Bhat [119b] were able to separate several vitamin A-related compounds, including 13-*cis*- and all-*trans*-retinoic acid. In their procedure a reversed-phase column was eluted with acetonitrile-potassium phosphate buffer (pH 7.2) mixtures. The order of elution under these conditions was 4-oxoretinoic acid, retinyl phosphate, 13-*cis*- and all-*trans*-retinoic acid, retinol, retinal, retinyl acetate, anhydro-retinol, and retinyl palmitate. However, again, no baseline separation was obtained between the 13-*cis* and the all-*trans* isomers of retinoic acid.

13-*cis*-Retinoic acid has been shown to be less toxic than the parent all-*trans* form in the treatment of severe acne [142-144]. In view of these applications, a pharmacological study and a method for quantitation were developed by Besner et al. in 1980 [55a]. After denaturation with methanol and pH adjustment the extraction was performed with hexane-methylenechloride-isopropanol (80:19:1). Using all-*trans*-retinyl acetate as internal standard, levels of 13-*cis*-retinoic acid down to 25 ng/ml were quantitated. Shelley et al. [55b] described an HPLC assay for 13-*cis*- or all-*trans*-retinoic acid. They simply diluted the serum with acetonitrile and an aliquot of the supernatant was analyzed on a C_{18} column. Under these conditions they were not able to separate 13-*cis*- and all-*trans*-retinoic acid. Recently, Vane et al. [55c] developed a simultaneous determination of 13-*cis*-retinoic acid and of its major metabolite, 4-oxo-13-*cis*-retinoic acid in human blood on a reversed-phase system using a gradient system.

4.4 Further Applications

Liquid chromatographic systems are often successfully applied to the separation and identification of different vitamin A metabolites. In these particular cases, reversed-phase chromatography is the method of choice. With this phase system, metabolites, which are usually more polar, elute before the parent compound. Epoxy derivatives were demonstrated in rat intestines after a single dose of retinoic acid. Extraction of the homogenate with methanol-chloroform and gel chromatography on Sephadex LH-20 and DEAE-Sephadex A-25 columns served as a cleanup step before the reversed-phase chromatography [145]. Mass spectrometric analysis of the collected fraction revealed an oxyderivative, tentatively identified as 5,8-oxyretinoic acid; however, it is very probable that the initial metabolite is 5,6-epoxyretinoic acid. The 5,8-oxy derivative is likely to be produced from the initial metabolite under the acidic conditions of the isolation procedure [146].

Two other metabolites were separated from hamster liver microsomal preparations. They were identified as 4-hydroxy- and 4-ketoretinoic acid by mass spectrometric analysis. The LC separations were performed either on a Partisil ODS-2 column eluted with a gradient of acetonitrile-1% ammonium acetate (from 2:98 to 75:25) or on a Partisil-10-SAX using acetonitrile-1 mM ammonium acetate (73:27) as eluting solvent [147,148].

Separation and purification of retinyl glucuronides by use of LC are also reported by Zile et al. and by Frolik et al. [149,150]. Extracts of bile were purified on Sephadex LH-20 and the polar peak of this column was applied to a reversed-phase system developed with 70% methanol in water containing 10 mM ammonium acetate. This solvent system is extremely interesting in these studies as it can be totally evaporated without salt rest and redissolution for a further step in the analysis gives no problem at all.

4.5 Conclusions

The number of applications of LC in either quantitative work or in biochemical studies has increased tremendously during recent years. The use of smaller column dimensions and better packing materials allowed excellent separations and detection even in the subnanogram range. For the near future the LC-MS combination looks very promising for the identification of unknown structures in metabolization studies.

PART B: PROVITAMIN A (CAROTENOIDS)

1. INTRODUCTION

1.1 Background

Carotenoids are widespread in the photosynthetic tissues of green plants [151]. These pigments participate in light harvesting [152] and protect against harmful radiation, probably via quenching of singlet oxygen [153]. Animals are unable to biosynthesize carotenoids but, with a few exceptions, assimilate them through their diet [154]. They may deposit them unaltered in their tissues [154] or, as crustaceans [155] and birds [156] do, metabolize them further, usually to keto and hydroxy derivatives. In higher animals a limited number of carotenoids are intestinally cleaved to yield retinal [154,157]. These vitamin A precursors are commonly designated with the collective term *provitamin A*.

This review is only concerned with the latter type of carotenoids and does not include an exhaustive literature survey on chromatographic methods for photosynthetic pigment profiling in plants. Reference to papers in this area will only be made to illustrate the principles that apply to chromatography of carotenoids in general.

1.2 Chemistry

The principal carotenoids with provitamin A activity have been listed in recent reviews [154,157]. Structural formulas of the four major vitamin A precursors, i.e., β-carotene, α-carotene, β-cryptoxanthin, and γ-carotene are depicted in Fig. 1.15. Although cis isomers have been detected in plants [158], most natural carotenoids occur in the more stable all-trans configuration. Cis isomers possess lower vitamin A activity than the all-trans form [154,157]. The official IUPAC nomenclature for carotenoids is rather complex [159] and of little concern in this context. Such common terms as carotenes and xanthophylls are used throughout to designate carotenoid hydrocarbons and oxygenated carotenoids, respectively.

1.3 Occurrence

Fruits and vegetables are the major dietary sources of provitamin A carotenoids [154,157]. These yellow to orange pigments, including β-carotene and β-8'-apocarotenal, are also frequently used as colorants

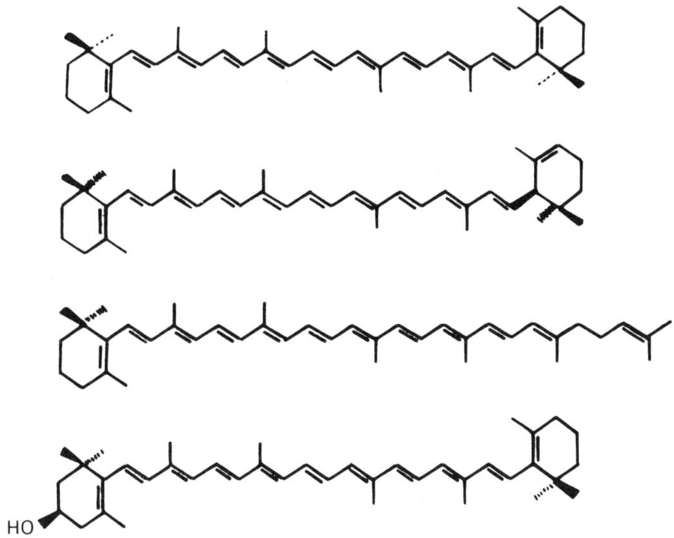

1

2

3

4

Figure 1.15 Structural formulas of the principal provitamins A. (1) β-carotene; (2) α-carotene; (3) γ-carotene; (4) β-cryptoxanthin.

in a variety of foods [160]. In humans, carotenoid blood and tissue levels are a mere reflection of eating habits [160] because, unlike retinol, β-carotene cannot be mobilized from specific body deposition sites, e.g., the liver, when required.

1.4 Biochemical Importance

β-Carotene has maximal provitamin A activity because both halves of its molecule can be converted to retinal [154,157]. This process, which is catalyzed by the intestinal enzyme β-carotene-15,15'-dioxygenase, involves cleavage of the central double bond. Alternatively, a stepwise degradation to retinal has been suggested [157]. Carotenoids containing one β end group, e.g., α-carotene, γ-carotene, and β-cryptoxanthin, theoretically possess 50% of the provitamin A activity of β-carotene, whereas derivatives without a β end group, such as lycopene, are biologically inactive. The biopotencies of several carotenoids relative to β-carotene have been tabulated by Bauernfeind et al. [154] and Simpson [157].

Recently, some hypotheses claim that β-carotene, in addition to its nutritional role, may also participate in non-vitamin A-related processes, such as protection against cancer [161]. However, so far no conclusive evidence proving such a role is available.

1.5 Analytical Considerations

"Classic" carotenoid assays were mostly based on spectrophotometry, often in conjunction with open column chromatography [162,163]. Non-chromatographic approaches fail to distinguish between the individual carotenoids contributing to the total absorbance in the sample [162]. The presence of non-provitamin A pigments in biological materials will lead to significant overestimations of biopotency. Hence, chromatography is an indispensable tool for carotenoid determination. In view of the instability of these pigments, sample pretreatment should be as mild and simple as possible [162,163]. The polyene chain is particularly vulnerable to oxidation and cis/trans isomerization, the latter process being catalyzed by light, heat, and acids [162,164]. Modern chromatographic techniques, especially liquid chromatography, permit sample preparation to be kept to a minimum. The older complex sample cleanup schemes, though still in use, involving saponification, partitioning between organic solvents, open column chromatography, and thin-layer chromatography [162,163], are gradually becoming obsolete.

2. GAS CHROMATOGRAPHY

Gas chromatography has found very limited application to carotenoids, the major obstacle being the thermolability of the polyene chain. Yet this technique, in combination with mass spectrometry, remains potentially useful for identification purposes. Taylor and Ikawa have thoroughly reviewed the state of the art in this field [165]. To increase the volatility and reduce the instability of carotenoids, perhydro derivatives can be prepared. Hydrogenation of the polyene chain is usually carried out with hydrogen gas and platinum dioxide as a catalyst. Still, however, perhydrocarotenoids require high temperatures (up to 300°C) to be eluted within a reasonable time. Xanthophylls containing hydroxyl groups may be acylated or silylated prior to chromatography. Among the different stationary phases tested by Taylor and Ikawa [165], SE-52 (2% on GasChrom Q) proved superior. These authors listed the retention characteristics of a great number of carotenoids for four different column systems. Elution orders were readily predictable from the respective structures. Surprisingly, perhydroketocarotenoids turned out to be less volatile than their corresponding perhydrohydroxy counterparts. It is worth noting that even positional isomers could be resolved on packed columns. A SE-52 column was also successfully coupled with an AEI MS 30 mass spectrometer. On the basis of the 24 eV electron impact spectra of the individual gas chromatographic peaks, several carotenoids of bacterial origin could be identified.

3. LIQUID CHROMATOGRAPHY

3.1 General Considerations

Since the pioneering work of Sweeney and Marsh [166] and Stewart and Wheaton [167] in the early seventies, modern liquid chromatography has superseded other analytical methods for carotenoids. The latter authors were the first to resolve over 20 xanthophylls and carotenes, including some geometrical isomers, in 4-5 hr runs, using either zinc carbonate or magnesium oxide as adsorbents. Sweeney and Marsh [166] used mixtures of magnesium oxide-Hyflo Super-Cel or magnesium hydroxide-calcium hydroxide, well-established supports for open column chromatography, to separate carotene geometrical isomers in vegetables. However, unlike Stewart and Wheaton [167], they did not monitor the column effluent on-line.

On modern microparticulate materials, the above separations would only take a few minutes. Apart from speed, other factors account for the present popularity of liquid chromatography for carotenoid separation as well. HPLC no longer suffers from the drawbacks of open column chromatography, which was characterized by long exposure of the labile polyenes to oxygen, light, active adsorbents, and solvents, nor does it require derivatization or elevated analysis temperatures. The presence of the polyene chain ensures excellent detectability with modern absorbance detectors.

An HPLC system suitable for carotenoids should meet the following requirements.

1. Separation of individual carotenes and xanthophylls in one run.
2. Separation of positional isomers, such as the pairs α-carotene-β-carotene and lutein-zeaxanthin, respectively (the latter is less critical in this context because lutein and zeaxanthin lack provitamin A activity).
3. Separation of geometrical (cis/trans) isomers.

Normal-phase and reversed-phase chromatography have their own specific applications and are often complementary. Isocratic as well as gradient elution systems have been reported.

3.2 Survey of Systems

3.2.1 Normal-Phase Systems

Solvents used for adsorption chromatography of carotenoids include silica [168-172], alumina [173-175], magnesia [176], and lime (calcium hydroxide) [177]. On silica separation distinct product classes are obtained, but individual components within a group may remain unresolved. For example, carotenes have been differentiated in acyclic (e.g., lycopene), monocyclic (e.g., γ-carotene) and bicyclic (e.g., α- and β-carotene) derivatives, with the latter positional isomers co-eluting as one peak [168]. Similar difficulties are often encountered in separating the corresponding hydroxylated derivatives lutein and zeaxanthin [169]. In general, selectivity on silica is determined by the number and the nature of the functional groups as well as the overall steric configuration (position of the double bonds) of the molecule. Thus, silica is the adsorbent of choice for the separation of cis/trans isomers [168,172] and diastereoisomers [168]. An example of a separation of geometrical isomers of astaxanthin diacetate on silica is shown in Fig. 1.16. However, this support has been suspected of catalyzing on-column carotenoid degradation, probably as a result of too strong interactions with active sites [178]. This unfortunate phenomenon precludes quantitative work on silica, particularly with cis isomers, which during chromatography tend to isomerize back to the more stable all-trans form [179].

Alumina and silica usually have quite different selectivities for carotenoids. Not only does alumina display a higher retention for the nonpolar carotenes, but it is also capable of separating α- and β-carotene [173]. Furthermore, resolution of complex mixtures of geometrical isomers has been achieved on alumina as well, but analysis times were long, i.e., exceeding 1 hr [175]. Careful control of the water content of the mobile phase and the temperature was crucial to obtain reproducible results.

Less common adsorbents, such as magnesia, zinc carbonate, or calcium carbonate, have been successfully used in open column chromatography [180,181]. These materials often display a unique selectivity, especially for cis/trans isomers. In general, however, they are insufficiently stable to be packed in high-performance columns [175]. Recently, however, an excellent separation of β-carotene geometrical isomers was reported on microparticulate (10 μm) lime (calcium hydrooxide) [177].

Polar bonded phases may be valuable alternatives if carotenoid degradation on active adsorbents occurs. For example, nitrile (cyano) phases have proved useful for the separation of carotene diastereoisomers [182] and for profiling microbial carotenoids [183].

3.2.2 *Reversed-Phase Systems*

Reversed-phase (RP) materials have a number of distinct advantages to silica for carotenoid analysis. Solute-support interactions on nonpolar bonded phases involve weaker forces, thus minimizing the risk for on-column artefact formation [157,178]. Retention in RP is little affected by small variations in the mobile phase composition. For example, no significant effect is seen from slight changes in water content. Compared with silica, reversed-phase materials show enhanced

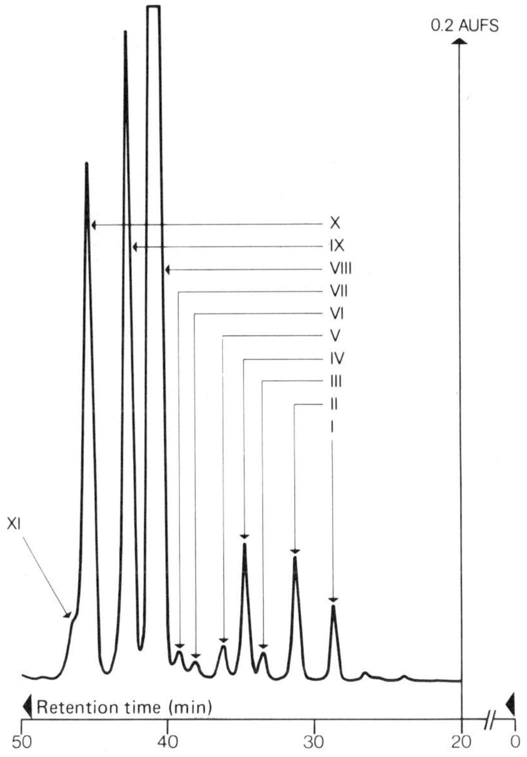

Figure 1.16 Chromatogram of a mixture of cis/trans isomers of astaxanthin diacetate. Chromatographic conditions: column, 50 × 0.32 cm packed with 5 μm Lichrosorb Si 60; mobile phase, hexane-ethyl acetate-acetonitrile (88:10:2); flow rate, 0.8 ml/min; detection, 473 nm. Peak identification: I, 9,9'-di-*cis*; II, 9,13'-di-*cis*; III, 9,15-di-*cis*; IV, 9,13-di-*cis*; V, 13,13'-di-*cis*; VI, 13,15-di-*cis*; VII, unidentified; VIII, all-*trans*; IX, 9-*cis*; X, 13-*cis*; XI, 15-*cis*. (From Ref. 172.)

stability and improved reproducibility. Hence, reversed-phase chromatography is preferable for quantitative work. This applies particularly to the extremely unstable *cis*-carotenoids. Unfortunately, nonpolar materials do not lend themselves very well to the separation of geometrical isomers. Very often, the trans peak is only partially resolved from a composite total cis peak [184], although there are examples of separation of individual cis isomers on RP as well [185]. In contrast, positional double-bond isomers, such as α- and β-carotene, can be readily distinguished on octadecyl-type (RP 18) phases [186-192], a basic requirement for any HPLC method to be acceptable for application to foods and serum.

Strong mobile phases, consisting of pure acetonitrile [187] or, more commonly, mixtures of acetonitrile with chloroform [186,190,192], dichloromethane [189,193], or tetrahydrofuran (THF) [191] are necessary to elute the nonpolar carotenes from RP 18 columns. Under those conditions, however, most RP materials exhibit too low retention for the more polar derivatives. As a result, concomitant chromatography of xanthophylls and carotenes requires gradient elution [178,194-196]. Sufficient retention for the polar xanthophylls is ensured by initial semiaqueous eluting conditions. The eluotropic strength is then gradually increased to end up with a nonaqueous mobile phase, usually containing a strong solvent component, which finally elutes the carotenes. However, use of semiaqueous mobile phases may have a deteriorating effect on the chromatographic efficiency and peak shape, because of poor sample solubility. This has been observed, for example, in connection with reversed-phase chromatography of K vitamins, another class of highly lipophilic compounds [197]. Switching from nonaqueous to semiaqueous conditions led to a considerable drop in plate count and a dramatic distortion of peak shape. Even the addition of as little as 1-2% of water had a marked effect.

The alternative remedy is to select a "less selective" column material, i.e., one that displays a relatively stronger retention for xanthophylls, but for the nonsubstituted carotenes maintaining a degree of retention similar to that on "too selective" materials. The result will be that nonaqueous mobile phases will now become useful to elute the whole spectrum of carotenoids, without the need for gradient elution. Zorbax ODS is one such unique material that permits isocratic elution of compounds of widely divergent polarity, even in extreme cases such as the concomitant chromatography of retinol (20 carbon atoms, hydroxylated) and β-carotene (40 carbon atoms, unsubstituted) [189]. The influence of different eluent parameters on retention and selectivity in nonaqueour reversed-phase chromatography (NARP) of carotenoids and retinoids has been systematically studied [189]. The concentration of the nonpolar modifier, e.g., dichloromethane, chloroform, or THF, in the mobile phase merely controls retention and has little effect on selectivity, except for THF. On the other hand, solvents with hydrogen-binding capability, such as methanol, profoundly affect selectivity and hence resolution. This was demonstrated using β-carotene and echinenone as model compounds [Fig. 1.17). A similar beneficial effect of methanol has been reported previously in connection with NARP of other fat-soluble vitamins [198]. The gain in resolution is even more pronounced for the critical pair retinol-retinyl propionate [189]. Increasing concentrations of methanol significantly decrease retention of the hydrogen donor retinol, but the capacity ratio of the nonhydrogen donor retinyl propionate is only slightly affected. Fig. 1.18 shows a separation of nine carotenoid standards using NARP conditions. Total chromatographic analysis time is 16-32 min. This system is now under extensive

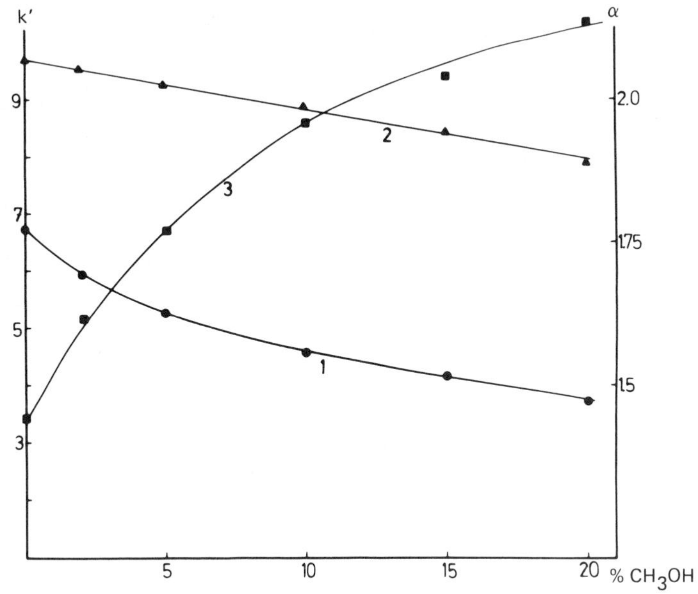

Figure 1.17 Capacity ratios (k') of echinenone (1), β-carotene (2), and the selectivity factor α (3) versus percentage (% v/v) of methanol in the eluent (20%, v/v dichloromethane in acetonitrile). (Reprinted with permission from Ref. 189. Copyright 1983 American Chemical Society.)

investigation as a basis for carotenoid quantitation in biological materials of vegetal, animal, and human origin [189].

In general, gradient elution has a number of disadvantages, the most notable the long equilibration periods needed between runs [157, 189]. Yet, Simpson developed some useful gradient elution conditions that apparently do not suffer from the above drawback [157]. This author has routinely used a RCM-C18 5 μm cartridge, eluted with a gradient of water-methanol (3:97) to tetrahydrofuran in methanol (10:90). Total analysis time for the whole range of carotenoids is 30 min, and a backwashing of only 10 min is sufficient to give satisfactory reproducibility. It is interesting to note that an isocratic system based on the strong eluting conditions (THF-methanol, 10:90) of this system is capable of resolving the principal provitamins A, i.e., β-cryptoxanthin, α-carotene, and β-carotene, as well as lycopene, in only 17 min [157].

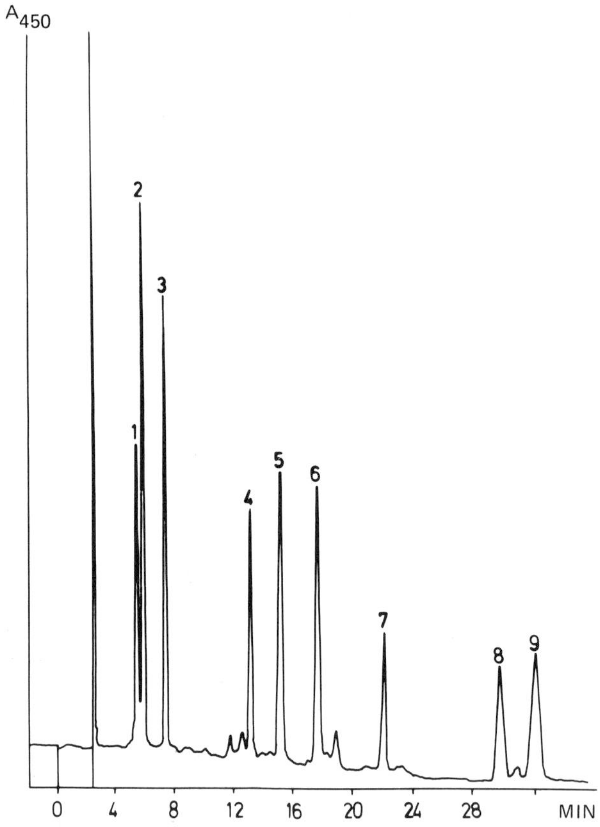

Figure 1.18 Separation of carotenoid standards on a Zorbax ODS (5 μm) column (25 × 0.46 cm). The mobile phase consisted of acetonitrile-dichloromethane-methanol (70:20:10, v/v). Flow rate was 1 ml/min and detection was at 450 nm. Peak identification: (1) lutein; (2) zeaxanthin; (3) canthaxanthin; (4) β-cryptoxanthin; (5) echinenone; (6) lycopene; (7) torulene; (8) α-carotene; (9) β-carotene. (Reprinted with permission from Ref. 189. Copyright 1983 American Chemical Society.)

3.2.3 Gel Permeation Chromatography

Gel permeation chromatography may be of interest when compounds of considerably different size, such as β-carotene (40 carbons), retinyl palmitate (36 carbon atoms), and retinyl acetate (22 carbon atoms), are to be separated. There is an example of such separation in the literature [199]. A μStyragel column eluted with dichloromethane containing

0.001% triethylamine was used. The whole separation process takes 24 min. This approach found an application for the determination of the above components in oil and margarine. Its value for separating closely related carotenoids, however, is negligible.

3.3 Detection

Carotenoids strongly absorb light in the 400-500 nm region. Absorption detection at these wavelengths is particularly favorable in terms of selectivity (little interference from biological materials) and sensitivity (molar absorption coefficients of 10^5 and more). So far, other detectors have not found any application. Although they are not fluorescent, carotenoids are electro-chemically active [200], so that amperometric or coulometric detection should be feasible.

A major problem that is to be faced is identification of chromatographic peaks in complex pigment patterns. On-line recording of the characteristic absorption spectra may provide useful structural information. For example, *cis*-carotenoids display, in addition to the main band around 450 nm, a cis peak in the long wavelength UV region [185, 201]. The hypsochromic shift of the main maximum after acid treatment of unknown peaks is characteristic for 5,6-epoxycarotenoids (conversion of 5,6-epoxides to furanoids) [185]. Modern scanning detectors and especially the recently marketed photodiode array detectors are strong analytical tools for identification of carotenoids. If not available, wavelength ratioing in connection with multiple injections remains a valuable alternative [185].

3.4 Applications

3.4.1 Foods

All vitamin A in the blood and tissues of humans and animals is directly or indirectly derived from dietary carotenoids. Nutritional surveys conducted in underdeveloped countries have demonstrated that a moderate daily intake of green leaf vegetables would lead to a total eradication of the incidence of hypovitaminosis A and its accompanying nutritional blindness (xerophthalmia) [202]. Food analysis in general is of growing interest because of the current awareness among consumers about nutritional requirements and dietary standards and the increasing weight of regulations imposed by governments on food producers and processors. The determination of provitamin A is a useful part of food quality control programs.

In most vegetables and fruits, α- and β-carotene comprise up to 90% of provitamin A activity [188,203], except in citrus fruits, in which β-cryptoxanthin is the predominating vitamin A precursor. γ-Carotene only contributes a few percentage by weight to the total provitamin A content. Food processing may lead to alterations in the carotenoid pattern, particularly because of heat-catalyzed trans → cis conversion [203]. As mentioned earlier, *cis*-carotenoids possess lower provitamin A activity than the corresponding *trans* forms. The picture is further complicated by the presence of significant amounts of carotenoids that lack provitamin A activity [157], typical examples being lycopene, lutein, and zeaxanthin. Thus, the biopotency of tomatoes may be at least 10-fold overestimated as a result of the presence in excess of lycopene, when nonchromatographic assays such as spectrophotometry are used [157]. In citrus juice, cryptoxanthin predominates, which is only 50% as active as the minor constituent β-carotene.

From the above considerations, two essential requirements for the "ideal" chromatographic method for carotenoid determination in foods emerge: differentiation between α- and β-carotene and at the same time separation of "active" from such "inactive" carotenoids as lycopene. If possible, the method should also be capable of resolving cis and trans isomers. The first requirement virtually precludes the use of other systems but reversed phase, alumina being a possible alternative.

Reversed-phase chromatography has undoubtedly become the method of choice in this regard. It forms the basis of methods for carotenoid determination in tomatoes [186], carrots [192], paprika [204], citrus juice [205,206], and a variety of other fruits and vegetables [157,188]. Figure 1.19 shows a typical chromatogram of a tomato extract. The separation between β-carotene and the inactive lycopene, is clear-cut. α-Carotene, though not detected in tomatoes, is well resolved from β-carotene in this system as well [186]. Several authors list the carotenoid content of a number of selected fruits and vegetables as determined by HPLC and, for reasons of comparison, spectrophotometry [157,188]. The latter technique yields, as predicted, erroneously high provitamin A contents [157]. In paprika, geometrical isomers of capsanthin (9-*cis* and 13-*cis*) have been detected and conveniently separated on reversed phase [204].

As an alternative to reversed phase, magnesium oxide [176,207] and alumina [173,174] have been used for the analysis of orange juice, citrus juice, and such processed foods as pizza.

Sample pretreatment usually consists of a crude lipid extraction, using, e.g., acetone or tetrahydrofuran, possibly followed by saponification [186,204], and reextraction with an organic solvent, such as diethyl ether. There is a tendency toward simplification of sample preparation, especially by omitting the saponification step [188,192]. The crude extract is often even directly injected on the column. The selective detection as 450 nm prevents potentially interfering compounds from showing up in the chromatograms.

3.4.2 Human Serum

Although there is still controversy about the relevance of plasma carotenoid levels as an indicator of vitamin A status [208], β-carotene

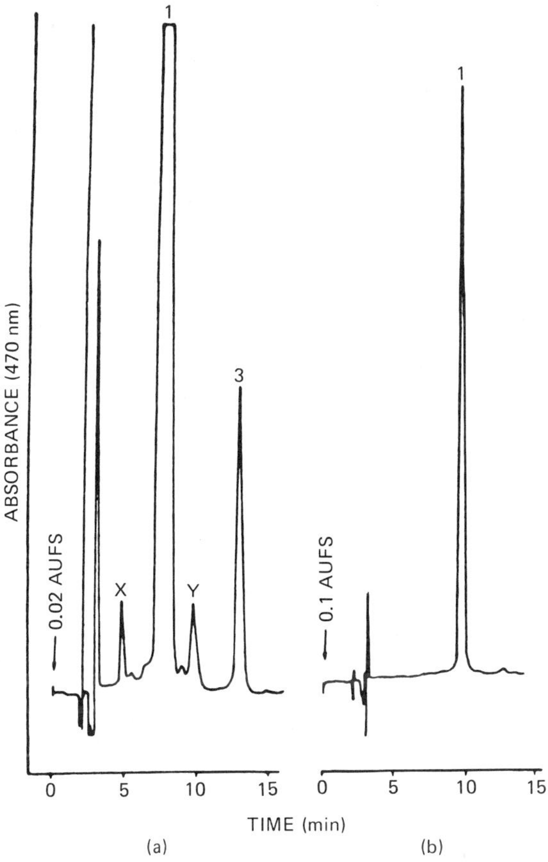

Figure 1.19 Chromatogram of a tomato extract. Chromatographic conditions: column, Partisil-5/ODS, 5 μm; eluent, chloroform-acetonitrile (8:92, v/v); flow rate, 2 ml/min; temperature, ambient; detection, 470 nm. (a), 50 μl injected; (b), 10 μl injected. Peak identification: (1) lycopene; (3) β-carotene; x and y, unknown. (From Ref. 186.)

and to a lesser extent other provitamins A are now commonly determined in human serum as part of nutritional assessment programs [209]. The International Vitamin A Consultative Group (IVACG) has recently given guidelines for the interpretation of hypo- or hypercarotenemia, with or without associated abnormal serum retinol levels [209]. In general, low β-carotene levels reflect either inadequate dietary intake or intestinal malabsorption. Hypercarotenemia may result from medical use of β-carotene in the treatment of photosensitivity [210].

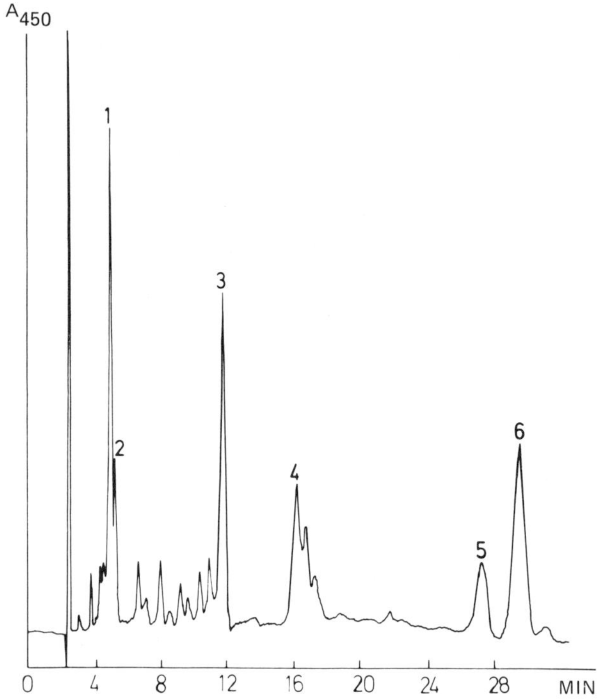

Figure 1.20 Carotenoid profile of a human serum extract. Column, 25 × 0.46 cm, packed with 5 μm Zorbax ODS. The mobile phase consisted of acetonitrile-dichloromethane-methanol (70:20:10, v/v). Flow rate was 1 ml/min and detection was at 450 nm. Peak identification: (1) lutein; (2) zeaxanthin; (3) β-cryptoxanthin; (4) lycopene; (5) α-carotene; (6) β-carotene. (Reprinted with permission from Ref. 189. Copyright 1983 American Chemical Society.)

The carotenoid pattern in serum of healthy humans is very complex (Figure 1.20), the predominating chromatographic peaks attributable to β-carotene, β-cryptoxanthin, α-carotene, lycopene, and lutein. Very recently, a few assays for serum carotenoids have been published [190, 191,193]. They all focus on the determination of α- and/or β-carotene, and one also includes lycopene [190]. Other provitamins A, such as β-cryptoxanthin, though readily detectable in serum (Fig. 1.20), are not determined. All three methods have the nonaqueous reversed-phase conditions in common, as well as similar sample pretreatment. The latter consists of an initial deproteinization with methanol [190], ethanol [193], or a mixture of tetrahydrofuran and ethyl acetate [191]. An aliquot of the organic layer may either be directly injected on the chromatographic column [190,191], or alternatively, the whole organic layer may be evaporated to dryness and the residue reconstituted with an organic solvent before injection [193]. Detection limits vary from 80 [193] to 10 μg/liter [191], with sample volumes ranging from 100 to 500 μl.

The method of Driskell et al. [193] is unique in that it is the only one to use a carotenoid as an internal standard (Fig. 1.21). Dimethyl-β-carotene, a synthetic derivative, seems to meet the criteria for an ideal internal standard. It is a close analog of β-carotene, exhibits a similar degree of lipophilicity and hence extractibility, does not occur in nature, and, most important, elutes after β-carotene, which of the natural C_{40} carotenoids is the one most retained on reversed phase. Any internal standard eluting before β-carotene could potentially coincide with one of the peaks of the complex serum chromatogram. Use of an internal standard should be highly recommended anyway, because, in the absence of compensatory measures, the above assays, involving handling of small volumes of organic solvents, will be subject to considerable variability.

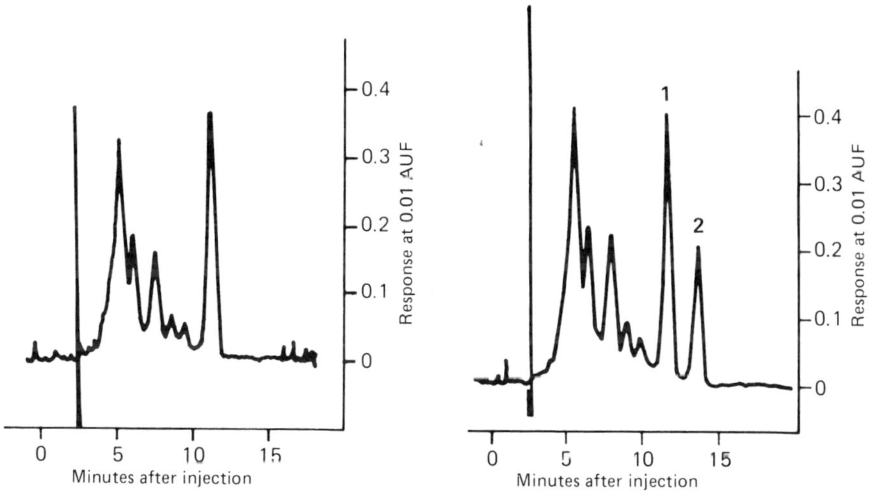

Figure 1.21 Liquid chromatogram of a human serum extract containing no internal standard (*left*) or containing the internal standard, dimethyl-β-carotene (*right*). Chromatographic conditions: column, 23 × 0.39 cm, packed with 10 μm μBondapak C_{18}. Eluent: acetonitrile-dichloromethane (89:11, v/v). Flow rate, 2 ml/min. Detection was at 450 nm. Peak 1 is β-carotene; peak 2 is the internal standard dimethyl-β-carotene. (From Ref. 193. Reprinted with permission from *Clinical Chemistry*.)

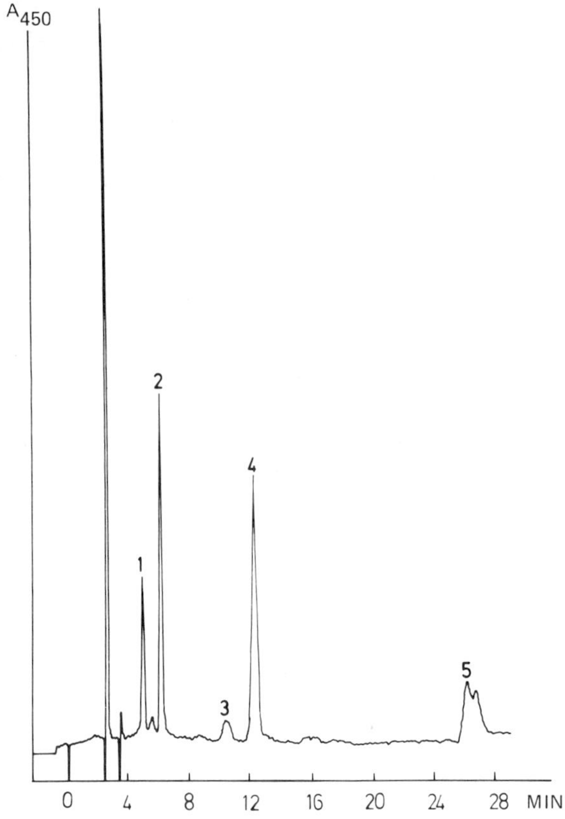

Figure 1.22 Carotenoid profile of adult Artemia (strain not specified) (acetone homogenate). Chromatographic conditions: column 25 × 0.46 cm, packed with Zorbax ODS 5 μm, eluent, acetonitrile-dichloromethane-methanol, 70:20:10, v/v; flow rate, 1 ml/min; detection, 450 nm. Peak identification: (1) zeaxanthin; (2) canthaxanthin; (3) β-cryptoxanthin; (4) echinenone; (5) β-carotene. (Reprinted from Ref. 189. Copyright 1983 American Chemical Society.)

A β-carotene-related compound, i.e., its colorless precursor phytoene, has also been determined in serum using HPLC [211].

3.4.3 Feeds

Like humans, animals depend on carotenoids as a primary source for their vitamin A supply. Bauernfeind et al. have exhaustively reviewed the occurrence and importance of provitamin A in animal feed [154]. Poultry feeds are often supplemented with carotenoids to en-

enhance skin and yolk pigmentation [212]. Similarly, salmon and trout pigmentation is strongly diet dependent [213]. A carotenoid pattern of Artemia, a very popular feed for fish, shrimp, and prawns, is shown in Fig. 1.22. Modern handbooks still recommend open column chromatography for the analysis of feeds [163]. Although no quantitative HPLC methods dealing exclusively with feeds have appeared as yet, the principles outlined for food analysis should equally apply to animal feeds as well. Only minor modifications in sample pretreatment might be necessary.

4. OPEN-COLUMN CHROMATOGRAPHY

Until the appearance of modern HPLC, open column chromatography in conjunction with off-line spectrophotometry was the most widely used method for carotenoid determination in biological materials [163]. It has now lost most of its significance, except as a (semi-)preparative cleanup step for pigment isolation from complex matrices. For the analytical separation of carotenoids it has been totally superseded by HPLC on microparticulate materials with on-line detection.

A great variety of adsorbents have found some application [214,215], of which magnesium oxide, often in combination with Hyflo Super Cel [216], is one of the most useful in terms of selectivity and compound stability. Most adsorbents are indeed more or less destructive of carotenoids because of high surface activity [217]. The risk on degradation is further enhanced by the long exposure to solvents, heat, and light during column development. Other disadvantages of open column chromatography include its poor resolving power, especially with regard to positional and geometrical isomers, and low sensitivity.

A typical analysis scheme based on this technique consists of successive elutions of one or more columns with solvent mixtures of increasing eluotropic strength [163]. Quantitation of individual carotenoids in collected fractions is subsequently carried out spectrophotometrically and concentrations of each component are calculated on the basis of molar absorption coefficients. Examples of this approach are the determination of carotenoids in fruits [216,218] and serum [219,220].

5. THIN-LAYER CHROMATOGRAPHY

TLC has always been of great value for the identification of carotenoids, particularly in photosynthetic tissues [214,221]. For quantitative work, it suffers from the same disadvantages as open column chromatography. A number of reviews deal with adsorbents and useful solvent combinations [214,222]. As in column chromatography, magnesium oxide [217, 221-224] and Hyflo Super Cel [217] are among the most valuable materials. Carotenoids have been determined in fruits [223], vegetables [217], and serum [225] by spectrophotometry following TLC fractionation.

6. CONCLUSIONS

Modern chromatography of carotenoids is becoming restricted to virtually one technique, i.e., liquid chromatography. The main reason for its exclusive position is that HPLC and especially reversed-phase chromatography guarantees a maximum degree of stability to these extremely labile compounds. HPLC separations are fast and can be conducted under very mild conditions, at room temperature and protected from light. As a result, the older assays based on spectrophotometry in conjunction with open column chromatography or thin-layer chromatography are gradually being abandoned. Other chromatographic techniques, such as TLC and GC (in combination with MS) remain of value, particularly for identification purposes. For quantitative work, however, modern HPLC has definitely become the method of choice.

ACKNOWLEDGMENTS

The authors gratefully acknowledge the support of their research on the subject treated in this chapter by the grants 20210, 3.0008.76, and 3.00011.81 from the National Foundation for Medical Research (FGWO) and by bursaries from the National Foundation of Scientific Research (NFWO) to Lambert and Nelis.

REFERENCES

1. IUPAC, Commission on the Nomenclature of Biological Chemistry, *J. Amer. Chem. Soc.*, *82*, 5581-5583 (1960).
2. H. R. Cama, F. D. Collins, and R. A. Morton, *Biochem. J.*, *50*, 48-59 (1951).
3. J. Boldingh, H. R. Cama, F. D. Collins, R. A. Morton, N.T. Gridgeman, O. Isler, M. Kofler, R. J. Taylor, A. S. Welland, and T. Bradbury, *Nature*, *168*, 598 (1951).
4. J. A. Olson, and V. Pungapongse, *Fed. Proc.*, *26*, 636 (1967).
5. D. E. Balke, and R. S. Becker, *J. Amer. Chem. Soc.*, *89*, 5061-5062 (1967).
6. M. Kofler, and S. H. Rubin, *Vitam. Horm.*, *18*, 315-339 (1960).
7. J. A. Olson, In *Newer Methods for Nutritional Biochemistry*, Vol. 2, (A. A. Albanese, ed.), Academic Press, New York, 1965, pp. 345-360.
8. O. Isler, H. Klaüi, and U. Solms, Vitamin A. III. Industrial preparation and production, in *The Vitamins* (W. H. Sebrell, Jr., and R. S. Harris, eds.), Academic Press, New York, 1967, pp. 101-113.
9. O. Isler, U. Solms, and J. Wursch, in *Fat Soluble Vitamins* (R. A. Morton, ed.), Pergamon Press, Oxford, 1970, p. 99.

10. V. Schwieter, and O. Isler, Vitamin A. II. Chemistry, in *The Vitamins*, Vol. 1 (W. H. Sebrell, Jr., and R. S. Harris, eds.), Academic Press, New York, 1967, pp. 5-101.
11. W. Stepp, *Biochem. Z.*, *22*, 452-460 (1909).
12. E. V. McCollum, and M. Davies, *J. Biol. Chem.*, *15*, 167-175 (1913).
13. E. V. McCollum, and N. Simmonds, *J. Biol. Chem.*, *32*, 181 (1917).
14. L. S. Fridericia, and E. Holm, *Amer. J. Physiol.*, *73*, 63-78 (1925).
15. D. S. Goodman, H. S. Huang, M. Kanai, and T. Shiratory, *J. Biol. Chem.*, *242*, 3543-3544 (1967).
16. H. H. Fidge, and D. S. Goodman, *J. Biol. Chem.*, *243*, 4372-4379 (1968).
17. D. S. Goodman, and J. A. Olson, In *Methods in Enzymology*, Vol. XV, *Steroids and Terpenoids*. New York, Academic Press, 1969, pp. 462-475.
18. J. S. K. David, and J. Ganguly, *Indian J. Biochem.*, *4*, 14-17 (1967).
19. H. H. Fidge, T. Sheraton, J. Ganguly, and D. S. Goodman, *J. Lipid Res.*, *9*, 103-109 (1968).
20. S. Mahadevan, D. S. Deshmukh, and J. Ganguly, *Biochem. J.*, *93*, 499-503 (1964).
21. D. S. Goodman, *Fed. Proc.*, *38*, 2501-2503 (1979).
22. D. S. Goodman, in *Trace Components of Plasma: Isolation and Clinical Significance* (G. A. Jamieson, and J. J. Greenwalt, eds.), Alan R. Liss, New York, 1976, pp. 313-330.
23. R. D. Zachman, and J. A. Olson, *Nature*, *201*, 1222-1223 (1964).
24. K. Lippel, and J. A. Olson, *J. Lipid Res.*, *9*, 168-175 (1968).
25. M. H. Zile, H. K. Schnoes, and H. F. DeLuca, *Proc. Nat. Acad. Sci. U.S.*, *77*, 3230-3233 (1980).
26. A. M. McCormick, J. L. Napoli, H. K. Schnoes, and H. F. DeLuca, *Biochemistry*, *17*, 4085-4090 (1978).
27. W. K. Sietsema, and H. F. DeLuca, *Biochem. Biophys. Res. Commun.*, *90*, 1091-1097 (1979).
28. C. A. Frolik, A. B. Roberts, T. E. Tavela, P. P. Roller, D. L. Newton, and M. B. Sporn, *Biochemistry*, *18*, 2092-2097 (1979).
29. A. B. Roberts, L. C. Lamb, and M. B. Sporn, *Arch. Biochem. Biophys.*, *199*, 374-382 (1980).
30. C. A. Frolik, P. P. Roller, A. B. Roberts, and M. B. Sporn, *J. Biol. Chem.*, *255*, 8057-8062 (1980).
31. C. A. Frolik, B. N. Swanson, L. L. Dart, and M. B. Sporn, *Arch. Biochem. Biophys.*, *208*, 344-352 (1981).
32. M. H. Zile, R. V. Inhorn, and H. F. DeLuca, *J. Nutr.*, *110*, 2225-2230 (1980).
33. G. Wald, *Vitam. Horm.*, *18*, 417 (1960).

34. T. Moore, in *The Vitamins*, Vol. 1, 2nd ed. (W. H. Sebrell, Jr., and R. S. Harris, eds.), Academic Press, New York, 1967, p. 245.
35. D. McLaren, in *The Vitamins*, Vol. 1, 2nd ed. (W. H. Sebrell, Jr., and R. S. Harris, eds.), Academic Press, New York, 1967, p. 267.
36. R. S. Harris, in *The Vitamins*, Vol. 1, 2nd ed. (W. H. Sebrell, Jr., and R. S. Harris, eds.), Academic Press, New York, 1967, p. 3.
37. W. Bollag, *Eur. J. Cancer, 8*, 689-693 (1972).
38. H. Mayer, W. Bollag, R. Hanni, and R. Ruegg, *Experientia, 34*, 1105-1119 (1978).
39. M. B. Sporn, N. M. Dunlop, D. L. Newton, and J. M. Smith, *Fed. Proc., 35*, 1332-1338 (1976).
40. M. B. Sporn, and R. A. Squire, *Science, 195*, 487-489 (1977).
41. A. K. Verma, H. M. Rice, B. G. Shapas, and R. K. Boutwell, *Cancer Res., 38*, 793-801 (1978).
42. G. L. Peck, T. G. Olsen, F. W. Yoder, J. S. Strauss, D. T. Downing, M. Pandya, D. Butkus, and J. Battandries, *N. Engl. J. Med., 300*, 329-333 (1979).
43. M. L. Reed, J. Stanley, F. Stengel, J. L. Shupack, and D. M. Benjamin, *Arch. Dermatol., 115*, 605-608 (1979).
44. W. O. Landen, and R. R. Eitenmiller, *J. Ass. Offic. Anal. Chem., 62*, 283-289 (1979).
45. W. O. Landen, *J. Ass. Offic. Anal. Chem., 63*, 131-136 (1980).
46. K. Aitzetmuller, J. Pilz, and R. Tasche, *Fette, Seifen, Anstrichm., 81*, 40-43 (1979).
47. R. Fankel, and K. Schmid, *Landwirtsch. Forsch., 33 II*, 283-289 (1976).
48. K. Ranfft, and H. Rückemann, *Z. Lebensm.-Unters. Forsch., 166*, 13-14 (1978).
49. H. K. Biesalski, W. Ehrenthal, G. Hafner, and O. Harth, *GIT Fachz. Lab. Suppl. Chromatographie*, 4-8 (1981).
50. M. G. De Ruyter, and A. P. De Leenheer, *Clin. Chem., 24*, 1920-1923 (1978).
51. M. G. M. De Ruyter, and A. P. De Leenheer, *Clin. Chem., 22*, 1593-1595 (1976).
52. F. G. Pilkiewicz, M. J. Pettei, A. P. Yudd, and K. Nakanishi, *Exp. Eye Res., 24*, 421-423 (1977).
53. G. W. T. Groenendijk, W. J. de Grip, and F. J. M. Daemen, *Biochim. Biophys. Acta, 617*, 430-438 (1980).
54. C. V. Puglisi, and J. A. F. De Silva, *J. Chromatogr., 152*, 421-430 (1978).
55a. J. G. Besner, and R. Leclaire, *J. Chromatogr. Biomed. Appl., 183*, 346-351 (1980).
55b. R. Shelley, J. C. Price, H. W. Jun, D. E. Cadwallader, and A. C. Capomacchia, *J. Pharm. Sci., 71*, 262-264 (1982).

55c. F. M. Vane, J. K. Stoltenborg, and C. J. L. Bugge, *J. Chromatogr. Biomed. Appl.*, *227*, 471-484 (1982).
56. M. G. De Ruyter, W. E. Lambert, and A. P. De Leenheer, *Anal. Biochem.*, *98*, 402-409 (1979).
57. E. Stahl, in *Thin Layer Chromatography* (E. Stahl, ed.), Springer-Verlag, Berlin, and Academic Press, New York, 1969, pp. 52-85.
58a. F. H. Carr, and E. A. Price, *Biochem. J.*, *20*, 497-502 (1926).
58b. J. B. Neeld, and W. N. Pearson, *J. Nutr.*, *79*, 454-462 (1963).
59. C. Von Planta, V. Schwieter, L. Chopard-dit-Jean, R. Ruegg, M. Kofler, and O. Isler, *Helv. Chim. Acta*, *45*(2), 548-561 (1962).
60. A. P. De Leenheer, and M. G. M. De Ruyter, *Anal. Biochem.*, *63*, 169-174 (1975).
61. D. Oesterhelt, M. Meentzen, and L. Schuhmann, *Eur. J. Biochem.*, *40* , 453-463 (1973).
62. L. Y. Jan, *Vision Res.*, *15*, 1081-1086 (1975).
63. H. Parizkova, and J. Blattna, *J. Chromatogr.*, *191*, 301-306 (1980).
64. R. M. McKenzie, M. L. McGregor, and E. C. Nelson, *J. Label. Compounds Radiopharm.*, *15*, 265-278 (1977).
65. T. N. R. Varma, T. Panalaks, and T. K. Murray, *Anal. Chem.*, *36*, 1864-1865 (1964).
66. Y. K. Fung, R. G. Rahwan, and R. A. Sams, *J. Chromatogr.*, *147*, 528-531 (1978).
67. S. R. Targan, S. Merrill, and A. D. Schwabe, *Clin. Chem.*, *15*, 479-486 (1969).
68. J. Kahan, *J. Chromatogr.*, *30*, 506-513 (1967).
69. K. Yagishita, P. R. Sundaresan, and G. Wolf, *Nature*, *203*, 410-412 (1964).
70. K. Lippel, and J. A. Olson, *J. Lipid Res.*, *9*, 168-175 (1968).
71. K. Lippel, and J. A. Olson, *J. Lipid Res.*, *9*, 580-586 (1968).
72. S. Baczyk, K. Bazanowska, W. Jakubowska, and I. Sobisz, *Fresenius Z. Anal. Chem.*, *255*, 132-133 (1971).
73. D. E. M. Lawson, and E. Kodicek, *Nutr. Proc.*, 8th Int. Congr. 158-158 (1969).
74. P. E. Dunagin, and J. A. Olson, *Anal. Chem.*, *36*, 756-759 (1964).
75. P. E. Dunagin, and J. A. Olson, *Methods Enzymol.*, *15*, 289-301 (1969).
76. T. W. Fenton, H. Vogtmann, and D. R. Clandinin, *J. Chromatogr.*, *77*, 410-412 (1973).
77. K. K. Carroll, and D. C. Herting, *J. Amer. Oil Chem. Soc.*, *41*, 473-474 (1964).
78. T. Ninomiya, K. Kikodoro, M. Horiguchi, and N. Higosaki, *Bitamin*, *27*, 349 (1963).
79. M. Vecchi, W. Vetter, W. Walther, S. F. Jermstad, and G. W. Schutt, *Helv. Chim. Acta*, *50*, 1243-1248 (1967).

80a. M. Vecchi, J. Vesely, and G. Oesterhelt, *J. Chromatogr.*, *83*, 447-453 (1973).
80b. J. Van Antwerp, and J. Lepore, *J. Liquid Chromatogr.*, *5*(3), 571-584 (1982).
81. C. C. Sweeley, R. Bentley, M. Mahita, and W. W. Wells, *J. Amer. Chem. Soc.*, *85*, 2497-2507 (1963).
82. J. Eagles, W. M. Laird, R. Self, and R. L. M. Synge, *Biomed. Mass Spectr.*, *1*, 43 (1974).
83. D. R. Hughes, P. Rietz, W. Vetter, and G. A. Pitt, *J. Int. Vitaminol. Nutr.*, *46*, 231-234 (1976).
84. P. E. Dunagin, R. D. Zachman, and J. A. Olson, *Biochim. Biophys. Acta*, *124*, 71-85 (1966).
85. T. C. Chiang, *J. Chromatogr.*, *182*, 335-340 (1980).
86. L. R. Snyder, J. J. Kirkland, *Introduction to Modern Liquid Chromatography*, 2nd ed., John Wiley & Sons, New York, 1979.
87. P. A. Bristow, *Liquid Chromatography in Practice, HETP*, Cheshire, 1976.
88. R. P. W. Scott, *Contemporary Liquid Chromatography*, John Wiley & Sons, New York, 1976.
89. D. L. Saunders, *J. Chromatogr.*, *125*, 163-177 (1976).
90. K. Tsukida, A. Kodama, M. Ito, M. Kawamoto, and K. Takahashi, *J. Nutr. Sci. Vitaminol.*, *23*, 263-264 (1977).
91. D. C. Egberg, J. C. Heroff, and R. H. Potter, *J. Agr. Food Chem.*, *25*, 1127-1132 (1977).
92. D. B. Dennison, J. R. Kirk, *J. Food Sci.*, *42*, 1376-1379 (1977).
93. R. Fankel, *Mikrochim. Acta*, *1*, 359-361 (1978).
94. J. E. Paanakker, and G. W. T. Groenendijk, *J. Chromatogr.*, *168*, 125-132 (1979).
95. A. P. De Leenheer, and M. G. M. De Ruyter, *Experientia*, *36*, 31-33 (1980).
96. M. Zile, and H. F. DeLuca, *Anal. Biochem.*, *25*, 307-316 (1968).
97. L. R. Snyder, *J. Chromatogr. Sci.*, *16*, 223-234 (1978).
98. W. Gartner, H. Hopf, W. E. Hull, D. Oesterhelt, D. Schentzow, and P. Towner, *Tetrahedron Lett.*, *21*, 347-350 (1980).
99. K. Tsukida, R. Masahara, and M. Ito, *J. Chromatogr.*, *192*, 395-401 (1980).
100. C. D. B. Bridges, S.-L. Fong, and R. A. Alvarez, *Vision Res.*, *20*, 355-360 (1980).
101. J. P. Rotmans, and A. Kropf, *Vision Res.*, *15*, 1301-1302 (1975).
102. W. H. Waddell, R. Crouch, K. Nakanishi, and N. J. Turro, *J. Amer. Chem. Soc.*, *98*, 4189-4192 (1976).
103. M. Denny, and R. S. H. Liu, *J. Amer. Chem. Soc.*, *99*, 4865-4867 (1977).
104. P. J. Bauer, and P. Carl, *J. Amer. Chem. Soc.*, *99*, 6850-6855 (1977).
105. K. Tsukida, A. Kodama, and M. Ito, *J. Chromatogr.*, *134*, 331-336 (1977).

106. A. Maeda, Y. Shichida, and T. Yoshizawa, *J. Biochem.*, *83*, 661-663 (1978).
107. K. Tsukida, A. Kodama, and M. Ito, *J. Nutr. Sci. Vitaminol.*, *24*, 593-596 (1978).
108. K. Tsukida, M. Ito, and A. Kodama, *J. Nutr. Sci. Vitaminol*, *24*, 143-148 (1978).
109. A. Maeda, Y. Shichida, and T. Yoshizawa, *Biochemistry*, *18*, 1449-1453 (1979).
110. T. Ebrey, R. Govindjee, B. Honig, E. Pollak, W. Chan, R. Crouch, A. Yudd, and K. Nakanishi, *Biochemistry*, *14*, 3933-3941 (1975).
111. W. H. Waddell, P. M. Dawson, D. L. Hopkins, K. L. Rach, M. Uemura, J. L. West, *J. Liquid Chromatogr.*, *2*, 1205-1218 (1979).
112. D. F. Tomkins, and R. J. Tscherne, *Anal. Chem.*, *46*, 1602-1604 (1974).
113. H. Steuerle, *J. Chromatogr.*, *206*, 319-326 (1981).
114. F. D. Grain, F. J. Lotspeich, and R. F. Krause, *J. Lipid Res.*, *8*, 249-254 (1967).
115. A. M. McCormick, J. L. Napoli, and H. F. DeLuca, *Anal. Biochem.*, *86*, 25-33 (1979).
116a. A. P. De Leenheer, and W. E. Lambert, *Int. J. Vitamin. Nutr. Res.*, *51*, 180-181 (1981).
116b. A. P. De Leenheer, W. E. Lambert, and I. Claeys, *J. Lipid Res.*, *23*, 1362-1367 (1982).
117. A. Kleiner-Bossaler, and H. F. DeLuca, *Arch. Biochem. Biophys.*, *142*, 371-377 (1971).
118. A. B. Roberts, and H. F. DeLuca, *Biochem. J.*, *102*, 600-605 (1967).
119a. C. A. Frolik, L. L. Dart, and M. B. Sporn, *Biochim. Biophys. Acta*, *663*, 329-335 (1981).
119b. P. R. Sundaresan, and P. V. Bhat, *J. Lipid Res.*, *23*, 448-455 (1982).
120. F. M. Rabel, *J. Chromatogr. Sci.*, *18*, 394-408 (1980).
121. H. Colin, and G. Guiochon, *J. Chromatogr.*, *141*, 289-312 (1977).
122. R. M. McCormick, and B. L. Karger, *J. Chromatogr.*, *199*, 259-273 (1980).
123. J. N. Thompson, and W. B. Maxwell, *J. Ass. Offic. Anal. Chem.*, *60*, 766-771 (1977).
124a. A. P. De Leenheer, V. O. De Bevere, M. G. De Ruyter, and A. E. Claeys, *J. Chromatogr. Biomed. Appl. 162*, 408-413 (1979).
124b. G. L. Catignani, and J. G. Bieri, *Clin. Chem.*, *29*, 708-712 (1983).
125. M. H. Bui-Nguyen, and B. Blanc, *Experientia*, *36*, 374-375 (1980).
126. N. A. Parris, *J. Chromatogr.*, *157*, 161-170 (1978).

127. W. O. Landen, *J. Chromatogr.*, *211*, 155-159 (1981).
128. K. Abe, K. Ishibashi, M. Ohmae, K. Kawabe, and G. Katsui, *Vitamins*, *51*, 275-280 (1977).
129. P. V. Bhat, H. T. Co., and A. Lacroix, *J. Chromatogr.*, *260*, 129-136 (1983).
130. C. A. Frolik, T. E. Tavela, and M. B. Sporn, *J. Lipid Res.*, *19*, 32-37 (1978).
131. A. M. McCormick, J. L. Napoli, and H. F. DeLuca, *Methods Enzymol.*, *67F*, 220-233 (1980).
132. R. C. Williams, J. A. Schmit, and R. A. Henry, *J. Chromatogr. Sci.*, *10*, 494-501 (1972).
133. J. W. Dolan, J. R. Gant, N. Tanaka, R. W. Giese, and B. L. Karger, *J. Chromatogr. Sci.*, *16*, 616-622 (1978).
134. M. Eriksson, T. Eriksson, and B. Sörensen, *Acta Pharm. Suec.*, *15*, 274-281 (1978).
135. H. Cohen, and M. Lapointe, *J. Agr. Food Chem.*, *26*, 1210-1213 (1978).
136. A. B. Roberts, M. D. Nichols, C. A. Frolik, D. L. Newton, and M. B. Sporn, *Cancer Res.*, *38*, 3327-3332 (1978).
137. A. C. Ross, *Anal. Biochem.*, *115*, 324-330 (1981).
138. M. G. De Ruyter, and A. P. De Leenheer, *Anal. Chem.*, *51*, 43-46 (1979).
139. R. M. McKenzie, M. L. McGregor, and E. C. Nelson, *J. Label. Compounds Radiopharm.*, *XV*, 265-278 (1981).
140. C. A. Frolik, T. E. Tavela, G. L. Peck, and M. B. Sporn, *Anal. Biochem.*, *86*, 743-750 (1978).
141. R. M. McKenzie, D. M. Hellwege, M. L. McGregor, N. L. Rockley, P. J. Riquetti, and E. C. Nelson, *J. Chromatogr.*, *155*, 379-387 (1978).
142. G. L. Peck, T. G. Olsen, F. W. Yoder, J. S. Strauss, D. T. Downing, M. Pandya, D. Butkus, and J. Arnaud-Battandier, *N. Engl. J. Med.*, *300*, 329-333 (1979).
143. M. L. Reed, J. Stanley, F. Stengel, J. L. Shupack, and D. M. Benjamin, *Arch. Dermatol.*, *115*, 605-608 (1979).
144. J. S. Strauss, A. M. Stranieri, L. N. Farrell, and D. T. Downing, *J. Invest. Dermatol.*, *74*, 66-67 (1980).
145. J. L. Napoli, A. M. McCormick, H. K. Schnoes, and H. F. DeLuca, *Proc. Nat. Acad. Sci. U.S.*, *75*, 2603-2605 (1978).
146. A. M. McCormick, J. L. Napoli, H. K. Schnoes, and H. F. DeLuca, *Arch. Biochem. Biophys.*, *192*, 577-583 (1979).
147. C. A. Frolik, A. B. Roberts, T. E. Tavela, P. P. Roller, D. L. Newton, and M. B. Sporn, *Biochem.*, *18*, 2092-2097 (1979).
148. A. B. Roberts, L. C. Lamb, and M. B. Sporn, *Arch. Biochem. Biophys.*, *199*, 374-383 (1980).
149. M. H. Zile, H. K. Schnoes, and H. F. DeLuca, *Proc. Nat. Acad. Sci. U.S.*, *77*, 3230-3233 (1980).

150. C. A. Frolik, B. N. Swanson, L. L. Dart, and M. B. Sporn, *Arch. Biochem. Biophys.*, *208*, 344-352 (1981).
151. T. W. Goodwin, *The Biochemistry of the Carotenoids*, Vol. 1, *Plants*, Chapman and Hall, London, 1980.
152. P. Mathis and C. C. Schenck, in *Carotenoid Chemistry and Biochemistry* (G. Britton and T. W. Goodwin, eds.), Pergamon, Oxford, 1982, pp. 339-347.
153. N. I. Krinsky, *Pure Appl. Chem.*, *51*, 649 (1979).
154. J. C. Bauernfeind, C. R. Adams, and W. L. Marusich, in *Carotenoids as Colorants and Vitamin A Precursors* (J. C. Bauernfeind, ed.), Academic Press, New York, 1981, pp. 563-743.
155. R. Castillo, G. Nègre-Sadargues, and R. Lenel, in *Carotenoid Chemistry and Biochemistry* (G. Britton and T. W. Goodwin, eds.), Pergamon, Oxford, 1982, pp. 211-214.
156. A. H. Brush, in *Carotenoids as Colorants and Vitamin A Precursors* (J. C. Bauernfeind, ed.), Academic Press, New York, 1981, pp. 539-562.
157. K. L. Simpson, *Proc. Nutr. Soc.*, *42*, 7 (1983).
158. G. Tóth and J. Szabolcs, *Phytochemistry*, *20*, 2411 (1981).
159. T. W. Goodwin, *The Biochemistry of the Carotenoids*, Vol. 1, *Plants*, Chapman and Hall, London, 1980, pp. 2-8, and refs. 19 and 20 therein.
160. H. Klaüi and J. C. Bauernfeind, in *Carotenoids as Colorants and Vitamin A Precursors* (J. C. Bauernfeind, ed.), Academic Press, New York, 1981, pp. 47-317.
161. K. Peto, R. Doll, J. D. Buckley, and M. B. Sporn, *Nature*, *290*, 201 (1981).
162. G. Arroyave, C. O. Chichester, H. Flores, J. Glover, L. A. Mejia, J. A. Olson, K. L. Simpson, and B. A. Underwood, in *Biochemical Methodology for the Assessment of Vitamin A Status*, A Report of the International Vitamin A Consultative Group (IVACG), Washington, 1982, pp. 77-84.
163. E. De Ritter and A. E. Purcell, in *Carotenoids as Colorants and Vitamin A Precursors* (J. C. Bauernfeind, ed.), Academic Press, New York, 1981, pp. 815-882.
164. L. Zechmeister, *Cis-Trans Isomeric Carotenoids, Vitamin A and Arylpolyenes*, Springer, Vienna, 1962, pp. 46-57.
165. R. F. Taylor and M. Ikawa, *Methods Enzymol.*, *67*, 233-261 (1980).
166. J. P. Sweeney, and A. C. Marsh, *J. Ass. Offic. Anal. Chem.*, *53*, 937 (1970).
167. I. Stewart and T. A. Wheaton, *J. Chromatogr.*, *55*, 325 (1971).
168. A. Fiksdahl, J. T. Mortensen, and S. Liaaen-Jensen, *J. Chromatogr.*, *157*, 111 (1978).

169. S. K. Hajibrahim, P. J. C. Tibbetts, C. D. Watts, J. R. Maxwell, G. Eglinton, H. Colin, and G. Guiochon, *Anal. Chem.*, *50*, 549 (1978).
170. H. Stransky, *Z. Naturforsch.*, *33C*, 836 (1978).
171. J. K. Abaychi, and J. P. Riley, *Anal. Chim. Acta*, *107*, 1 (1979).
172. G. Englert, and M. Vecchi, *Helv. Chim. Acta*, *63*, 1711 (1980).
173. S. K. Reeder, and G. L. Park, *J. Ass. Off. Anal. Chem.*, *55*, 595 (1975).
174. B. S. Kamel, and M. Bueno, *Lebensm.-Wiss. Technol.*, *13*, 134 (1980).
175. M. Vecchi, G. Englert, R. Maurer, and V. Meduna, *Helv. Chim. Acta*, *64*, 2746 (1981).
176. I. Stewart, *J. Ass. Off. Anal. Chem.*, *60*, 132 (1977).
177. K. Tsukida, K. Saiki, T. Takii, and Y. Koyama, *J. Chromatogr.*, *245*, 359 (1982).
178. T. Braumann, and L. H. Grimme, *Biochim. Biophys. Acta*, *637*, 8 (1981).
179. H. J. C. F. Nelis, and A. P. De Leenheer, unpublished results (1982).
180. L. Zechmeister, and A. Polgar, *J. Amer. Chem. Soc.*, *64*, 1856 (1942).
181. J. W. Porter, and F. B. Zscheibe, *Arch. Biochem.*, *10*, 537 (1946).
182. M. Vecchi, G. Englert, and H. Mayer, *Helv. Chim. Acta*, *65*, 1050 (1982).
183. F. T. Gillan, and R. B. Johns, *J. Chromatogr. Sci.*, *21*, 34 (1983).
184. H. J. C. F. Nelis, and A. P. De Leenheer, unpublished results (1983).
185. Z. Matus, M. Baranyai, G. Tóth, and J. Szabolcs, *Chromatographia*, *14*, 337 (1981).
186. M. Zakaria, K. Simpson, P. R. Brown, and A. Krstulovic, *J. Chromatogr.*, *176*, 109 (1979).
187. H. Pfander, H. Schurtenberger, and V. R. Meyer, *Chimia*, *34*, 179 (1980).
188. R. J. Bushway, and A. M. Wilson, *Can. Inst. Food Sci. Technol. J.*, *15*, 165 (1982).
189. H. J. C. F. Nelis, and A. P. De Leenheer, *Anal. Chem.*, *55*, 270 (1983).
190. C. R. Broich, L. E. Gerber, and J. W. Erdman, *Lipids*, *18*, 253 (1983).
191. Y.-M. Peng, J. Beaudry, D. S. Alberts, and T. P. Davis, *J. Chromatogr.*, *273*, 410 (1983).
192. Y.-P. C. Hsieh, and M. Karel, *J. Chromatogr.*, *259*, 515 (1983).
193. W. J. Driskell, M. M. Bashor, and J. W. Neese, *Clin. Chem.*, *29*, 1042 (1983).
194. K. Eskins, C. R. Scholfield, and H. J. Dutton, *J. Chromatogr.*, *135*, 217 (1977).

195. K. Eskins, and H. J. Dutton, *Anal. Chem.*, *51*, 1885 (1979).
196. T. Braumann, and L .H. Grimme, *J. Chromatogr.*, *170*, 264 (1979).
197. M. F. Lefevere, Ph.D. thesis, State University of Gent, Belgium, 1983.
198. W. O. Landen, *J. Chromatogr.*, *211*, 155 (1981).
199. W. O. Landen, and R. R. Eitenmiller, *J. Ass. Offic. Anal. Chem.*, *62*, 283 (1979).
200. V. G. Mairanovsky, A. A. Engovatov, N. T. Ioffe, and G. I. Samokhvalov, *J. Electroanal. Chem.*, *66*, 123 (1975).
201. L. Zechmeister, Cis-Trans *Isomeric Carotenoids, Vitamins A and Arylpolyenes*, Springer, Vienna, 1962, pp. 25-45.
202. A. Sommer, *Nutritional Blindness*, Oxford University Press, Oxford, 1982, pp. 219-222.
203. J. P. Sweeney, and A. C. Marsh, *J. Amer. Diet. Ass.*, *59*, 238 (1971).
204. M. Baranyai, Z. Matus, and J. Szabolcs, *Acta Aliment.*, *11*, 309 (1982).
205. G. Calabro, G. Micali, and P. Curro, *Atti-Conv. Naz. Olii Essenz. Sui. Deriv. Agrum.*, 7, 171 (1978); *CA*, *93*, 166194m (1980).
206. G. Calabro, G. Micali, and P. Curro, *Essenze Deriv. Agrum.*, *48*, 359 (1978); *CA*, *93*, 6195j (1980).
207. I. Stewart, *J. Agr. Food Chem.*, *25*, 1132 (1977).
208. P. Le François, S. Chevassus-Agnes, and A. Makhtar Ndiaye, *Am. J. Clin. Nutr.*, *34*, 434 (1981); J. A. Olson, *Am. J. Clin. Nutr.*, *34*, 435 (1981).
209. G. Arroyave, C. O. Chichester, H. Flores, J. Glover, L. A. Mejia, J. A. Olson, K. L. Simpson, and B. A. Underwood, in *Biochemical Methodology for the Assessment of Vitamin A Status*, International Vitamin A Consultative Group (IVACG), Washington, 1982, pp. 72-76.
210. M. M. Mathews-Roth, in *Carotenoids as Colorants and Vitamin A Precursors* (J. C. Bauernfeind, ed.), Academic Press, New York, 1981, pp. 755-785.
211. C. V. Puglisi and J. A. F. de Silva, *J. Chromatogr.*, *120*, 457 (1976).
212. W. L. Marusich, and J. C. Bauernfeind, in *Carotenoids as Colorants and Vitamin A Precursors* (J. C. Bauernfeind, ed.), Academic Press, New York, 1981, pp. 319-462.
213. K. L. Simpson, T. Katayama, and C. O. Chichester, in *Carotenoids as Colorants and Vitamin A Precursors* (J. C. Bauernfeind, ed.), Academic Press, New York, 1981, pp. 516-528.
214. H. H. Strain, and W. A. Svec, *Adv. Chromatogr.*, *8*, 119 (1969).

215. E. De Ritter and A. E. Purcell, in *Carotenoids as Colorants and Vitamin A Precursors* (J. C. Bauernfeind, ed.), Academic Press, New York, 1981, pp. 829-831.
216. D. B. Rodriguez, L. C. Raymundo, T. C. Lee, K. L. Simpson, and C. O. Chichester, *Ann. Bot., 40*, 615 (1976).
217. A. K. Baloch, K. A. Buckle, and R. A. Edwards, *J. Chromatogr., 139*, 149 (1977).
218. S. E. Gebhardt, E. R. Elkins, and J. Humphrey, *J. Agr. Food Chem., 25*, 629 (1977).
219. N. I. Krinsky, D. G. Cornwell, and J. L. Oncley, *Arch. Biochem. Biophys., 73*, 223 (1958).
220. D. S. McLaren, W. W. C. Read, Z. L. Awdeh, and M. Tchalian, *Methods Biochem. Anal., 15*, 1 (1967).
221. R. Sadowski, and W. Wojcik, *J. Chromatogr., 262*, 455 (1983).
222. H. R. Bolliger, and A. König, in *Thin-layer Chromatography* (E. Stahl, ed.), Springer, Berlin, 1969, pp. 266-273.
223. J. Gross, *Chromatographia, 13*, 572 (1980).
224. V. Schwieter, H. R. Bolliger, L. H. Chopard-dit-Jean, G. Englert, M. Kofler, A. König, C. Von-Planta, R. Ruegg, W. Vetter, and O. Isler, *Chimia, 19*, 294 (1965).
225. S. R. Targan, S. Merrill, and A. D. Schwabe, *Clin. Chem., 15*, 479 (1969).

2

VITAMIN D: CHOLECALCIFEROL, ERGOCALCIFEROL, AND HYDROXYLATED METABOLITES

*GLENVILLE JONES**

Research Institute, The Hospital for Sick Children, Toronto, Ontario, Canada

D. A. SEAMARK, D. J. H. TRAFFORD, and H. L. J. MAKIN

Department of Chemical Pathology, The London Hospital Medical College, London, England

---

**Present affiliation:* Departments of Biochemistry and Medicine, Queen's University, Kingston, Ontario, Canada

1. INTRODUCTION

The demonstration over a decade ago [1] of the conversion of vitamin D_3 into 25-hydroxyvitamin D_3 (25-OH-D_3) triggered a dramatic surge of interest in the basic biochemistry and physiology of vitamin D. The resultant elucidation of the metabolism and role of vitamin D in calcium homeostasis [1-5] has led to a change of emphasis from the study of the parent vitamin itself to the study of its metabolites. Older methods of bioassay, colorimetric and ultraviolet (UV) absorption assay, have been replaced by newer techniques based upon chromatography, radioligand procedures, and mass spectrometry. Present methods separate and measure a wide variety of metabolites and chemical analogs instead of a heterogeneous "vitamin D" group. The increased sensitivity and specificity offered by the newer assays have enabled the measurement of picogram quantities of vitamin D metabolites, thereby permitting these techniques to be applied to the analysis of vitamin D and its metabolites in physiological fluids. Chemical synthesis of vitamin D metabolites for clinical trials has led to improvements in the separation and analysis of the vitamin D family of compounds, which has also been beneficial to analytical chemists in the pharmaceutical and food industries.

1.1 Basic Formulas

Vitamin D and its metabolites are a group of 9,10-seco-steroids with the basic structure shown in Fig. 2.1. The A ring of the steroid nucleus is rotated about the C6 position. The conjugated cis-triene system gives rise to the characteristic UV-absorption spectrum of vitamin D (molar extinction coefficient $\varepsilon = 18{,}300$; $\lambda_{max} = 265$ nm; $\lambda_{min} = 228$ nm). The D_3 series has a side chain derived from cholesterol; the D_2 series has a side chain derived from ergosterol containing an additional C22(23) double bond and a C24 methyl group.

1.2 Chemistry

Vitamin D_2 and vitamin D_3 are derived by photoirradiation from their respective 5,7-diene sterol precursors: 7-dehydrocholesterol or ergosterol. The intact sterol precursor, known as a provitamin, undergoes photolysis when exposed to UV light of wavelength 280-320 nm to yield a variety of photoirradiation products, the principal ones being previtamin D, tachysterol, and lumisterol, the structures of which are

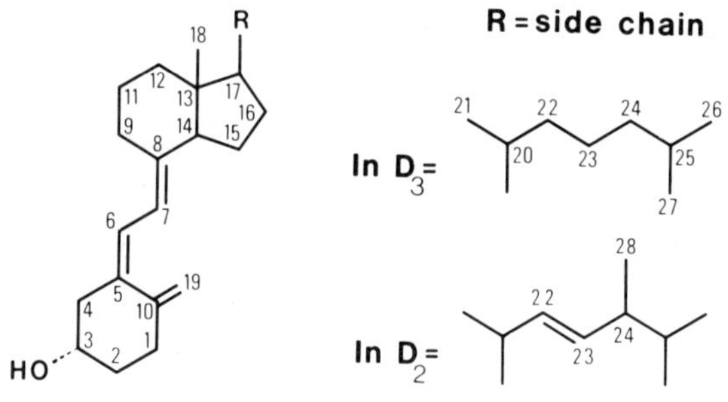

Figure 2.1 Basic formula of vitamin D compounds.

illustrated in Fig. 2.2. The previtamin D undergoes spontaneous thermal rearrangement to vitamin D [6]. Vitamin D has a number of other closely related isomers and derivatives that have little biological activity but can be formed during synthesis, derivatization, or handling.

The isolation and identification of the metabolites of vitamin D have been followed by a keen interest in the chemical synthesis of the D vitamins. Clinical use of these new compounds has led to more efficient synthesis [7]. Almost without exception the synthesis of vitamin D metabolites involves the preparation of a suitably hydroxylated or substituted provitamin, conversion of this to the equivalent previtamin, followed by thermal isomerization to the vitamin D. Beside the natural hydroxylated metabolites described below, a series of fluorinated vitamin D compounds (e.g., 24-difluoro-25-OHD_3), as well as 25-aza and 19-hydroxylated derivatives of vitamin D, have been synthesized for potential use as antivitamins.

1.3 Biochemistry

Vitamin D_3 is synthesized in the skin from 7-dehydrocholesterol by a process closely resembling the photochemical process in the test tube [8]. Previtamin D_3 made by UV light exposure undergoes thermal isomerization to vitamin D_3, which is selectively removed by the vitamin D-binding globulin (DBP) of the blood [8]. Alternatively, preformed vitamin D_3 of D_2 in the diet is absorbed from the intestine into lymph, enters the circulation, and is rapidly taken up by the liver [9]. Vitamin D_3 is converted in the liver to 25-OH-D_3 [1] (see Fig. 2.3). Two enzymes are present in the liver to carry out 25-hydroxylation and are located in the endoplasmic reticulum and mitochondrion, respectively [10,11]. The former appears to be the specific 25-hydroxylase enzyme with an apparent K_m for 25-OH-D_3 of 5.6×10^{-8} M [12], a limited binding capacity, and tight regulation [11]. The mitochondrial enzyme,

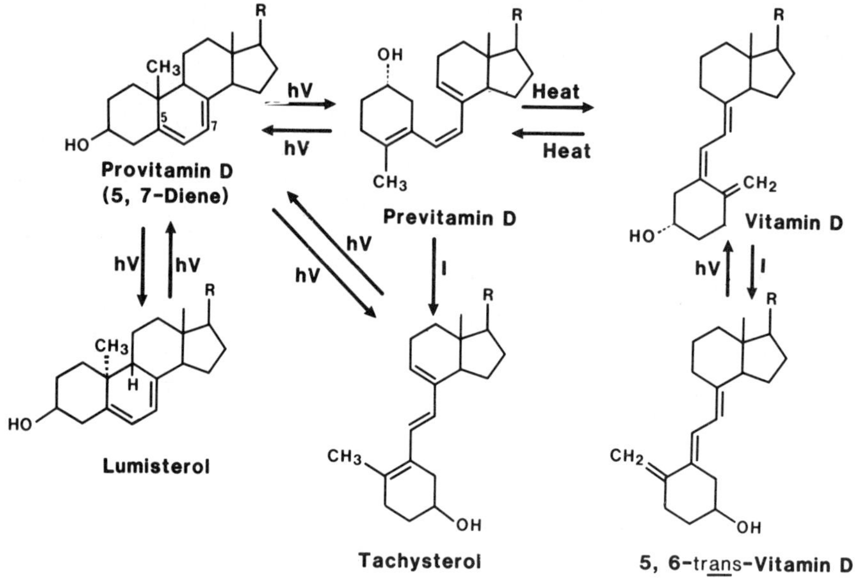

Figure 2.2 Irradiation products of provitamin D.

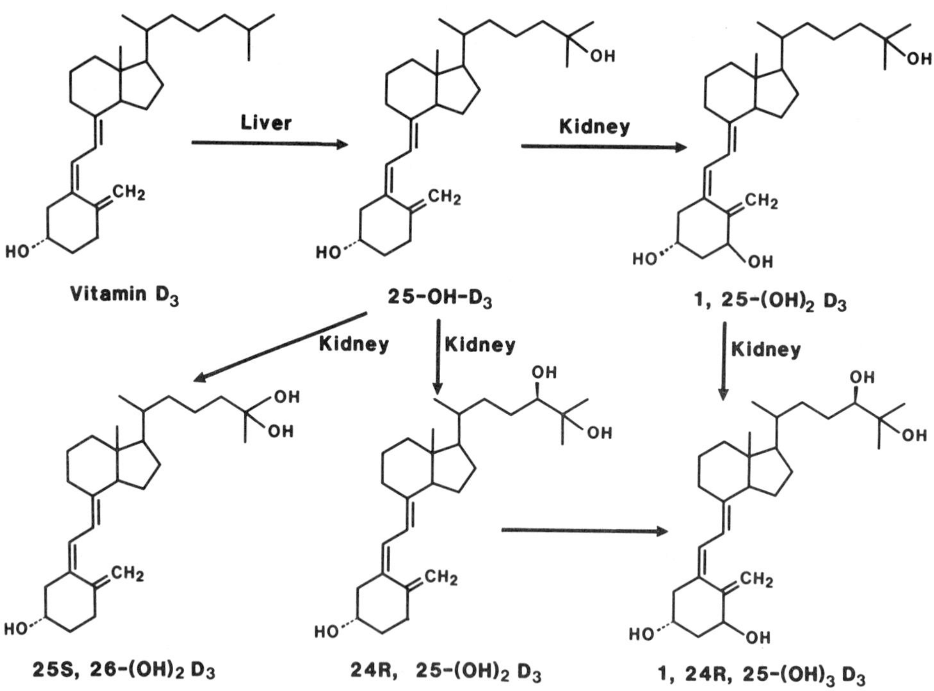

Figure 2.3 Metabolism of vitamin D_3.

on the other hand, appears to be nonspecific and unregulated but with a high capacity [12], perhaps utilizing cholesterol as its normal substrate. In vivo regulation of the level of 25-OH-D_3 is thus incomplete and 25-OH-D_3 is the circulating vitamin D metabolite in highest concentration.

25-Hydroxyvitamin D_3 is metabolized by the kidney (see Fig. 2.3) into a series of dihydroxylated compounds, the important ones being 1α,25-dihydroxyvitamin D_3 [1α,25-$(OH)_2D_3$] [13,14] and 24R,25-dihydroxyvitamin D_3 [24(*R*),25-$(OH)_2D_3$] [15,16]. 1α,25-Dihydroxyvitamin D_3 is believed to be the active form of vitamin D necessary to raise serum calcium and phosphate [2] and is at least 10 times more biologically active in this respect than vitamin D_3. 1α,25-Dihydroxyvitamin D_3 raises serum calcium and phosphate by stimulating intestinal calcium and phosphate absorption [17-19] and, along with parathyroid harmone (PTH), by stimulating bone resorption [20]. It may also play a role in calcium and phosphate reabsorption by the kidney. Since 1α,25-$(OH)_2D_3$ is such a potent hormone, it is not surprising to find that its synthesis is tightly controlled. The kidney 1-hydroxylase [21] is stimulated by low concentrations of calcium [22], phosphate [23], or parathyroid hormone [24] and inhibited by 1α,25-$(OH)_2D_3$ [25]. Thus, the circulating level of 1α,25-$(OH)_2D_3$ in the plasma rarely exceeds 100 pg/ml in any mammalian species [4]. The synthesis of 24(*R*),25-$(OH)_2$-D_3 is a detoxication product of vitamin D_3 [2,3]. The synthesis of 24(*R*),25-$(OH)_2D_3$ is stimulated by 1α,25-$(OH)_2D_3$ administration [27], and 24(*R*),25-$(OH)_2D_3$ is the major dihydroxylated metabolite in the plasma of normocalcemic mammals. A role for 24(*R*),25-$(OH)_2D_3$ in bone mineralization has been suggested by some authors [28,29]. 1,24(*R*),25-Trihydroxyvitamin D_3 is the product of the administration of 1α,25-$(OH)_2D_3$ to a normocalcemic animal, and its function is unknown [30]. The metabolism of vitamin D_2 parallels that of vitamin D_3 in most mammalian species with the synthesis of 25-OH-D_2 [31], 1,25-$(OH)_2D_2$ [32] (Fig. 2.4), and 24(*R*),25-$(OH)_2D_2$ [33].

The 1α,25-$(OH)_2D_3$ undergoes side-chain cleavage [34] to calcitroic acid [35] (Fig. 2.4), whereas 24(*R*),25 $(OH)_2D_3$ probably is converted to cholecalcioic acid [36]. The kidney is the likely site of the formation of two other vitamin D metabolites, 25-OH-D_3-26,23-lactone [37] and 25(*S*),26-$(OH)_2D_3$ [38], shown in Figs. 2.3 and 2.4.

Many other minor metabolites are being reported, but little is known beyond their structure. These include 1,25(*S*),26-$(OH)_3D_3$; 1α,25-$(OH)_2D_3$-26-lactone; 24-OH-D_2; 25,26-$(OH)_2D_2$; vitamin D_3 sulfate; and 5,6-*trans*-25-OH-D_3 [39-44].

1.4 Application

The analysis of vitamin D and its metabolites is worthy of attention for many reasons.

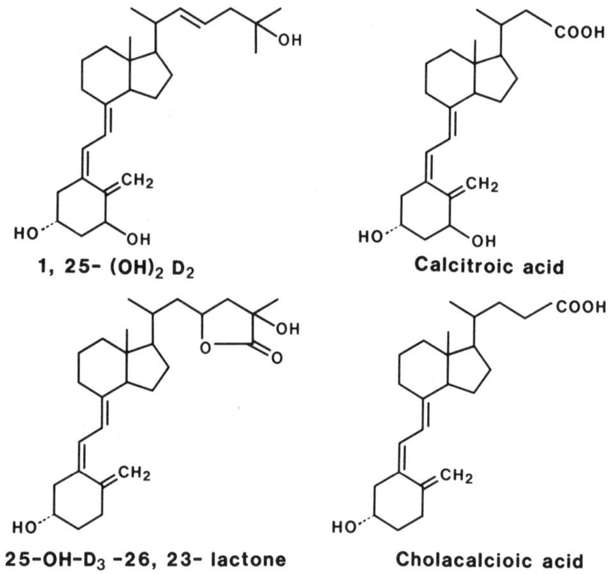

Figure 2.4 Other biological metabolites of vitamin D.

The measurement of the level of vitamin D metabolites provides important knowledge about the etiology, pathogenesis, and treatment of diseases involving disturbances of calcium and phosphorus metabolism.

The analysis of vitamin D compounds in drug formulations, such as multivitamin tablets, injectable solutions, and new metabolite preparations, is required to ensure the purity, stability, and potency of the compounds.

There is at present tremendous interest in elucidating the full details of the metabolism and mode of action of vitamin D. The need for unequivocal identification of increasing numbers of metabolites in animal studies in vivo and in vitro is placing increased demands on the resolving capabilities of modern chromatographic techniques and the sensitivity of detection and measurement.

Part of the interest in the study of the vitamin D stems from the availability of pure radioisotope-labeled derivatives of vitamin D and its metabolites and chemically synthesized metabolites. Present techniques of quality control can accurately determine purity and specific activity in radioactively labeled vitamin D, and chemical synthesis has been greatly aided by chromatographic techniques able to resolve stereoisomers.

Last but not least, the food industry has now introduced liquid chromatographic (LC) and gas chromatography-mass spectrometry (GC-MS) methods for vitamin D analysis of the traditional vitamin D-supplemented human foods. Milk and margarine provide the analyst with the task of measuring nanogram quantities of vitamin D in a matrix containing milligram or gram quantities of fat. Animal feeds also contain added vitamin D. It is likely that, if vitamin D metabolites are used in the beef and dairy science industry on any major scale, there will be a need for accurate methods for the analysis of vitamin D and its metabolites in tissues and milk.

Attempts will be made in this chapter to describe examples of the methods used to solve the many problems of analysis of vitamin D and its metabolites.

2. PREPARATION OF SAMPLES FOR ANALYSIS

The need for sample extraction, concentration, and partial purification prior to final analysis naturally depends upon the nature of the sample matrix and the method of quantitation used. Minimal prepurification is required for concentrated samples of synthetic vitamin D compounds during chemical synthesis. An example is given in Sec. 3 of such an analysis involving separation of the photoirradiation isomers of ergosterol. At the other extreme, the analysis of vitamins in foodstuffs involves saponification, extraction, and concentration prior to quantitation. The analysis of trace quantities of vitamin D metabolites, e.g., $1,25\text{-}(OH)_2D_3$, in human plasma requires the extraction of picogram quantities, purification by column chromatography, and, finally, radioligand assay. Much attention is now being paid to sample preparation, as a significant reduction in analysis time can be achieved by optimizing purification procedures.

2.1 Saponification

Where samples contain significant quantities of lipid, e.g., in the analysis of such foodstuffs as milk and margarine, a saponification step using KOH or $NaHCO_3$ is usually included to eliminate the bulk of the neutral lipid (e.g., triglyceride and phospholipid) prior to extraction. There is also the possibility that conjugates or esters of vitamin D might be released by saponification.

2.2 Extraction

Since vitamin D and its metabolites are fat-soluble sterols, partition into organic solvents provides significant purification by removal of water-soluble contaminants. Solvent systems for extraction fall into two basic categories (Table 2.1).

1. Total lipid extraction, e.g., methanol-chloroform-water (2:1:0.8) [45] or ethanol water (9:1) [46]

TABLE 2.1 Extraction Procedures for Vitamin D Compounds

Solvent	Metabolite(s)	Reference
Methanol-chloroform (2:1)	General	45
Methanol-methylene chloride (2:1)	General	83
Ethyl acetate-cyclohexane (1:1)	General	47
Hexane-isopropanol (1:2)	General	48
Ethanol	D, 25-OH-D	52
Hexane (after saponification)	D	63
Ether	25-OH-D	53
Methanol-water (Sep-Pak)	25-OH-D	59
Dichloromethane	24,25-$(OH)_2D_3$, 1,25-$(OH)_2D_3$	50
Dichloromethane on Extrelut	24,25-$(OH)_2D_3$, 1,25-$(OH)_2D_3$	51

2. Selective lipid extraction, e.g., ether, ethylacetate-cyclohexane (1:1), hexane, dichloromethane, or hexane-isopropanol (1:2)

Clearly, though solvents in the second class are desirable to minimize contamination of vitamin D extracts, they post problems in that they tend to provide efficient extraction for one particular metabolite or group of metabolites (e.g., the dihydroxylated metabolites) and poorly extract those of a different polarity. When simultaneous analysis of several vitamin D metabolites with a wide range of polarities is requried, a total lipid extraction [45] may be necessary.

The high efficiency of the Bligh and Dyer technique [45] for total lipid extraction is probably due to the formation of a monophasic dispersion of sample in extracting solvent, followed by return to the classic two-phase system by addition fo extra chloroform and saturated KCl. All hydroxylated nonacidic vitamin D metabolites can be extracted qualitatively by this technique. Substitution of methylene chloride for chloroform avoids the known carcinogenicity of chloroform and reduces evaporation times by taking advantage of the lower boiling point of methylene chloride. The use of ethanol is a rapid method for precipitation of protein and extraction of vitamin D metabolites but becomes increasingly impractical as sample volumes exceed 1 ml.

Two-phase liquid-liquid extraction of samples is the preferred choice of sample preparation in the majority of applications. Vitamin D and its photoisomers are relatively nonpolar and can be extracted with hexane or other low-polarity solvents, providing the matrix is broken down by saponification, etc. Acid hydrolysis must be avoided due to the tendency to dehydrate the 1-hydroxyl and to give rise to isomerization. In the biochemical areas of analysis, all types of saponification are avoided, chiefly because of concern about the stability of vitamin D metabolites but also because of lack of convenience. More polar solvents, such as modified hexane mixtures, cyclohexane-ethyl acetate (1:1) [47], hexane-isopropanol (1:2) [48], or hexane-isopropanol-*n*-butanol (93:3:4) [49], are employed for the extraction of metabolites of vitamin D and to break protein-bound vitamin D complexes. Horst et al. [42] use ether followed by methanol-methylene chloride (1:3) in a lengthy but ingenious attempt to analyze a series of vitamin D_2 and D_3 metabolites simultaneously. Their general scheme for the extraction of plasma and subsequent analysis is shown in Fig. 2.5. Care must be taken when using ether, since this solvent may contain peroxides that can react with the cis-triene of vitamin D.

Methylene chloride is frequently used for the extraction of 24(*R*),25-$(OH)_2D_3$ and 1,25-$(OH)_2D_3$ from plasma. Though the use of methylene chloride was originally as a two-phase system with plasma [50], one modification is to adsorb the plasma onto Kieselguhr packed in a column (commercial version, Extrelut, Merck, Darmstadt, West Germany) and elute by passing methylene chloride through the column [51]. A total of 80-90% extraction of both dihydroxylated compounds is obtained, and the amount of lipid extracted is so small as to make additional cleanup steps unnecessary.

An additional postextraction prechromatographic step that can be used is to backwash the extract with a low ionic strength weakly basic, pH 9-10, buffer, such as phosphate. This tends to remove any acidic lipid components poorly ionized at neutral pH levels.

2.3 Minicolumns

The use of an inexpensive, disposable, gravity flow minicolumn as a penultimate step prior to LC serves several useful purposes.

1. It protects the expensive LC column from lipid, particulate matter, etc.
2. It eliminates remaining neutral and highly polar lipid contaminants, further reducing the lipid load applied to the LC.
3. It can be used to fractionate the vitamin D metabolites into major classes for subsequent analysis using different solvent systems.

Minicolumns have disadvantages, too.

1. Batch-to-batch variability, requiring frequent calibration checks.

PLASMA (3-5 ml) + ^{3}H METABOLITES

Diethylether (2X)

Methanol:methylene chloride (1:3)

Backwash w/0.1 m $NaPO_4$ (pH 10.5)

Sephadex LH-20
(1:1:9 Chloroform:methanol:hexane)

Vitamin D

Lipidex 5000
(5:95 Chloroform:hexane)

HPLC on Zorbax SiL
(0.25:99.75 Isopropanol:methylene chloride)

HPLC on Zorbax ODS
(4:96 Water:methanol)

Quantitate both D_2 and D_3
Forms by UV absorbance

25-OHD

Sephadex LH-20
(50:50 Hexane:chloroform)

or

Lipidex 5000
(10:90 Chloroform:hexane)

HPLC on Zorbax SiL
(4:96 Isopropanol hexane)

Quantitative both D_2 and D_3
Forms by UV absorbance

Quantitate both D_2 and D_3
forms simultaneously by
competitive protein
binding assay

Dihydroxyvitamin D

HPLC on Zorbax SiL
(11:89 Isopropanol:hexane)

$24,25\text{-}(OH)_2D$ Complex $25,26\text{-}(OH)_2D_3$ $1,25\text{-}(OH)_2D$

Quantitate both D_2 and D_3
forms simultaneously by
competitive protein
binding assay

HPLC on Zorbax SiL
(2:98 Isopropanol:methylene chloride)

$25\text{-}OHD_3\text{-}26,23$ Lactone $24,25\text{-}(OH)_2D_2$ $24,25\text{-}(OH)_2D_3$ $25,26\text{-}(OH)_2D_2$

Quantitated individually using competitive protein binding assay

2. The minicolumn sometimes provides unwanted resolution of D_2 and D_3 metabolites, particularly when using Sephadex-LH 20 and hydroxyalkoxypropyl Sephadex (HAPS), giving misleading results when using radioactive vitamin D metabolites as internal standards, unless larger fractions are collected to include both D_2 and D_3 metabolites.
3. Losses can occur without achieving any substantial purification of the sample.
4. Adsorbants used in the minicolumns can be dirty and introduce interference into subsequent analyses.

2.3.1 Silicic Acid

This packing has been used for some time and resolves vitamin D and 25-OH-D fractions [52,53], and it has been incorporated into a method for LC assay of 25-OH-D [54,55]. Recoveries with silicic acid can be low, and care must be taken to select the appropriate particle size to permit adequate flow and good performance. Koshy [56] recommends use of columns containing 2 g silica gel 60 (230-400 mesh, E. Merck, Darmstadt, West Germany). 25-Hydroxyvitamin D can be eluted with ether-ethyl acetate (9:1).

Sep-Pak SIL cartridges (Waters Associates, Milford, Mass.) permit the use of high-efficiency silica particles and offer minimal variability. The cartridges have been used by Hollis and colleagues [57] to demonstrate the lack of vitamin D sulfate in human milk and by Adams et al. [58] to fractionate classes of vitamin D metabolites in human plasma. A recent evaluation of the use of Sep-Pak SIL as a means of preliminary fractionating vitamin D and its metabolites has been published [58]. Hollis et al. [57] demonstrated the use of these cartridges for the rapid separation of vitamin D and vitamin D sulfate. The plastic coating of the Sep-Pak is made of virgin polyethylene with minimal or zero amounts of plasticizers. Early versions of the Sep-Pak SIL were not completely devoid of UV-absorbing impurities, and as a result numerous interfering peaks were introduced into the sample during "purification." Sep-Pak-C18 is the reverse-phase analog of the silica cartridge and uses a bonded octadecasilane (C18) phase. It can also be used to fractionate plasma in the reverse manner to that described by Adams et al. [58]. Dabek et al. [59], Turnbull et al. [60], and Fraher et al. [61] have developed methods for 25-OH-D_2 and 25-OH-D_3 using Sep-Pak C18 for both extraction [60,61] and preliminary purification [59] of the sample. The method of Dabek et al. [59] utilizes the properties of methanol as both an extraction solvent and as a mobile phase.

2.3.2 Alumina

This adsorbant is particularly useful for the purification of vitamin D from the unsaponifiable lipid of milk, margarine, and infant formulas

[62,63]. Neutral alumina (AG7, Biorad Labs, Richmond, Cal.) is deactivated prior to use, and vitamin D is eluted with 1% isopropanol in hexane.

2.3.3 Sephadex-LH 20 (Pharmacia, Uppsala, Sweden)

This gel was first used with the solvent systems hexane-chloroform (50:50-35:65) [64] for vitamin D and its metabolites. In fact, without Sephadex-LH 20, it is doubtful if the early metabolites would have been purified sufficiently for MS identification. Later solvent systems based upon hexane-methanol-chloroform (9:1:1) gave much sharper peaks and permitted reduction of column size to 15 × 0.6 cm bed dimension [50]. A sample elution pattern for vitamin D metabolites in plasma is shown in Fig. 2.6 [42]. Sephadex-LH 20, although widely used for preliminary purification of the hydroxylated metabolites of vitamin D, provides poor resolution of vitamin D and neutral lipid and is rarely used for analysis of vitamin D itself.

2.3.4 Hydroxyalkoxypropyl Sephadex-LH20

Coupling of a C11-C14 or C15-C18 chain in the form of an olefin epoxide (Nedox 114 or 1518, Ashland Chemical Co., Peoria, Ill.) to Sephadex-LH 20 generates a gel, HAPS, which is far less polar than

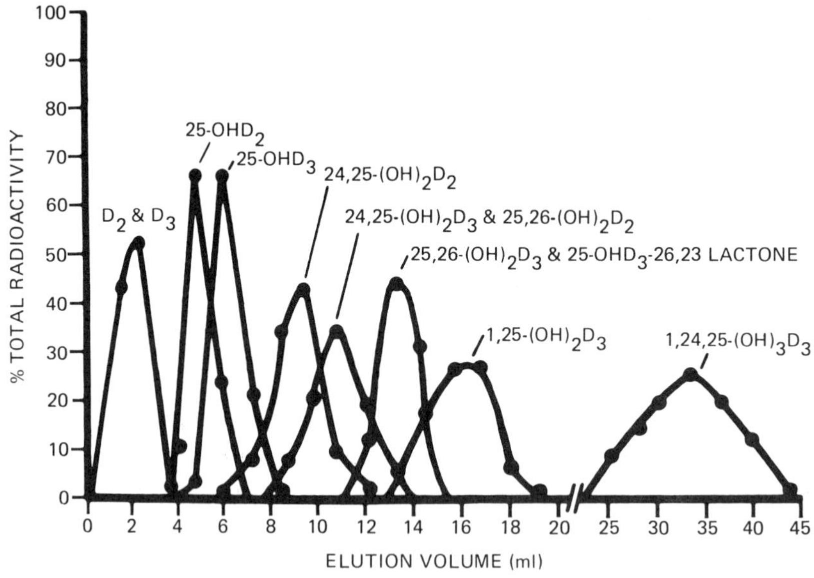

Figure 2.6 Elution pattern of vitamin D metabolites on a 0.6 × 15.5 cm Sephadex-LH 20 column developed with hexane-chloroform-methanol (9:1:1). (From Ref. 42.)

its precursor [65]. This permits the use of solvents such as hexane, which swell the gel and increase the retention of nonpolar vitamin D metabolites, such as vitamin D itself and its photoisomers [32]. Thompson et al. [66] used HAPS instead of alumina to purify vitamin D from unsaponifiable material of milk samples. HAPS is now commercially available under the name Lipidex 5000 and is increasingly used in place of LH 20. Both Shepard et al. [67] and Horst et al. [42] have incorporated it into their schemes for simultaneous vitamin D metabolite assay. Jones et al. [32] used the solvent hexane-chloroform (95:5) to separate vitamin D_2 from its photoisomers on a HAPS column.

2.3.5 *Celite*

This support for liquid-liquid partition chromatography provides good resolution and excellent recoveries but the technique is somewhat difficult to carry out reproducibly. The stationary phase methanol-water (80:20) has a tendency to be stripped by the mobile phase if temperature or other conditions fluctuate. Celite 545 (Johns-Manville Product Co., New York, N.Y.) was used by Koshy [56]. It is likely to be replaced by the bonded phase packings, such as RP C18 or Sep-Pak C18, which do not lose stationary phase. Celite chromatography as described is a straight-phase system, whereas Sep-Pak C18 is a reversed-phase system. Results would be similar, but metabolites would be eluted in reverse order.

2.3.6 *TLC*

This technique is mainly used for prepurification of saponified samples for gas-liquid chromotography (GLC) and has been used for the separation of vitamin D from cholesterol [68] and in combination with GLC for the estimation of 25-OH-D_3 in chick plasma [68]. Care must be taken to avoid oxidation of vitamins on the TLC plate during drying of the applied sample and during removal of the plate from the tank. Such oxidation can be minimized by running plates at 0°C in an atmosphere of nitrogen. Thin-layer chromatography (TLC) is relatively inexpensive and yet offers high resolution and has the advantage that many samples can be purified using one TLC run. Vieth et al. [69] used TLC to separate tritiated products in biochemical studies of renal 25-OH-D-hydroxylases.

2.4 Other Procedures

One of the most common problems with LC is the presence of particulate matter in the samples for injection. Unchecked, this ultimately leads to clogging of the injector or, worse, the inlet frit of the column. Changing the inlet frit of the column is a procedure to be avoided by the inexperienced chromatographer if at all possible, since disturbance of the bed may occur, unless great care is taken, in which case column efficiency is probably reduced. Thus, filtering of samples prior to injection is recommended. Such a procedure is described by Jones [70]. Use of a relatively inexpensive guard column may also give greater protection to the main analytical column. Waters Associates (Milford, Mass.) market a syringe fitted with a Swinney filter holder containing a Millipore 0.45 μm Teflon filter (called an organic sample clarification kit) but its use may lead to losses. Alternatively, samples for LC analysis can be transferred to a 5 ml conical screw-capped vial (Reactivial; Pierce Chemical Co., Rockford, Ill.) and centrifuged at low speed (around 500× *g* for 5 min), although this is not always a completely satisfactory procedure.

It is our experience at the Hospital for Sick Children in Toronto that the principal determinant of the lifetime of an LC column used in the assay of plasma and kidney perfusate samples is not the lipid content of the sample but the mechanical and physical abuse the column takes. Thus, specific columns will withstand heavy lipid loads very well without major diminution in column efficiency provided the adsorbed lipids can subsequently be eluted. Columns that permit washing with highly polar solvents such as methanol, water and even 0.1 N nitric acid (Zorbax-SIL, Dupont Instruments, Wilmington, Del.) allow their use as preparative columns (without sample cleanup) prior to LC analysis on a second LC system [70]. Washing the column at the end of each day of analysis is sufficient. Jones [70] used silica adsorption LC followed by reverse-phase LC on octadecasilane-bonded phase (Fig. 2.7). A similar technique, but reversing the order of the LC columns, C18-LC followed by silica LC, has been used successfully by Okano and colleagues [71]. The use of reversed-phase Sep-Pak C18 cartridges followed by straight-phase LC on Zorbax-SIL [60] is a similar approach. The main advantage is the use of the resolving power of both types of LC (straight phase and reverse phase). Clearly, this is advantageous over a combination of open column and one LC step. It also permits greater automation of LC and overcomes the disadvantages of minicolumns alluded to in Sec. 2.3. The disadvantages are the need for expensive instrumentation, and that samples are run serially rather than in parallel, and the LC column needs frequent washes. An alternative procedure [72] for plasma 25-OH-D used two separate straight-phase Zorbax-SIL columns but interpolated a simple specific chemical reaction (formation of isotachysterol isomers [73]) between the two LC separations.

3. LIQUID CHROMATOGRAPHIC ANALYSIS

3.1 General Considerations

LC analysis of vitamin D has only become practically feasible over the last decade with the tremendous improvements in LC instrumentation. Major advances in detector sensitivity and column particle technology

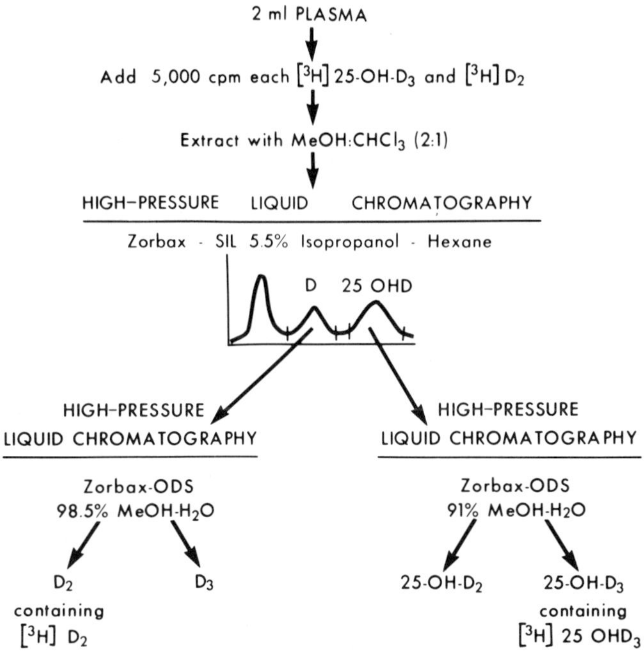

Figure 2.7 Scheme for the analysis of vitamins D_2 and D_3 and 25-OH-D_2 and 25-OH-D_3 in human plasma by LC with UV detection. (From Ref. 70, ©1978, Clinical Chemistry.)

have made possible the detection of as little as 1 ng vitamin D under ideal conditions. Extra column peak broadening has been minimized by the design of eddy-free flow connectors and injectors and use of minimum dead-space tubing. The objective of all good LC analyses is to maximize resolution R, which is defined by the equation

$$R = \frac{1}{4} \underbrace{\left(\frac{\alpha - 1}{\alpha}\right)}_{\text{Selectivity}} \underbrace{\left(\sqrt{N}\right)}_{\text{Plates}} \underbrace{\left(\frac{k'}{1 + k'}\right)}_{\text{Retention}}$$

N is determined by the size and nature of the particles used to make the column and by the quality of the packing procedure. Most vitamin D applications utilize particles in the 5-10 μm size range and values of N in excess of 10,000 per 25 cm column should be expected. The selectivity factor (α) and capacity factor (k') can be altered by changing the chemical characteristics and the strength of the solvent, respectively. Examples of the effect of increasing R by changing either α or k' are given below.

The nature of the type of analysis to be carried out and the design of the LC system may be important considerations in the selection of the type and size of the column to be used since they may influence N. Some practical rules or considerations are given below.

1. Use a column size appropriate to the volume of the injection, weight of solute, etc.
2. Use small volume injections since these are theoretically best for sharp peaks (i.e., large N), but practical considerations of sample handling may dictate a compromise to a larger injection volume.
3. Use mobile phase as the sample solvent to avoid peak artefacts.
4. Standardize injection volumes for more reproducible results since change in injection volume will influence the retention time and hence peak height-area ratio. If peak areas are used for quantitation instead of peak heights, changes in injection volume have little effect.

Whereas a 25 × 0.21 cm LC column may be ideal for analysis of repeated 2-5 μl injections containing microgram quantities of a synthetic vitamin D metabolite using UV detection, it is not sensitive enough for the analysis of 1,25-$(OH)_2D_3$ in human plasma. The purification of 1,25-$(OH)_2D_3$ from 2-5 ml plasma by LC is therefore followed by a radioligand assay, and in such a case, good LC is required, with injection of the whole extract and minimal sample losses. Choice of a 25 × 0.62 cm silica column allows large injections to be made with minimal losses of sample, tolerance of large lipid loads in the sample, and minimal effects of extracolumn band broadening. In addition, if use of an automatic injection system (Fig. 2.8, e.g., WISP, Waters Associates, Milford, Mass.) is required with the capability of injecting from minimum

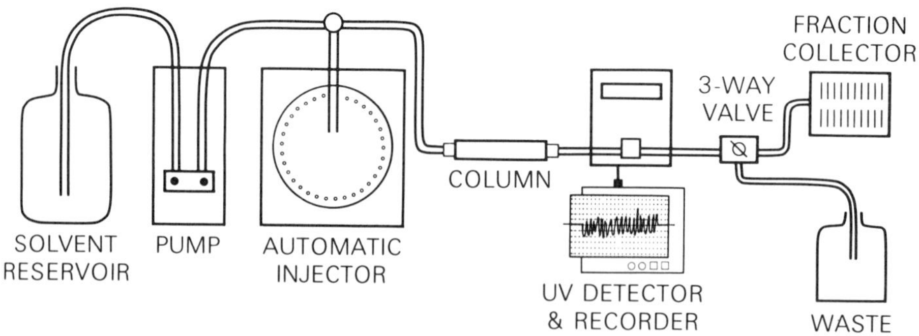

Figure 2.8 Typical LC apparatus used for the analysis of vitamin D and its metabolites in human plasma. (Modified from Ref. 47.)

volume inserts, The WISP can be programmed to make 200 μl injections from a total volume of 210 μl, thus losing only 5% of the sample (the WISP has a 6 μl dead space). In this example, the chromatography of biological extracts makes it preferable to have a precolumn filter (e.g., Rheodyne Inlet Filter) or a precolumn to protect the inlet end of the LC column, since direct injection of extracts from large volumes of sample increases the likelihood of column contamination. Though these devices help to prolong column life, they can also cause extracolumn band broadening especially if they are poorly installed or badly packed. Larger columns (25 × 0.62 cm) are less affected by these protective devices. The mobile phase is a rich source of contaminants over the life time of a column, and a preinjector filter or precolumn is therefore again recommended for prolonging column life.

3.2 Columns and Solvents

3.2.1 Silica

Several microparticulate (5-6 μm) spherical silicas are now available: Lichrospher, Zorbax-SIL, and Spherisil. The advantage of these packings is that they pack more uniformly than irregularly shaped particles, and have less tendency to settle with time and pressure.

3.2.1.1 Isopropanol-Hexane Mixtures. These mixtures provide a good solubility for the D vitamins and have been widely applied to the resolution of a variety of compounds ranging in polarity from photoirradiation mixtures of D to 1,25,25-$(OH)_3D_3$. Isocratic mixtures of 2.5% isopropanol in hexane separate vitamins D_2 and D_3 from 24-OH-D_2, 24-OH-D_3, 25-OH-D_2, and 25-OH-D_3 (Fig. 2.9). Isopropanol-hexane (10:90) provides resolution of most of the known vitamin D_3 metabolites [74] in a single chromatographic run (Fig. 2.10), and can also be useful for vitamin D_2 metabolites (Fig. 2.11). These early separations can now be achieved on a single 25 × 0.46 or 25 × 0.62 cm column using similar solvent systems. The use of ternary solvent systems, incorporating methanol into the standard isopropanol-hexane mixture [75], greater minimizes peak tailing, particularly of 24(R),25-$(OH)_2D_3$. Figure 2.12 shows the results of using a hexane-isopropanol-methanol (87:10:3) system solvent on the separation of 25-OH-D_3, 24(*R*),25-$(OH)_2D_3$, and 1,25-$(OH)_2D_3$. The polarity of the solvent can be reduced while keeping the proportion of isopropanol-methanol approximately the same.

3.2.1.2 Methanol-Methylene Chloride Mixtures. Ikekawa and Koizumi [76] have described a gradient solvent system of 0.02-6% methanol in methylene chloride to resolve a mixture of dihydroxy metabolites on Zorbax-SIL (25 × 0.21 cm) (Fig. 2.13). This solvent system

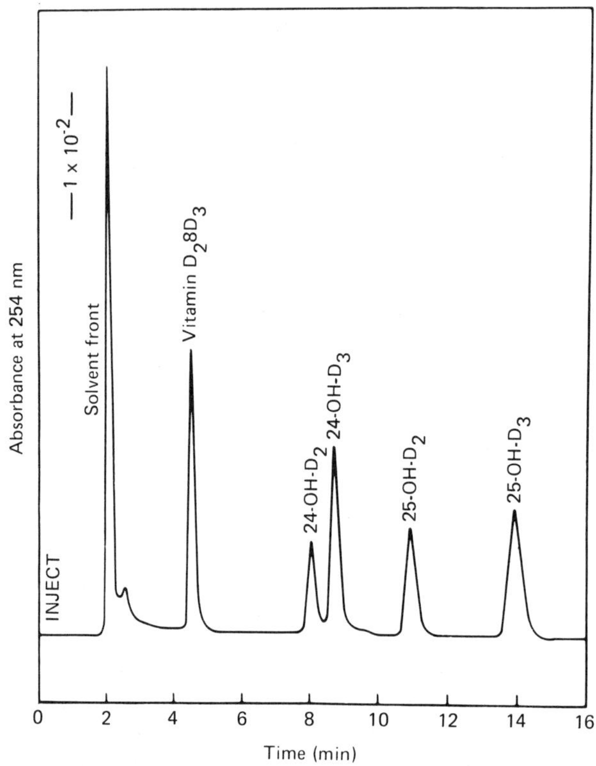

Figure 2.9 LC separation of vitamin D_2 and D_3, 24-OH-D_2, 24-OH-D_3, 25-OH-D_2, and 25-OH-D_3, using Zorbax-SIL and 2.5% isopropanol in hexane. (From Ref. 74.)

gives higher theoretical plate count N than does the isopropanol-hexane-methanol mixture under otherwise identical conditions. However, the low viscosity of methylene chloride-methanol mixtures can lead to bubble problems with some solvent delivery systems. Replacing some of the methylene chloride with hexane and using a hexane-methylene chloride-methanol mixture (90:10:10) overcomes some of the viscosity problems, but poor miscibility and poor sample solubility develop. The hexane-methylene chloride-methanol (90:10:10) mixture beautifully separates D_2 and D_3 metabolites (Fig. 2.14), but not D_2 and D_3 themselves.

Figure 2.15 illustrates the influence of small changes in solvent composition (α) on the resolution of 1,25-$(OH)_2D_2$ and 1,25-$(OH)_2D_3$ using the same column (Zorbax-SIL, 25× 0.62 cm) while keeping the retention approximately the same.

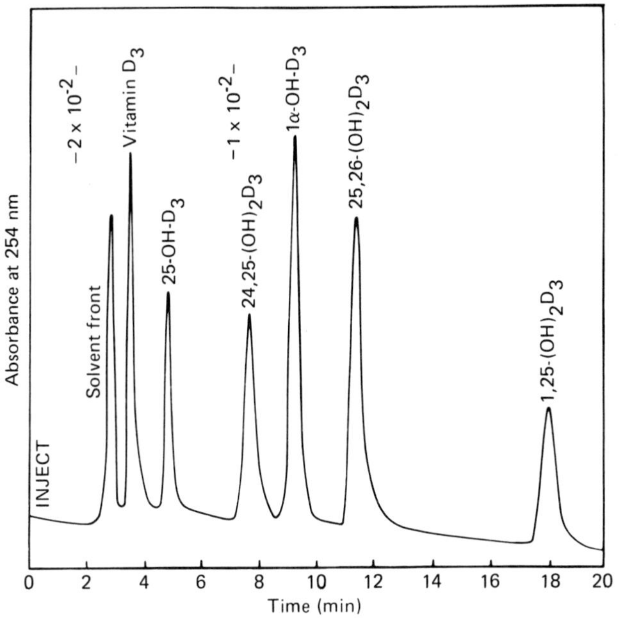

Figure 2.10 LC separation of vitamin D_3 and its metabolites on Zorbax-SIL using the solvent system 10% isopropanol in hexane. (From Ref. 74.)

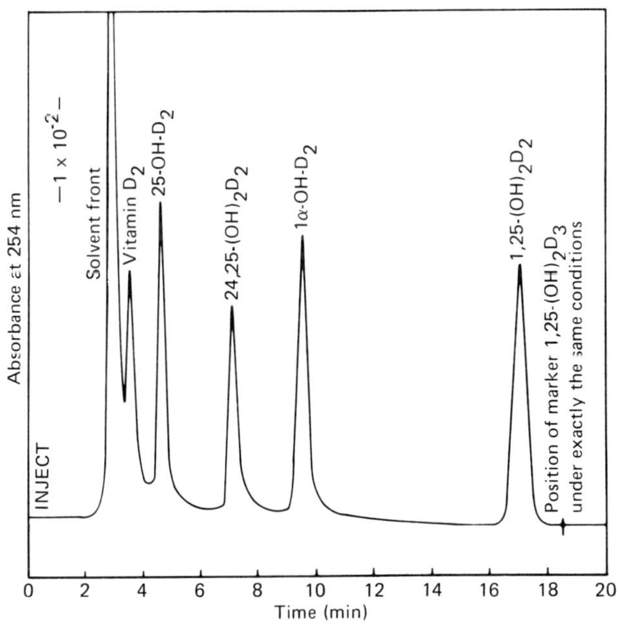

Figure 2.11 LC separation of vitamin D_2 and its metabolites on Zorbax-SIL using the solvent system 10% isopropanol in hexane. (From Ref. 74.)

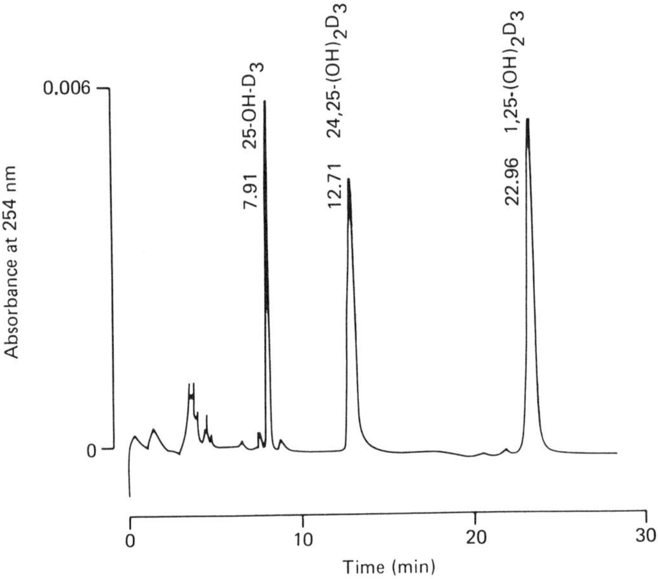

Figure 2.12 LC separation of 25-OH-D_3, 24(*R*),25-$(OH)_2D_3$, and 1,25-$(OH)_2D_3$ on Zorbax-SIL (25 cm × 6.2 mm) using the solvent system hexane-isopropanol-methanol (87:10:3). (Redrawn after Ref. 75.)

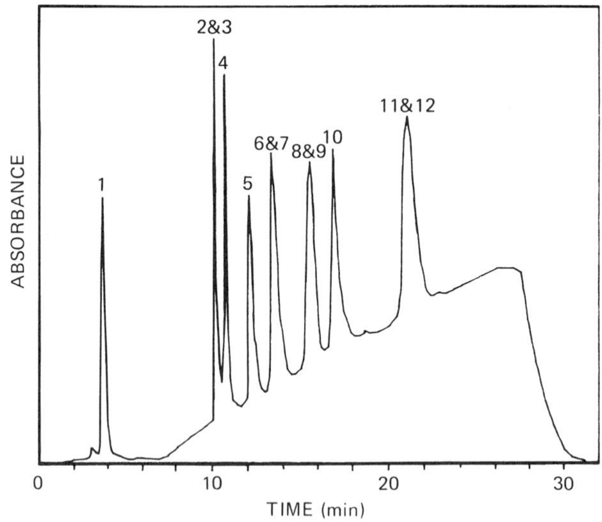

Figure 2.13 LC separation of vitamin D_3 metabolites on Zorbax-SIL using a gradient of 0.02-6% methanol in methylene chloride. Peaks: (1) vitamin D_3; (2, 3) 24(*R*)-OH-D_3 and 24(*S*)-OH-D_3; (4) 25-OH-D_3; (5) 1α-OH-D_3; (6, 7) 24(*R*),25-$(OH)_2D_3$ and 24(*S*),25-$(OH)_2D_3$; (8, 9) 1α-24(*R*)-$(OH)_2D_3$ and 1α,24(*S*)-$(OH)_2D_3$; (10) 1α,25-$(OH)_2D_3$; (11, 12) 1α,24(*R*),25-$(OH)_3D_3$ and 1α,24(*S*),25-$(OH)_3D_3$. (Modified from Ref. 76.)

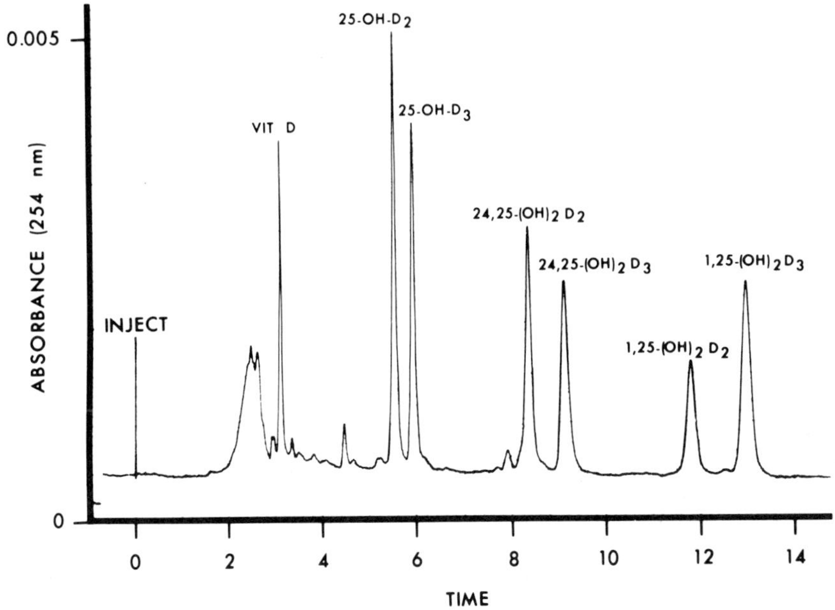

Figure 2.14 LC separation of vitamin D_2 and D_3 metabolites on Zorbax-SIL using the solvent system hexane-methanol-methylene chloride (80:10:10). (From Ref. 75.)

One further solvent system that should be mentioned which separates 25-OH-D_3-26,23-lactone and 24(*R*),25-$(OH)_2D_3$, is a methylene chloride-isopropanol (96.5:3.5) mixture [77]. The lactone runs with a retention time almost identical to 24,25-$(OH)_2D_3$ with isopropanol-hexane mixtures but elutes with 25-OH-D_3 with a solvent mixture rich in methylene chloride.

3.2.2 Octadecasilane (ODS or C18) Bonded Phases

Matthews et al. [78] were first to report LC of vitamin D metabolites using the pellicular packing ODS-Permaphase and a gradient of methanol and water. With the advent of microparticulate bonded phases, such as μBondapak C18 and Zorbax-ODS, many analysts have successfully used reverse-phase LC for many compounds. However, in the vitamin D field, their use has been sparing for all but the separation of vitamins D_2 and D_3 (Fig. 2.16). Mixtures of methanol-water or acetonitrile-methanol-water [77] are preferred for vitamin D and its metabolites. Acetonitrile greatly reduces the viscosity of the solvent mixture and results in lower back pressures. The limited use of reverse-phase packings stem from the aversion of vitamin D analysts to aqueous mobile

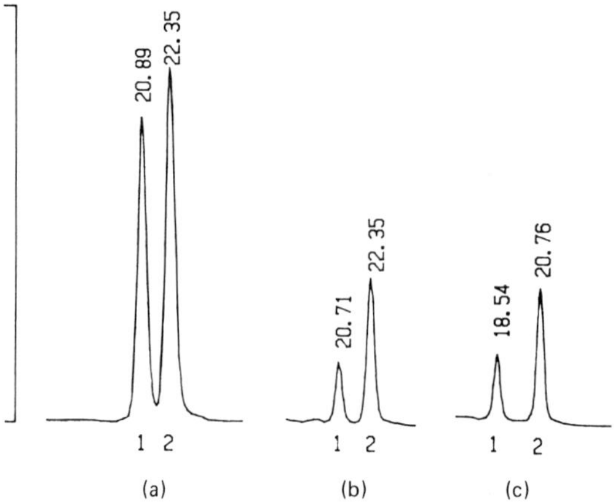

Figure 2.15 LC separation of 1,25-$(OH)_2D_2$ (peak 1) and 1,25-$(OH)_2D_3$ (peak 2) on Zorbax-SIL using (a) hexane-isopropanol (85:15); (b) hexane-isopropanol-methanol (87:10:3); and (c) hexane-ethanol-chloroform (80:10:10). Retention times in minutes are shown above peaks. Note the effect of changing solvent selectivity α on resolution. (Modified from Ref. 75.)

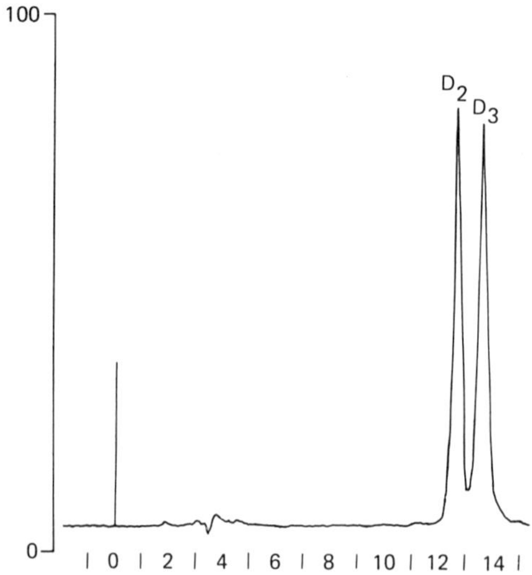

Figure 2.16 LC separation of vitamins D_2 and D_3 on Zorbax-ODS using the solvent system methanol-water (98.5:1.5). (From G. Jones, unpublished results.)

phases, particularly when competitive binding assay of collected fractions is used for quantitation, since aqueous phases are difficult to evaporate, tend to dissolve support material, and, since they contain dissolved oxygen, may give rise to oxidation of the 5,7-diene system. Second, the resolution of the dihydroxylated compounds is not as good as that achieved using silica (see Fig. 2.17). Plate counts on bonded-phase columns are generally lower, pressures are higher, and column lifetimes are shorter than systems using straight-phase silica LC.

3.2.3 Other Packings

There has been limited use of other packings. Straight-phase bonded packings offer excellent recoveries but often the solvent strength has to be weakened significantly in comparison to that for silica in order to conserve adequate retention. μBondapak-CN, Zorbax-CN [79], and Nucleosil 10-NO2 [49] have been used with some success.

3.3 Methods of Detection

3.3.1 Ultraviolet Absorbance (Fixed or Variable Wavelength)

Variable wavelength detectors are useful for photoirradiation mixtures (see Sec. 3.4.3), where they take advantage of the differences in the UV spectra of the various photoisomers. Normally little is gained in the analysis of vitamin D metabolites, since ε254 is 90% of ε265 and usually variable wavelength detectors are noisier at comparable sensitivity than fixed wavelength detectors due to an inherently weaker lamp. But variable wavelength detectors can have a considerable value, for example, when isotachysterol derivatives are used for LC, since these isomers at 290 nm give approximately twice the absorbance of unisomerized vitamin D metabolites at 264 nm [80]. Isotachysterols have been used to improve the sensitivity and specificity of an LC system for plasma 25-OH-D_2 and 25-OH-D_3. A simple method for 25-OH-D_2 and 25-OH-D_3 in plasma has been described using LC of isotachy sterols and monitoring at 301 nm, where absorbance for vitamin D metabolites is minimal but where isotachysterols still absorb significantly [60].

3.3.2 Radioactivity Detector

One manufacturer (Berthold, West Germany) has developed an extremely sensitive radioactivity monitor to measure radioisotopes flowing through a flow cell placed between two photomultiplier tubes. Early models were bulky and used a spirally arranged Teflon tube between two plates as the flow cell. A post-LC column mixer was used to mix effluent with scintillation fluid. The mixture flowed through the flow

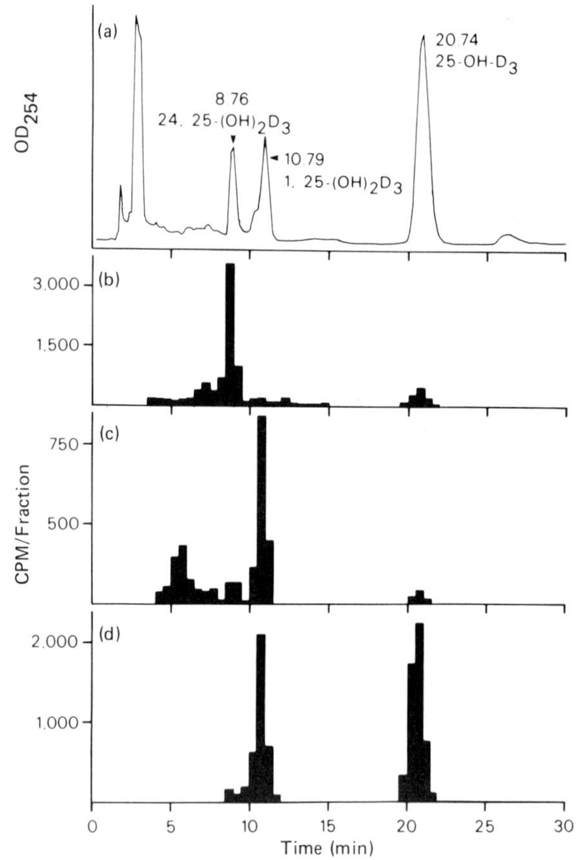

Figure 2.17 LC separation of in vitro generated metabolites of [^{3}H]25-OH-D_3 on μBondapak-C_{18} using the solvent system methanol-water (85:15). Fetal kidney tissue was homogenized and mitochondria prepared, and these were incubated with [^{3}H]25-OH-D_3 to produce two major metabolites, [^{3}H]1,25-$(OH)_2D_3$ and [^{3}H]24,25-$(OH)_2D_3$. The extract was chromatographed initially on Zorbax-SIL, as in Fig. 2.12 and the metabolite regions isolated and rechromatographed as above. (a) non-radioactive standards; (b) [^{3}H]24,25-$(OH)_2D_3$ region; (c) [^{3}H]1,25-$(OH)_2D_3$ region; and (d) radioactive standards. (From M. Rawlins, S. W. Kooh, and G. Jones, unpublished results.)

cell, where counting efficiencies for ^3H were high but residence time short. Results are shown in Fig. 2.18.

Later models (e.g., LB503HS) have reduced lead shielding and use highly sensitive electronics as before. The liquid scintillation fluid and mixer have been replaced by a flow cell of 300 μl volume containing a solid glass scintillator. Adsorption of lipids to the scintillator is minimal, but counting efficiency for ^3H is in the 2-5% range. Nevertheless, peaks are much sharper than when liquid scintillator is used (Fig. 2.19). The detector provides excellent on-line counting efficiencies for [^{14}C]25-OH-D_3 of around 80%.

3.3.3 *Off-Line Radioligand Assays*

One of the most common uses of LC in the analysis of clinical samples, where UV sensitivity is at its limit, is in the preparation of purified extracts containing individual metabolites (or groups of metabolites) for competitive protein-binding assay or radioimmunoassay. Specific binding proteins, such as vitamin D-binding globulin (DBP; VDBG), the blood carrier protein [46], or target tissue receptors for 25-OH-D_3 or 1,25-$(OH)_2D_3$, can be used in these systems (reviewed in Ref. 81). Antibodies have been raised in sheep and rabbits [47,82] and various radioimmunoassays described [81].

3.4 Applications

3.4.1 *Vitamin D and 25-OH-D Measurement in Plasma*

Vitamin D and 25-OH-D can be measured in plasma by LC with UV monitoring at 254 nm as described by Shepard et al. [67], Horst et al. [42], Jones [70], and Lambert et al. [83]. Schemes shown in Figs. 2.5 and 2.7 illustrate the different approaches used to purify the sample prior to LC. Examples of chromatograms from Ref. 70 are shown in Figs. 2.20 and 2.21.

24,25-Dihydroxyvitamin D_3 can also be measured by LC with UV monitoring at 254 nm, but identification of the small peak is tentative. The potentially interfering 25-OH D_3-26,23-lactone can be rapidly eluted with methylene chloride-isopropanol mixtures. An LC method for plasma 24,25-$(OH)_2D_3$ has recently been described [84] using two Zorbax-SIL LC steps.

3.4.2 *Analysis of Enzymic Products in Physiological Buffers*

This is a relatively straightforward application of the technique of LC to separate and identify radioactively labeled peaks by comigration with authentic standards. Two examples have already been shown in Figs. 2.17 and 2.18. Unfortunately, although this is one of the potentially useful applications of LC analysis, it is also one of the most

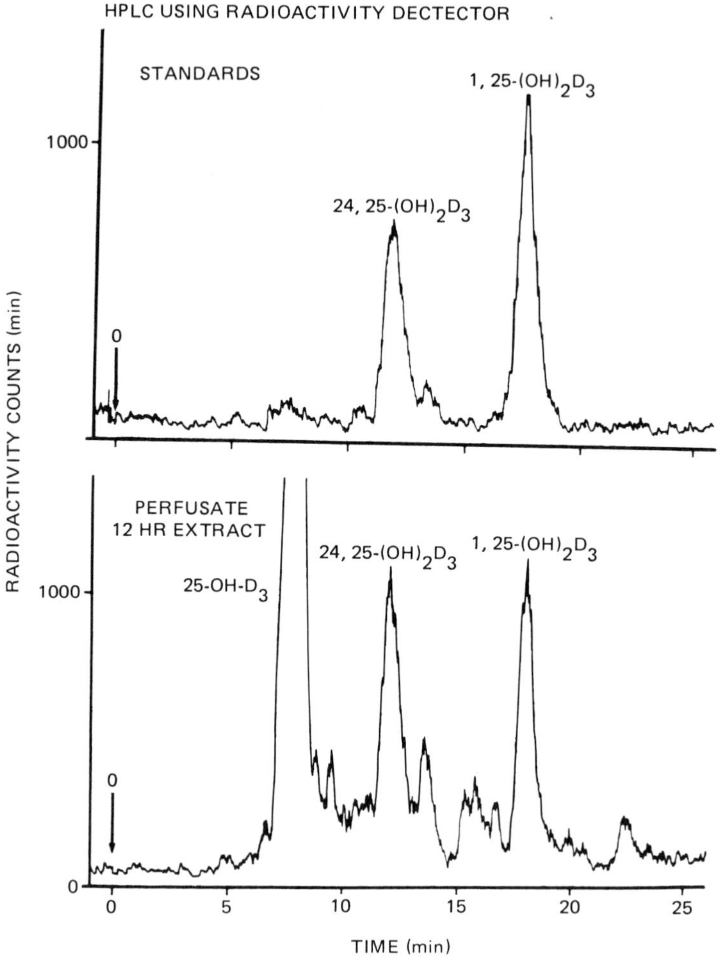

Figure 2.18 LC using an on-line liquid scintillation counter for detecting radioactive (^3H or ^{14}C) peaks of vitamin D. The example illustrates the metabolic picture obtained from the analysis of an incubation of [^3H]25-OH-D_3 with an isolated perfused rat kidney. (From Ref. 79.)

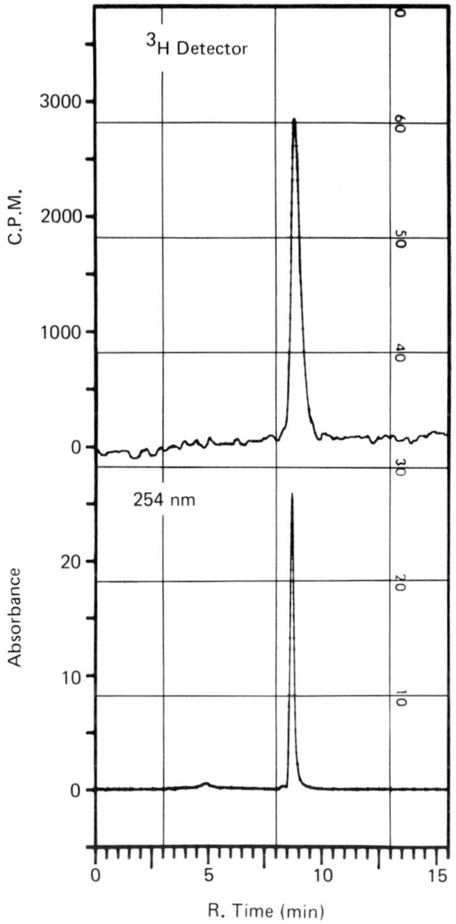

Figure 2.19 LC of [^{3}H]25-OH-D (400,000 dpm) as in Fig. 2.12, using a radioactivity detector (Berthold). Upper panel: ^{3}H detector. Lower panel: Waters 440 UV detector at 254 nm. Note the slight peak broadening effect of the 300 μl flow cell.

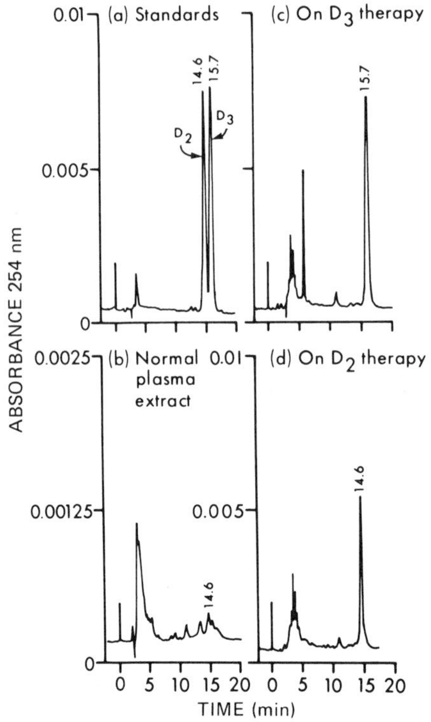

Figure 2.20 LC analysis of vitamin D fractions from human plasma using the scheme shown in Fig. 2.8. (a) Standards (D_2, 112 ng; D_3, 117 ng). (b) Plasma extract of normal adult. (c) Plasma extract of child treated with 50,000 IU vitamin D_3 daily. (d) Plasma extract of child treated with 100,000 IU vitamin D_2 daily. Fractions b, c, and d were purified by Zorbax-SIL LC before the reverse-phase LC shown here. Numbers above peaks represent retention times in minutes. Flow rate = 1.5 ml/min; solvent system, methanol-water (98.5:1.5); k'_{D_2} = 6.3; k'_{D_3} = 6.85; α = 1.087; N = 4720; R = 1.19. Column: Zorbax-ODS (6.2 mm × 25 cm). (From Ref. 70, © 1978, Clinical Chemistry.)

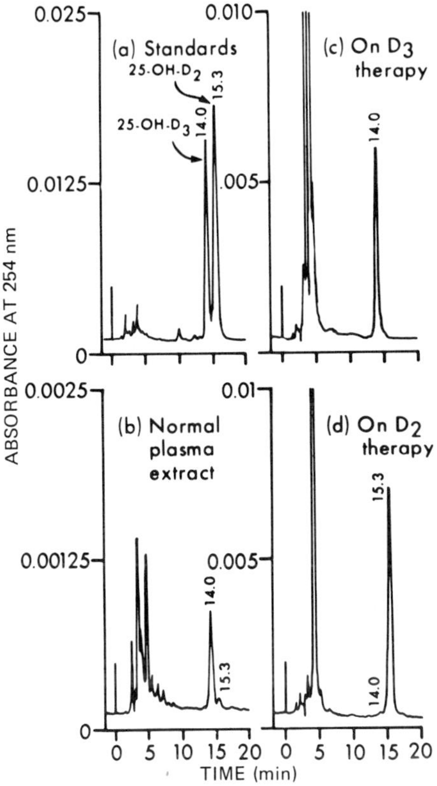

Figure 2.21 LC analysis of 25-OH-D fractions from human plasma using the scheme shown in Fig. 2.8. (a) Standards (25-OH-D_3, 250 ng; 25-OH-D_2, 345 ng). (b) Plasma extract of normal adult. (c) Plasma extract of child treated with 50,000 IU vitamin D_3 daily. (d) Plasma extract of child treated with 100,000 IU vitamin D_2 daily. Fractions b, c, and d were purified by Zorbax-SIL LC before the reverse-phase LC shown here. Numbers above peak represent retention times, in minutes. Flow rate = 1.5 ml/min; solvent system, methanol-water (91:9, v/v); $k'_{25\text{-OH-D}_3} = 6.80$; $k'_{25\text{-OH-D}_2} = 6.05$; $\alpha = 1.12$; N = 4314; R = 1.51. Column: Zorbax ODS (6.2 mm × 25 cm). (From Ref. 70, © 1978, Clinical Chemistry.)

abused. It is not valid to assume that, because a radioactive compound, e.g., peak B of Fig. 2.17, comigrates with the standard in a single LC system, it is unequivocally identified. Too often it is believed that liquid chromatography always produces "high-performance" results. Clearly, comigration depends on all factors in the R_s equation, and several solvent and column systems should be tested before firm identification of unknown peaks is made. Interpolation of specific chemical transformation (such as isotachysterol formation), changing the chromatographic properties of the peaks under investigation, also enhances specificity.

3.4.3 Photoirradiation Products

Figures 2.22 and 2.23 show LC runs of the photoisomers of vitamin D following UV irradiation of ergosterol. Each trace shows $\lambda_{254\ nm}$ and $\lambda_{280\ nm}$ detector responses, illustrating the different absorbance properties of the different isomers. Previtamin D_2 and tachysterol$_2$ form the major products but after heating overnight pre-D_2 isomerizes to vitamin D_2, which is difficult to resolve from tachysterol$_2$. Figure 2.23 illustrates the same irradiation mixture as that in Fig. 2.22, but the chromatographic separation has been altered by substitution of 0.25% methanol in methylene chloride for 1% isopropanol in hexane. Resolution of tachysterol2 and vitamin D_2 is marginally improved. Other secondary solvents, such as amyl alcohol, have been used to change α.

Liquid chromatography of stereoisomeric mixtures with naturally synthesized metabolites has led to the identification of at least four structures as 24(*R*),25-$(OH)_2D_3$, 24(*R*),25-$(OH)_2D_2$, 1,24(*R*),25-$(OH)_3D_3$, and 25(*S*),26-$(OH)_2D_3$. The separation of 24(*R,S*),25-$(OH)_2D_2$ is shown in Fig. 2.24.

3.4.4 Pharmaceutical Preparations

Aside from the specially formulated vitamin D metabolite preparations now being produced by Leo (1α-OH-D_3), Phillips-Duphar (vitamin D_2 and D_3), Upjohn (25-OH-D_3), and Hoffmann-La Roche [1α,25-$(OH)_2D_3$ and 24(*R*),25-$(OH)_2D_3$], the major type of vitamin D analysis in the pharmaceutical industry is measurement of vitamin D in multivitamin preparations and vitamin D resin-containing powders. Methodology involves a reverse-phase chromatographic step followed by separation of Δ4,6-cholestadienol(internal standard), vitamin D, and previtamin D. Concentrations of vitamin D represent the sum of previtamin D and vitamin D in the mixture [85]. An example of analysis of dry multivitamin concentrates after saponification and extraction is shown in Fig. 2.25.

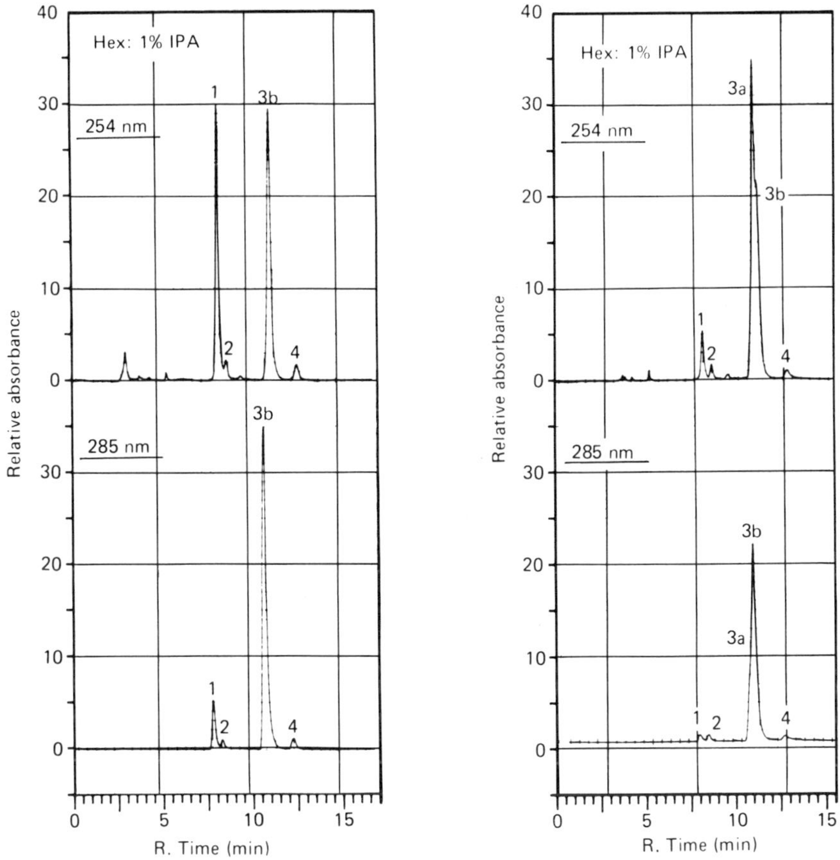

Figure 2.22 LC of mixture of photoisomers of ergosterol. Ergosterol (1 mg/ml) was irradiated with UV light from a quartz mercury vapor lamp screened by a vicor tube (Ace Glass, Inc., Vineland, N.J.). The hexane solution (450 ml) was irradiated in a photochemical reaction vessel equiped with double water jackets and a vent for passing a stream of N through the solution. Samples were drawn off for analysis every 5 min, and the analyses illustrated represent the steady-state mixture at 30 min. (a) Analyzed immediately. Conditions: Zorbax-SIL (25 cm × 6.2 mm). Solvent: hexane-1% isopropanol; 2.0 ml/min; 6900 kPa. Detectors: Upper trace, fixed wavelength; 254 nm; 100% = 0.5 Absorbance units. Lower trace, variable wavelength; 285 nm; 100% = 2.0 Absorbance units. Sample: 100 μl reaction mixture. Peak identity: (1) Previtamin D_2, (2) lumisterol$_2$, (3a) vitamin D_2, (3b) tachysterol$_2$, (4) ergosterol. (b) Analyzed following heating of reaction mixture. The mixture of photoisomers was heated in a screw-capped vial at 60°C under nitrogen for 8 hr. Conditions for analysis as in (a).

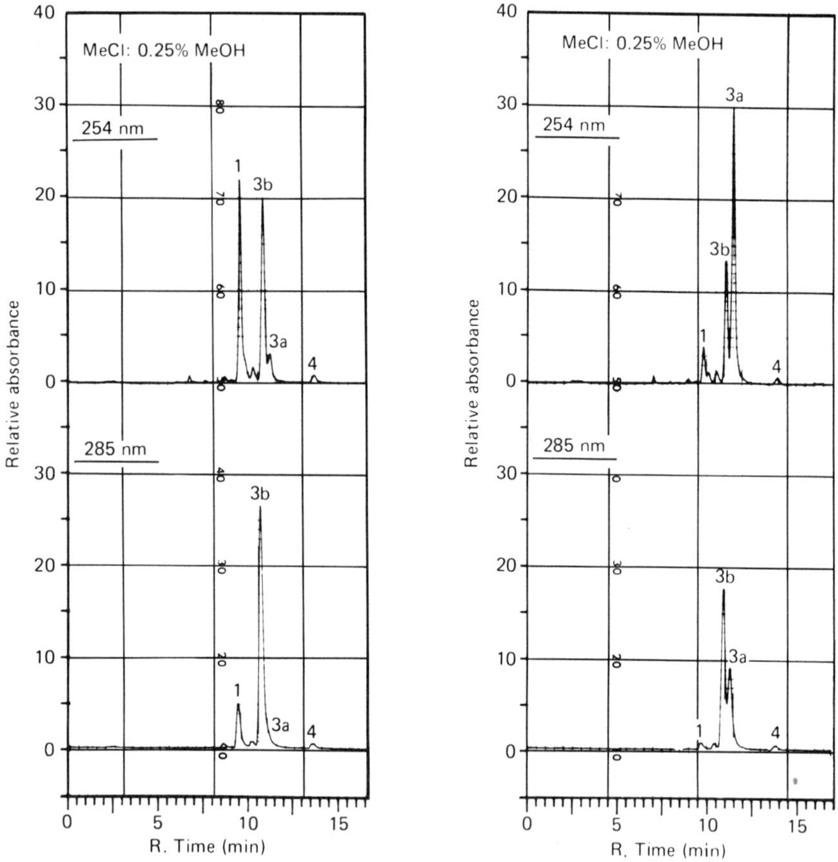

Figure 2.23 LC of mixture of photoisomers of ergosterol. As in Fig. 2.22, except the sample was evaporated and redissolved in methylene chloride-0.25% methanol prior to analysis. The LC solvent system was changed to methylene chloride-0.25% methanol. (a) Analyzed immediately. Condition: Zorbax SIL (25 cm × 6.2 mm). Solvent: Methylene chloride-0.25% methanol; 1.0 ml/min; 5520 kPa. Detectors: Upper trace, fixed wavelength; 254 nm; 100% = 1.0 Absorbance units. Lower trace, variable wavelength; 285 nm; 100% = 2.0 Absorbance units. Sample: 50 μl reaction mixture. Peak identity: (1) previtamin D_2, (2) lumisterol$_2$, (3a) vitamin D_2, (3b) tachysterol$_2$, (4) ergosterol. Note the reversal and better resolution of vitamin D_2 and tachysterol$_2$ with this solvent system. (b) Analyzed following heating of reaction mixture. Conditions of analysis as in (a) above.

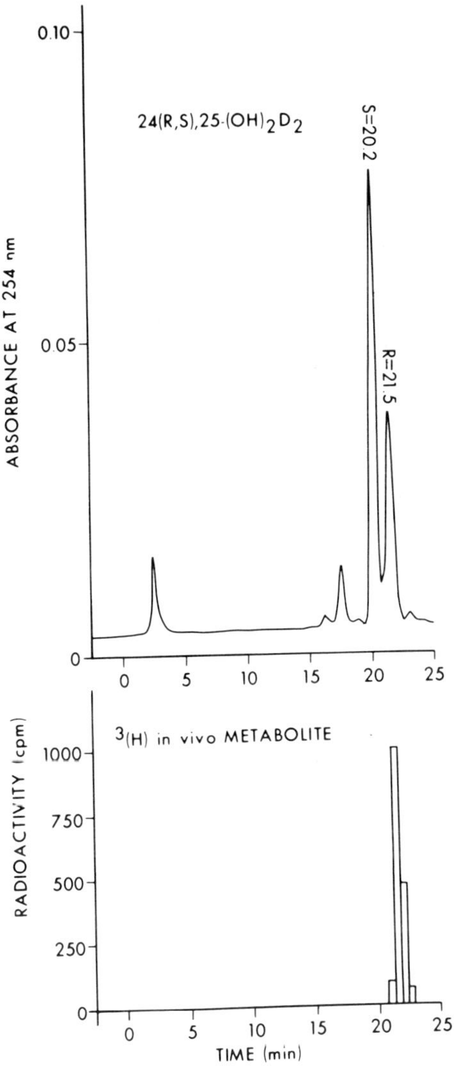

Figure 2.24 LC separation of stereoisomers of 24(*R*,*S*),25-$(OH)_2D_2$. (Reprinted with permission from Ref. 33, © 1979, American Chemical Society.)

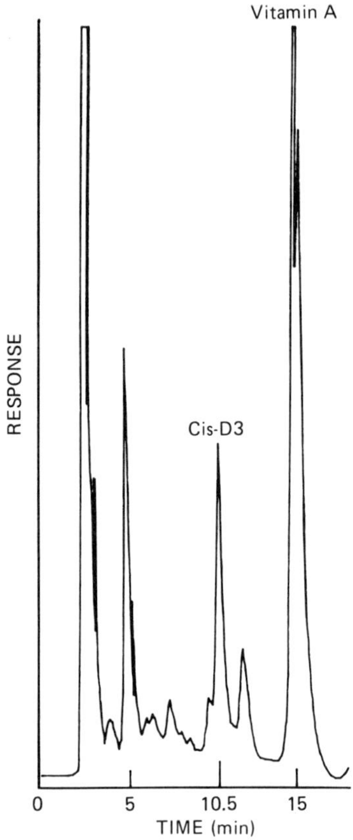

Figure 2.25 LC analysis of vitamin D_3 in a multivitamin concentrate after saponification and extraction. The concentrate contained 200,000 IU D_3 (5 mg) per gram. Note the good resolution of vitamins D and A. (From H. Hofsass, in Symposium Proceedings of the Association of Vitamin Chemists, Oak Brook, Illinois, March 30, 1979. Reproduced courtesy of Waters Associates.)

3.4.5 *Foodstuffs*

As stated earlier, current methodology for margarine involves saponification, selective extraction, and alumina column chromatography followed by a combination of silica, 5 μm, and reversed-phase Radial Pak-C18 separation of vitamins D_2 and D_3. See Fig. 2.26 for the type of separation. Milk analysis can be achieved without the alumina step.

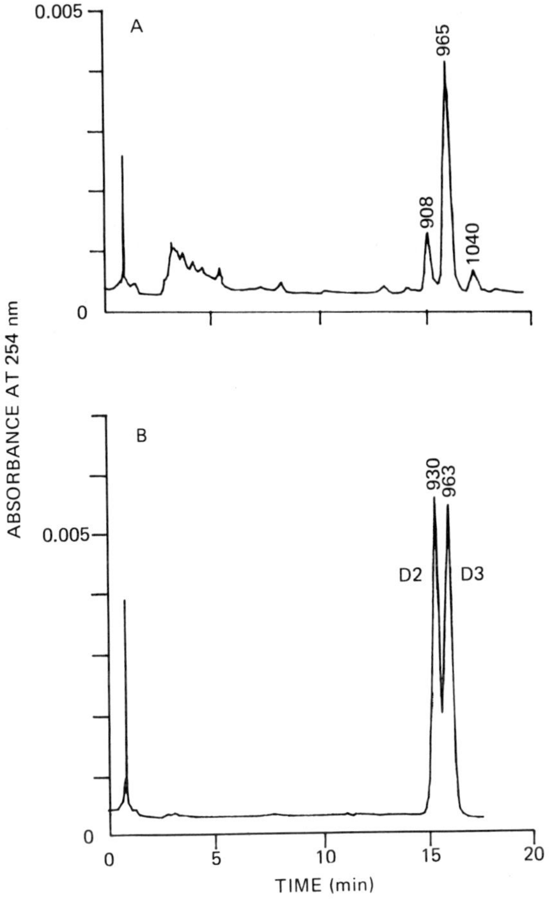

Figure 2.26 (A) LC analysis of vitamin D_3 in milk after saponification, extraction, and alumina chromatography. (B) LC of standards of D_2 and D_3 under identical conditions. Retention times are marked above the peaks in seconds. Spherisorb ODS (25 cm × 4.6 mm, 10 μm). Solvent system: methanol-acetronitrile (1:9). Flow rate: 1 ml/min. Detection: 254 nm. The milk sample contained 24 IU D_3 per 100 ml according to area under peak at 965 sec. (Modified from Ref. 63.)

4. GAS CHROMATOGRAPHIC ANALYSIS

4.1 B-ring Closure

A prerequisite for the use of gas-liquid chromatography is that the solute should be in the vapor phase during the process of partition between the mobile gas phase and the stationary liquid phase. For compounds of molecular weight in excess of around 200 daltons, this requires analysis to be carried out at elevated temperatures, usually in excess of 100°C. At these temperatures, vitamin D and its hydroxylated metabolites undergo thermal rearrangement involving B-ring closure. In the naturally occurring steroid, the 19-methyl attached to C10 is always β and is trans with respect to the 9α-hydrogen. When B-ring closure occurs in the vitamin D series, two isomers are formed, both of which are cis across the 9-10 carbon-carbon bond (i.e., C19β, 9β: isopyro-; and C19α, 9α: pyro-). The formation of these isomers was originally described by Askew and his colleagues in 1932 [86], who heated vitamin D in the test tube to around 190°C. Although isomerization took place rapidly at this temperature and was complete in around 4 hr, ring closure has been shown to occur at all temperatures in excess of 125°C [87]. It is irreversible and the ratio of pyro and isopyro remains at 2:1. This reaction occurs in the absence of oxygen, and it is now thought that the immediate precursor of these isomers is the previtamin (see Fig. 2.27).

At the temperatures at which GLC is carried out, therefore, vitamin D and its metabolites are all converter irreversibly and quantitatively to a 2:1 mixture of pyro and isopyro isomers that separate in all the GLC systems so far examined [68,73,88,89]. The formation of these two isomers during GLC was first demonstrated by Ziffer et al. [90].

4.2 Derivatives for GLC

As with all GLC systems for steroid analysis, the stability and volatility of vitamin D and its metabolites can be increased by the formation of derivatives on the polar hydroxyl groups. Free underivatized hydroxyl groups may give rise to adsorption during chromatography [91] and nonlinearity of detector response. The 25-hydroxyl group in vitamin D is sterically hindered, and if vitamin D and particularly its polyhydroxylated metabolites are injected into a GC without derivatization, broad peaks are obtained, indicating some degree of adsorption. In addition, dehydration of the 25-hydroxyl can occur, giving rise to two extra peaks in addition to the pyro and isopyro isomers. This dehydration is variable, but 100% conversion to the 25-dehydro derivative can sometimes be achieved by introducing powdered glass at the top of the GC column, thus allowing use of the dehydro peaks for quantitation.

Formation of derivatives on the polar hydroxyl groups of vitamin D and metabolites, although it improves stability and volatility, does not

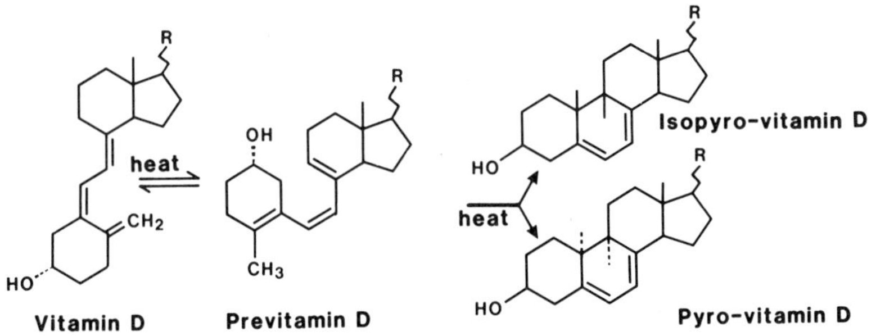

Figure 2.27 Thermal transformation of vitamin D to pyro and isopyro isomers.

overcome the problem of the B-ring closure and the formation of the two pyro and isopyro peaks. The formation of these isomers is quantitative and either or both peaks can be used for measurement [92-96], but the separation of a number of metabolites in a single run can be complicated by the multiplicity of peaks. Figure 2.28a illustrates the separation of trimethylsilyl ether derivatives on a conventional packed column. In this separation, the 25-hydroxyl has been derivatized by the formation of a 25-trimethylsilyl ether. Because the 25-hydroxyl is sterically hindered, it is difficult to derivatize, and so far, in our hands, only one reagent has proved satisfactory for the low levels present in plasma—trimethylsilylimidazole (TMSI). Improvement in separation can be achieved by using capillary columns, and Fig. 2.28b shows the separation of a persilylated mixture, similar to that used in Fig. 2.28a, on a 25 cm wall-coated capillary column.

In attempts to simplify the chromatographic profile obtained, a considerable amount of study has been undertaken into the formation of derivatives of vitamin D that were thermally stable and did not isomerize during GLC. Nair and DeLeon [89], using trifluoroacetic anhydride, formed "acetate esters" and suggested that although pyro and isopyro derivatives were still formed during GLC, they did not separate as acetates. This is now known to be incorrect, and it is recognized that the use of trifluoroacetic anhydride gave rise to the acetate of the isotachysterol isomer of vitamin D, which because of the rearrangement of the A ring is resistant to thermal cyclization (see Fig. 2.29). Injection of isotachysterol isomers onto GLC columns gives rise, therefore, to single peaks. Adsorption and dehydration of the 25-hydroxyl group still occur, however, and the use of derivatives is still necessary. 1α-Hydroxylated derivatives of vitamin D so not undergo this isomerization, probably because the suggested reaction mechanism for this reaction (illustrated in Fig. 2.29) involves the intermediate formation of isovitamin D with a Δ^1 double bond [97]. The presence of a 1α-hydroxyl

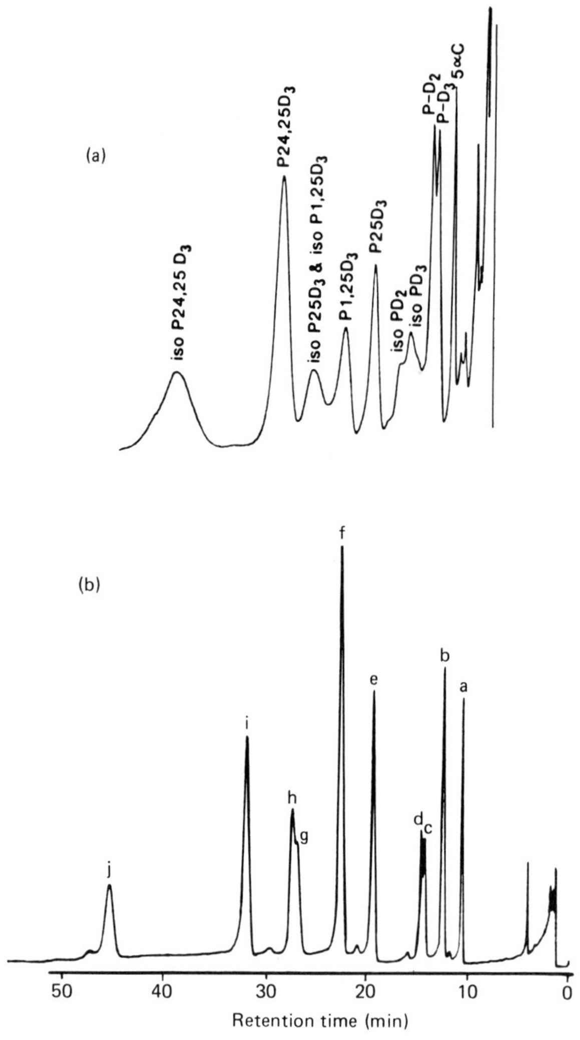

Figure 2.28 Gas chromatography of trimethylsilyl ether derivatives of vitamin D and various metabolites using (a) conventional packed column and (b) wall-coated glass capillary column. (a) Separation by GLC of the trimethylsilyl ether derivatives of ergocalciferol, cholecalciferol, and some metabolites of cholecalciferol. P = pyro, isoP = isopyro, 5αC = 5α-cholestane, D_2 = ergocalciferol, D_3 = cholecalciferol, $25D_3$ = 25-hydroxycholecalciferol, $1,25D_3$ = 24,25-dihydroxyvitamin D_3. Gas chromatography was carried out on OV1 at 230°C using total ion current detection. (From Ref. 73.) (b) Total ion current chromatogram of a mixture of the TMS ethers of five cholecalciferols showing peaks corresponding to cyclized isomers: vitamin D_3 (pyro, a, and isopyro, c), 1α-hydroxyvitamin D_3 (peak I, b, and peak II, d), 25-hydroxyvitamin D_3 (peak I, e, and peak II, g), 1α,25-dihydroxyvitamin D_3 (peak I, f, and peak II, h), and 24(*R*),25-dihydroxyvitamin D_3 (peak I, i, and peak II, j). (From Ref. 116.)

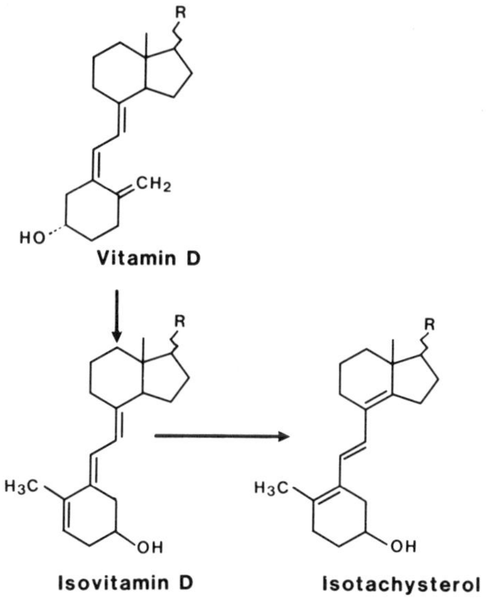

Figure 2.29 Formation of the isotachysterol isomers of vitamin D. (From Ref. 97.)

may well inhibit the formation of this intermediate. Figure 2.30 shows the GLC separation of the same mixture as that illustrated in Fig. 2.28a after formation of isotachysterol isomers.

A wide variety of reagents has been used to effect this isomerization: acetyl chloride [89], trifluoroacetic anhydride [89], penta- and heptafluorovutyryl chlorides [89], and heptafluorobutyric anhydride [98,99]. These reagents not only give rise to the formation of the isotachysterol isomers but at the same time form derivatives on the hydroxyl group. Acetyl chloride esterifies all groups, including the sterically indered 25-hydroxyl, whereas anhydrides do not esterify the 25-hydroxyl.

Other reagents that have been used to form the isotachysterol isomer, without concomitant ester formation, include boron trifluoride [100], antimony trichloride [101], and a number of acidic reagents [102]. None of these reagents were found to be of value when forming the isotachysterol isomers of the very small amounts of vitamin D and its metabolites present in human plasma. Seamark et al. [73] described a new procedure that gave quantitative conversion down to 1 ng levels using hydrochloric acid gas in chloroform or methylene chloride. At levels below 10 ng, carrier vitamin D_2 was added to protect against destruction. This simple HCl procedure, which takes only 5 min allows isomerization without formation of esters, thus allowing a choice of derivative without the interpolation of a hydrolysis step.

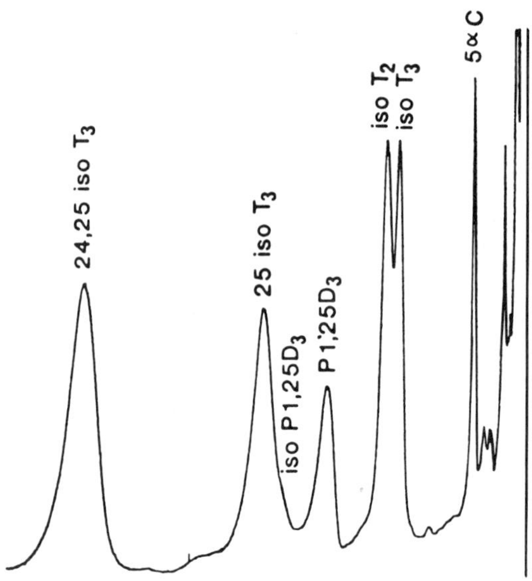

Figure 2.30 Gas chromatography of trimethylsilyl ether derivatives of isotachysterol isomers of vitamin D and various metabolites. Separation by GLC of the trimethylsilyl ether derivatives of ergocalciferol, cholecalciferol, and some metabolites of cholecalciferol, after isomerization to isotachysterol derivatives. P = pryo, isoP = isopyro, isoT_2 - isotachysterol$_2$, isoT_3 - isotachysterol$_3$, 5αC = 5α-cholestane, D_2 = ergocalciferol, D_3 = cholecalciferol, 25D_3 = 25-hydroxycholecalciferol, 1,25D_3 = 1α,25-dihydroxycholecalciferol. Gas chromatography was carried out on OV1 at 230°C using total ion current detection. This is the same mixture separated before isomerization in Fig. 2.28a. (From Ref. 73.)

Although these isomers give single peaks during GLC, formation of derivatives improves linearity of detector response and prevents destruction and adsorption in the same way as for isomerized vitamins. Underivatized 25-hydroxyisotachysterols still show dehydration. Figure 2.31 shows an example of this during the GLC spearation of methoxy derivatives of isotachysterol isomers using conventional packed columns. It is not clear whether this dehydration occurs during GLC or, as suggested by De Leenheer and Cruyl [103], during formation of the methoxy derivative or during the isomerization itself [104] using antimony trichloride. The quantitative formation of methoxy derivatives requires careful control and so far has only been examined using microgram amounts of vitamin D and 25-OH-D [103].

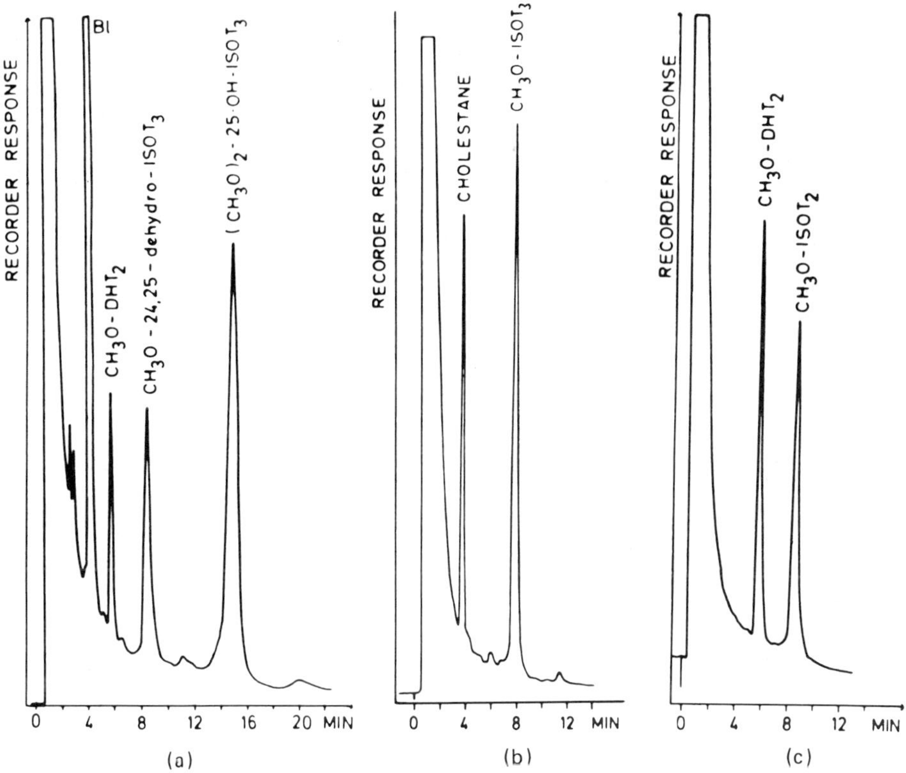

Figure 2.31 GLC of methoxy derivatives of isotachysterol isomers of vitamin D_2, vitamin D_3, and 25-OH-D_3. (a) Gas-liquid chromatography (FID) of 3β-methoxydihydrotachysterol$_2$ (CH_3O-DHT_2: 1 μg) and derivatized 25-hydroxyvitamin D_3 (4 μg), i.e., 3β-methoxy-24,25-dehydroisotachysterol$_3$ (CH_3O-24,25-dehydro-$ISOT_3$) and 3β,25-bis-methosyisotachystcrol$_3$ [$(CH_3O)_2$-25-OH-$ISOT_3$]. (b) GLC (FID) of cholestane (0.75 μg) and 3β-methoxyisotachysterol$_3$ (CH_3-$ISOT_3$, 3 μg). (c) GLC (FID) of 3β-methoxydihydrotachysterol$_2$ (CH_3O-DHT_2, 4 μg) and 3β-methoxyisotachysterol$_2$ (CH_3O-$ISOT_2$, 4 μg). GLC conditions: A 2% QF-1 column is operated at an oven temperature of 225°C and at a linear velocity of nitrogen (carrier gas) of 6 cm/sec. (From Ref. 109.)

4.3 Removal of Cholesterol

Unlike LC, in which the presence of large amounts of cholesterol in biological extracts can largely be ignored since cholesterol does not absorb in the UV, the presence of unesterified cholesterol during GLC is a major problem. It is present in human plasma in very much larger concentrations (50 mg per 100 ml) than vitamin D_3 (1 μg per 100 ml) and has very similar GLC properties to vitamin D_3. The removal of cholesterol prior to quantitation by GLC is therefore essential, but it has unfortunately proved to be extremely difficult to find a completely satisfactory method, although a number of procedures, such as TLC [68], digitonin precipitation [98,99], and LC on silver nitrate-impregnated columns [92], have been described.

4.4 Flame Ionization and Electron Capture Detectors

Although it is clear that vitamin D and its metabolites can be separated by GLC, their measurement in biological fluids has been hampered by the lack of a suitable detector with the required sensitivity and specificity. Two common detectors for gas chromatography are in use today, the flame ionization detector (FID) and the electron capture detector (ECD), and a number of methods using FID have been described for the estimation of vitamin D in pharmaceutical preparations or animal foodstuffs [105-107]. However, the FID is not sufficiently sensitive for the estimation of vitamin D metabolites in human plasma. The ECD, however, is sufficiently sensitive, and GLC methods have been described in which very low levels of vitamin D (5-10 nm; Refs. 108 and 109) using pure standards, and 25-OH-D_3 (50 pg; Ref. 68) using chick plasma, have been detected. However, the electron capture detector, although sufficiently sensitive, lacks specificity, and its use for the analysis of vitamin D in human plasma has so far not been described.

4.5 Mass Spectrometry

4.5.1 MS as a Detector

Linking the gas chromatograph to a mass spectrometer, which can be used both for identification and as a detector, gives comparable sensitivity to the ECD but provides greatly enhanced specificity. Low-resolution mass spectrometers are not as sensitive as high-resolution machines but are at the moment much cheaper to purchase and operate. Conventional packed columns with carrier gas flow rates around 40 ml/min can be linked to the mass spectrometer via separator systems, which might be expected to be a possible source of destruction and/or loss of sensitivity. Capillary columns can be linked to the mass spectrometer directly, since the carrier gas flow is low. Apart from its use as a detector on the end of a gas chromatograph, the mass spectrometer has found considerable use in the vitamin D field as a means of establishing the identity of isolated metabolites (e.g., Refs. 22 and 110-113) and for establishing the purity of biological extracts (see Fig. 2.32). Jones [70], for example, has also used mass spectrometry as a means of establishing the purity of a 25-OH-D_3 fraction isolated from plasma in an LC method for the measurement of 25-OH-D_2 and 25-OH-D_3.

The mass spectrometer can be used as a detector for gas chromatography in two says. One relies on the ability of the MS to focus onto specific ion fragments with time. This can be done in a conventional magnetic sector machine by altering the accelerating voltage so that

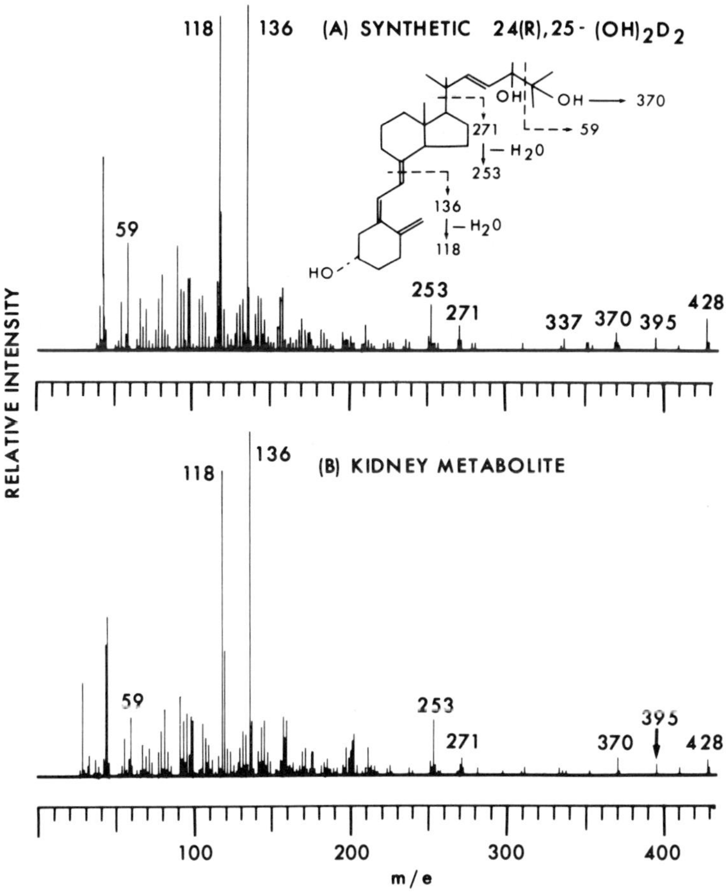

Figure 2.32 Mass spectra of 24(*R*),25-$(OH)_2D_2$ after LC. (A) Chemically synthesized. (B) Biologically generated. (Reprinted with permission from Ref. 33, © 1979, American Chemical Society.)

different ions are sequentially focused onto the electron multiplier. Quadrapole instruments which have not been used in the vitamin D field as yet, operate on a different principle. This process of mass fragmentography (MF) is the most sensitive mode of operation, but some GC-MS apparatus is attached to data handling systems that can be used to provide quantitative information in a different fashion to MF. The effluent from the GLC is scanned every 5-10 sec and the mass spectra are stored in the data core. At the end of the GLC run, these data can be recalled and chromatograms can be constructed for any ion fragment of choice. Such a system (mass chromatography) is not as sensitive as MF but provides a very valuable method of establishing the ion fragment to monitor for the very best sensitivity and specificity.

4.5.2 *Derivatives for MS*

The use of the mass spectrometer as a gas chromatographic detector places an extra requirement on the derivatives of vitamin D used for this analysis. Not only must they have satisfactory GLC properties but in addition they should fragment in the mass spectrometer to give a suitable ion of high m/z, preferably specific to the metabolite of interest. There are various methods of producing ion fragments, but in the vitamin D field electron impact (EI) has been most commonly used. Chemical ionization (CI) has been used, with ammonia as the reagent gas, in the GC-MS separation and identification of vitamin D_3 and previtamin D_3 after irradiation of the skin of vitamin D-deficient rats by ultraviolet light [114].

The most extensive evaluation of the mass spectrometric properties of vitamin D and its metabolites have been carried out in Lisboa's laboratory using open tubular capillary column GLC and a conventional low-resolution mass spectrometer. Trimethysilylated derivatives of D_3, 25-OH-D_3, their isotachysterol derivatives, and 1,25-$(OH)_2D_3$ have been examined by mass chromatography [104]. Increased intensity of the molecular ion in the mass spectrum of the isotachysterol$_3$ trimethysilyl (TMS) ether compared with that obtained from either the pyro- or isopyrovitamin D_3 trimethylsilyl ether was observed. This has also been described by Seamark et al. [115] and comparative values are given in Table 2.2. Thus, not only does the formation of isotachysterol derivatives give single peaks during GLC, but it also increases the sensitivity of mass fragmentography. Mass spectrometric data and retention times of these TMS ether derivatives on OV101-coated capillary columns have been obtained [116].

In addition, studies have been carried out using t-butyldimethylsilyl (t-BDMS) ether derivatives, which have been shown to give greatly enhanced intensity of the $(M - 57)^+$ ion fragment for steroids [117] and acetates. Retention time data have been obtained for these derivatives and for mixed derivatives containing the 25-trimethylsilyl ether [116].

TABLE 2.2 Comparison of the Peak Areas of ITS-TMS and pyro-D_3-TMS + Isopyro-D_3-TMS Prepared from Equal Amounts of D_3 as a Measure of Relative Ion Intensity[a]

Compound	"Area"
ITS_3-TMS	0.924
Pyro-D_3-TMS	0.585
Isopyro-D_3-TMS	0.175
ITS/(pyro + isopyro)	1.22
ITS/pyro	1.58
ITS/isopyro	5.28

[a]ITS = isotachysterol$_3$, D_3 = vitamin D_3. The isotachysterol derivative was prepared by the method of Seamark, Trafford, and Makin [73] and analyzed by mass fragmentography. The molecular ion (*m/z* 456) was monitored in each case, as was the ion $(M - 57)^+$ of cholesteryl-*t*-BDMS, used as an internal standard.

An interesting derivative for 24,25-$(OH)_2D_3$ has been described by Lisboa and Halket [116], who formed a cyclical boronate across the adjacent 24- and 25-hydroxyl groups as previously described for steroids [118-120]. Halket et al. [121] obtained mass spectrometric and retention times for methyl- and *n*-butylboronate-3-trimethylsilyl ether derivatives of 24,25-OH-D_3 and found that the use of the cyclic boronate derivative stablized the molecule and gave rise to considerably enhanced intensity of ions of high mass-charge ratio. Figure 2.33 compares the normalized mass spectra of 24,25-$(OH)_2D_3$-TMS with that obtained from the methylboronate TMS derivative showing the greatly increased mass fragments at *m/z* 381 $(M - 131)^+$, the fragment probably obtained by A-ring cleavage containing C2, C4, and C3-O-TMS and *m/z* 407, $[M - 90 + 15]^+$. Trimethylsilyl ether derivatives have been widely used for GC-MS, but as can be seen from Fig. 2.33, they are not ideal derivatives for mass fragmentography in the vitamin D field, since intensities of ion fragments of high mass are very low. The mass spectra of TMS ethers of 25-hydroxylated vitamin D_3 metabolites all contain a base peak at *m/z* 131 (the fragment obtained by side-chain cleavage, between C24 and C25, containing C26, C27, and C25-O-TMS). 26 Hydroxylation increases the mass of the base peak to *m/z* 219

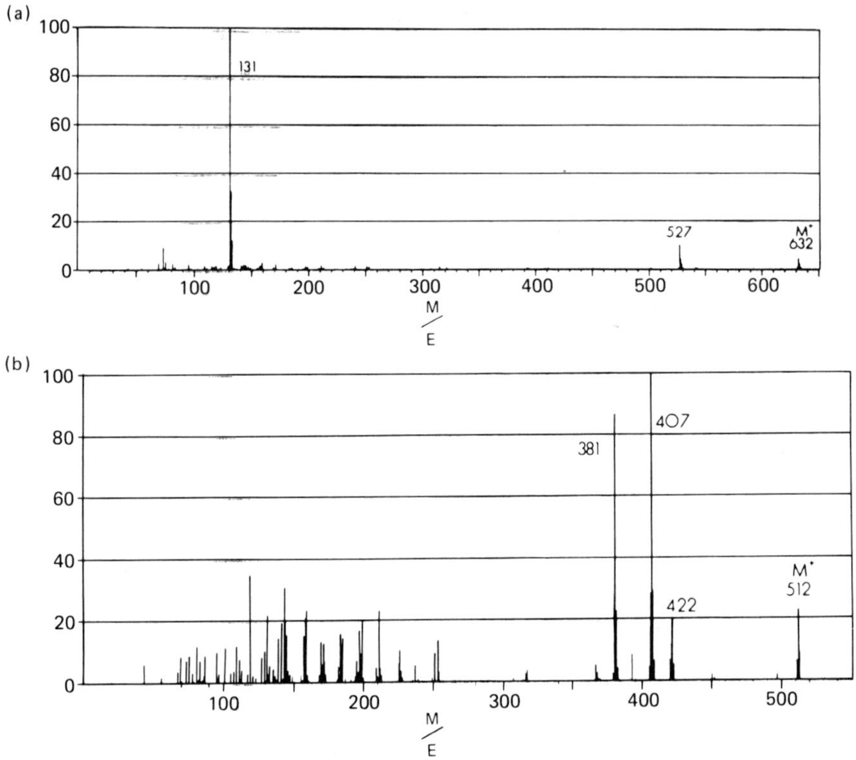

Figure 2.33 Normalized mass spectra of two derivatives of 24,25-$(OH)_2D_3$ during GC-MS on a 50 m OV-17 capillary column at 260°C. In each case the pyro peak was scanned. (a) 24(*R*),25-Dihydroxycholecalciferol-tris-trimethylsilyl ether. (b) 3-Trimethylsilyl ether-24,25-methylboronate ester derivative of 24(*R*),25-dihydroxycholecalciferol. (From Ref. 121.)

because of the presence of the extra 26-trimethylsilanol group. Use of other derivatives, such as the cyclic boronates, can, as illustrated for 24,25-$(OH)_2D_3$, greatly the intensities of higher mass fragments more suitable for use in MF analysis. Both mass spectra in Fig. 2.33 were obtained using the pyro peak, since formation of these derivatives does not prevent the thermal cyclization of the B ring.

Although information about the behavior of these derivatives of vitamin D in the mass spectrometer has been obtained, cyclic boronate derivatives have so far not yet been used in the analysis of vitamin D or its metabolites in biological fluids. Halket and Lisboa [122] in a brief communication did illustrate a mass fragmentograph of 25-OH-D_3-TMS extracted from plasma, although no quantitative data were given.

4.6 Vitamin D Estimation in Plasma Using GC-MS

The first mass fragmentographic assay of a vitamin D metabolite in human plasma was described for 25-OH-D_3 using [26-$^2H_3$]25-OH-D_3 as an internal standard [95]. Stable isotope-labeled 25-OH-D_3 was synthesized by the route shown in Fig. 2.34, and after purification a preparation containing 92% trideuterated molecules was obtained. Then, 2.5 ml serum was extracted with chloroform-methanol, the 25-OH-D_3 fraction eluted from a LH 20 column, and the TMS derivative formed and analyzed by GC-MS using a conventional packed column (1% SE30). The ion fragment *m/z* 131 (for 25-OH-D_3) and 134 (for [$^2H_3$]25-OH-D_3) were monitored. Because the mass fragments monitored were so low, interference was occasionally encountered, so that the *m/z* 131 channel was obscured and it was not possible to estimate 25-OH-D_3. In such cases, alternative ions could not be monitored since their intensities were too low. However, it was suggested that the 25-OH-D_3-TMS derivative could be resolved from the interfering substances by reducing the column temperature.

De Leenheer and Cruyl [98,99] described a method for plasma D_3 in which the interfering cholesterol was removed by digitonide precipitation and D_3 was purified using Lipidex 5000. After formation of isotachysterol$_3$ heptafluorobutyrate using heptafluorobutyric anhydride, mass fragmentography was carried out monitoring the molecular ion (M^+ = 580). Dihydrotachysterol$_2$ was used as the internal standard. A similar procedure for assaying plasma D_3 using added D_2 as an internal standard was described by Bjorkhem and Larsson [96]. Cholesterol

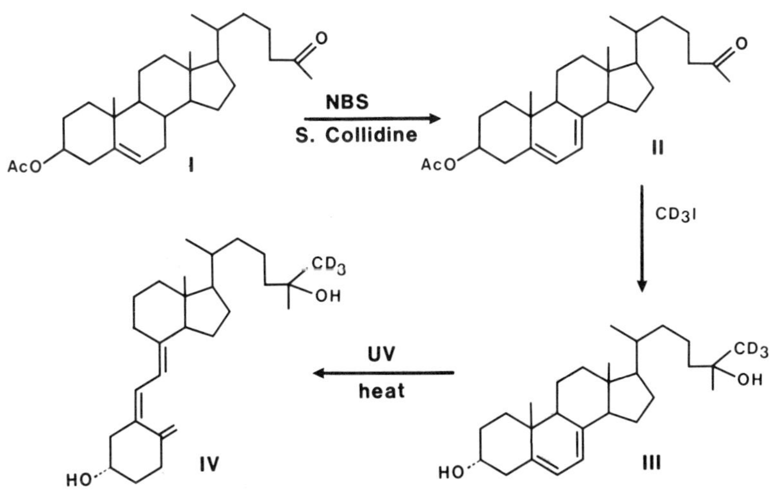

Figure 2.34 An outline of a synthetic route for the synthesis of 26-trideuterated 25-OH-D_3. (From Ref. 95.)

was removed by TLC and trimethylsilyl ether derivatives were formed. Although the base peak was at *m/z* 351 ($[M - 90 + 15)]^+$), it was considered better to monitor smaller peaks at *m/z* 325 (and 337 for D_2). Again, pyro and isopyro peaks were formed during chromatography.

The first use of capillary columns in the GC-MS of vitamin D derivatives was by Zagalak et al. [92], who described a procedure for D_3 and 25-OH-D_3 in human serum. A number of deuterated vitamin D metabolites were synthesized and used as internal standards. TLC was followed by LC using different columns for 25-OH-D_3 and D_3. Trimethylsilyl ethers were formed and the pyro peaks monitored using the molecular ion (*m/z* 456).

1,25-Dihydroxyvitamin D_3 has also been assayed [93] using GC-MS. After extraction, LH 20 chromatography and straight-phase followed by reverse-phase LC, trimethylsilyl ethers were formed and analyzed on a packed column (1.5% SE30) monitoring the fragment *m/z* 131, derived from side-chain cleavage, in a similar fashion to that described previously for 25-OH-D_3 [95]. This assay required 20 ml serum and used $[26^2H_3]1,25\text{-}(OH)_2D_3$ as an internal standard.

In an attempt to develop a mass fragmentographic "profile" for the major metabolites of vitamin D in plasma, a method was described [115] in which D_2, 25-(OH)-D_2, 25-OH-D_3, and 24,25-OH-D_3 were assayed. After extraction and Lipidex 5000 chromatography, isotachysterol isomers were formed using the HCl method of Seamark et al. [73]. Trimethylsilyl ethers were formed using TMSI and mass fragmentography carried out, monitoring the molecular ions of each metabolite.

These methods and the values obtained in plasma using mass fragmentography are summarized in Table 2.3. Bjorkhem's group have summarized their work in this area [123] and have used their methods for a study of the seasonal variation in plasma D_3 and 25-OH-D_3 [124].

4.7 In Vitro Studies Using GC-MS

Apart from the use of GC-MS for the analysis of plasma levels of vitamin D and its metabolites, Bjorkhem's group have applied mass gragmentographic methods to the study of in vitro metabolism, investigating the properties of a mitochondrial 25-hydroxyl [125,126] and a microsomal 25-hydroxylase [127]. In these studies the previous published methods were modified and mass fragmentographic assays were developed using *t*-butyldimethylsilyl derivatives of 25-OH-D_3 with a trimethylsilyl ether on C25 and the molecular ion *m/z* 586 was monitored. The $(M - 57)^+$ ion, found in the steroids, was not found to be prominent, although the $[M - (90 + 57)]^+$ was slightly more intense than the M^+ ion. Capillary column GC-MS has been used by Shackleton et al. [128] in a study of metabolites of vitamin D in the urine.

TABLE 2.3 Plasma Levels of Vitamin D and Metabolites Estimated Using Gas-Liquid Chromatography

Metabolite	Level[a]	Assay details[b]	Reference
25-OH-D_3	17 and 25[c]	EC, PC, ITS-TFA	68
25-OH-D_3	27± 10 (23)	MF, PC, TMS, SID	95
D_3	5-11 (4)†	MF, PC, ITS-HFV, DHT_2	94
D_3	1.6 (19)*	MF, PC, TMS, D_2IS	96
	0-6.3†	MF, PC, TMS, D_2IS	96
1,25-$(OH)_2D_3$	55 ± 10 pg/ml (15)	MF, PC, TMS, SID	93
25-OH-D_3	23* Aug (12)	MF, PC, TMS, SID	124
	11* Apr (12)		
D_3	4* Aug (7)	MF, PC, TMS, D_2IS	124
	2* Apr (7)		
	26* Jun (7)		
D_3	8.1 ± 4.8 (15)	MF, PC, ITS-TMS, D_2IS	115
25-OH-D_3	17.9 ± 7.4 (18)		
25-OH-D_2	0.9 ± 0.2 mg/ml (3)		
24,25-$(OH)_2D_3$			

[a]Mean values—ng/ml unless otherwise noted—± standard deviation (number of patients) are recorded, except where * mean and † range are recorded. Values measured in August (Aug), April (Apr), and June (Jun).
[b]EC, electron capture; PC, packed column; ITS, isotachysterol; TFA, trifluoroacetate; TMS, trimethylsilyl ether; MF, mass fragmentography; SID, stable isotope dilution; D_2IS, D_2 as an internal standard; HFB, heptafluorobutyrate.
[c]Two values in chick plasma.
Source: From Ref. 81.

TABLE 2.4 Synthesis of Some Deuterated Analogs of Various Vitamin D Metabolites for Use in GC-MS

Metabolite	Carbon atom on which ^2H attached	Number of ^2H molecules per mole	Reference
D_3	26	3	128
25-OH-D_3	26	3	128
1,25-$(OH)_2D_3$	26	3	128
24,25-$(OH)_2D_3$	26	3	128
25,26-$(OH)_2D_3$	26	3	128
1,24,25-$(OH)_3D_3$	26	3	128
25-OH-D_3	26	3	95,123
1,25-$(OH)_2D_3$	26	3	93,123
D_3	3,4	2	92
D_3	2,3,4,6	6	92
25-OH-D_3	26	3	92

4.8 Synthesis of Deuterated Standards

It is clear that the best mass fragmentographic assays utilize stable isotope-labeled anlogs for internal standards. The use of nonradioactive standards does away with the requirement for liquid scintillation counting and simplifies the assay and provides a greater degree of precision. Table 2.4 summarizes some of the deuterated standards that have been produced so far and used in mass fragmentographic analysis of vitamin D and its metabolites [92,93,123,129]. Using stable isotopes as internal standards, stable isotope dilution mass fragmentographic (SID-MF) assays can be developed, providing definitive procedures against which the specificity of many of the commonly used competitive protein binding, radioreceptor, or radioimmunoassays may be assessed.

ACKNOWLEDGMENTS

Some of the work described in this chapter was supported by the Medical Research Council of Canada (Grant No. MT-5777) to Jones, and by grants from the Wellcome Trust and Research Advisory Committee of the London Hospital to Makin. The authors wish to acknowledge the technical assistance of F. Palma in some of this work.

REFERENCES

1. J. W. Blunt, H. F. DeLuca, and H. K. Schnoes, *Biochemistry, 7*, 3317 (1968).
2. H. F. DeLuca, *Vitamin D: Metabolism and Function*, Springer-Verlag, Berlin, 1979, pp. 1-80.
3. H. F. DeLuca, *Ann. Rev. Physiol., 43*, 199 (1981).
4. M. R. Haussler and T. A. McCain, *N. Engl. J. Med., 297*, 974 (1977).
5. A. W. Norman, in *Vitamin D, The Calcium Homeostatic Steroid Hormone*, Academic Press, London, 1979. pp. 1-450.
6. E. Havinga, *Experientia, 29*, 1181 (1973).
7. W. G. Salmond, in *Vitamin D: Basic Research and Its Clinical Application* (A. W. Norman, K. Schaefer, D. V. Herrath, H.-G. Grigoleit, J. W. Coburn, H. F. DeLuca, E. B. Mawer, and T. Suda, eds.), Walter de Gruyter, Berlin, 1979, p. 25.
8. M. F. Holick, N. M. Richtand, S. C. McNeill, S. A. Holick, J. E. Frommer, J. W. Henley, and J. T. Potts, *Biochemistry, 18*, 1003 (1979).
9. G. Ponchon and H. F. DeLuca, *J. Clin. Invest., 48*, 1273 (1969).
10. M. Horsting and H. F. DeLuca, *Biochem. Biophys. Res. Commun., 36*, 251 (1969).
11. M. H. Bhattacharyya and H. F. DeLuca, *Arch. Biochem. Biophys., 160*, 58 (1974).
12. T. Suda, N. Horiuchi, M. Fukishima, Y. Nishii, and E. Ogata, in *Vitamin D: Biochemical, Chemical and Clinical Aspects Related to Calcium Metabolism* (A. W. Norman, K. Schaefer, J. W. Coburn, H. F. DeLuca, D. Fraser, H. G. Grigoleit, D. V. Herrath, eds.), Walter de Gruyter, Berlin, 1977, p. 201.
13. D. E. M. Lawson, D. R. Fraser, E. Kodicek, H. R. Morris, and D. H. Williams, *Nature, 230*, 228 (1971).
14. M. F. Holick, H. K. Schnoes, and H. F. DeLuca, *Proc. Nat. Acad. Sci. U.S., 68*, 803 (1971).
15. M. F. Holick, H. K. Schnoes, H. F. DeLuca, R. W. Gray, I. T. Boyle and T. Suda, *Biochemistry, 11*, 4251 (1973).
16. J. C. Knutson and H. F. DeLuca, *Biochemistry, 13*, 1543 (1974).
17. J. F. Myrtle, and A. W. Norman, *Science, 171*, 79 (1971).
18. M. R. Haussler, D. W. Boyce, E. T. Littledike, and H. Rasmussen, *Proc. Nat. Acad. Sci. U.S., 68*, 177 (1971).
19. T. C. Chen, L. Castillo, M. Korycka-Dahl, and H. F. DeLuca, *J. Nutr., 104*, 1056 (1974).
20. Y. Tanaka and H. F. DeLuca, *Arch. Biochem. Biophys., 146*, 574 (1971).
21. D. R. Fraser and E. Kodicek, *Nature, 228*, 764 (1970).
22. U. Trechsel, J. A. Eisman, J. A. Fischer, J. P. Bonjour, and H. Fleisch, *Am. J. Physiol., 239*, E119 (1980).
23. Y. Tanaka and H. F. DeLuca, *Arch. Biochem. Biophys., 154*, 566 (1973).

24. M. Garabedian, M. F. Holick, H. F. DeLuca, and I. T. Boyle, *Proc. Nat. Acad. Sci. U.S.*, *69*, 1673 (1972).
25. R. W. Gray, J. L. Omdahl, J. G. Ghazarian, and H. F. DeLuca, *J. Biol. Chem.*, *247*, 7528 (1972).
26. R. M. Shepard and H. F. DeLuca, *Arch. Biochem. Biophys.*, *202*, 43 (1980).
27. Y. Tanaka and H. F. DeLuca, *Science*, *183*, 1198 (1974).
28. P. Bordier, H. Rasmussen, P. Marie, L. Miravet, J. Gueris, and A. Ryckwart, *J. Clin. Endocrinol. Metab.*, *46*, 284 (1978).
29. D. Goodwin, D. Noff, and S. Edelstein, *Nature*, *276*, 517 (1978).
30. M. F. Holick, A. Kleiner-Bossaller, H. K. Schnoes, P. M. Kasten, I. T. Boyle, and H. F. DeLuca, *J. Biol. Chem.*, *248*, 6691 (1973).
31. T. Suda, H. K. Schnoes, H. F. DeLuca, and J. W. Blunt, *Biochemistry*, *8*, 3515 (1969).
32. G. Jones, H. K. Schnoes, and H. F. DeLuca, *Biochemistry*, *14*, 1250 (1975).
33. G. Jones, A. Rosenthal, D. Segev, Y. Mazur, F. Frolow, Y. Halfon, D. Rabinovich, and Z. Shakked, *Biochemistry*, *18*, 1094 (1979).
34. D. Harnden, R. Kumar, M. F. Holick, and H. F. DeLuca, *Science*, *193*, 493 (1976).
35. R. P. Esvelt, H. F. Schnoes, and H. F. DeLuca, *Biochemistry*, *18*, 3977 (1979).
36. H. F. DeLuca and H. K. Schnoes, in *Vitamin D: Basic Research and Its Clinical Application* (A. W. Norman, K. Schoefer, D. V. Herrath, H.-G. Grigoleit, J. W. Coburn, H. F. DeLuca, D. B. Mawer, and T. Suda, eds.), Walter de Gruyter, Berlin, 1979, p. 445.
37. J. K. Wichman, H. F. DeLuca, H. K. Schnoes, R. L. Horst, R. M. Shepard, and N. A. Jorgensen, *Biochemistry*, *18*, 4775 (1979).
38. J. J. Partridge, S.-J. Shiney, N. K. Chadha, E. G. Baggiolini, J. F. Blount and M. R. Uskokovic, *J. Amer. Chem. Soc.*, *103*, 1253 (1981).
39. Y. Tanaka, H. K. Schnoes, C. M. Smith, and H. F. DeLuca, *Arch. Biochem. Biophys*, *210*, 104 (1981).
40. N. Ohnuma, K. Bannai, H. Yamaguchi, Y. Hashimoto, and A. W. Norman, *Arch. Biochem. Biophys.*, *204*, 387 (1980).
41. G. Jones, H. K. Schnoes, L. Levan, and H. F. DeLuca, *Arch. Biochem. Biophys.*, *202*, 450 (1980).
42. R. L. Horst, E. T. Littledike, J. L. Riley, and J. L. Napoli, *Anal. Biochem.*, *116*, 189 (1981).
43. N. LeBoulch, C. Gulat-Marnay, and Y. Raoul, *J. Vitam. Nutr. Res.*, *44*, 167 (1974).
44. R. Kumar, S. Nagubandi, I. Jardine, J. Londowski, and S. Bollman, *Amer. Soc. Bone Mineral Res.*, *51A*, (1981).
45. E. G. Bligh and W. J. Dyer, *Can. J. Biochem.*, *37*, 911 (1957).

46. R. E. Belsey, H. F. DeLuca, and J. T. Potts, *J. Clin. Endocrinol. Metab.*, *38*, 1046 (1974).
47. R. Bouillon, P. DeMoor, E. G. Baggiolini, and M. R. Uskokovic, *Clin. Chem.*, *26*, 562 (1980).
48. M. T. Parviainen, K. E. Savolainen, P. H. Korhonen, E. M. Alhava, and J. K. Visakorpi, *Clin. Chim. Acta*, *114*, 233 (1981).
49. N. J. M. Jongen, W. J. F. Vander Vijgh, H. J. J. Willems, and J. C. Netelenbes, *Clin. Chem.*, *27*, 444 (1981).
50. J. A. Eisman, A. J. Hamstra, B. E. Kream, and H. F. DeLuca, *Science*, *193*, 1021 (1976).
51. R. S. Mason, D. Lissner, C. Reek, and S. Posen, in *Vitamin D: Basic Research and Its Clinical Application* (A. W. Norman, K. Schoefer, D. V. Herrath, H.-G. Grigoleit, J. W. Coburn, H. F. DeLuca, D. B. Mawer, and T. Suda, eds.), Walter de Gruyter, Berlin, 1979, p. 243.
52. R. Belsey, H. F. DeLuca, and J. T. Potts, *J. Clin. Endocrinol. Metab.*, *33*, 554 (1971).
53. E. E. Delvin, M. Dussualt, and F. H. Glorieux, *Clin. Biochem.*, *13*, 106 (1980).
54. K. T. Koshy and A. L. VanDerSlik, *Anal. Biochem.*, *74*, 282 (1976).
55. T. J. Gilbertson and R. P. Stryd, *Clin. Chem.*, *23*, 1700 (1977).
56. K. T. Koshy, *Meth. Enzymol.*, *67*, 357 (1980).
57. B. W. Hollis, B. A. Roos, H. H. Draper, and P. W. Lambert, *J. Nutr.*, *111*, 384 (1981).
58. J. S. Adams, T. Clemens, and M. F. Holick, *J. Chromatogr.*, *226*, 198 (1981).
59. J. T. Dabek, M. Harkonen, O. Wahlroos, and H. Adlercreutz, *Clin. Chem.*, *27*, 1346 (1981).
60. H. Turnbull, D. J. H. Trafford, and H. L. J. Makin, *Clin. Chem. Acta*, *120*, 65 (1982).
61. L. J. Fraher, G. Jones, T. L. Clemens, S. Adami, and J. L. H. O'Riordan, *Acta Endocrinol. (KBH.)*, *97*, Suppl. 243, Abstract 8 (1981).
62. S. K. Henderson and A. F. Wickroski, *J. Ass. Offic. Anal. Chem.*, *61*, 1130 (1978).
63. J. N. Thompson, G. Hatina, W. B. Maxwell, and S. Duval, *J. Ass. Offic. Anal. Chem.*, *65*, 624 (1982).
64. M. F. Holick and H. F. DeLuca, *J. Lipid Res.*, *12*, 460 (1971).
65. J. N. Ellingboe, E. Nystrom, and J. Sjovall, *J. Lipid Res.*, *11*, 266 (1970).
66. J. N. Thompson, W. B. Maxwell, and M. L'Abbe, *J. Ass. Offic. Anal. Chem.*, *60*, 998 (1977).
67. R. M. Shepard, R. L. Horst, A. J. Hamstra, and H. F. DeLuca, *Biochem. J.*, *182*, 55 (1979).
68. D. Sklan, P. Budowski, and M. Katz, *Anal. Biochem.*, *56*, 606 (1978).

69. R. Vieth, D. Fraser, and G. Jones, *Anal. Chem.*, *50*, 2150 (1978).
70. G. Jones, *Clin. Chem.*, *24*, 287 (1978).
71. T. Okano, N. Mizuno, S. Shida, N. Takahashi, T. Kobayashi, E. Kuroda, S. Kodama, T. Matsuo, *J. Nutr. Sci. Vitaminol.*, *27*, 43 (1981).
72. D. J. H. Trafford, D. A. Seamark, H. Turnbull, and H. L. J. Makin, *J. Chromatogr.*, *226*, 351 (1981).
73. D. A. Seamark, D. J. H. Trafford, and H. L. J. Makin, *J. Steroid Biochem.*, *13*, 1057 (1980).
74. G. Jones and H. F. DeLuca, *J. Lipid Res.*, *16*, 448 (1975).
75. G. Jones, *J. Chromatogr. Biomed. Appl.*, *221*, 27 (1980).
76. N. Ikekawa and N. Koizumi, *J. Chromatogr.*, *119*, 227 (1976).
77. R. L. Horst, *Biochem. Biophys. Res. Commun.*, *89*, 286 (1979).
78. E. W. Matthews, P. G. H. Byfield, K. W. Colston, I. M. A. Evans, L. S. Galante, and I. MacIntyre, *FEBS Lett.*, *48*, 122 (1974).
79. A. M. Rosenthal, G. Jones, S. W. Kooh, and D. Fraser, *Amer. J. Physiol.*, *239*, E12 (1980).
80. D. A. Seamark, D. J. H. Trafford, P. G. Hiscocks, and H. L. J. Makin, *J. Chromatogr.*, *197*, 271 (1980).
81. D. A. Seamark, D. J. H. Trafford, and J. L. J. Makin, *J. Steroid Biochem.*, *14*, 111 (1981).
82. T. L. Clemens, G. N. Hendry, R. F. Graham, E. G. Baggiolini, M. R. Uskokovic and J. L. H. O'Riordan, *Clin. Sci. Mol. Med.*, *54*, 329 (1978).
83. P. W. Lambert, P. B. DeOreo, B. W. Hollis, I. Y. Fu, D. J. Grinsberg, and B. A. Roos, *J. Lab. Clin. Med.*, *98*, 536 (1981).
84. B. E. Dreyer and D. B. P. Goodman, *Anal. Biochem.*, *114*, 37 (1981).
85. E. de Vries, F. J. Mulder, and B. Borsje, *J. Ass. Offic. Anal. Chem.*, *64*, 61 (1981).
86. F. Askew, R. B. Bourdillon, H. M. Bruce, R. K. Callow, J. Philpot, and T. A. Webster, *Proc. Roy. Soc.*, *Series B*, *107*, 76 (1932).
87. B. Pelc and D. H. Marshall, *Steroids*, *31*, 23 (1978).
88. P. P. Nair, C. Bucana, S. DeLeon, and D. A. Turner, *Anal. Chem.*, *37*, 631 (1965).
89. P. P. Nair and S. DeLeon, *Arch. Biochem. Biophys.*, *128*, 663 (1968).
90. H. Ziffer, W. J. A. Vanden Heuvel, E. O. A. Haahti, and E. C. Horning, *J. Amer. Chem. Soc.*, *82*, 6411 (1960).
91. H. L. J. Makin and D. J. H. Trafford, in *Methodological Surveys in Biochemistry*, Vol. 7 (E. Reid, ed.), Ellis Horwood, Chichester, 1978, p. 312.
92. B. Zagalak, H. D. L. Curtius, R. Foschi, G. Wipf, V. Redweik, and M. J. Zagalak, *Experientia*, *34*, 1537 (1978).
93. I. Bjorkhem, I. Holmberg, T. Kristansen, and J. L. Pedersen, *Clin. Chem.*, *25*, 584 (1979).
94. J. M. Halket and B. P. Lisboa, *Acta Endocrinol. (Kbh.)*, *87*, Suppl. 251, p. 120 (1978).

95. I. Bjorkhem and I. Holmberg, *Clin. Chim. Acta, 68*, 215 (1976).
96. I. Bjorkhem and A. Larsson, *Clin. Chim. Acta, 88*, 559 (1978).
97. T. Kobayashi and A. Adachi, *J. Nutr. Sci. Vitaminol. (Tokyo), 19*, 311 (1973).
98. A. P. De Leenheer and A. A. M. Cruyl, in *Quantitative Mass Spectrometry in Life Sciences* (A. P. De Leenheer and R. R. Roncucci, eds.), Elsevier, Amsterdam, 1977, p. 165.
99. A. P. De Leenheer and A. A. M. Cruyl, *Anal. Biochem, 91*, 293 (1978).
100. H. H. Inhoffen, K. Bruckner, and R. Grundel, *Chem. Bericht., 87*, 1 (1954).
101. T. K. Murray, K. C. Day, and E. Kodicek, *Biochem. J., 98*, 293 (1966).
102. T. Kobayashi and A. Adachi, *J. Nutr. Sci. Vitaminol. (Tokyo), 19*, 303 (1973).
103. A. P. De Leenheer and A. A. M. Cruyl, *J. Chromatogr. Sci., 14*, 434 (1976).
104. J. M. Halket and B. P. Lisboa, in *Recent Developments in Mass Spectrometry in Biochemistry and Medicine*, Vol. 1, (A. Frigerio, ed.), Plenum Publishing, New York, 1978, p. 457.
105. T. K. Murray, P. Erdody, and T. Panalaks, *J. Ass. Offic. Anal. Chem., 51*, 839 (1968).
106. D. O. Edlund, *Methods Enzymol., 67*, 343 (1980).
107. T. Kobayashi, *Methods Enzymol., 67*, 347 (1980).
108. P. W. Wilson, D. E. M. Lawson, and E. J. Kodicek, *J. Chromatogr., 39*, 75 (1969).
109. A. P. De Leenheer and A. A. M. Cruyl, *Methods Enzymol., 67*, 335 (1980).
110. N. Ohnuma, K. Bannai, H. Yamaguchi, Y. Hashimoto, and A. W. Norman, *Arch. Biochem. Biophys, 204*, 387 (1980).
111. L. W. LeVan, H. K. Schnoes, and H. F. DeLuca, *Biochemistry, 20*, 222 (1981).
112. Y. Takasaki, T. Suda, S. Yamada, H. Takayama, and Y. Nishii, *Biochemistry, 20*, 1681 (1981).
113. T. A. Reinhardt, J. L. Napoli, D. C. Beitz, E. T. Littledike, and R. L. Horst, *Biochem. Biophys. Res. Commun., 99*, 302 (1981).
114. T. Okano, K. Mizuno, N. Matsuyama, N. Nobuhara, and T. Kobayashi, *Rec. Trav. Chim. Pays-Bas, 98*, 253 (1979).
115. D. A. Seamark, D. J. H. Trafford, and H. L. J. Makin, *Clin. Chim. Acta, 106*, 51 (1980).
116. B. P. Lisboa and J. M. Halket, in *Recent Developments in Chromatography and Electrophoresis* (A. Frigerio and L. Renoz, eds.), 1979, pp. 141-162.
117. G. Phillipou, D. A. Bigham, and R. F. Seamark, *Steroids, 26*, 516 (1975).

118. C. J. W. Brooks and B. S. Middleditch, *Clin. Chim. Acta, 34*, 145 (1971).
119. C. J. W. Brooks and D. J. Harvey, *J. Chromatogr., 54*, 193 (1971).
120. T. A. Baillie, C. J. W. Brooks, and B. S. Middleditch, *Anal. Chem., 44*, 30 (1972).
121. J. M. Halket, I. Ganschow, and B. P. Lisboa, *J. Chromatogr., 192*, 434 (1980).
122. J. M. Halket and B. P. Lisboa, *Acta Endocrinol. (Kbh.), 87*, Suppl. 215, 120 (1978).
123. I. Bjorkhem, I. Holmberg, J. Kristiansen, A. Larsson, and J. I. Pedersen, in *Vitamin D: Basic Research and Its Application* (A. W. Norman, K. Schaefer, D. V. Herrath, H.-G., Grigoleit, J. W. Coburn, H. F. DeLuca, E. B. Mawer, and T. Suda, eds.), Walter de Gruyter, Berlin, 1979, p. 183.
124. I. Holmberg and A. Larsson, *Clin. Chim. Acta, 100*, 173 (1980).
125. I. Bjorkhem and I. Holmberg, *J. Biol. Chem., 253*, 842 (1978).
126. I. Bjorkhem and I. Holmberg, *J. Biol. Chem., 254*, 9518 (1980).
127. I. Bjorkhem, R. Hansson, I. Holmberg, and K. Wikvall, *Biochem. Biophys. Res. Commun., 90*, 615 (1979).
128. C. H. L. Shackleton, E. Roitman, and J. J. Whitney, *Steroid Biochem., 12*, 521 (1980).
129. J. O. Whitney, C. H. L. Shackleton, C. G. Edmonds, A. L. Burlingame, and C. F. Piel, in *Stable Isotopes (Proceedings of the Third International Conference)* (E. R. Klein and P. D. Klein, eds.), Academic Press, New York, 1979.

3

VITAMIN E: TOCOPHEROLS AND TOCOTRIENOLS

HANS J. NELIS, VEERLE O. R. C. DE BEVERE, and ANDRÉ P. DE LEENHEER

Laboratoria voor Medische Biochemie en vor Klinische Analyse, Rijksuniversiteit Gent, Gent, Belgium

1. INTRODUCTION

Since Evans and Burr demonstrated the presence of an unknown dietary "factor X" in natural foodstuffs [1], vitamin E has become a collective term for a series of compounds exhibiting more or less alike biological activity [2]. Although the exact biochemical function of α-tocopherol, the predominant form in nature, has yet to be elucidated, most evidence so far justifies its classification among the vitamins. Chromatography played an important role in the isolation and characterization of several homologous and isomeric vitamin E-like compounds. It was soon recognized that there were quantitative differences in the biological activities of those various forms. The complexity of biological matrices usually does not permit direct determination of vitamin E without a preliminary lipid fractionation step. As a result, chromatography has emerged as an essential part of most vitamin E assays. It is the aim of this chapter to rationalize this trend, to demonstrate the superiority of chromatography-based to classic colorimetric and fluorimetric vitamin E assays, and to review the existing literature in this area, with special emphasis on applications in biological media.

2. CHEMISTRY

The most simple synthetic vitamin E-like derivative, which is, however, devoid of biological activity, is *dl*-tocol or *dl*-2-methyl-2(*R*)[4'(*R*),8'(*R*), 12'-trimethyltridecyl]-6-chromanol. Tocopherols can be regarded as substituted tocols. The four naturally occurring tocopherols, *d*-α-, *d*-β-, *d*-γ-, and *d*-δ-tocopherol, differ in the number and position of methyl groups on the chromanol ring, as shown in Table 3.1. Similarly, four derivatives with an unsaturated side chain (tocotrienols) consisting

TABLE 3.1 Structures of Tocopherols

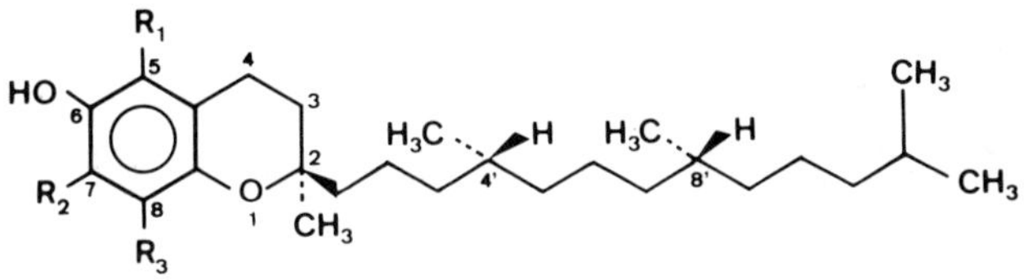

Compound	R_1	R_2	R_3
α-Tocopherol	CH_3	CH_3	CH_3
β-Tocopherol	CH_3	H	CH_3
γ-Tocopherol	H	CH_3	CH_3
δ-Tocopherol	H	H	CH_3
5,7-Dimethyltocol[a]	CH_3	CH_3	H
Tocol[a]	H	H	H

[a]Synthetic compound.

of three double bonds are accordingly designated α-, β-, γ-, and δ-tocotrienol (Table 3.2). All those eight vitamin E forms were isolated from vegetable oils and other plant materials [3].

Since the tocopherol molecule contains three asymmetric centers, i.e., at C2, C4', and C8', different stereoisomers are possible. The natural compounds have been shown to possess the 2(*R*),4'(*R*),8'(*R*) configuration, whereas the synthetic substances are 2(*S*),4'(*R*),8'(*R*). Tocotrienols occur in the all-trans configuration.

3. OCCURRENCE

Vegetable materials and oils, nuts and seeds in particular, are rich sources of vitamin E. Since animals and humans are unable to synthesize this vitamin, they have to rely on an external supply through their diet. Consequently, products of animal origin are relatively poor in this vitamin. *d*-α-Tocopherol, biologically the most potent derivative, is the predominant form in plasma and tissues. In humans, almost 88% of total vitamin E consists of the latter homolog, whereas *d*-β-tocopherol (2%) and the γ analog (10%) account for the rest of the activity. Until now, no detectable amounts of δ-tocopherol or tocotrienols have been found in humans or animals under physiological conditions. More stable *dl*-tocopheryl acetate and succinate esters have been synthesized and

TABLE 3.2 Structures of Tocotrienols

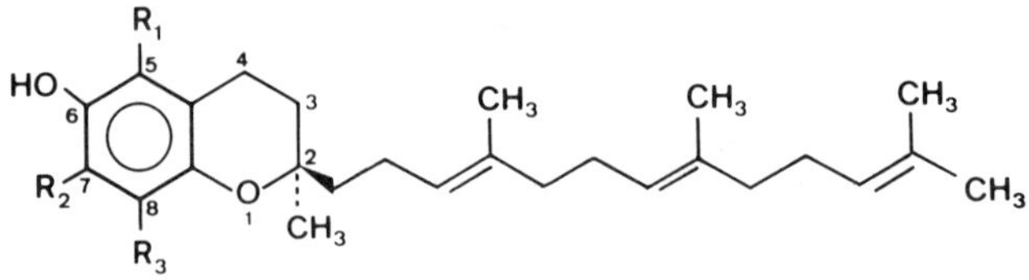

Compound	R_1	R_2	R_3
α-Tocotrienol	CH_3	CH_3	CH_3
β-Tocotrienol	CH_3	H	CH_3
γ-Tocotrienol	H	CH_3	CH_3
δ-Tocotrienol	H	H	CH_3

are commonly used to fortify foods and feeds. The recommended daily intake of vitamin E can be estimated at 5-15 mg (expressed as α-tocopheryl acetate).

4. BIOCHEMICAL IMPORTANCE

All tocopherols and tocotrienols are potent antioxidants, but their relative biological activities are not predictable on the basis of this chemical property. To a certain extent, the action of vitamin E in vivo can be understood in terms of its ability to prevent lipid peroxidation [3]. For example, it is well established that vitamin E plays a significant role in the maintenance of red cell viability by counteracting free radical-mediated peroxidation processes [4-6]. The compound reportedly inhibits platelet aggregation by blocking the formation of the endoperoxide that triggers this process [7,8]. Vitamin E has also been shown to affect certain mitochondrial functions [9], as well as microsomal mixed-function oxidase systems [10], which suggests a possible role in the electron transport chain. Its close interaction with the selenium-containing glutathione peroxidase further indicates active involvement in oxidative processes [10,11].

However, there is growing evidence that more specific cellular functions, unrelated to antioxidant properties, are involved [3]. The wide diversity of vitamin E-deficiency symptoms in certain animal species cannot be explained solely by the antioxidant theory [3]. Regulatory functions in nucleic acid and protein metabolism have been postulated, but so far there is still insufficient evidence supporting this hypothesis

[12]. At present, a unifying theory that fits all effects associated with vitamin E deficiency is still lacking.

5. ANALYTICAL ASPECTS

5.1 The Need for Assaying Vitamin E

Most vitamin E assays focus on the determination of the compound in either its primary sources, or blood and tissues of animals and humans, who assimilate the vitamin secondarily through dietary consumption. Analysis of tocopherols and tocotrienols in primary sources, e.g., vegetable oils, is desirable to detect losses that may occur during refinement processes. The marked susceptibility of those compounds to oxidation makes them particularly vulnerable. Foods, such as cereals, and feeds are often enriched with α-tocopheryl acetate in order to ensure adequate vitamin E supply to their consumers. A determination of "total" tocopherol, i.e., naturally present tocopherol plus added esters, may be relevant to assess the nutritional value of certain foodstuffs. The vitamin E status of humans and animals is generally reflected by their α-tocopherol plasma and tissue levels. In humans, subnormal plasma concentrations may indicate a deficiency state, and may prompt to therapeutic supplementation with synthetic material. As an example, the high incidence of vitamin E deficiency-related anemia in premature infants warrants close monitoring of their plasma tocopherol levels [13-15]. Determination of vitamin E levels in pathological states will further contribute to a better understanding of this vitamin's biochemical function. Also, effective quality control of vitamin E-containing pharmaceuticals, e.g., multivitamin preparations, requires suitable analytical methodology.

5.2 The Need for Chromatography

As an older review article by Bunnell indicated, even during the early years of vitamin E research a number of chromatographic systems became available as an alternative and/or complement to colorimetric or fluorimetric methods [16]. Current colorimetric assays take advantage of a nonspecific property of vitamin E, i.e., its ability to reduce ferric to ferrous ions [17,18]. Consequently, interferences from other reducing substances present in biological matrices is to be expected, unless an extensive sample cleanup step is included. This is, however, complicated by the fact that vitamin E is only a minor constituent in most products of vegetable and animal origin. Its isolation from an excess of lipids, such as sterols and triglycerides, requires more or less extensive fractionation procedures, of which chromatography is by far the most powerful in terms of resolution, selectivity, and speed. The chromatographic step may either precede "off-line" quantitation of the compounds of interest, e.g., by colorimetry or gas-liquid chromatography (GLC), or, as in modern liquid chromatography (LC), may itself be utilized for quantitation as well.

Separation and individual measurement of the various homologs and isomers is often desirable. Those forms show pronounced quantitative differences in biological activity and, hence, contribute to a different degree to the overall vitamin E supply. They also differ in their ability to protect natural products, e.g., vegetable oils, from oxidative degradation. Although α-tocopherol is by far the most active form, assays should be capable of distinguishing between the various analogs. In addition, knowledge of the relative abundance of the various forms can be utilized for identification and purity assessment, e.g., to detect possible adulteration of vegetable oils. These considerations offer a rationale for the systematic use of a chromatographic step in tocopherol and tocotrienol assays.

5.3 Analytical Challenges

Three major problems must be faced in connection with vitamin E assays in biological materials. First, the instability of tocopherols implicates some limitations with regard to sample preparation. Oxidative destruction of the molecule is readily catalyzed by heat, light, alkaline pH conditions, and metal ions. Therefore, extreme care should be exercised during isolation and fractionation. A critical step in this respect is the rather harsh saponification procedure (see below) as well as, in some cases, the chromatography itself. For example, the risk of oxidative losses is considerably enhanced in thin-layer chromatography (TLC), in which compounds are spread out on a large surface and remain exposed to air for prolonged times. Second, as already mentioned, major interferences in tocopherol assays are caused by the presence of other lipids in excess, particularly cholesterol. A large variety of approaches have been devised to remove this troublesome endogenous substance, often at the expense of assay simplicity and speed.

The most challenging goal for analytical chemists dealing with vitamin E has always been the separation of positional isomers, i.e., β- and γ-tocopherol and, to a lesser extent, β- and γ-tocotrienol. Up to now, the majority of literature procedures fail to meet this challenge. Only in recent years has LC enabled a fast and reproducible separation of both components.

5.4 Sample Preparation

The extent of sample pretreatment is largely dependent upon the selectivity and resolving power of the chromatographic technique and the method used for final quantitation. Colorimetric and gas chromatographic vitamin E assays, as opposed to LC, have always suffered from the need for elaborate lipid fractionation steps, involving thin-layer or column chromatography. Modern liquid chromatography has succeeded in reducing sample pretreatment to its absolute minimum. Usually, a simple solvent extraction is sufficient in this case. A "standard" analysis scheme for assays not involving LC can be outlined as follows.

1. Physical treatment of the sample: Most samples, such as oils, plasma, and foods, do not require any physical treatment, e.g., grinding. However, in case of plant and animal tissues, a preliminary homogenization step is mandatory.
2. Lipid extractions: A first lipid extraction is commonly performed with such organic solvents as chloroform, acetone, diethyl ether, or ethanol. Very often, this step is omitted. Instead, one may proceed immediately to the saponification step.
3. Saponification is helpful to remove interfering triglycerides. At the same time it liberates tocopherol from any esters that may be present. Essentially, the method involves refluxing the sample with ethanolic potassium hydroxide for variable times. At this point, serious precautions should be taken to avoid oxidative degradation of vitamin E. Air should be excluded from the reaction system, which can be accomplished by working in a nitrogen atmosphere. It is also common practice to add antioxidants, such as pyrogallol or sodium ascorbate.
4. Reextraction of lipids: After saponification, the mixture is diluted with water and reextracted with a nonpolar solvent, such as hexane or, occasionally, diethyl ether. The organic layer is isolated and evaporated to dryness and the residue reconstituted with an organic solvent.

6. GAS CHROMATOGRAPHY

6.1 General Analytical Considerations

Gas-liquid chromatography has been widely used as an analytical tool for the determination of vitamin E analogs [19]. Yet, certain physicochemical properties of these compounds do not seem to make GLC the technique par excellence for this purpose. Their low volatility, in connection with their considerable molecular weight, accounts for high temperatures and long analysis times, inherent in most available GLC systems. Furthermore, in view of their instability, the possibility of degradation at elevated temperatures should always be considered.

Sample preparation for GC analysis of vitamin E is more critical than with other chromatographic techniques. Interferences of lipids present in the unsaponifiable fraction of biological materials are more likely to occur since, unlike in liquid chromatography, universal GLC detectors (flame ionization detectors, FID) do not provide an extra means of selectivity in addition to the column. An unfortunate phenomenon associated with most of the GC systems described is the coelution of α-tocopherol with cholesterol. Given the huge excess of the latter in biological materials, more than one step is often needed to remove this major interference. As a result, sample pretreatment prior to GLC often becomes tedious and time consuming.

Little effort is involved in separating homologous α-, γ-, and δ-tocopherols. In contrast, few existing GLC systems lend themselves to the differentiation of the positional isomers β-, and γ-tocopherol. Even recent methods fail to meet this challenge.

In the following sections, literature GLC procedures for vitamin E are critically evaluated on the basis of three criteria, i.e., the relative expedience and effectiveness of sample pretreatment, with emphasis on cholesterol removal, the ability to separate positional isomers, and the assay's overall performance (analysis time, temperature, reproducibility, and sensitivity).

6.2 Survey of Existing Systems

6.2.1 *Derivatization*

Although tocopherols can be chromatographed as such, derivatization contributes to reducing retention times and analysis temperatures. Peak shape and selectivity are often found to improve as well. Most authors recommend preparation of trimethylsilyl (TMS) ethers [20-38] or acetates [34,39-43]. A variety of other derivatives, including propionates [34,44], butyrates [45,46], trifluoroacetates [27,34,47,48], and pentafluoropropionates [34], have been occasionally employed. Table 3.3 gives details about common derivatization procedures. The historical merits of quinone derivatives [29,34,49] should be acknowledged in that their use, in conjunction with a special binary liquid phase, represents the first successful attempt to separate β- and γ-tocopherol [29]. This is further discussed in Sec. 6.2.5.1.

6.2.2 *Working Conditions*

Since Nicolaides in 1960 reported that α- and γ-tocopherol were partially resolvable on a silicone gum rubber column [50], a large number of similar silicone-type liquid phases have been tested on their capability to chromatograph tocopherols and tocotrienols. As inert supports, silanized Gas Chrom, Chromosorb, or Celite have become most popular. Underivatized tocopherol can be readily chromatographed on nonpolar to medium polar stationary phases, such as SE 30 [39, 40,51-66], OV-1 [67-69], OV-17 [48,70,71], and SE 52 [72]. It is our personal experience that the use of more polar phases, including SP-2401, XE 60, FFAP, and OV-25, invariably results in moderate to severe tailing of the tocopherol peak [48]. Recently, however, a 3% Silar

TABLE 3.3 Derivatization Procedures for Tocopherols

Reagents[a]	Reaction temperature (°C)	Reaction time	Estimated amount of tocopherol to be derivatized	Reference
I. TMS ethers				
Pyridine-HMDS-TMCS (10:9:6, v/v) 0.01 ml	Ambient	—	<1 μg	20-22
Pyridine-HMDS-TMCS (10:2:1, v/v) 0.1 ml	Ambient	15 min	25-50 μg	23
Pyridine-HMDS-TMCS (10:9:6, v/v) 0.1 ml	Ambient	15 min	2-200 μg	24-26
Chloroform, 0.5 ml; HMDS, 0.1 ml; TMCS, 6 drops	55	3 hr	50 μg	27
Chloroform, 5 ml; HMDS, 0.2 ml; TMCS, 5 drops	Ambient	12 hr	5 mg	28
Hexane, 0.5 ml; BSA, 0.5 ml	50	45 min	<100 μg	29
Hexane, 0.2 ml; BSA-TMCS (5:1, v/v), 0.05 ml; BSTFA, 0.25 ml	50	45 min	15-60 μg	30
Cyclohexane, 0.2 ml; BSA, 0.2 ml	Ambient	15 min	100-200 μg	31
Pyridine, 0.05 ml; BSA, 0.05 ml; TMCS, 0.025 ml	90	1 hr	5-20 μg	32
BSA, 0.1 ml	Ambient	15 min	50-75 μg	33
BSA, 0.5 ml	Ambient	12 hr	<100 μg	34
Hexane, 0.1 ml; MSTFA (BSA), 0.1 ml	50	30 min	Up to 1 mg	35
Pyridine-BSTFA (1:1, v/v), 1 ml or tetrahydrofuran-BSTFA (1:1, v/v), 1 ml	Ambient	—	10-20 μg	36
Pyridine, 0.015 ml; BSTFA-TMCS (99:1, v/v), 0.015 ml	Ambient	10 min	Picogram range	37

TABLE 3.3 (*Continued*)

Reagents[a]	Reaction temperature (°C)	Reaction time	Estimated amount of tocopherol to be derivatized	Reference
BSTFA 0.5 ml	Ambient	30 min	10 mg	38
II. Acetates				
Pyridine-acetic anhydride (2:1, v/v), 1 ml	Ambient	12 hr	1 mg	39
Pyridine-acetic anhydride (75:25, v/v), 5 ml	100	30 min	300 mg	40,41
Pyridine-acetic anhydride vapors	Ambient	5 hr	0.25-5 μg	42
Pyridine, 20 μl; acetic anhydride, 100 μl	50	15 min	0.5-20 μg	43
Glacial acetic acid-acetic anhydride (1:1, v/v)	95	3 hr	100 μg	34
III. Propionates				
Pyridine, 0.5 ml; propionyl chloride, 6 drops	95	3 hr	100 μg	34
Pyridine-propionic anhydride (2:1, v/v), 3 ml	75	30 min	0.5 mg	44
IV. Butyrates				
Pyridine-butyric anhydride (2:1, v/v), 8 ml	130	10 min	?	45
Pyridine-butyric anhydride (2:1, v/v), 3 ml	130	15 min	0.5-1 mg	46
V. Trifluoroacetates				
Hexane, 0.3-0.4 ml; trifluoroacetic anhydride, 0.2-0.3 ml	50	1 hr	100 μg	27,34, 47
Hexane, 0.3 ml; trifluoroacetic anhydride, 0.2 ml	80	1 hr	100 μg	48

VI. Pentafluoropropionates				
Pyridine, 0.5 ml; pentafluoropropionic anhydride, 6 drops	95	3 hr	100 μg	34
VII. Quinones				
Dissolve sample in 1.0 ml ethanol Oxidize with 0.5 ml $FeCl_3$ solution (5% w/v in 1% hydrochloric acid) Add 1 ml α,α'-dipyridyl solution (1% w/v in ethanol)	50	10 min	100 μg	29,34
Dissolve sample in 0.6 ml chloroform-ethanol (1:5, v/v) Oxidize with 0.25 ml $FeCl_3$ solution (5% w/v in 1% hydrochloric acid) Add 0.5 ml α,α'-dipyridyl solution (1% w/v in ethanol)	50	10 min	~2 μg	49

[a]Abbreviations: HMDS, hexamethyldisilazane; TMCS, trimethylchlorosilane; BSA, bis-(trimethylsilyl)-acetamide; MSTFA, *N*-methyl-*N*-trimethylsilyltrifluoroacetamide; BSTFA, *N*,*O*-bis(trimethylsilyl)-trifluoroacetamide.

10C column was used as part of a gas chromatographic-mass spectrometric (GC-MS) assay of tocopherol [37]. A highly polar SP 2340 phase proved also useful for the separation of tocopherol diastereoisomers [38]. Table 3.4 lists GLC working conditions reported in the literature up to 1970. A survey of papers published after 1970 is presented in Table 3.5. Although, as indicated in the tables, many studies included β- and γ-tocopherol, this does not necessarily mean that both compounds were resolved from each other.

6.2.3 Detection

In earlier years, most gas chromatographic work on vitamin E was done using instruments equipped with an argon ionization detector. Two variants of this device, containing either a ^{226}Ra [51,53,54,60] or a ^{90}Sr source [28,29,34,39,47], gained wide popularity. More recently, however, both detector types have been almost completely replaced by the more reliable flame ionization detector, which has now found general acceptance for the gas chromatography of vitamin E. A Japanese group reported promising results with a ^{63}Ni electron capture detector (ECD) for the determination of ubiquinones and tocopherols [65]. Unfortunately, no sensitivity data were given in their paper, so that the merits of this alternative approach can hardly be evaluated. In view of the relatively high tocopherol concentrations occurring in most biological materials, conventional detectors usually provide the required sensitivity (detection of microgram amounts). For this reason, there has been little need so far for more sophisticated techniques, e.g., mass spectrometry with selected ion monitoring. Exceptionally, sensitivity in the picogram range as well as enhanced specificity may be required, as exemplified by a recently published tocopherol assay in lung tissues [37]. To this end, a GC-MS method, based on selected ion monitoring and involving the use of deuterated tocopherol as an internal standard, was developed. Ions at m/z 502 (molecular ion of tocopherol trimethylsilyl ether) and 515 (molecular ion of tocopherol-d_{13}-TMS ether) were simultaneously monitored using 20 eV in the electron impact (EI) mode.

Quantitative GC-MS has further been employed as part of a biopharmaceutical study to monitor α-tocopherol and α-tocopheryl acetate plasma levels in animals following parenteral administration of the latter compound [73]. Ions at m/z 502 (α-tocopherol-TMS ether) and 430 (α-tocopheryl acetate and the internal standard, α-tocopheryl propionate) were followed. More frequently, combined GC-MS has been applied for qualitative purposes, such as the identification of tocopherol in vegetable oils [33], milk [71], tissues [27], and pharmaceutical preparations [76]. De Leenheer and coworkers characterized a serum peak, tentatively ascribed to α-tocopherol via LC, by collecting the corresponding fraction and recording a 20 eV EI mass spectrum after gas chromatography on a 1% OV-1 column [69].

TABLE 3.4 Gas Chromatographic Working Conditions for Tocopherol and Tocotrienol Analysis, Literature Published Before 1970

Compounds investigated	Stationary phase, column dimensions	Temperature (°C)	Carrier gas and flow rate (ml/min)	Derivatization	Reference (year)
α-, β-, γ-, δ-Tocopherol, α-, β-tocotrienol, tocol, methyltocols, dimethyltocols	4% SE 30 on Celite, 100/120 mesh, 1.2 m × 4 mm	250	Ar, 40	None or acetates	39 (1962)
α-, γ-, δ-Tocopherol	1% SE 30 on Chromosorb W, 80/100 mesh, 1.8 m × 4 mm	235	Ar	—	51 (1963)
α-, β-, γ-, δ-Tocopherol, tocol, methyltocols, dimethyltocols	5% SE 30 on Celite, 100/120 mesh, 2 m × 4 mm	235	Ar, 40	—	52 (1962)
α-, γ-Tocopherol, α-tocopheryl acetate	5% SE 30 on Gas Chrom P, 100/140 mesh, 2.4 m × 3 mm	260	Ar	—	53 (1964)
α-, β-, γ-, δ-Tocopherol, α-tocopheryl acetate	3% SE 30 on Gas Chrom P, 100/140 mesh, 1.8 m × 3 mm	205	Ar	—	54 (1964)
α-, β-, γ-, δ-Tocopherol, α-tocopheryl acetate	1.5% SE 30 on Chromosorb W, 60/80 mesh, 2.25 m × 4 mm or 4.5 m × 4 mm	250	N_2, 80-100	None, acetates, or TMS ethers	40 (1966)
α-Tocopherol	5% SE 30 on Gas Chrom P, 100/120 mesh, 2.4 m × 4 mm	250	75	—	56 (1967)
α-Tocopherol	3.8% UCW-982 on Diatoport S, 80/100 mesh, 1.2 m × 3 mm	260	He, 25	—	55 (1969)

TABLE 3.4 (*Continued*)

Compounds investigated	Stationary phase, column dimensions	Temperature (°C)	Carrier gas and flow rate (ml/min)	Derivatization	Reference (year)
α-, β-, γ-, δ-Tocopherol, α-, β-, γ-, δ-tocotrienol, tocol, methyltocols, dimethyltocols	2% SE 30 on Anakrom, 110/120 mesh, 4.5 m × 2 mm	235	He, 15-20	TMS ethers	23 (1967)
α-, β-, γ-, δ-Tocopherol, α-, β-, γ-tocotrienol	4% SE 30 on Gas Chrom Z, 100/120 mesh, 1.5 m × 4 mm	250	N_2, 40	—	57 (1968)
α-Tocopheryl acetate	3% SE 30 on Gas Chrom Q, 100/120 mesh, 2 m × 3 mm	235	He, 50	—	76 (1968)
α-, β-, γ-, δ-Tocopherol	3.5% SE 30 on Gas Chrom Q, 1.8 m × 3.5 mm	225	He, 150	—	58 (1968)
α-Tocopherol, α-tocopheryl acetate, α-tocopheryl succinate	10% SE 30 on Aeropak 30, 100/120 mesh, 1.22 m × 3 mm	245	N_2, 40	—	59 (1968)
α-, β-, γ-, δ-Tocopherol, α-, β-, γ-, δ-tocotrienol, methyltocols, dimethyltocols	1.7% SE 30 on Gas Chrom Q, 100/120 mesh, 4.5 m × 2 mm	235	He, 30	TMS ethers	25 (1969)
α-, β-, γ-, δ-Tocopherol	1.5% SE 52 on Gas Chrom P, 100/120 mesh, 1.8 m × 6 mm	210	Ar	—	72 (1966)
α-, γ-, δ-Tocopherol	2% XE 60 on Celite 545, 60/80 mesh, 0.265 m × 0.2 mm	250	N_2, 30	Acetates	41 (1968)

α-, β-, γ-, δ-Tocopherol, tocol, methyltocols, dimethyltocols	5% Apiezon N on Celite, 100/120 mesh, 2 m × 4 mm	260	Ar, 60	—	52 (1962)
α-, β-, γ-, δ-Tocopherol, α-, β-, γ-, δ-tocotrienol, tocol, methyltocols, dimethyltocols	0.5% Apiezon L on Anakrom, 110/120 mesh, 4.5 m × 2 mm	235	He, 15-20	TMS ethers	23 (1967)
α-, β-, γ-, δ-Tocopherol, α-, β-, γ-, δ-tocotrienol, methyltocols, dimethyltocols	0.4% Apiezon L on Gas Chrom Q, 100/120 mesh, 4.5 m × 2 mm	235	He, 30	TMS ethers	25 (1969)
α-, β-, γ-, δ-Tocopherol, α-tocopheryl acetate	3% QF-1 on Gas Chrom P, 100/140 mesh, 1.8 m × 3 mm	195	Ar	—	54 (1964)
α-, β-, γ-, δ-Tocopherol, α-, β-tocotrienol, tocol, methyltocols, dimethyltocols	5% QF-1 on Celite, 100/120 mesh, 1.2 m × 4 mm	240	Ar, 50	None or acetates	39 (1962)
α-, β-, γ-, δ-Tocopherol, α-, β-, γ-, δ-tocotrienol, methyltocols, dimethyltocols	1.6% OV-17 on Gas Chrom Q, 100/120 mesh, 4.5 m × 2 mm	235	He, 30	TMS ethers	25 (1969)
α-, β-, γ-, δ-Tocopherol, α-tocopherylquinone, α-tocopheryl hydroquinone	10% SE 52 + 10% nitrile polysiloxane on Gas Chrom P, 60/80 mesh, 1.8 m × 5 mm	228	Ar, 75	None or TMS ethers	28 (1963)

TABLE 3.4 (*Continued*)

Compounds investigated	Stationary phase, column dimensions	Temperature (°C)	Carrier gas and flow rate (ml/min)	Derivatization	Reference (year)
α-, β-, γ-, δ-Tocopherol, tocol, methyltocols, dimethyltocols	10% SE 52 + 5% XE 60 on Gas Chrom P, 80/100 mesh, 1.8 m × 5 mm	220	Ar, 200	TMS ethers, quinones, acetates, propionates, trifluoroacetates, pentafluoropropionates	29 (1966) 34 (1967)
α-Tocopherol	10% SE 52 + 5% XE 60 on Gas Chrom P, 80/100 mesh, 1.8 m × 5 mm	220	Ar, 100	Trifluoroacetates	47 (1969)

TABLE 3.5 Gas Chromatographic Working Conditions for Tocopherol and Tocotrienol Analysis, Literature Published After 1970

Compounds investigated	Stationary phase, column dimensions	Temperature (°C)	Carrier gas and flow rate (ml/min)	Derivatization	Reference (year)
α-, β-, γ-, δ-Tocopherol	1% SE 30 on Gas Chrom Q, 1.8 m × 3.5 mm	235	He, 170	Butyrates	45 (1970)
α-Tocopherol, α-tocopheryl acetate	5% SE 30 on an unspecified support, 100/120 mesh, 2.44 m × 4 mm	270-285	Ar	—	60 (1970), 61 (1971)
α-, γ-Tocopherol	3% SE 30 on Gas Chrom Q, 80/100 mesh, 1.8 m × 6 mm	232	Ar, 30	—	62 (1972)
α-Tocopherol	10% SE 30 on Chromosorb W-AW-DMCS, 1.8 m × 6 mm	270	N_2	—	63 (1972), 64 (1973)
α-Tocopheryl acetate	5% SE 30 on Chromosorb W-AW-DMCS, 1.8 m × 3 mm	235	N_2	—	74 (1972)
α-, β-, γ-, δ-Tocopherol	2.5% SE 30 on Chromosorb G, 1.8 m × ? mm	225	N_2	TMS ethers	31 (1973)
α-Tocopherol	5% SE 30 on Gas Chrom Q, 60/80 mesh, 1.2 m × 3 mm	245	N_2, 45	—	65 (1973)

TABLE 3.5 (*Continued*)

Compounds investigated	Stationary phase, column dimensions	Temperature (°C)	Carrier gas and flow rate (ml/min)	Derivatization	Reference (year)
α-, β-, γ-, δ-Tocopherol, α-tocotrienol, α-tocopheryl quinone	3% SE 30 on Gas Chrom Q, 100/120 mesh, 1.5 m × ? mm	230	N_2, 30	TMS ethers	35 (1976)
α-Tocopherol	5% SE 30 on Chromosorb W-AW-DMCS, 1.8 m × 3 mm	270	N_2	—	66 (1977)
α-, β-, γ-, δ-Tocopherol	3% SE 30 on Gas Chrom Q, 100/120 mesh, 1.8 m × 5 mm	240	He, 60	Butyrates	46 (1977)
α-Tocopherol	3% SE 30 on Gas Chrom Q, 100/120 mesh, 2 m × 4 mm	242	He, 70	Acetates	42 (1978)
α-, β-, γ-, δ-Tocopherol, α-, β-, γ-, δ-tocotrienol	2% SE 30 on Anakrom AW, 110/120 mesh, 5 m × ? mm	235	He, 15-20	TMS ethers	26 (1978)
α-, β-, γ-, δ-Tocopherol, α-, β-, γ-, δ-tocotrienol	Open tubular column coated with SE 30, 50 m × 0.4 mm	240	H_2	TMS ethers	26 (1978)
α-Tocopherol, α-tocopheryl acetate, α-tocopheryl succinate	3% OV-1 on Chromosorb W, 60/80 mesh, 1.5 m × 2 mm	260	He, 54	—	67 (1971)
α-, β-, γ-, δ-Tocopherol	3% OV-1 on Gas Chrom Q, 2.7 m × ? mm	235	N_2	TMS ethers	31 (1973)
α-, β-, γ-, δ-Tocopherol	3% OV-1 on Supelcoport, 80/100 mesh, 1.8 m × 3 mm	240	He, 80	TMS ethers	30 (1973)

α-, β-, γ-, δ-Tocopherol	2% OV-1 on Gas Chrom Q, 100/120 mesh, 2 m × 3 mm	265	N_2, 40	—	68 (1974)
α-Tocopherol	5% OV-1 on Gas Chrom Q, 80/100 mesh, 1.8 m × 2 mm	250	He, 30	—	69 (1978)
α-, β-, γ-Tocopherol, α-tocopheryl acetate	3% OV-1 on Gas Chrom Q, 80/100 mesh, 2.5 m × 2 mm	250	He	—	48
α-, β-, γ-, δ-Tocopherol, α-, β-, γ-, δ-tocotrienol	Open tubular column coated with OV-17, 20 m × 0.4 mm	240	H_2	TMS ethers	26 (1978)
α-, β-, γ-Tocopherol, α-tocopheryl acetate	3% OV-17 on WHP, 80/100 mesh, 1.8 m × 2 mm	240	He	—	48 (1977)
α-Tocopherol	3% OV-17 on Gas Chrom Q, 2 m × 4 mm	235	—	—	70 (1976)
α-, β-, γ-, δ-Tocopherol	2% OV-17 on Chromosorb W-AW-DMCS, 80/100 mesh, 2 m × 3 mm	235	N_2, 40	—	71 (1975)
α-Tocopherol, α-tocopheryl acetate	3% OV-17 on Gas Chrom Q, 80/100 mesh, 0.5 m × 3 mm	285	He, 30	TMS ethers	73 (1979)
α-, β-, γ-, δ-Tocopherol	1% SE 52 on Chromosorb W-AW-DMCS, 2.44 m × 4 mm	255	He, 70	Propionates	44 (1974)
α-, γ-Tocopherol	0.5% Apiezon L on Gas Chrom Q, 100/120 mesh, 4.5 m × ? mm	250	He, 25	TMS ethers	20 (1971)
α-Tocopherol	0.3% Apiezon L on Gas Chrom Q, 100/120 mesh, 4.5 m × 3 mm	245	He, 25	TMS ethers	21 (1978)
α-Tocopherol	5% Apiezon L on Gas Chrom Q, 80/100 mesh, 1 m × 2.2 mm	260	He, 40	TMS ethers	36 (1980)

TABLE 3.5 (*Continued*)

Compounds investigated	Stationary phase, column dimensions	Temperature (°C)	Carrier gas and flow rate (ml/min)	Derivatization	Reference (year)
α-Tocopherol	3% JXR on Gas Chrom Q, 100/120 mesh, 1.8 m × 2 mm	260	N_2, 30	Acetates	43 (1980)
α-Tocopherol	3% QF-1 on Supelcoport, 80/100 mesh, 1.8 m × 3 mm	230	Ar, 50-70	Quinones	49 (1970)
α-, β-, γ-, δ-Tocopherol, α-tocotrienol, α-tocopherylquinone	10% SE 52 + 5% XE 60 on Gas Chrom P, 80-100 mesh, 3 m × ? mm	235	N_2, 50	TMS ethers	35 (1976)
α-Tocopherol, α-tocopheryl acetate	1.5% OV-101 on Chrom GHP, 100/120 mesh, 1.8 m × 2 mm	260	He	—	48 (1977)
α-, β-, γ-, δ-Tocopherol, α-, β-, γ-, δ-tocotrienol	2% MS-550/HP on Chromosorb W-AW-DMCS, 80/100 mesh, 1 m × 4 mm	235	N_2, 45	—	75 (1979)
α-, β-, γ-, δ-Tocopherol	PZ-176, open tubular column, 32 m × 0.25 mm	180 → 2°C/min	N_2	TMS ethers	32 (1975)
α-Tocopherol	3% Silar-10C on Gas Chrom Q, 100/120 mesh, 2 m × 2 mm	230	He	TMS ethers	37 (1981)
Diastereoisomers of mono-, di-, and trimethyltocols	Open tubular column coated with SP 2340, 115 m × 0.25 mm	195	H_2	TMS ethers	38 (1981)

6.2.4 Quantitation

Quantitation in gas chromatography is based on peak height or area measurements. This can be done either manually, e.g., by triangulation, or by electronic integration. Ideally calibration should involve analysis of samples, supplemented with known amounts of tocopherol. Standard curves can be constructed by plotting peak heights or areas versus the amount of tocopherol added to the sample. Those curves show a substantial intercept, which depends on the amount of endogenously present tocopherol. Satisfactory precision and accuracy are to be expected only if at the same time a suitable internal standard (IS) is carried through the entire procedure. Advantages of this approach in chromatography can no longer be ignored. The internal standard compensates for losses during sample pretreatment (extraction of samples and evaporation of organic solvents containing the extracts), as well as for variations in injection volume and conditions during a chromatographic run. The above principles for quantitation remain valid except that now peak height or area ratios (product/IS) are determined instead of absolute heights or areas. To ensure maximal compensation, use of a structural analog is desirable. A closely related component, e.g., a homolog, will most likely approach the physicochemical properties, such as partition coefficient of the compound to be assayed.

Proper chromatographic resolution between the two is obviously a requirement. Table 3.6 lists some of the internal standards that have been proposed for tocopherol analysis in plasma, tissues, and red cells. Such compounds as 5,7-dimethyltocol [20-22,30], tocol, or α-tocopheryl propionate [31,70] are likely to meet the above criteria. Structurally unrelated substances, e.g., octacosane [23,43], cholestane [51], squalane [58], stigmasterol [36], cholesteryl isovalerate [45,46], cetyl palmitate [44], didecyl pimelate [25], dotriacontane [59,68,70], and hexadecyl stearate [75], may be acceptable from a chromatographic standpoint but it is questionable whether they provide proper compensation for variabilities during sample pretreatment.

Very often, simplified approaches based on "response factors" are used as an alternative to the standardization procedure outlined above [25,44-46,58,75].

6.2.5 Applications

6.2.5.1 Separation of Isomers. As early as 1962, Wilson and coworkers showed that GC separation of vitamin E analogs takes place according to the number of methyl groups present on the chromanol moiety in their structures [39]. These authors were able to differentiate tocopherols in three groups consisting of mono-, di-, and trimethyltocols, respectively. Calculated separation factors, as correlated with the number of methyl groups on the ring, permitted prediction of relative retention volumes for the different homologs. Kofler et al. [52]

TABLE 3.6 GC-Based Vitamin E Assays in Biological Materials of Animal or Human Origin

Biological matrix, sample size	Isolation method	Lipid fractionation step	Internal standard	Detection limit	Chromatographic analysis time (min)	Reference
Animal tissues, 0.5-1.5 g	Saponification, hexane extraction	Column chromatography on alumina-Celite-zinc carbonate	Cholestane	<1 μg/g	20	51
Beef heart mitochondria	Saponification, hexane extraction	Column chromatography on silicic acid	—	—	20	47
Red blood cells (humans, rats), 0.5 ml	Deproteinization with methanol, hexane extraction	Oxidation to quinones, benzene extraction, TLC	[^{3}H]Tocopherol	—	10	49
Human plasma, 0.1 g	Deproteinization with ethanol, petroleum ether extraction	TLC	5,7-Dimethyltocol	—	25	20
Rat plasma	Saponification, hexane extraction	Column chromatography on Celite-digitonin	—	—	27	61
Human plasma (3 ml) or red blood cells (3 ml)	Saponification, hexane extraction	TLC	5,7-Dimethyltocol	0.03 μg[a]	22	30
Human plasma, 1 ml	Extraction with isopropanol-isooctane-sulfuric acid	—	—	<0.15 μg/ml	55	32

Human plasma, 0.1 ml; human platelets, 0.5 ml	Saponification, petroleum ether extraction	—	5,7-Dimethyltocol	0.2 μg/ml	40	21
Human plasma, 0.1-1 ml	Deproteinization with ethanol, hexane extraction	Column chromatography on Celite-digitonin	Octacosane	0.2 μg/ml	6	43
Human red blood cells, 0.5 ml	Saponification, petroleum ether extraction	Column chromatography on silicic acid	5,7-Dimethyltocol	0.2 μg/ml	30	22
Liver, 2 g	Saponification, diethyl ether-hexane extraction	Column chromatography on Kieselgel 60	Stigmasterol	0.5 μg/g	40	36
Human serum, 0.5 ml	Saponification hexane extraction	TLC	—	—	22	42
Lung tissue	Extraction with silylating agent (15 μl) + pyridine (15 μl)	—	α-Tocopherol-d_{13}	<200 pg[a]	—	37

[a]Absolute amount injected.

reported a first partial separation of β- and γ-tocopherol on a 5% Apiezon N (on Celite 100/120 mesh) column and demonstrated the superiority of this phase over SE 30. A similar incomplete separation of the two isomers was achieved by Slover et al., using a 0.5% Apiezon L liquid phase [23]. These workers introduced a major improvement by preparing tocopherol silyl ethers, which proved to yield more symmetrical peaks than the underivatized compounds. In an attempt to separate isomeric dimethyltocols (β-tocopherol, γ-tocopherol, and 5,7-dimethyltocol), Nair et al. tested the applicability of a binary stationary phase, consisting of 10% SE 52 and 5% XE 60 on Gas Chrom Q (80/100 mesh) [29]. The β- and γ-tocopherol were indistinguishable from each other, either as free compounds or as TMS ethers, but could be separated from 5,7-dimethyltocol. However, conversion of tocopherols to their quinones resulted in complete separation of the two positional isomers β- and γ-tocopherol. Unfortunately under these conditions 5,7-dimethyltocol and 5,8-dimethyltocol (β-tocopherol) appeared as a single peak in the chromatogram. The authors extended their study to the separation of three isomeric monomethyltocols as well [34]. To this end, they prepared a variety of derivatives, including acetates, propionates, trifluoroacetates, pentafluoropropionates, quinones, and silyl ethers, of which the latter gave superior results in terms of resolution.

So far, few authors have applied this approach for the determination of β- and γ-tocopherol in biological materials. Rather, a complementary step, thin-layer chromatography (TLC), is used to differentiate the isomeric forms, whereas separate quantitation is then carried out in two consecutive GC runs [30]. Recently, capillary GC has been shown to yield satisfactory resolution of both isomers without the need for preparing quinones [26,32]. Shen-Nan Lin and Horning resolved all four tocopherols as their TMS ethers using a 32 m open tubular column coated with a polar PZ-176 liquid phase [32]. Figure 3.1 illustrates the resolution obtainable in a 55 min run. Similar results were accomplished by a French group, who reported separation of all tocopherols and tocotrienols in a single run on a 20 m OV-17 capillary column [26]. Surprisingly, these authors managed to reduce the analysis time to only 15 min.

The unique resolution attainable with capillary GC is further illustrated by its ability to yield separation of various tocopherol stereoisomers. α-Tocopherol isomers produced by different synthetic routes have been readily resolved on a 115 m × 0.25 mm glass capillary column coated with SP 2340 [38]. Racemic mixtures of other mono- and dimethyltocols could likewise be differentiated as well [38].

6.2.5.2 Determination of Vitamin E in Biological Materials of Animal or Human Origin. Application of gas chromatography to vitamin E determination in biological materials requires more or less complex sample preparation. Saponification is still frequently used as a first step,

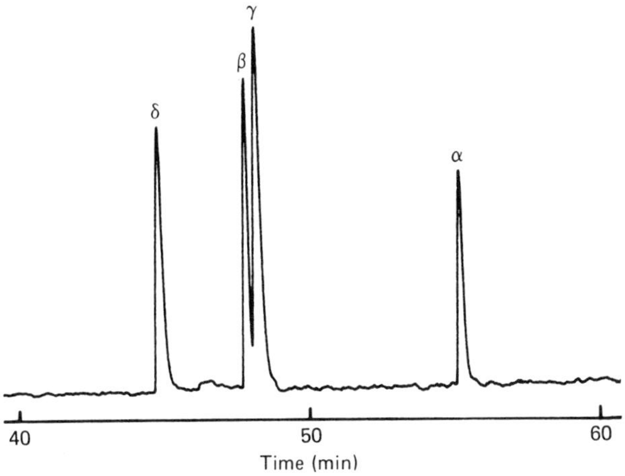

Figure 3.1 Separation of α-, β-, γ-, and δ-tocopherol as trimethylsilyl ethers. Operating conditions: temperature programmed from 180°C at 2°C/min; 32 m × 0.25 mm ID open tubular glass capillary column, coated with polar phase PZ-176; carrier gas, nitrogen; initial column pressure, 5 psi; split ratio, 10:1. (From Ref. 32.)

although deproteinization followed by a simple double phase extraction with a nonpolar solvent is often sufficient to isolate the lipid fraction. Table 3.6 compares various literature procedures for tocopherol assay in plasma, red blood cells, and tissues. Special attention is paid to the potential interference of cholesterol. Unlike capillary columns, packed columns usually lack sufficient resolving power to separate α-tocopherol from an excess of cholesterol.

As exemplified in Table 3.6, various thin-layer and column chromatographic procedures, as well as digitonin precipitation, have been devised to remove interfering sterols. The effect of efficient sample cleanup on the final chromatogram is illustrated in Fig. 3.2. The left part (Fig. 3.2a) represents a trace of a human plasma extract after passage through a digitonin-Celite column, whereas Fig. 3.2b depicts the same extract not subjected to any purification procedure [43]. In the latter case the α-tocopherol peak coincides with the leading edge of the huge cholesterol acetate peak and consequently becomes difficult to quantitate.

Lehmann [21] developed an assay for plasma α-tocopherol that apparently did not suffer from cholesterol interference. Both α-tocopherol and the internal standard were distinctly separated from the excess of cholesterol (Fig. 3.3). However, obviously this separation could only be accomplished at the expense of chromatographic analysis time (35 min as opposed to 6.0 min in another study [43]). Capillary

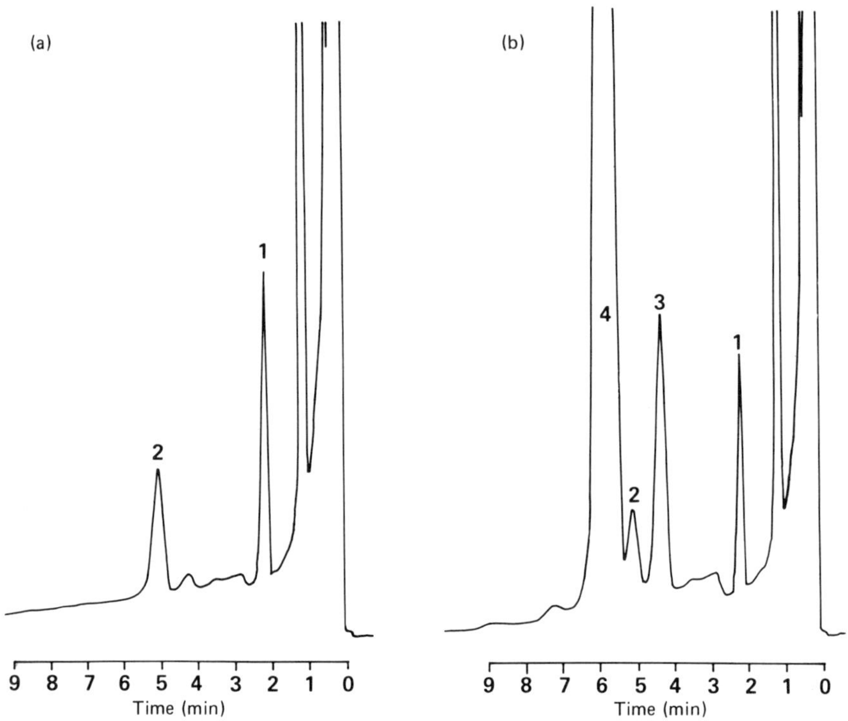

Figure 3.2 Chromatogram of an extract of normal plasma subjected to digitonin-Celite column chromatography (a) and not treated on a digitonin-Celite column (b). Chromatographic conditions: 180 × 0.2 cm column packed with 3% JXR on Gas Chrom Q (100/120 mesh); oven temperature, 260°C; carrier gas, nitrogen; flow rate, 30 ml/min. Peak identification: (1) octacosane, (2) α-tocopheryl acetate; (3) cholesterol; (4) cholesteryl acetate. (From Ref. 43.)

GC allows analysis of crude plasma extracts, without the need for saponification, TLC, or any other sample cleanup step. Again, however, a considerable time (65 min) is required to resolve α-tocopherol from cholesterol and other lipids [32].

Some overall performance criteria of various assays, such as chromatographic analysis time and sensitivity, are also presented in Table 3.6. The advantages of an internal standard have been discussed previously. If properly selected, it should have a favorable effect on both precision and accuracy.

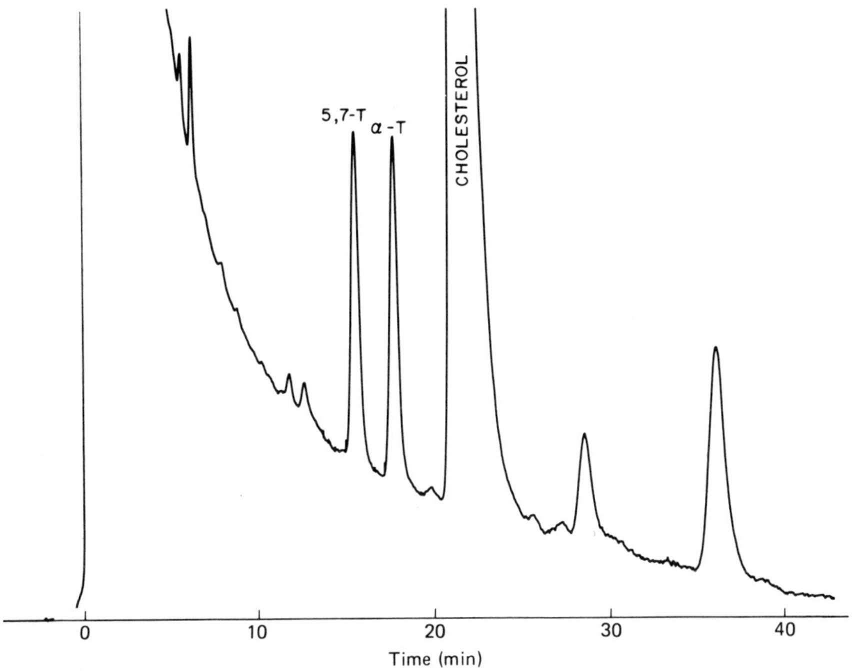

Figure 3.3 Chromatogram of an extract of rat plasma. Chromatographic conditions: 450 × 0.3 cm column packed with 0.3% Apiezon L on Gas Chrom Q, 100/120 mesh; oven temperature, 245°C; carrier gas, helium; flow rate, 25 ml/min. Peak identification: 5,7-T, 5,7-dimethyltocol TMS ether; α-T, α-tocopherol TMS ether. (From Ref. 21.)

6.2.5.3 Determination of Vitamin E in Biological Materials of Vegetable Origin. As a biological matrix, foods, feeds, and vegetable oils differ from animal products in that, besides the predominant α-tocopherol and, to a lesser extent, β- and γ-tocopherol, other analogs, i.e., tocotrienols, are present in significant amounts. In addition, foods and feeds are often fortified with α-tocopheryl acetate, which does not occur in nature. Generally extensive sample cleanup is necessary since vegetable materials contain a large number of potentially interfering substances, such as chlorophylls, xanthophylls, sterols, and carotenoids. A typical analysis scheme consists of a saponification step, which is mandatory in this case, extraction with a nonpolar solvent, a lipid fractionation step, derivatization, and GC separation. Purification procedures primarily meant to remove sterols include TLC [23,25,35,75], column chromatography on florisil [23,72] or aluminium oxide [31,57,63,64,74,77], digitonin precipitation [31,70], and digitonin-

Celite chromatography [31,70]. Some authors managed to simplify the assay by omitting this step [44-46,58,66].

GC systems for the analysis of vitamin E in plant materials can be evaluated in terms of their ability to separate the various homologs and isomers, as indicated by their relative retention times (Table 3.7). A satisfactory separation of all naturally occurring tocopherols and tocotrienols has only been accomplished on an open tubular column coated with OV-17 [26]. Substitution of OV-17 by SE 30 resulted in overlapping of γ-tocopherol and β-tocotrienol [26]. Packed columns fail to resolve β- and γ-tocopherol [30,36,40,43-46,58,75], as well as β- and γ-tocotrienol [75]. An Apiezon L phase yields partial separation between positional isomers but at the expense of long analysis times [23, 25].

Most assays in foods and feeds do not distinguish between free α-tocopherol and its acetate ester [31,63,66,68], unless the saponification step is omitted [64]. Ranfft reported a determination of α-tocopherol acetate in feeds with a detection limit of about 2 ng [74]. If a differentiation of the ester and the naturally occurring α-tocopherol is required, the content of the latter can be calculated by subtracting the tocopheryl acetate concentration from the total tocopherol level [64]. The latter determination obviously involves saponification to hydrolyze the ester.

6.2.5.4 Determination of Vitamin E in Pharmaceutical Preparations. In pharmaceutical preparations, vitamin E is usually present in the form of its more stable acetate or succinate ester. The first paper pertaining to the analysis of vitamin E in pharmaceuticals was published in 1964 by Libby and Sheppard [53]. These authors developed a GC separation of α-tocopherol, α-tocopheryl acetate, and γ-tocopherol. Later, the study was extended to α-tocopheryl succinate as well [56]. The esters were separately quantitated in addition to the total tocopherol determination after saponification of the sample.

Classic lipid extractions are applicable to pharmaceutical preparations. However, hydrophilic dosage forms, e.g., gelatin capsules, require preliminary treatment in aqueous medium [70]. Kovensky and Day described a method that was claimed to have the advantage of increased speed [67]. However, no separation between tocopherol and either of its esters was achieved so that separate quantitation was carried out. A comprehensive study describing assays of α-tocopherol and its esters in a wide variety of multivitamin preparations was published by Mahn and coworkers [59]. These workers found a cleanup step on an alumina column useful to obtain sufficiently pure extracts. Thus, their method could be extended beyond the usual tablets and liquids to include other products, such as multivitamin creams. Bowman and West compared a GC assay with the Emmerie-Engel approach for the determination of α-tocopheryl acetate in multivitamin preparations

TABLE 3.7 Separation of Tocopherols and Tocotrienols in Samples of Vegetable Origin[a]

Column	Temperature (°C)	Derivatization	Relative retention times								Reference
			α-T	β-T	γ-T	δ-T	α-T_3	β-T_3	γ-T_3	δ-T_3	
1% SE 30 on Gas Chrom Q, 1.8 m × 3.5 mm	235	Butyrates	1.00	0.81	0.81	0.61	—	—	—	—	45
3% SE 30 on Gas Chrom Q, 100/120 mesh, 1.8 m × 5 mm	240	Butyrates	1.00	0.82	0.82	0.63	—	—	—	—	46
3.5% SE 30 on Gas Chrom Q, 1.8 m × 3.5 mm	225	—	1.00	0.80	0.81	0.60	—	—	—	—	58
2% MS 550HP on Chromosorb W, 80/100 mesh, 1 m × 4 mm	235	—	1.00	0.79	0.79	0.58	1.58	1.25	1.25	0.93	75
0.5% Apiezon L on Anakrom, 110/120 mesh, 4.5 m × 2 mm	235	TMS ethers	1.00	0.59	0.61	0.44	1.33	0.78	0.82	0.59	23
1% SE 52 on Chromosorb W-AW-DMCS, 2.44 mm × 4 mm	255	Propionates	1.00	0.84	0.84	0.67	—	—	—	—	44
10% SE 52 + 5% XE 60 on Gas Chrom P, 80/100 mesh, 3 m × 5 mm	220	TMS ethers	1.00	0.66	0.66	0.51	1.42	—	—	—	35
OV-17 open tubular column, 20 m × 0.4 mm	240	TMS ethers	1.00	0.64	0.66	0.50	1.71	1.09	1.13	0.86	26
SE 30 open tubular column, 50 m × 0.4 mm	240	TMS ethers	1.00	0.68	0.70	0.54	1.31	0.89	0.92	0.70	26

[a] T, tocopherol; T_3, tocotrienol.

[76]. As they pointed out, the major disadvantage of the colorimetric assay lies in the fact that it does not distinguish between the esters and free tocopherol.

Most procedures employ internal standards for quantitation. Unlike structurally unrelated components, such as hexadecyl palmitate [67], dotriacontane [59], and 3-(*p*-methoxyphenyl)-2,4,5-triphenylcyclopentadienone [76], α-tocopheryl propionate, as proposed by Christie et al. [31], seems to be a most appropriate choice.

A number of collaborative studies have been carried out under the auspices of the Association of Official Analytical Chemists, in which GC methods were compared for the analysis of vitamin E in pharmaceutical preparations [78-80]. The most recent official AOAC method meets the requirements for a NF (National Formulary) Compendial method [80]. Previously such studies had led to the replacement of the colorimetric determination by a GC assay as the official AOAC method, which was essentially adopted from the work of Pillsbury [56] and Mahn et al. [59].

7. LIQUID CHROMATOGRAPHY

7.1 General Considerations

Classic column chromatography is characterized by low efficiency, poor sensitivity, and long analysis times. The use of large particle supports in connection with low flow rates accounts for this poor performance. In the vitamin E area, the technique is virtually worthless for the separation of individual tocopherols and tocotrienols. Its merely historical merit lies in the application as a crude lipid fractionation step prior to colorimetric or gas chromatographic tocopherol quantitation in biological samples.

Since the introduction of microparticulate packing materials for high-performance liquid chromatography (HPLC), classic column chromatography is becoming more and more obsolete. Unprecedented improvements in column efficiency, resolution, sensitivity, and speed have been achieved so that LC has definitely emerged as the method of choice among chromatographic techniques. Its major advantages, particularly for the analysis of labile products with low volatility, are well documented: compounds can be chromatographed at room temperature, and in general no derivatization is required. Only in recent years, though, has this technique superseded its less versatile GC and TLC counterparts for vitamin E analysis. A paper by Thompson et al. [81] can be considered a precursor of modern LC analysis of vitamin E in biological materials. For the first time, a chromatographic column (hydroxyalkoxypropyl Sephadex) was coupled with on-line spectrofluorimetric detection, in order to assay tocopherols in tissues, foods, and serum.

However, real "high performance" was not yet reached, as reflected from the long chromatographic analysis time, 90-120 min, and the limited sensitivity (minimum detectable amount about 1 μg). This is not surprising in view of the large particle size of the Sephadex material. No on-line quantitation was attempted. Instead, peaks were collected and tocopherols determined "off-line" by fluorimetry. Yet, a major breakthrough was made in dealing with some of the basic challenges posed by vitamin E analysis: resolution of positional isomers, separation of tocopherols and tocotrienols in a single run, and determination of realistic concentrations in biological matrices.

Although initially LC systems for vitamin E were based on pellicular supports [82-86], their use has now been completely abandoned in favor of their microparticulate equivalents. The low efficiency and sample capacity of the former render them unacceptable for modern chromatographic work. Their historical merits should be acknowledged insofar as they indicated useful eluent combinations for later application on microparticulate supports.

In recent years microparticulate materials have formed the basis of numerous assays of tocopherols and tocotrienols in plasma, blood cells, tissues, foods, feeds, and vegetable oils. Absorbance and fluorescence detectors now permit quantitation of levels down to the low nanogram range, and provide a marked selectivity in addition to the column separation. As a result, sample preparation has been tremendously simplified.

7.2 Survey of Existing Systems

Tocopherol can be readily chromatographed on both polar and nonpolar stationary phases. In view of its pronounced lipophilic character, unfavorable chromatographic phenomena, e.g., low efficiency and peak tailing, which are often associated with polar molecules, are rarely encountered.

Reversed phase offers some practical advantages, e.g., column stability, reproducibility of retention times, and fast equilibration, and, hence, is preferable for routine analysis of biological samples. In addition, it is ideally suited to separate homologs, since retention on this type of support is essentially governed by the number of carbon atoms in a molecule. The failure of reversed phase to resolve positional isomers can be understood on this basis as well.

In the early seventies, a few papers appeared on the separation of fat-soluble vitamins using pellicular reversed-phase columns, gradient elution, and elevated temperatures [82,83]. Meanwhile, those supports have been replaced by microparticulate materials, particularly with octadecylsilane (ODS) bonded chains. The growing popularity of the latter can be judged from Table 3.8, which gives a survey of reversed-phase systems that have been reported for the LC of vitamin E analogs.

TABLE 3.8 Reversed-Phase Liquid Chromatographic Systems for Vitamin E

Column	Mobile phase	Flow rate (ml/min)	Detection (nm)	Application	Reference (year)
Pellicular supports					
Permaphase ODS, 100 × 0.21 cm	Gradient, 50°C, water → methanol	0.9	UV 254	Vitamins in general	82 (1971)
Permaphase ODS, 100 × 0.21 cm	Gradient, 70°C, water → methanol	2	UV 254	Separation of fat-soluble vitamins	83 (1972)
Permaphase ODS, 100 × 0.21 cm	Gradient, 50°C, methanol-water (80:20, v/v) → methanol-water (85:15, v/v)	0.7	UV 254	Separation of tocopherols (standard mixture)	83 (1972)
Zipax HCP, 100 × 0.21 cm	Methanol-water-phosphoric acid (77:23:0.1, v/v)	1	UV 254	α-Tocopheryl succinate	83 (1972)
BondaPak C_{18}/ Corasil 37-50 μm, 175 × 0.2 cm	Methanol-water (97.5:2.5, v/v)	0.56	Fluorescence: λ_{exc}, 295; λ_{em}, 340	α-, β-, γ-, δ-Tocopherol	87 (1978)
BondaPak Phenyl/ Corasil, 37-50 μm 100 × 0.2 cm	Methanol-water (85:15, v/v)	0.40	Fluorescence: λ_{exc}, 295; λ_{em}, 340	α-, β-, γ-, δ-Tocopherol	87 (1978)
Microparticulate Supports					
MicroPak CH 10 10 μm, 25 × 0.21 cm	Methanol-water (97:3, v/v)	1	UV 285	α-Tocopheryl acetate in pharmaceuticals	88 (1977)

MicroPak CH 10 10 μm, 25 × 0.21 cm	Methanol-water (98:2, v/v)	1	UV 265	Liposoluble vitamins in pharmaceuticals	89 (1978)
μBondaPak C18 10 μm, 30 × 0.39 cm	Methanol-water (98:2, v/v or 96:4 v/v) or methanol	1	UV 280	α-Tocopheryl acetate in feeds	90 (1977)
μBondaPak C18 10 μm, 30 × 0.42 cm	Methanol-water (95:5, v/v)	1	UV 280	Liposoluble vitamins in feeds	91 (1978)
μBondaPak C18 10 μm, 30 × 0.39 cm	Methanol-water (95:5, v/v)	3	UV 280 or fluorescence: λ_{exc}, 296; λ_{em}, 330	α-Tocopherol in feedstuffs	92 (1979)
μBondaPak C18 10 μm, 30 × 0.39 cm	Methanol-water (97:3, v/v)	3	UV 280 or fluorescence: λ_{exc}, 296; λ_{em}, 330	α-Tocopherol in serum of cattle, pigs	93 (1979)
μBondaPak C18 10 μm, 30 × 0.39 cm	Methanol-water (95:5, v/v)	2.5	UV 280	Simultaneous determination of α-tocopherol and retinol in plasma, red blood cells	94 (1979)
μBondaPak C18 10 μm, 30 × 0.39 cm	Methanol-water (95:5, v/v)	2	Fluorescence: λ_{exc}, 205; λ_{em}, 340	α-Tocopherol in plasma and red blood cells	95 (1979)
μBondaPak C18 10 μm, 30 × 0.39 cm	Gradient: methanol-water (91.5:8.5, v/v) → methanol	Flow progr.	UV 280 UV 254	Liposoluble vitamins in pharmaceuticals	96 (1979)

TABLE 3.8 (*Continued*)

Column	Mobile phase	Flow rate (ml/min)	Detection (nm)	Application	Reference (year)
μBondaPak C18 10 μm, 30 × 0.39 cm, coupled with μBondaPak Phenyl 10 μm, 30 × 0.39 cm	Gradient: methanol-water (70:30, v/v) → methanol	Flow progr.	UV 280 UV 254	Liposoluble vitamins in pharmaceuticals	96 (1979)
μBondaPak C18 10 μm, 30 × 0.39 cm	Methanol-water (90:10, v/v)	2	UV 280	α-Tocopherol in fish liver	97 (1980)
μBondaPak C18 10 μm, 30 × 0.39 cm	Methanol-water (95:5, v/v)	1	UV 280	α-Tocopherol and α-tocopheryl acetate in feeds	98 (1980)
μBondaPak C18 10 μm, 30 × 0.39 cm	Methanol-water (95:5, v/v)	3	Fluorescence: λ_{exc}, 296; λ_{em}, 330	α-, β-, γ-, δ-Tocopherol, α-, β-, γ-tocotrienol in feeds	99 (1980)
RSIL C18 10 μm, 25 × 0.46 cm	Methanol, 40°C	2	UV 292	α-Tocopherol in human serum	69 (1978)
RSIL C18 HL 10 μm, 25 × 0.46 cm	Methanol, 40°C	2	UV 292	Simultaneous determination of α-tocopherol and retinol in human serum	100 (1979)

Nucleosil C18 10 μm, 25 × 0.46 cm	Methanol-pyridine (99.9: 0.1, v/v) + 0.05% M sodium perchlorate	0.75	Electrochemical +0.7 V versus Ag/AgCl	α-Tocopherol in serum	101 (1979)
Nucleosil C18 10 μm, 20 × 0.5 cm	Methanol-water (98:2, v/v) + 0.25 mM TBAH[a]	2	UV 280 UV 254	α-Tocopherol in brain	102 (1981)
Spherisorb ODS 10 μm, 25 × 0.32 cm	Methanol-water (90:10, v/v)	0.8	UV 280	α-Tocopherol and retinol in feeds	103 (1978)
C18 column 10 μm, 25 cm	Methanol-1% ammonium carbonate in water (95:5, v/v)	5	UV 292	α-Tocopherol in feeds	104 (1978)
Zorbax ODS 7 μm, 50 × 0.46 cm	Gradient: methanol-ethyl acetate (86:14, v/v) → acetonitrile	Flow progr.	UV wavelength programming	Liposoluble vitamins in infant formulas, dairy products	105 (1980)
Zorbax ODS, 6 μm, 25 × 0.46 cm	Acetonitrile-methylene chloride (70:30, v/v)	1	UV 280	Liposoluble vitamins	106 (1981)
Lichrosorb RP8 10 μm, 25 × 0.46 cm	Gradient, 35°C: acetonitrile-water (80:20, v/v) → acetonitrile	2	UV 295	Liposoluble vitamins in foods	107 (1979)
Vydac C8 10 μm, 25 × 0.32 cm	Acetonitrile-water (95:5, v/v), acetic acid at pH 4	1.5	Fluorescence: λ_{exc}, 295; λ_{em}, 330; UV 308	α-, β-, γ-, δ-Tocopherol in foods	108 (1979)

[a]TBAH, tetrabutylammonium hydroxide.

Nearly all eluents consist of methanol as a basis, with small amounts of water added to control retention. Recently, nonaqueous reversed-phase chromatography has emerged as a promising tool for the separation of nonpolar compounds [106]. The rationale behind this concept is that neat organic eluents ensure optimal solubility of the solutes and thus prevent partial precipitation that may occur in the presence of water. Even small amounts of water often have a pronounced deteriorating effect on the peak shape of lipophilic compounds.

For this reason, the use of reversed-phase materials with short alkyl chains and lower carbon content should be discouraged in view of the necessity to incorporate substantial amounts of water in the eluent to produce sufficient solute retention. Nonaqueous reversed-phase chromatography is only possible on materials displaying higher retentivity than conventional supports. Zorbax ODS (du Pont) meets this criterion [109,110]. Although its carbon content apparently does not exceed that of other ODS phases, this support requires much stronger eluents in order to yield comparable retention times, as obtained with other reversed-phase materials. So far, two studies have taken advantage of this unique property to chromatograph a variety of liposoluble vitamins, including α-tocopherol on Zorbax ODS using mixtures of methylene chloride, acetonitrile, and methanol [106] or methanol and ethyl acetate as the eluent [105].

Unlike reversed-phase, straight-phase chromatography on a polar stationary phase allows fast and easy differentiation of positional isomers. This is in agreement with the well-known stereochemical selectivity of silicic acid for isomers. Van Niekerk was the first to report a successful separation of β- and γ-tocopherol [85]. This author employed a pellicular Corasil II column, eluted with diisopropyl ether-hexane (5:95, v/v). The same mobile phase proved directly applicable to microparticulate silica columns, which, as predicted, yielded superior performance in terms of efficiency, resolution, and analysis time [111]. This system has now almost become standard for the differentiation of β- and γ-tocopherol [88,111-119]. Diisopropyl ether was later shown to be substitutive by less hazardous solvents, e.g., isopropanol [120-126].

Similarly, the first separation of all eight natural tocopherols and tocotrienols was accomplished on a Corasil II column (Fig. 3.4) [86]. Again, substitution of the pellicular by a microparticulate column and use of a modified eluent resulted in a dramatic improvement [124-126], as can be judged from Fig. 3.5. The same separation was now completed in 15 min as opposed to 80 min on the pellicular column (Fig. 3.4) (δ-tocotrienol not included). Table 3.9 lists chromatographic conditions that have proved useful in connection with silica columns. Polar bonded phases have found limited application for vitamin E liquid chromatography. They are generally inferior to silica with regard to the resolution of positional isomers. Matsuo noted that both nitrile and

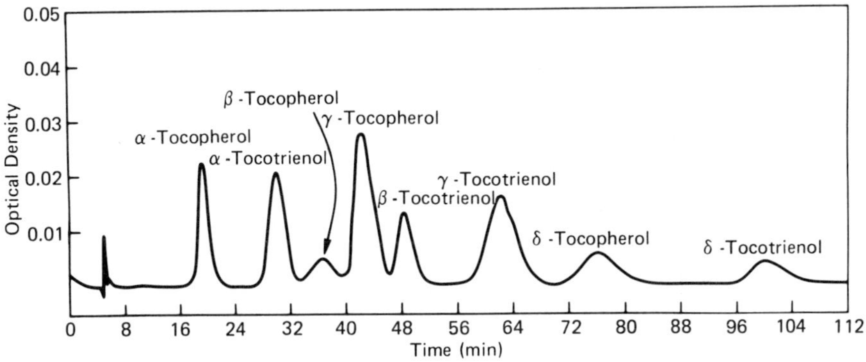

Figure 3.4 Separation of tocopherols and tocotrienols. Chromatographic conditions: 200 × 0.2 cm stainless steel column, packed with 37-50 μm Corasil II; eluent, hexane-tetrahydrofuran (99.5:0.5, v/v); flow rate, 1 ml/min; detection, UV 254 nm. (From Ref. 86.)

amino phases gave incomplete separation of β- and γ-tocopherol [111]. This observation has been confirmed by our own results as far as the nitrile support is concerned [48]. Contrarily, Jansson and coworkers obtained satisfactory resolution between the two positional isomers on an amino column using hexane-isopropanol as the eluent [117]. However, application to serum presented an additional problem in that a large peak interfered with the elution position of γ-tocopherol. Recently, Westerberg found the same support useful both for the separation of isomeric dimethyltocols and for the quantitative determination of α-tocopherol in rat brain [102].

7.3 Detection

To detect tocopherols in column eluates, advantage can be taken of either their UV characteristics (λ_{max} = 292 nm), their native fluorescence, or their electrochemical properties. Although tocopherols display relatively poor ultraviolet (UV) absorbance (ε = 3530 cm^2/mmol at 292 nm), detection at this or a near wavelength, using a variable wavelength detector, in general yields sufficient sensitivity [69,88,104,107, 111,120,122,129,130,135]. In our hands, α-tocopherol could be accurately quantitated at levels down to 0.6 μg/ml [69]. The minimum detectable amount of pure α-tocopherol under those conditions was estimated at 60 ng [69]. Reduction of the column dimensions from 25 × 0.46 cm to 15 × 0.32 cm yielded another gain in sensitivity by a factor of approximately 2. The performance of a UV detector, being concentration

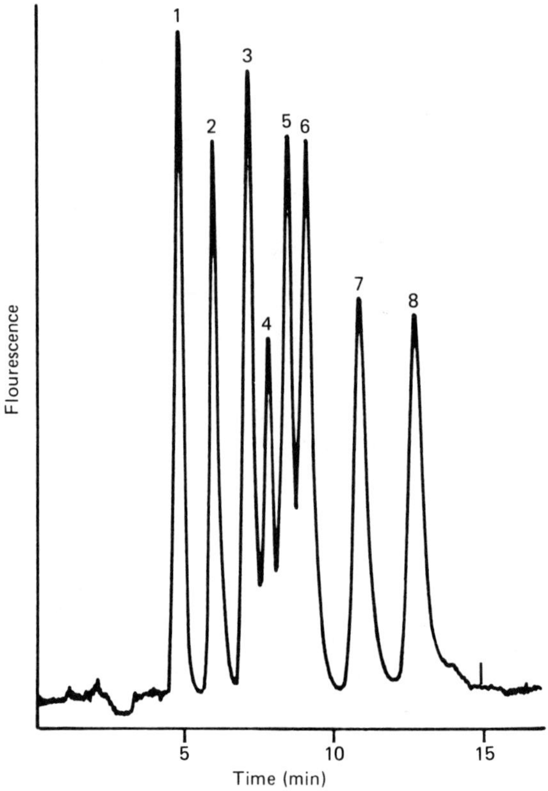

Figure 3.5 Separation of tocopherols and tocotrienols. Chromatographic conditions: 25 × 0.46 cm stainless steel column packed with 5 μm Lichrosorb Si 60; eluent, hexane-diethyl ether (95:5, v/v); flow rate, 2 ml/min; detection: fluorescence, λ_{exc} 295 nm; λ_{em} 330 nm. Peak identification: (1) α-tocopherol; (2) α-tocotrienol; (3) β-tocopherol; (4) butylhydroxyanisole (BHA); (5) γ-tocopherol; (6) β-tocotrienol; (7) γ-tocotrienol; (8) δ-tocopherol. (From Ref. 124.)

rather than mass sensitive, is indeed primarily dependent on solute dilution during chromatography. Consequently, as rationalized by Karger et al. [136], column size greatly affects the detectability of compounds.

Alternatively, several authors recommend detection at 280 nm, using fixed wavelength instruments [86,90-94,97,98,102,103,117,121,128,131]. A better signal-to-noise ratio may partly compensate for the loss in absolute sensitivity, caused by working at this nonoptimal wavelength.

Superior performance in terms of selectivity and sensitivity may be readily expected from fluorescence detection. Since tocopherols exhibit

TABLE 3.9 Silica-Based HPLC Systems for Vitamin E

Column	Mobile phase	Flow rate (ml/min)	Detection (nm)	Application	Reference (year)
Pellicular Supports					
Vydac, TM, 100 × 0.2 cm	Pentane-tetrahydrofuran (97.5:2.5, v/v)	0.15	UV 254	Separation of fat-soluble vitamins	84 (1972)
Corasil II 37-50 μm, 150 × 0.2 cm	Hexane-diisopropyl ether (95:5, v/v)	1.5	Fluorescence: λ_{exc}, 295; λ_{em}, 340	α-, β-, γ-, δ-Tocopherol in vegetable oils	85 (1973)
Corasil II 37-50 μm, 200 × 0.2 cm	Hexane-tetrahydrofuran (99.5:0.5, v/v)	1	UV 280, UV 254	α-, β-, γ-, δ-Tocopherol, α-, β-, γ-, δ-tocotrienol in cereals	86 (1974)
Corasil II 37-50 μm, 180 × 0.2 cm	Hexane-methanol (99.4:0.6, v/v)	0.5	Fluorescence: λ_{exc}, 295; λ_{em}, 320	α-Tocopherol in rat brain	127 (1977)
Corasil II 37-50 μm, 175 × 0.2 cm	Hexane-methanol (99.2:0.8, v/v)	0.43	Fluorescence: λ_{exc}, 295; λ_{em}, 340	α-, β-, γ-, δ-Tocopherol, standards	87 (1978)
Corasil I, 60 × 0.2 cm	Hexane-diisopropyl ether (96:4, v/v)	1	UV 280	α-Tocopherol in plasma	128 (1978)
Microparticulate supports					
Jascopack WC-03-500, 50 × 0.23 cm	Hexane-diisopropyl ether (98:2, v/v)	0.8	Fluorescence: λ_{exc}, 298; λ_{em}, 325	α-, β-, γ-, δ-Tocopherol: standards, in serum, in vegetable oils, in liver	112-115 (1975-1976)

TABLE 3.9 (*Continued*)

Column	Mobile phase	Flow rate (ml/min)	Detection (nm)	Application	Reference (year)
Jascopack WC-03-500, 50 × 0.23 cm	Hexane-diisopropyl ether (99.5:0.5, v/v)	0.5-0.7	Fluorescence: λ_{exc}, 298; λ_{em}, 325	α-, β-, γ-, δ-Tocopherol in red blood cells	116 (1977)
MicroPak SI-10, 10 μm, 50 × 0.24 cm	Hexane-methylene chloride-isopropanol (98.8:1.1:0.1, v/v)	1	UV 297	α-Tocopherol standard	129 (1974)
MicroPak SI-10, 10 μm, 25 × 0.22 cm	Gradient (1%/min), hexane → hexane-diisopropyl ether (95:5, v/v)	0.67	UV 297	α-, β-, γ-, δ-Tocopherol in cell cultures	111 (1977)
MicroPak SI-10, 10 μm, 25 × 0.21 cm	Hexane-diisopropyl ether (95:5, v/v)	1	UV 298	α-, β-, γ-, δ-Tocopherol standards	88 (1977)
MicroPak SI-10, 10 μm, 25 × 0.21 cm	Hexane-1-chlorobutane (80:20, v/v)	1	UV 285	Acetates of α-, β-, γ-, δ-tocopherol standards	88 (1977)
Lichrosorb SI-60, 5 μm, 25 × 0.32 cm	Hexane-isopropanol (99.8:0.2, v/v), hexane-diethyl ether (95:5, v/v)	2	Fluorescence: λ_{exc}, 290; λ_{em}, 330	α-, β-, γ-, δ-Tocopherol, α-, β-, γ-, δ-tocotrienol in foods, tissues	124-126 (1978-1979)
Lichrosorb SI-60, 7 μm, 25 × 0.46 cm	Isooctane-isopropanol (99.5:0.5, v/v)	1	UV 296	α-, β-, γ-, δ-Tocopherol standards	130 (1980)

Spherisorb 5 μm, 25 × 0.2 cm	Hexane-isopropanol (99.75:0.25, v/v)	0.8	UV 280	α-, β-, γ-, δ-Tocopherol standards	121 (1979)
μPorasil 10 μm, 30 × 0.4 cm	Hexane-chloroform (85:15, v/v)	1.5	UV 280	α-Tocopheryl acetate in cereals	131 (1979)
μPorasil 10 μm, 30 × 0.4 cm	Isooctane-isopropanol (99:1, v/v)	1	UV 280	α-Tocopherol in foods	132 (1979)
μPorasil, ?	Chloroform	—	UV 254	α-, β-, γ-, δ-Tocopherol in vegetable oils	133 (1979)
μPorasil 10 μm, 30 × 0.4 cm	Hexane-diisopropyl ether (92:8, v/v)	2.5	Fluorescence: λ_{exc}, 295; λ_{em}, 370; UV 280	α-, β-, γ-, δ-Tocopherol in serum	117 (1980)
μPorasil 10 μm, 30 × 0.4 cm	Hexane-isopropanol (98.5:1.5, v/v)	2	UV 295	α-, β-, γ-, δ-Tocopherol in vegetable oils	122 (1979)
RSIL 5 μm, 15 × 0.32 cm	Hexane-isopropanol (99.4:0.6, v/v), 30°C	1	UV 294	α-, β-, γ-, δ-Tocopherol in serum	120 (1979)
RSIL 10 μm, 25 × 0.3 cm	Hexane-ethyl acetate (97:3, v/v)	1.5	Fluorescence: λ_{exc}, 303; λ_{em}, 328	α-, β-, γ-, δ-Tocopherol α-, β-, γ-, δ-tocotrienol in vegetable oils	134 (1980)
Zorbax Sil	Hexane-diisopropyl ether (98:2, v/v)	0.9	Fluorescence: λ_{exc}, 298; λ_{em}, 325	α-, β-, γ-, δ-Tocopherol in foods	118 (1979)

TABLE 3.9 (*Continued*)

Column	Mobile phase	Flow rate (ml/min)	Detection (nm)	Application	Reference (year)
Zorbax Sil 10 μm, 25 × 0.21 cm	Hexane-diisopropyl ether (87.5:12.5, v/v)	1	Electrochemical +0.7 V versus Ag/AgCl	α-, β-, γ-, δ-Tocopherol in wheat germ oil	119 (1981)
Partisil-10 PAC, 10 μm, 25 × 0.46 cm	Hexane-methylene chloride-isopropanol (70:30:0.2, v/v)	1	UV 292	α-Tocopherol in feeds	135 (1980)
Partisil PXS-10, 10 μm, 25 × 0.46 cm	Hexane-methanol (99.7:0.3, v/v)	2.39	Fluorescence: λ_{exc}, 295; λ_{em}, 340	α-, β-, γ-, δ-Tocopherol standards	87 (1978)
Partisil PXS-5, 5 μm, 25 × 0.46 cm	Hexane-isopropanol (98.6:1.4, v/v)	1.2	Fluorescence: λ_{exc}, 200; λ_{em}, 340	α-, β-, γ-, δ-Tocopherol in plasma, liver, oils, platelets	123 (1981)

native fluorescence (λ_{exc}, 205 and 295 nm; λ_{em}, 330 nm), no derivatization is necessary. Using the long wavelength excitation maximum, a minimum detectable amount of 9 ng (21 pmol) of pure α-tocopherol was reportedly reached [117].

Fluorescence detection has been adopted by numerous authors in connection with tocopherol assay in serum and plasma [93,113,117], red blood cells [116], liver [115], brain [127], vegetable oils [112,134], and foods and feeds [99,108,118,124-126]. Hatam and Kayden achieved a 20-fold gain in sensitivity by exciting tocopherol at 205 nm rather than at 295 nm [95]. Recently, this approach has been used again for the quantitation of tocopherol in plasma, platelets, oils, and liver [123]. A striking example of the enhanced selectivity of fluorescence in comparison with absorbance detection is illustrated in Fig. 3.6. In Fig. 3.6a a chromatogram representing extracts of bovine and pig plasma does not show any appreciable fluorescent peaks beyond the elution position of α-tocopherol (4 min), whereas a corresponding UV trace at 280 nm (Fig. 3.6b) reveals a large β-carotene peak in the pig plasma with a retention time of about 19 min.

Electrochemical LC detectors have not yet experienced a real breakthrough, although for selected compounds they are capable of affording a sensitivity not attainable with any other detector type. Tocopherols are oxidizable at a glassy carbon electrode at potentials approximating +0.7 V (versus Ag/AgCl reference electrode). Both straight- and reversed-phase systems, even in combination with nonaqueous eluents, are compatible with this detector type, provided that the mobile phase contains sufficient amounts of supporting electrolyte. This can be achieved either by adding salts to the chromatographic solvent or by mixing the column effluent "postcolumn" with an electrolyte solution. The first approach has been found useful in reversed-phase chromatography, since several electrolytes are readily soluble in methanol-based eluents. As much as 0.05 M sodium perchlorate could be incorporated in a methanol-pyridine (99.9:0.1, v/v) mixture [101]. This eluent combination in conjunction with a C_{18} column formed the basis of a liquid chromatographic assay, using electrochemical detection, for tocopherols and vitamin K-related compounds in blood [101]. Tocopherol amounts as low as 50 pg were detectable. Comparison was made with a similar LC serum assay but using fluorescence detection. No significant differences between the two sets of results were found.

Similarly, a straight-phase system was developed and applied to the determination of tocopherols in wheat germ oil [119]. The polarity of the mobile phase, consisting of diisopropyl ether-hexane (12.5:87.5, v/v), was too low to allow incorporation of substantial amounts of salts. To overcome this problem, the electrolyte solution consisting of 0.1 M sodium perchlorate in ethanol-methanol (1:9, v/v) was mixed postcolumn with the column eluate. The detection limit of the method was estimated at 1 ng.

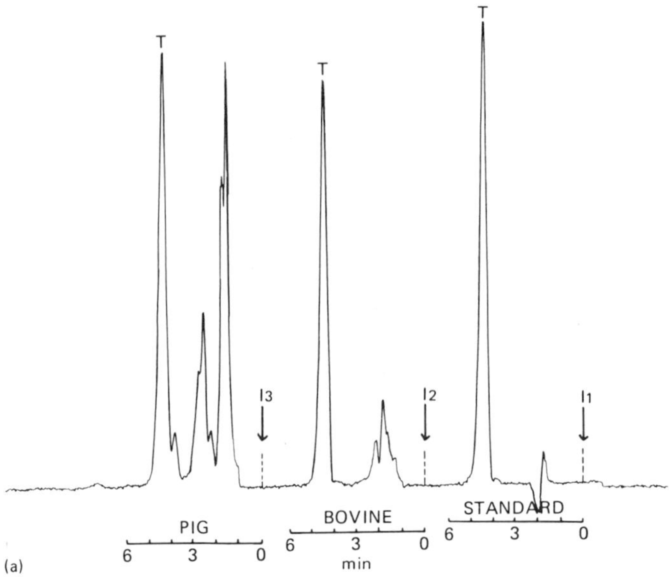

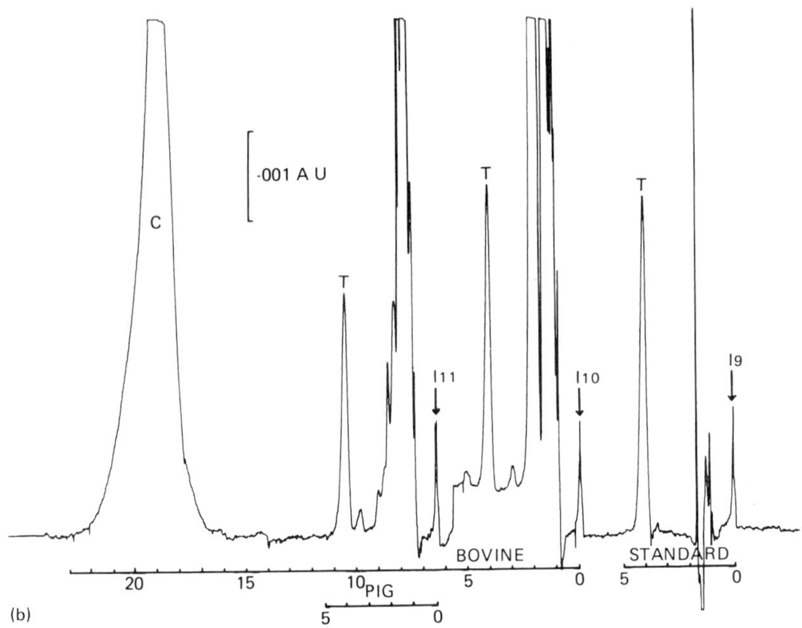

Figure 3.6 Comparison of fluorescence (a) and ultraviolet (b) detection of α-tocopherol in extracts of pig and bovine plasma. Chromatographic conditions: 30 × 0.39 cm stainless steel column packed with 10 μm μBondapak C18; eluent, methanol-water (97:3, v/v); flow rate, 3 ml/min. Fluorescence detection: λ_{exc} 296 nm; λ_{em} 330 nm; UV detection at 280 nm. Peak identification: T, α-tocopherol; C, β-carotene. (From Ref. 93.)

7.4 Quantitation

Most calibration principles outlined in the gas chromatography section are essentially valid for liquid chromatographic assays as well. The external standard method involves processing of samples containing known concentrations of the compound of interest, without addition of an internal standard. To construct standard curves, absolute peak heights or areas are plotted versus concentration. The choice of either of the two approaches is mainly determined by instrumental and chromatographic factors. For example, it is well known that peak areas are more affected by flow rate changes than peak heights, whereas the opposite is true in case of fluctuations in solvent composition [137]. Excellent guidelines in this respect have been given by Snyder and Kirkland [137]. Numerous tocopherol assays have relied on external standardization using peak height [92,93,95,97,98,103,112,131,135] or area measurements [122,124,127,135]. The use of modern sampling valves has largely eliminated imprecision caused by poor sample injection, so that the majority of analytical errors are now attributable to variabilities during sample pretreatment, e.g., extraction. Hence, precision is likely to improve by use of an appropriate internal standard.

More than in GC, the internal standard has indeed found general acceptance in modern LC. Apart from compensating for losses during sample handling, it also provides a means to correct for variations of chromatographic parameters, e.g., solvent composition and flow rate. The suitability of different internal standards proposed for vitamin E analysis can be readily judged in view of the eight requirements formulated by Snyder and Kirkland [138]. A basic criterion that encompasses all the others is structural similarity to the compound of interest.

Tocol is by far the most widely used internal standard in vitamin E assays [69,100,101,115-117,119,120]. Although this compound is very likely to behave physicochemically like α-tocopherol (Snyder's third criterion), there is a considerable polarity difference between the two analogs since their structures differ by three methyl groups. As a result, both compounds do not elute close to each other in some chromatographic systems [117] (criterion 2). Some authors found tocol unstable during saponification [95] (criterion 7). The synthetic 5,7-dimethyltocol, as used by Westerberg et al. in their assay of brain tocopherols [102], might be a better choice, but unfortunately this compound is not readily available. Some authors have used α-tocopheryl acetate as an alternative [94,102,128]. Its usefulness is, however, limited to reversed-phase systems. Despite the availability of tocopherol analogs, compounds not bearing any chemical relationship to vitamin E are still in use. Examples of "bad" internal standards include substances like naphthalene [111], 2-hydroxybiphenyl [88], and 2,4,6-trimethylphenol [88].

7.5 Applications

7.5.1 General Considerations

The resolving power of LC in conjunction with the unique selectivity of absorbance and fluorescence detection has allowed considerable simplification of sample workup as part of liquid chromatographic vitamin E assays. A typical analysis scheme consists of a double-phase solvent extraction, isolation and evaporation of the organic layer, reconstitution of the residue with a small solvent volume and injection of an aliquot on the column. In some cases, part of the organic supernatant may even be directly injected, if this is compatible with the mobile phase [102]. In reversed-phase chromatography, the residue should be redissolved in water-miscible organic solvents, such as methanol or, preferably, in the chromatographic solvent itself.

Saponification is generally omitted, which contributes to minimizing risks of oxidative product losses. With few exceptions, tocopherol extraction from biological materials of animal or human origin is based upon partitioning in a nonpolar solvent, e.g., hexane, usually preceded by a protein precipitation step [69,93-95,113,115,117,120,123, 128]. In the case of vegetable products, e.g., oils, sample handling is even often reduced to a direct injection of the sample onto the liquid chromatographic column [85,122,126,134].

7.5.2 Determination of α-Tocopherol in Serum and Red Blood Cells

Although as early as in 1975 Abe and coworkers had reported on the simultaneous determination of α-, β-, and γ-tocopherol in serum [113] and red blood cells [116] by adsorption chromatography with fluorescence detection (see below), the first papers specifically dealing with practically useful α-tocopherol assays were published by De Leenheer et al. [69] and Nilsson et al. [128]. The former authors introduced some important novelties by chromatographing α-tocopherol on reversed-phase rather than on silica columns and, unlike the Japanese workers, by UV monitoring the column eluate at 292 nm. Advantages of reversed-phase supports for the analysis of biological samples have been emphasized before. Variable wavelength UV detectors are still less troublesome and more convenient than their fluorescence counterparts. De Leenheer's procedure involves deproteinization of 200 μl serum, containing 2 μg tocol as an internal standard, with an equal volume of ethanol, duplicate extraction with 0.5 ml hexane, evaporation of the organic layer to dryness, and reconstitution of the residue with 200 μl methanol. A 100 μl aliquot is then injected on a RP 18 column, eluted with methanol at slightly increased temperature (40°C). Linearity of standard curves based on peak height ratio measurements extended from at least 5 to 25 mg/liter. Extraction recoveries were nearly quantitative, whereas within-day and day-to-day precision (coefficient of variation, CV) averaged 2.3% ($\overline{x}$ = 10.1 mg/liter, n = 24) and 3.2% ($\overline{x}$ = 10.8 mg/liter, n = 20), respectively. The detection limit of this method was estimated at 0.6 mg/liter, which is far beyond the normal concentration range. This figure, together with the small sample volume required, illustrates the excellent sensitivity affordable with absorbance detection at 292 nm. Figure 3.7 depicts a representative chromatogram of a human serum extract, obtained with this procedure. Chromatographic analysis time is only 6 min. The only non-vitamin E-like lipid constituent showing up in the chromatograms is retinol, the majority of other plasma lipids, e.g., cholesterol and triglycerides, not detectable at 292 nm.

The method of Nilsson et al. [128] differs thoroughly from the above in its chromatographic part. A straight-phase system, based on a pellicular Corasil silica column, was employed. Consequently, chromato-

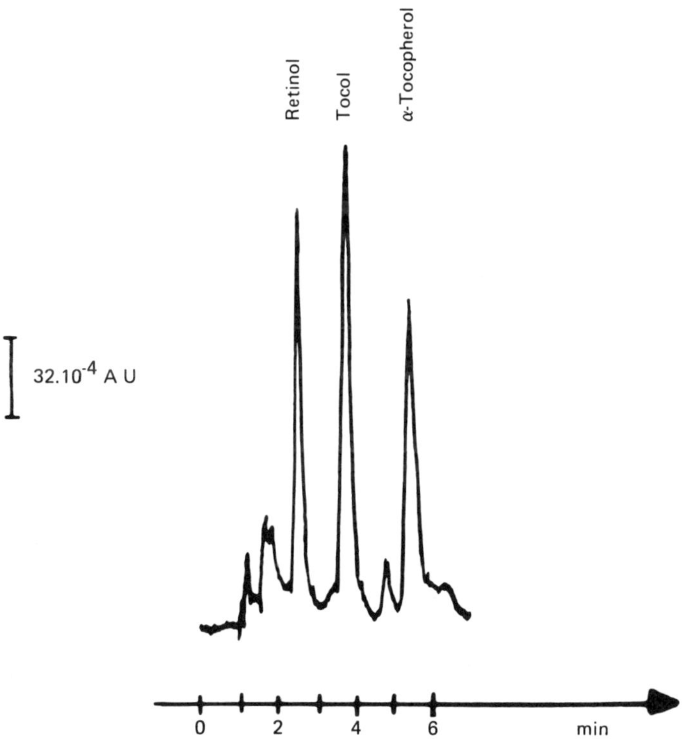

Figure 3.7 Chromatogram of a human serum extract. Chromatographic conditions: 25 × 0.46 cm stainless steel column, packed with 10 μm RSIL C18; eluent, methanol; flow rate, 2 ml/min; oven temperature, 40°C; detection at 292 nm. (From Ref. 100.)

graphic efficiency remains below current standards, as reflected, e.g., from the inability to separate γ- from β-tocopherol and other interferences. α-Tocopheryl acetate was selected as an internal standard, and detection was carried out at 280 nm. The authors claim a detection limit of 3.6 ng pure α-tocopherol under those conditions. However, in view of reported sensitivity data, obtained at the optimal wavelength of 292 nm—minimum detectable amount about 60 ng [69]—this figure does not seem to be realistic. That a pellicular column was used, which affords poor efficiency and sensitivity, certainly does not add any further to its credibility.

Later studies consistently abandoned the use of silica in favor of reversed-phase supports, which are preferable unless the other tocopherols are to be determined simultaneously. In the latter case, fluorescence detection is often indicated in order to ensure detectability of the minor β species. This extra sensitivity is generally not required in assays exclusively focusing on α-tocopherol. However, in certain animals, e.g., cattle and pigs, extremely low circulating levels of the latter may be associated with many disorders. To deal with this additional problem, McMurray and Blanchflower developed a method for α-tocopherol determination both in normal and in clinically affected animals [93]. Fluorescence detection was found superior to UV detection in terms of sensitivity and selectivity (Fig. 3.6). Although no internal standard was used, excellent precision (CV = 5.1% at $\bar{x}$ = 0.96 and 2.20 mg/liter, n = 12) was obtained. Further enhancement of sensitivity using fluorescence detection may be achieved by utilizing excitation at 205 nm instead of at the common long wavelength (295 nm) [95,123]. Under those conditions, Hatam and Kayden were able to reduce the sample size to 50-100 μl (human) plasma, the absolute detection limit of α-tocopherol being 1 ng [95]. The method could also be extended with minor modifications to red blood cells, leukocytes, and platelets. The authors discuss the potential use of an internal standard but fail to offer an acceptable explanation for not using one. Although α-tocopherol levels in red blood cells are significantly lower than in plasma, UV detection is capable of providing sufficient sensitivity. This was demonstrated by Bieri et al., who applied a procedure originally devised for the simultaneous determination of vitamin A and E in plasma (see Sec. 7.5.5) to erythrocytes [94]. Sample pretreatment was essentially the same as for plasma and involved deproteinization of 0.5 ml of cell suspension with methanol, lipid extraction with *n*-heptane, and use of α-tocopheryl acetate as an internal standard. The 280 nm detector was operated at its maximum sensitivity, i.e., 0.005 AUFS (Absorption Units Full Scale).

In exceptional cases in which still higher sensitivity is required, electrochemical detection may be tried. A reversed-phase liquid chromatographic determination of α-tocopherol in plasma using amperometric detection has been described [101]. Sensitivity of this assay is in the picogram range, and no interferences were noted.

7.5.3 Determination of α-, β-, and γ-Tocopherol in Serum and Red Blood Cells

Plasma levels of α-, β-, and γ-tocopherol differ by several orders of magnitude, which complicates a simultaneous determination of the three species. α-Tocopherol is found in a 10- to 100-fold excess with respect to the other two analogs. Hence, it is nearly impossible to find an adequate amount of internal standard that would cover this wide concentration range. For this reason, De Leenheer et al. recommend separate assays of α-tocopherol and the pair β- and γ-tocopherol, respectively [120]. Except for the chromatographic part, the β- and γ-tocopherol assay is based upon the same analytical principles (e.g., extraction procedure) as its α-tocopherol counterpart. It is obvious that only silica columns, or possibly some polar bonded phase materials, can be considered for the chromatography of the two positional isomers, since they are not separable on reversed-phase supports. In this case α-tocopheryl acetate is not useful as an internal standard, as tocopherol esters show little or no retention in straight-phase systems.

The first successful attempt to quantitate the three analogs in one run in serum [113] and red blood cells [116] was reported by Abe and coworkers. These authors used a 50 × 0.2 cm Jascopack silica column and fluorescence detection. However, the overall performance of the system was poor (low efficiency and long analysis time) so that the method can no longer compete with more recent assays. Data on precision and sensitivity are lacking in the paper. Tocol was used as an internal standard only in the red blood cell assay. Two major improvements were introduced by De Leenheer et al. [120]. First, use of a short, high-efficiency 5 μm silica column (15 × 0.32 cm, RSIL) considerably enhanced the speed and the sensitivity of the method. A chromatographic run, including elution of the internal standard tocol, was now completed in less than 10 min (Fig. 3.8) as compared with 30 min in Abe's system. The β- and γ-tocopherol could be readily separated. Second, application of the method for routine analysis was greatly facilitated by the use of a variable wavelength detector, operated at 294 nm. Detection limits for α-, β-, and γ-tocopherol were 30, 40, and 40 ng, respectively. Sample size ranged from 100 (α-tocopherol) to 500 μl (β- and γ-tocopherol).

New methods continue to appear in this area. They all have the same general outline in common, but they usually differ from each other in some quantitative aspects. For example, a recently developed assay claims improved sensitivity [123]. Although no actual data were given, this can be inferred indirectly from the reduced sample size required (0.2 ml plasma). This gain in sensitivity may be partly ascribed to the use of a 5 μm instead of 10 μm column, as well as to the superior performance of the fluorescence detector used (e.g., a higher energy output of its light source).

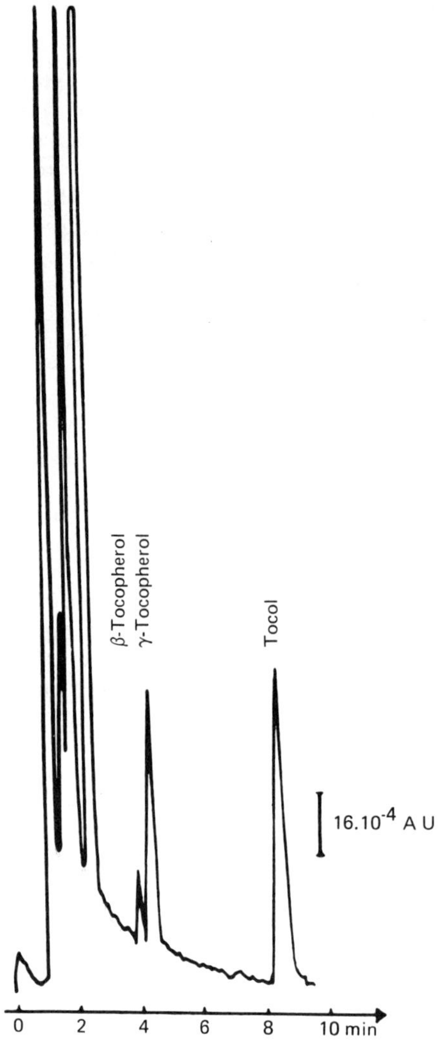

Figure 3.8 Chromatogram of a human serum extract containing β- and γ-tocopherol. Chromatographic conditions: 15 × 0.32 cm stainless steel column containing 5 μm RSIL (silica); eluent, hexane-isopropanol (99.4:0.6, v/v); flow rate, 1 ml/min); oven temperature, 30°C; detection at 294 nm. (From Ref. 120.)

7.5.4 Determination of Tocopherols in Organs and Tissues

Organs and tissues can be processed in a similar way as serum provided they are first homogenized. The homogenization mixture usually contains ethanol and hexane to ensure tocopherol extraction. Abe and coworkers were again the first to develop a method for the determination of tocopherol in rat liver [115], which was a slight modification of their previous serum assay [113]. The homogenate was extracted with hexane after deproteinization with ethanol. Other authors recommend saponification prior to extraction [123]. Chromatography is performed on silica columns but so far has failed to reveal the presence of β- or δ-tocopherol in liver samples [115,123]. Similar procedures have been reported for α-tocopherol analysis in brain [102,127]. Vatassery and Hagen used a pellicular column, fluorescence detection, and external calibration [127]. Brain is extracted with hexane in the presence of ascorbic acid. An aliquot of the organic phase can be directly injected on the column. The paper by Westerberg et al. [102] is outstanding from the standpoint of both methodology and results. Homogenization of 100 mg of brain was carried out in a mixture of water, containing ascorbic acid, ethanol, and hexane. 5,7-Dimethyltocol was added as an internal standard. After centrifugation of the mixture, a 50 μl aliquot of the organic layer was directly injected on a μBondaPak-NH_2 column. Alternatively, the hexane layer was evaporated to dryness, the residue redissolved in methanol, and an aliquot injected on a reversed-phase column. The polar bonded phase was found to be superior in terms of selectivity. A disadvantage was the poor resolution between the internal standard and α-tocopherylquinone. However, no traces of this alleged degradation product or metabolite were detected in brain. This is consistent with observations by other authors about the absence of the quinone in serum [48] and red blood cells [94]. Bieri and coworkers demonstrated the quinone only when sample pretreatment was carried out in the absence of pyrogallol, indicating that its presence is probably an artefact [94].

7.5.5 Simultaneous Determination of α-Tocopherol and Retinol in Serum

A simultaneous determination of α-tocopherol and retinol is meaningful in view of the recently suggested biological interaction between the two vitamins. For example, a sparing action of vitamin E on the expenditure of vitamin A has been demonstrated in animals and children [139-141]. Existing liquid chromatographic systems for α-tocopherol assay [69] can serve as a basis for their simultaneous quantitation. Since both compounds are liposoluble, the classic tocopherol extraction procedure is useful for their isolation from serum. Simultaneous absorbance detection of retinol and α-tocopherol is necessarily a compromise. At 330 nm (λ_{max} of retinol) tocopherol is virtually nonabsorbing, whereas at the λ_{max} of tocopherol there is a fair amount of retinol absorbance left. De Leenheer et al. accordingly propose detection at 292 nm [100], whereas Bieri et al. selected 280 nm as the detection wavelength [94], which affords somewhat less sensitivity and specificity. Tocol is useful as an internal standard for both tocopherol and retinol [100]. Alternatively, two separate internal standards may be employed, e.g., α-tocopheryl acetate and retinyl acetate [94].

7.5.6 *Determination of Tocopherols and Tocotrienols in Vegetable Oils, Foods, and Feeds*

The pioneering paper by Van Niekerk [85], pertaining to the analysis of vitamin E in vegetable oils, witnesses the tremendous simplification sample pretreatment has undergone since the introduction of modern LC. Gas chromatographic assays had to include saponification and extensive lipid fractionation steps; LC now permits processing of vegetable oils without any prepurification whatsoever. Van Niekerk's procedure merely consists of a direct injection of 1-6 μl oil, dissolved in hexane, on the liquid chromatographic column. This remarkable success became achievable owing to the exceptional selectivity of fluorescence detection. Duggan et al. showed indeed that few natural compounds are similar to tocopherols in their fluorescence characteristics [142]. The direct injection approach forms the basis of several later assays in this area [122,126,134]. Individual tocopherols and tocotrienols have been determined in a variety of plant oils using silica columns and fluorescence [85,114,118,123,126,134] or absorbance detection [107,122]. Occasionally some assays still include saponification [123], but in general a single extraction step is sufficient.

Somewhat more complex sample pretreatment is usually required for the analysis of solid foods and feeds [86,90,92,98,99,103,107,124-126, 135]. Organic solvents, such as acetone [86], chloroform-ethanol [131], diethyl ether [92], perchloroethylene [98], isopropanol [124], or isooctane-dioxane [135], can be considered for a crude lipid extraction. Preliminary aqueous treatment of samples to dissolve the gelatin coating of supplemental α-tocopheryl acetate may be required [98]. Saponification before or after extraction is generally recommended. McMurray and Blanchflower, in their comparative study of tocopherol isolation techniques, found solvent extraction followed by saponification the most effective approach [99]. A very simple method for the determination of α-tocopheryl acetate in feed has been reported by Shaikh et al. [90]. The procedure involves methanol extraction followed by injection of an aliquot of the supernatant on a reversed-phase column. Thompson and Hatina determined the tocopherol and tocotrienol content of a variety of foods [124]. By using fluorescence detection, naturally occurring tocopherols and supplemental α-tocopheryl acetate could be easily differentiated, since the latter does not possess native fluorescence. Although silica columns are preferable in view of their ability to resolve positional isomers [86,124-126,131], reversed phase has been used as well, particularly for the quantitation of α-tocopheryl acetate [90,98], the simultaneous determination of vitamin A and E [103,108], and the concurrent chromatography of all fat-soluble vitamins [91,105, 107].

7.5.7 *Determination of Tocopherols in Pharmaceuticals*

The advantages of LC for vitamin E determination in various biological materials apply to pharmaceuticals as well. Most studies so far have focused on the simultaneous determination of all fat-soluble vitamins in multivitamin preparations [88,89,96,143]. In practice, reversed phase is recommended for routine determination of α-tocopheryl acetate, whereas the use of silica columns should be reserved for differentiation of various homologs and isomers, including their esters [88].

7.5.8 *Recent Developments in LC of Vitamin E*

Liquid chromatographic methods for vitamin E continue to proliferate. A number of recent papers are concerned with the determination of tocopherols in blood plasma or serum [144-153], platelets [147,153], erythrocytes [154], tissues [146,149,150], human foods [155-160], including edible oils [157-160], and animal feeds [149,156]. The state of the art, with particular emphasis on the analysis of foods and feeds, has been reviewed by Barnes and Taylor [161]. New methods further elaborate on the principles outlined above and generally entail further improvements in sensitivity, specificity, speed, and ease of sample preparation. Both absorbance and fluorescence detection are in use, although there is a tendency in favor of the latter technique. In any case, the wealth of literature that continues to appear illustrates the unique position that LC has acquired over the years for vitamin E determination.

7.5.9 *Classic Column Chromatography*

Classic column chromatography, though far inferior to modern LC, is still a valuable tool for fractionation of complex biological samples. The technique is simple and requires no sophisticated equipment. It is not capable of providing truly differentiated chromatographic patterns but it does act as a more or less effective sample cleanup step prior to colorimetric or gas chromatographic tocopherol quantitation.

Adsorbents that have gained wide popularity for vitamin E column chromatography include secondary magnesium phosphate [162-164], silicic acid [22,47,164,165], alumina [31,57,59,63,64,74,77,166-169], Florisil [23,72,164], and a mixture of alumina and zinc carbonate [51, 170]. The latter reportedly yields a unique selectivity for α-tocopherol with regard to other lipids. This phenomenon is in agreement with the successful tocopherol separations obtained on thin layers of this material as well as on zinc carbonate-impregnated paper (Section 9). Special properties have also been ascribed to secondary magnesium phosphate. Unlike most other adsorbents, this particular support tends to afford a fair resolution between individual tocopherols [164]. Partial separation into homologous groups was also achieved on Florisil [164].

The applications of column chromatography as part of vitamin E assays in biological materials, e.g., plasma, animal tissues, fish, milk, vegetable oils, red blood cells, and foods, are multiple. Tocopherols have been determined in column eluates using colorimetry [163-165,169-171], gas chromatography [22,31,51,57,59,63,64,72,74,77], or TLC plus colorimetry [172,173].

8. THIN-LAYER CHROMATOGRAPHY

8.1 Introduction

TLC continues to occupy an important position among analytical methods in the vitamin E field. Before the breakthrough of LC, two considerations underlay the search for new and more efficient TLC systems. First, as a part of assays in biological materials, TLC has been frequently used to separate tocopherols from an excess of other lipids, including sterols, carotenoids, and other fat-soluble vitamins. Second, TLC has been extensively studied as a possible tool to meet the persisting analytical challenge of separating the various tocopherols and tocotrienols occurring in nature. However, few systems have been successful in differentiating positional isomers.

At this point, paper chromatography deserves a mention, in that it was the first technique to yield separation of β- and γ-tocopherol [174]. Although years ago a paper chromatographic approach was adopted as an official standard method by the vitamin E panel of the Society of Analytical Chemistry in the United Kingdom [175], the technique has now been almost completely superseded by TLC and LC.

8.2 Survey of TLC Systems for Vitamin E

Silica gel is the adsorbent of choice for TLC of vitamin E derivatives. A large variety of useful solvent combinations has been reported. This area has been thoroughly reviewed by Bunnell [16] and, more recently, by Aratani and Asakawa [176]. Most systems fail to distinguish between isomeric dimethyltocols and -tocotrienols. Stowe was the first to achieve a complete separation of β- and γ-tocopherol in a one-dimensional system, using a five-component solvent mixture [177]. However, R_f values were dependent on the complexity of the sample applied and, hence, were not reproducible. Although a similar system was applied by other authors to the analysis of vegetable oils [178], Stowe's approach never gained wide popularity. The introduction of a two-dimensional technique, as proposed by Pennock and coworkers [179], afforded a major breakthrough for the separation of β- and γ-tocopherol. Development of the plates in the first dimension was originally carried out with chloroform and yielded separation of the different homologs. Differentiation of β and γ isomers was obtained in the second dimension with a mixture of diisopropyl ether and petroleum ether. This approach found wide application for the analysis of complex tocopherol and tocotrienol mixtures, e.g., in plant materials [179-183]. More recently, Lovelady again reported a simpler one-dimensional system, suitable for the resolution of the four natural tocopherols [184], which, in combination with GC, formed the basis of an assay in serum and red blood cells [30].

So far, one has to rely on a two-dimensional approach if one aims at the separation of all tocopherols and tocotrienols in one run. However, Müller-Mulot was reportedly able to resolve all those compounds with the exception of the pair β-tocotrienol and γ-tocopherol, in a one-dimensional silica gel-based system with a two-component solvent mixture [185]. A relative disadvantage of this procedure are the three successive developments recommended by the authors. Table 3.10 summarizes the conditions of existing TLC methods that have proved valuable to separate tocopherols and tocotrienols, and, in particular, the β and γ forms.

Besides silica, a number of other supports have been used, including alumina [172,187-189], Florisil [190], kieselguhr [190], secondary magnesium phosphate [190], and mixtures of zinc carbonate with alumina [190,191], or silica gel G [190,191], respectively. In general, those alternatives are inferior for the separation of positional isomers. However, they often provide a unique selectivity for the resolution of homologs. For example, optimal separation of α- and β-tocopherol was obtained either on zinc carbonate-alumina or zinc carbonate-silica gel layers [190]. This is in keeping with Green's observation that the resolving power of zinc carbonate-impregnated paper for this particular pair of tocopherols is rarely equaled by an other conventional thin-layer material [174].

8.3 Detection

In order to visualize tocopherols on thin layers, a fluorescent indicator, such as fluorescein, is commonly incorporated in the support. Tocopherols then appear as quenched dark spots on a fluorescent background. Apart from inspection under UV light, chromatograms can be sprayed with color reagents. Most popular in this respect is the Emmerie-Engel reagent, originally devised for in vitro colorimetric vitamin E assays [17]. Color development is based upon the ability of tocopherols to reduce ferric ions to ferrous ions, the latter forming a red-colored complex with α,α-dipyridine (λ_{max}, 520 nm) [17] or bathophenanthroline [18]. Both reagents are frequently used to detect

TABLE 3.10 Silica Gel-Based TLC Systems for the Separation of Homologous and Isomeric Tocopherols and Tocotrienols

Method (1- or 2-dimensional)	Solvent system	Comments	Reference
1D	Petroleum ether-diisopropyl ether-acetone-diethyl ether-glacial acetic acid (85:12:4:1:1, v/v)	R_f values dependent on the complexity of the sample applied	177
1D	Petroleum ether-diisopropyl ether-acetone-diethyl ether-glacial acetic acid (127:16:6:1.5:1.5, v/v)	—	178
1D	Cyclohexane-hexane-diisopropyl ether-ammonia (40:40:20:2, v/v)	Reproducible R_f values	30,184
1D	Hexane-ethyl acetate (92.5:7.5, v/v)	No separation of γ-tocopherol and β-tocotrienol, 3 successive developments	185
2D	1st dimension: chloroform; 2nd dimension: diisopropyl ether-petroleum ether (or hexane) (20:80, v/v)	No separation of γ-tocopherol and β-tocotrienol	179-183
2D	1st dimension: chloroform; 2nd dimension: hexane-isopropanol (4:1, v/v)		186

tocopherols on thin-layer plates [185,186,188,192,193]. Reaction of tocopherols with the stable free radical α,α-diphenyl-β-picrylhydrazyl is based on a similar reduction reaction [194]. This spray reagent affords acceptable sensitivity but is not more specific for tocopherols than the former. Phosphomolybdic acid has the advantage of differentiating β- and γ-tocopherol on a color basis. β-Tocopherol yields a reddish brown spot, whereas γ-tocopherol turns blue upon spraying [195]. The combination of phosphomolybdic acid with 2,7-dichlorofluorescein in ammoniacal medium enhances the stability of the color produced [184]. Ceric(IV)sulfate allows a similar color differentiation of β- and γ-tocopherol [185,195], as does antimony(V)chloride [195]. An excellent spray reagent is also Turnbull's blue, which is a mixture of potassium ferricyanide and ferric chloride [196,197]. Other, less common reagents include 2,6-dichloroquinonechlorimide [198,199], vanadium pentoxide in sulfuric acid [200], rhodamine B [191,201], and a Cu(I) complex reagent with bis-(diisopropyloxythiophosphor)disulfide [202].

8.4 Quantitation

8.4.1 "Off-Line" Quantitation

As indicated previously, the majority of the older tocopherol assays in complex biological materials include TLC, which merely serves as a lipid fractionation step prior to off-line (i.e., after scraping off and eluting the chromatographic spots), colorimetric, or gas chromatographic quantitation. In both cases, extensive sample purification is usually necessary. The classic colorimetric Emmerie-Engel approach, as well as its later modifications, is subject to interferences from other reducing substances and does not distinguish between individual tocopherols. In addition, the color formation may be inhibited in the presence of other lipids present in vegetable oils [122]. Gas chromatography often suffers from the inability to separate α-tocopherol from an excess of cholesterol (see Sec. 6).

After development of the TLC plates, tocopherol spots are visualized and the areas of interest scraped off. Compounds to be assayed are subsequently eluted with an organic solvent, e.g., ethanol. After centrifugation of the mixture, the supernatant is mixed with the Emmerie-Engel reagent. Alternatively, the tocopherol-containing eluate is evaporated to dryness, the residue derivatized, and an aliquot injected on a GC column. Bieri et al. combined the two approaches by using the Emmerie-Engel reaction to convert tocopherols to their quinones prior to TLC and finally GC [49].

8.4.2 Densitometry

The current status of densitometry for tocopherol quantitation on thin layers has been thoroughly reviewed [176]. In a photodensitometric tocopherol determination, transmission can be measured at 291 nm. Aratani applied this approach in conjunction with micro-TLC on silica gel-covered quartz plates [203]. Alternatively, tocopherols can be chemically converted to colored spots prior to densitometric scanning by using spray reagents, such as ceric(IV)sulfate-trichloroacetic acid [204], 2,6-dichloroquinonechlorimide [199], and the Emmerie-Engel reagent [193]. Hess et al. converted tocopherols to tocopherylquinones in situ by heating the TLC plates at 110-120°C for prolonged time periods [205]. Scanning at the new λ_{max} of the reaction product (270 nm) afforded enhanced sensitivity, as compared with the determination of the native compounds. So far, fluorodensitometry has found no application for tocopherol analysis [176].

Direct in situ analysis on thin-layer chromatograms suffers from a number of limitations. For example, problems related to the variability in thickness of the adsorbents should be recognized. Extensive information with regard to calibration and reproducibility of the technique is given by Aratani in the above-mentioned review article [176].

8.5 Applications

TLC in conjunction with gas chromatography or colorimetry has been used to determine tocopherols and tocotrienols in blood serum or plasma [30,186,201,205,206], red blood cells [49,201], liver [172,173,194, 207], vegetable oils [25,75,178,185,188,189,192-194,197,199,203,208, 209], leaves [210], algae [183,211], lichens [212], foods [172], and pharmaceutical preparations [213]. The TLC step greatly contributes to the reliability of vitamin E determinations in biological materials. Colorimetry in particular lacks the specificity required to distinguish between tocopherols and interferences from the biological matrix. Conditions of TLC systems, used as part of gas chromatographic or colorimetric vitamin E assays, are outlined in Tables 3.11 and 3.12. Despite the presence of a chromatographic step in the procedure, sample pretreatment remains complex, as opposed to the minimal cleanup required for LC-based assays. Very often, saponification and solvent extraction prior to the TLC step are inadequate. Additional fractionation, e.g., by column chromatography, may be required to obtain extracts sufficiently pure for colorimetry.

9. PAPER CHROMATOGRAPHY

Although among modern separation techniques paper chromatography is gradually losing its significance, some recent applications indicate that it continues to maintain a certain importance in the vitamin E area. The approach of Green et al. [174], who employed zinc carbonate-impregnated paper to separate tocopherols, might still be a valuable

TABLE 3.11 TLC Systems as Part of Gas Chromatographic Tocopherol Assays

Support	Mobile phase composition	Compounds involved	Application	Reference
Silica gel H	Benzene-methanol (98:2, v/v)	α-, β-, γ-, δ-Tocopherol, α-, β-, γ-tocotrienol	Foods	25
Silica gel G	Benzene	α-Tocopherol	Beef heart mitochondria	47
Silica gel H	Benzene-ethyl acetate (2:1, v/v)	α-, β-, γ-, δ-Tocopherol	Red blood cells	49
Silica gel HR	Benzene-methanol (99.25:0.75, v/v)	α-, γ-Tocopherol	Plasma	20
Silica gel G	Cyclohexane-hexane-diisopropyl ether-ammonia (40:40:20:2, v/v)	α-, β-, γ-, δ-Tocopherol	Plasma, red blood cells	30
Silica gel G	Cyclohexane-diethyl ether (8:2, v/v)	α-Tocopherol	Serum	42
Silica gel G	Benzene-ethyl acetate (96:4, v/v)	α-, β-, γ-, δ-Tocopherol, α-, β-, γ-, δ-tocotrienol	Palm oil	75

TABLE 3.12 Conditions of TLC Systems as Part of Colorimetric Tocopherol Assays

Support	Mobile phase composition	Compounds involved	Application	Reference
Silica gel G	Benzene	α-, β-, γ-, δ-Tocopherol	Serum	186
Silica gel G	Petroleum ether-chloroform (1:1, v/v)	α-, β-, γ-, δ-Tocopherol	Plasma	206
Silica gel G	Petroleum ether-diethyl ether-diisopropyl ether-acetone-acetic acid (127:1.5:16:6:1.5, v/v)	α-, β-, γ-, δ-Tocopherol	Vegetable oils	178
Silica gel G	Chloroform or benzene or petroleum ether-diethyl ether (5:1, v/v)	α-, β-, γ-, δ-tocopherol	Peanut oil	197
Silica gel G	1st dimension: chloroform; 2nd dimension: diisopropyl ether-petroleum ether (20:80, v/v)	α-, β-, γ-, δ-Tocopherol, α-, β-, γ-, δ-tocotrienol	Vegetable oils	180
Aluminum oxide	Carbon tetrachloride	α-, β-, γ-, δ-tocopherol	Vegetable oils	189
Aluminum oxide	Chloroform	α-, β-, δ-tocopherol	Oils, fats	188
Silica gel G	Carbon tetrachloride-chloroform (1:1, v/v)	α-Tocopherol (liver), α-, β-, γ-tocopherol (oils)	Rat liver vegetable oils	173
Silica gel G	Benzene-methanol (9:1, v/v) or cyclohexane-diethyl ether (8:2, v/v) and benzene-methanol (99:1, v/v)	α-, β-, γ-, δ-Tocopherol	Soybean oil, rat liver, oats	194

Aluminum oxide	Benzene-diethyl ether (50:50, v/v)	α-, γ-, δ-Tocopherol	Foods, oils, milk, liver	172
Silica gel F	Toluene	α-Tocopherol	Multivitamin products	213
Silica gel G	Hexane-ethyl acetate (92.5:7.5, v/v)	α-Tocopherol	Alfalfa	214
Silica gel G	Benzene	α-Tocopherol	Rat liver	207
Silica gel G	1st dimension: chloroform; 2nd dimension: diisopropyl ether-petroleum ether (20:80, v/v)	α-, β-, γ-, δ-tocopherol, α-, β-, γ-, δ-tocotrienol + esters	Latex	181
Silica gel F	Cyclohexane-diethyl ether (8:2, v/v)	α-Tocopherol	Foods, feeds	209
Silica gel G	Benzene-ethyl acetate (2:1, v/v), benzene-ethanol (99:1, v/v)	Total tocopherol	Plasma, red blood cells (rats)	201
Silica gel F	Hexane-ethyl acetate (92.5:7.5, v/v)	α-, β-, γ-, δ-Tocopherol, α-, β-, γ-, δ-tocotrienol	Vegetable oils	185

alternative when other methods fail. As mentioned above, zinc carbonate-containing adsorbents have been shown to display unique selectivity for vitamin E derivatives. The standard method of vitamin E assay, proposed by the vitamin E panel of the Analytical Methods Committee in England, utilized this approach as part of a two-dimensional system, employing reversed-phase chromatography on paraffin-impregnated paper in the second dimension [175]. The applicability of this approach for tocopherol assay in biological materials, e.g., in blood serum, was later demonstrated as well [215]. Recently, the same two-dimensional technique formed the basis of an improved method for the determination of α-tocopherol in sheep liver [216]. Although the usefulness of paper chromatography for this purpose was convincingly demonstrated, the technique remains time consuming (several hours of development) and relatively insensitive as compared with e.g., modern LC.

10. THE IDEAL CHROMATOGRAPHIC TECHNIQUE FOR TOCOPHEROL ASSAY

The ideal chromatographic assay for tocopherols in biological materials should possess the following distinct characteristics. It should be capable of separating tocopherols from an excess of other lipids (specificity), while at the same time effectively differentiating vitamin E analogs, i.e., homologs as well as positional isomers (selectivity and efficiency). In addition, the extent of sample pretreatment should be kept to an absolute minimum, particularly in view of the compounds' potential degradation under the influence of air and light. The assay's overall quantitative performance, as indicated by such criteria as accuracy, precision, and sensitivity, should be satisfactory.

A critical comparison of the three principal chromatographic techniques discussed in this chapter shows that only LC is able to satisfy all those demands.

10.1 Qualitative Aspects: Specificity, Selectivity, Chromatographic Efficiency, and Compound Stability

Vitamin E forms a minor constituent of the lipid fraction in most biological materials of animal and vegetable origin. A highly specific technique is therefore required to quantitate tocopherols in the presence of an excess of sterols, triglycerides, carotenoids, fat-soluble vitamins, higher alcohols, etc., while avoiding unnecessarily complex purification steps. Liquid chromatography combines a powerful resolving capacity and high efficiency with the unique selectivity of spectrophotometric or spectrofluorimetric detection. Interfering lipids may either be separated from the compounds of interest or, in most cases, remain undetected, thus rendering extensive sample pretreatment superfluous. GC suffers from the drawback of universal (flame ionization) detection, except when sophisticated detectors, e.g., a mass spectrometer, are used. The need to remove an excess of cholesterol negatively affects the simplicity of GC-based assays. In TLC, separation capacity and efficiency are limited, very specific spray reagents are lacking, and on-line quantitation is uncommon.

LC lends itself ideally to the resolution of the various vitamin E homologs and positional isomers. Such separation using conventional packed GC columns is only possible at the expense of long analysis times or involves the preparation of unusual derivatives (quinones). Alternatively, capillary columns may be used but this does not always preclude the need for long chromatographic runs to separate β- and γ-tocopherol.

TLC separation of positional isomers is only readily achievable using time-consuming and tedious two-dimensional approaches. In addition, TLC enhances the risk of oxidative losses by exposing tocopherols on large surface areas to light and air for prolonged times. On the other hand, LC with its short chromatographic runs and minimal sample pretreatment is less likely to cause such degradation.

10.2 Quantitative Aspects: Sensitivity, Accuracy, Precision, and Analysis Time

Tocopherol levels in biological materials are relatively high, i.e., of the order of micrograms per milliliter in blood and micrograms per gram in vegetable oils, so that analytical sensitivity problems are generally not a major concern. Existing TLC systems for tocopherols are less sensitive than their LC and GC counterparts and hence require larger sample quantities. Modern absorbance detectors for LC permit tocopherol quantitation in the nanogram range, whereas fluorescence and electrochemical detection can cover those cases in which increased sensitivity (and/or specificity) is required. In GC, the classic FID detector is satisfactory for most applications. If, occasionally, ultimate sensitivy is to be reached, selected ion-monitoring mass spectrometry is an obvious choice. This approach could form the basis of an absolute reference method.

The accuracy of an assay in general is largely dependent on sensitivity and specificity. From the specificity standpoint, LC is by far superior to the other chromatographic techniques (except GC-MS), which thus guarantees better accuracy as well. Although no systematic studies have been undertaken in this area, LC is likely to surpass TLC and GC in accuracy. The precision of most tocopherol assays in biological materials, using any of the three chromatographic approaches, is mostly acceptable. Coefficient of variation values of a few percent are common. LC- and, to a lesser extent, GC-based tocopherol assays usually include an internal standard, which favorably affects their reproducibility.

10.3 Conclusion

A survey of the most recently published analytical papers in the vitamin E area undoubtedly shows a marked trend in favor of LC. This technique is gradually emerging as the method of choice for vitamin E analysis. Several laboratories adopted it already as a routine method for the determination of tocopherols in blood and vegetable oils. This is merely another illustration of the steadily growing importance of LC as an analytical tool in general.

ACKNOWLEDGMENT

This work was supported by the National Fund for Scientific Research (NFWO) through a grant to the first author.

REFERENCES

1. M. Evans and G. O. Burr, *Proc. Nat. Acad. Sci. U.S.*, *11*, 373 (1925).
2. S. Kasparek, in *Vitamin E, a Comprehensive Treatise* (L. J. Machlin, ed.), Marcel Dekker, New York, 1980, p. 8.
3. P. B. McCay and M. M. King, in *Vitamin E, a Comprehensive Treatise* L. J. Machlin, ed.), Marcel Dekker, New York, 1980, p. 289.
4. J. Stocks, E. L. Offerman, C. B. Modell, and T. L. Dormandy, *Br. J. Haematol.*, *23*, 713 (1972).
5. G. R. Tudhope and J. Hopkins, *Acta Haematol.*, *53*, 98 (1975).
6. J. Bland, W. Canfield, T. Kennedy, J. Vincent, and R. Wells, *Physiol. Chem. Phys.*, *10*, 145 (1978).
7. M. Steiner, in *Tocopherol, Oxygen and Biomembranes* (C. De Duve and O. Hayaishi, eds.), Elsevier, Amsterdam, 1978, p. 143.
8. J. Svensson and T. Oki, *Int. J. Vitam. Nutr. Res.*, *48*, 250 (1978).
9. L. M. Corwin, in *Vitamin E, a Comprehensive Treatise* (L. J. Machlin, ed.), Marcel Dekker, New York, 1980, p. 332.
10. A. T. Diplock, *Vitam. Horm.*, *32*, 445 (1974).
11. H. H. Draper, in *Vitamin E, a Comprehensive Treatise* (L. J. Machlin, ed.), Marcel Dekker, New York, 1980, p. 276.
12. G. L. Catignani, in *Vitamin E, a Comprehensive Treatise* (L. J. Machlin, ed.), Marcel Dekker, New York, 1980, p. 318.
13. F. A. Oski and L. A. Barness, *J. Pediatr.*, *70*, 211 (1967).
14. S. Gross and D. K. Melhorn, *J. Pediatr.*, *85*, 753 (1974).
15. S. Gross, *Semin. Hematol.*, *13*, 187 (1976).
16. R. H. Bunnell, *Lipids*, *6*, 245 (1971).
17. A. Emmerie and C. Engel, *Rec. Trav. Chim.*, *57*, 1351 (1938).
18. C. C. Tsen, *Anal. Chem.*, *33*, 849 (1961).

19. A. J. Sheppard, A. R. Prosser, and W. D. Hubbard, *J. Amer. Oil Chem. Soc.*, *49*, 619 (1972).
20. J. Lehmann and H. T. Slover, *Lipids*, *6*, 35 (1971).
21. J. Lehmann, *Lipids*, *13*, 616 (1978).
22. J. Lehmann, *Lipids*, *15*, 135 (1980).
23. H. T. Slover, L. M. Shelley, and T. L. Burks, *J. Amer. Oil Chem. Soc.*, *44*, 161 (1967).
24. H. T. Slover, R. J. Valis, and J. Lehmann, *J. Amer. Oil Chem. Soc.*, *45*, 580 (1968).
25. H. T. Slover, J. Lehmann, and R. J. Valis, *J. Amer. Oil Chem. Soc.*, *46*, 417 (1969).
26. F. Mordret and A. M. Laurent, *Rev. Fr. Corps Gras*, *25*, 245 (1978).
27. P. P. Nair and Z. Luna, *Arch. Biochem. Biophys.*, *127*, 413 (1968).
28. P. P. Nair and D. A. Turner, *J. Amer. Oil Chem. Soc.*, *40*, 353 (1963).
29. P. P. Nair, I. Sarlos, and J. Machiz, *Arch. Biochem. Biophys.*, *114*, 488 (1966).
30. H. G. Lovelady, *J. Chromatogr.*, *85*, 81 (1973).
31. A. A. Christie, A. C. Dean, and B. A. Millburn, *Analyst*, *98*, 161 (1973).
32. Shen-Nan Lin and E. C. Horning, *J. Chromatogr.*, *112*, 465 (1975).
33. M. K. Govind Rao and E. G. Perkins, *J. Agr. Food Chem.*, *20*, 240 (1972).
34. P. P. Nair and J. Machiz, *Biochim. Biophys. Acta*, *144*, 446 (1967).
35. W. U. Dompert and H. Beringer, *Fette, Seifen, Anstrichm.*, *78*, 108 (1976).
36. A. W. Kormann, *J. Lipid Res.*, *21*, 780 (1980).
37. D. W. Thomas, R. M. Parkhurst, D. S. Negi, K. D. Lunan, A. C. Wen, A. E. Brandt, and R. J. Stephens, *J. Chromatogr.*, *225*, 433 (1981).
38. H. T. Slover and R. H. Thompson, *Lipids*, *16*, 268 (1981).
39. P. W. Wilson, E. Kodicek, and V. H. Booth, *Biochem. J.*, *84*, 524 (1962).
40. S. Ishikawa and G. Katsui, *J. Vitaminol.*, *12*, 106 (1966).
41. C. Kanno, K. Yamauchi, and T. Tsugo, *J. Dairy Sci.*, *51*, 1713 (1968).
42. M. Chiarotti and G. V. Giusti, *J. Chromatogr.*, *147*, 481 (1978).
43. G. Österlöf and A. Nyheim, *J. Chromatogr. Biomed. Appl.*, *183*, 487 (1980).
44. D. K. Feeter, *J. Amer. Oil Chem. Soc.*, *51*, 184 (1974).
45. J. P. Nelson, A. J. Milun, and H. D. Fisher, *J. Amer. Oil Chem. Soc.*, *47*, 259 (1970).
46. K. T. Hartman, *J. Amer. Oil Chem. Soc.*, *54*, 421 (1977).

47. M. M. Oliveira, W. V. Weglicki, A. Nason, and P. P. Nair, *Biochim. Biophys. Acta, 180*, 98 (1969).
48. V. O. R. C. De Bevere and A. P. De Leenheer, unpublished results, 1977.
49. J. G. Bieri, R. K. H. Poukka, and E. L. Prival, *J. Lipid Res., 11*, 118 (1970).
50. N. Nicolaides, *J. Chromatogr., 4*, 469 (1960).
51. J. G. Bieri and E. L. Andrews, *Iowa State J. Sci., 38*, 3 (1963).
52. M. Kofler, P. F. Sommer, H. R. Bolliger, B. Schmidli, and M. Vecchi, *Vitam. Horm., 20*, 407 (1962).
53. D. A. Libby and A. J. Sheppard, *J. Ass. Offic. Anal. Chem., 47*, 371 (1964).
54. K. K. Carroll and D. C. Herting, *J. Amer. Oil Chem. Soc., 41*, 473 (1964).
55. G. H. Dialameh and R. E. Olson, *Anal. Biochem., 32*, 263 (1969).
56. H. C. Pillsbury, A. J. Sheppard, and D. A. Libby, *J. Ass. Offic. Anal. Chem., 50*, 809 (1967).
57. G. S. Hall and D. L. Laidman, *Biochem. J., 108*, 465 (1968).
58. J. P. Nelson and A. J. Milun, *J. Amer. Oil Chem. Soc., 45*, 848 (1968).
59. F. P. Mahn, V. Viswanathan, C. Plinton, A. Menyharth, and B. Z. Senkowski, *J. Pharm. Sci., 57*, 2149 (1968).
60. A. J. Sheppard, A. R. Prosser, and W. D. Hubbard, *Methods Enzymol., 18* (part C), 356 (1971).
61. M. L. Quaife and L. Friedman, *Proc. Soc. Exp. Biol. Med., 133*, 849 (1970).
62. T. Gutfinger and A. Letan, *Lipids, 7*, 483 (1972).
63. K. Ranfft, *Z. Anal. Chem., 259*, 28 (1972).
64. K. Ranfft, *Landwirt. Forsch., 28*, 146 (1973).
65. H. Morimoto, I. Imada, T. Amano, M. Toyoda, and Y. Ashida, *Biochem. Med., 7*, 169 (1973).
66. H. Rückemann, *Landwirt. Forsch., 34*, 80 (1977).
67. B. J. Kovensky and R. I. Day, *J. Chromatogr. Sci., 9*, 442 (1971).
68. K. Abe, Y. Yuguchi, and G. Katsui, *Eiyo To Shokuryo, 27*, 329 (1974); *CA, 82*, 153888p (1975).
69. A. P. De Leenheer, V. O. De Bevere, A. A. Cruyl, and A. E. Claeys, *Clin. Chem., 24*, 585 (1978).
70. R. A. Wiggins, *Proc. Anal. Div. Chem. Soc., 13*, 133 (1976).
71. H. Kobayashi, C. Kanno, K. Yamauchi, and T. Tsugo, *Biochim. Biophys. Acta, 380*, 282 (1975).
72. J. Eisner, J. L. Iverson, and D. Firestone, *J. Ass. Offic. Anal. Chem., 49*, 580 (1966).
73. M. Nakanishi, K. Tsuchiya, K. Sakaguchi, and T. Fujita, *Yakugaku Zasshi, 99*, 1037 (1979); *CA, 92*, 34251x (1980).
74. K. Ranfft, *Landwirt. Forsch., 25*, 144 (1972).

75. P. W. Meijboom and G. A. Jongenotter, *J. Amer. Oil Chem. Soc.*, *56*, 33 (1979).
76. P. B. Bowman and W. E. West, *J. Pharm. Sci.*, *57*, 470 (1968).
77. P. Uebersax, *Schweiz. Landwirt. Forsch.*, *11*, 363 (1972).
78. A. J. Sheppard, W. D. Hubbard, and A. R. Prosser, *J. Ass. Offic. Anal. Chem.*, *52*, 442 (1969).
79. B. C. Rudy, F. P. Mahn, and B. Z. Senkowski, *J. Ass. Offic. Anal. Chem.*, *55*, 1211 (1972).
80. A. J. Sheppard and W. D. Hubbard, *J. Pharm. Sci.*, *68*, 98 (1979).
81. J. N. Thompson, P. Erdody, and W. B. Maxwell, *Anal. Biochem.*, *50*, 267 (1972).
82. J. A. Schmit, R. A. Henry, R. C. Williams, and J. F. Dieckman, *J. Chromatogr. Sci.*, *9*, 645 (1971).
83. R. C. Williams, J. A. Schmit, and R. A. Henry, *J. Chromatogr. Sci.*, *10*, 494 (1972).
84. G. J. Krol, C. A. Mannan, F. Q. Gemmill, G. E. Hicks, and B. T. Kho, *J. Chromatogr.*, *74*, 43 (1972).
85. P. J. Van Niekerk, *Anal. Biochem.*, *52*, 533 (1973).
86. J. F. Cavins and G. E. Inglett, *Cereal Chem.*, *51*, 605 (1974).
87. G. T. Vatassery, V. R. Maynard, and D. F. Hagen, *J. Chromatogr.*, *161*, 299 (1978).
88. T. Eriksson and B. Sörensen, *Acta Pharm. Suec.*, *14*, 475 (1977).
89. M. Eriksson, T. Eriksson, and B. Sörensen, *Acta Pharm. Suec.*, *15*, 274 (1978).
90. B. Shaikh, H. S. Huang, and W. L. Zielinski, *J. Ass. Offic. Anal. Chem.*, *60*, 137 (1977).
91. H. Cohen and M. Lapointe, *J. Agr. Food Chem.*, *26*, 1210 (1978).
92. C. H. McMurray and W. J. Blanchflower, *J. Chromatogr.*, *176*, 488 (1979).
93. C. H. McMurray and W. J. Blanchflower, *J. Chromatogr.*, *178*, 525 (1979).
94. J. G. Bieri, T. J. Tolliver, and G. L. Catignani, *Amer. J. Clin. Nutr.*, *32*, 2143 (1979).
95. L. J. Hatam and H. J. Kayden, *J. Lipid Res.*, *20*, 639 (1979).
96. S. A. Barnett and L. W. Frick, *Anal. Chem.*, *51*, 641 (1979).
97. S. S. O. Hung, Y. C. Cho, and S. J. Slinger, *J. Ass. Offic. Anal. Chem.*, *63*, 889 (1980).
98. S. Eriksen, *J. Ass. Offic. Anal. Chem.*, *63*, 1154 (1980).
99. C. H. McMurray, W. J. Blanchflower, and D. A. Rice, *J. Ass. Offic. Anal. Chem.*, *63*, 1258 (1980).
100. A. P. De Leenheer, V. O. R. C. De Bevere, M. G. M. De Ruyter, and A. E. Claeys, *J. Chromatogr. Biomed. Appl.*, *162*, 408 (1979).
101. S. Ikenoya, K. Abe, T. Tsuda, Y. Yamano, O. Hiroshima, M. Ohmae, and K. Kawabe, *Chem. Pharm. Bull.*, *27*, 1237 (1979).

102. E. Westerberg, M. Friberg, and B. Åkesson, *J. Liquid Chromatogr.*, *4*, 109 (1981).
103. P. Söderhjelm and B. Andersson, *J. Sci. Food Agr.*, *29*, 697 (1978).
104. H. Rückemann and K. Ranfft, *Z. Lebensm. Unters. Forsch.*, *166*, 151 (1978).
105. S. A. Barnett, L. W. Frick, and H. M. Bainc, *Anal. Chem.*, *52*, 610 (1980).
106. W. O. Landen, *J. Chromatogr.*, *211*, 155 (1981).
107. A. Mankel, *Deut. Lebensm. Rundsch.*, *75*, 77 (1979).
108. J. W. De Vries, D. C. Egberg, and J. C. Heroff, *Proceedings Symposium Analysis Foods and Beverages*, Vol. 2 (G. Charalambous, ed.), Academic Press, New York, 1979, p. 477.
109. N. A. Parris, *J. Chromatogr.*, *157*, 161 (1978).
110. LC Column report, Non-Aqueous Reversed Phase Chromatography, Du Pont Instruments, Wilmington, DE.
111. M. Matsuo and Y. Tahara, *Chem. Pharm. Bull.*, *25*, 3381 (1977).
112. K. Abe, Y. Yuguchi, and G. Katsui, *J. Nutr. Sci. Vitaminol.*, *21*, 183 (1975).
113. K. Abe and G. Katsui, *Vitamins (Japan)*, *49*, 259 (1975); *CA*, *83*, 110684x (1975).
114. K. Abe and G. Katsui, *Eiyo To Shokuryo*, *28*, 453 (1975); *CA*, *85*, 61507 (1976).
115. K. Abe, M. Ohmae, and G. Katsui, *Vitamins (Japan)*, *50*, 453 (1976); *CA*, *86*, 100661k (1977).
116. K. Ishibashi, K. Abe, M. Ohmae, K. Kawabe, and G. Katsui, *Vitamins (Japan)*, *51*, 415 (1977); *CA*, *87*, 196608t (1977).
117. L. Jansson, B. Nilsson, and R. Lindgren, *J. Chromatogr. Biomed. Appl.*, *181*, 242 (1980).
118. H. Fukuba, T. Miyoshi, and T. Tsuda, *J. Amer. Oil Chem. Soc.*, *56*, 189A (1979).
119. O. Hiroshima, S. Ikenoya, M. Ohmae, and K. Kawabe, *Chem. Pharm. Bull.*, *29*, 451 (1981).
120. A. P. De Leenheer, V. O. R. C. De Bevere, and A. E. Claeys, *Clin. Chem.*, *25*, 425 (1979).
121. C. C. Tangney, J. A. Driskell, and H. M. McNair, *J. Chromatogr.*, *172*, 513 (1979).
122. A. P. Carpenter, *J. Amer. Oil Chem. Soc.*, *56*, 668 (1979).
123. C. C. Tangney, H. M. McNair, and J. A. Driskell, *J. Chromatogr.*, *224*, 389 (1981).
124. J. N. Thompson and G. Hatina, *J. Liquid Chromatogr.*, *2*, 327 (1979).
125. J. N. Thompson, Proceedings Symposium Application of HPLC Methods for Determination of Fat Soluble Vitamins A, D, E, and K in Foods and Pharmaceuticals (Association of Vitamin Chemists), Chicago, 1978, p. 62.

126. J. N. Thompson, G. Hatina, and W. B. Maxwell, Proceedings Symposium Application of HPLC Methods for Determination of Fat Soluble Vitamins A, D, E, and K in Foods and Pharmaceuticals (Association of Vitamin Chemists), Chicago, 1978, p. 84.
127. G. T. Vatassery and D. F. Hagen, *Anal. Biochem.*, *79*, 129 (1977).
128. B. Nilsson, B. Johansson, L. Jansson, and L. Holmberg, *J. Chromatogr. Biomed. Appl*, *145*, 169 (1978).
129. D. Carr, *Anal. Chem.*, *46*, 743 (1974).
130. R. K. D. Tiebach and M. Schramm, *Chromatographia*, *13*, 403 (1980).
131. W. A. Widicus and J. R. Kirk, *J. Ass. Offic. Anal. Chem.*, *62*, 637 (1979).
132. J. P. Antalick, *Ann. Fals. Exp. Chim.*, *72*, 571 (1979).
133. H. E. Woziwodzki, U.S. Patent 4,122,094; *CA*, *90*, 53394w (1979).
134. P. Deldime, G. Lefebvre, Y. Sadin, and M. Wybauw, *Rev. Fr. Corps Gras*, *27*, 279 (1980).
135. H. Cohen and M. R. Lapointe, *J. Ass. Offic. Anal. Chem.*, *63*, 1254 (1980).
136. B. L. Karger, M. Martin, and G. Guiochon, *Anal. Chem.*, *46*, 1640 (1974).
137. L. R. Snyder and J. J. Kirkland (eds.), *Introduction to Modern Liquid Chromatography*, 2nd ed., Wiley, New York, 1979, p. 556.
138. L. R. Snyder and J. J. Kirkland (eds.), *Introduction to Modern Liquid Chromatography*, 2nd ed., Wiley, New York, 1979, p. 554.
139. E. Søndergaard, *Experientia*, *28*, 773 (1972).
140. M. K. Horwitt, *Amer. J. Clin. Nutr.*, *29*, 569 (1976).
141. V. Jagadeesan and V. Reddy, *Clin. Chim. Acta*, *90*, 71 (1978).
142. D. E. Duggan, R. L. Bowman, B. B. Brodie, and S. Udenfriend, *Arch. Biochem. Biophys.*, *68*, 1 (1957).
143. D. Thorburn Burns and C. Mackay, *J. Chromatogr.*, *200*, 300 (1980).
144. M. Leclercq and M. Bourgeay-Causse, *Rev. Inst. Pasteur Lyon*, *14*, 475 (1981).
145. H. Kato, H. Tanimura, and Y. Hikasa, *Arch. Jpn. Chir.*, *50*, 740 (1981); *CA*, *96*, 82141R (1982).
146. S. K. Howell and Y.-M. Wang, *J. Chromatogr. Biomed. Appl.*, *227*, 174 (1982).
147. J. Lehmann and H. L. Martin, *Clin. Chem.*, *28*, 1784 (1982).
148. W. J. Driskell, J. W. Neese, C. C. Bryant, and M. M. Bashor, *J. Chromatogr. Biomed. Appl.*, *231*, 439 (1982).
149. C. G. Rammell, B. Cunliffe, and A. J. Kieboom, *J. Liquid Chromatogr.*, *6*, 1123 (1983).
150. B. J. Zaspel and A. Saari Csallany, *Anal. Biochem.*, *130*, 146 (1983).

151. E. Leng Heng, D. Klein, and G. Debry, *Int. J. Vitam. Nutr. Res.*, *53*, 280 (1983).
152. G. L. Catignani and J. G. Bieri, *Clin. Chem.*, *29*, 708 (1983).
153. G. T. Vatassery, A. M. Krezowski, and J. H. Eckfeldt, *Amer. J. Clin. Nutr.*, *37*, 1020 (1983).
154. J. Lehmann and H. L. Martin, *Clin. Chem.*, *29*, 1840 (1983).
155. K. Hirai, S. Ishizawa, M. Miura, T. Hara, and K. Owada, *Osaka-shiritsu Daigaku Seikatsukagakubu Kiyo*, *29*, 9 (1981); *CA*, *97*, 90538M (1982).
156. U. Manz and K. Philipp, *Int. J. Vitam. Nutr. Res.*, *51*, 342 (1981).
157. W. Müller-Mulot, *Fette, Seifen, Anstrichm.*, *85*, 66 (1983).
158. C. Gertz and K. Herrmann, *Lebensm. Gericht. Chem.*, *36*, 53 (1982).
159. H. G. Brown and H. E. Snyder, *J. Amer. Oil Chem. Soc.*, *59*, 280 (1982).
160. P. Taylor and P. Barnes, *Chem. Ind.*, *20*, 722 (1981).
161. P. J. Barnes and P. W. Taylor, *Chem. N.Z.*, *45*, 176 (1981).
162. F. Bro-Rasmussen and W. Hjarde, *Acta Chem. Scand.*, *11*, 34, 44 (1957).
163. D. C. Herting and E.-J. E. Drury, *Amer. J. Clin. Nutr.*, *17*, 351 (1965).
164. M. W. Dicks-Bushnell, *J. Chromatogr.*, *27*, 96 (1967).
165. D. R. Erickson and W. L. Dunkley, *Anal. Chem.*, *36*, 1055 (1964).
166. D. J. Nazir and N. G. Magar, *Biochem. J.*, *90*, 268 (1964).
167. R. H. Bunnell, in *The Vitamins*, Vol. 6 (P. György and W. N. Pearson, eds.), Academic Press, New York, 1967, p. 261.
168. D. L. Laidman and G. S. Hall, *Methods Enzymol.*, *18* (part C), 349 (1971).
169. M. P. Grigor'eva and E. N. Stepanova, *Vopr. Pitan.*, *1*, 59 (1979); *CA*, *90*, 136322x (1979).
170. J. G. Bieri, C. J. Pollard, I. Prange, and H. Dam, *Acta Chem. Scand.*, *15*, 783 (1961).
171. K. R. Millar, *N.Z. J. Agr. Res.*, *14*, 142 (1971).
172. D. C. Herting and E.-J. E. Drury, *J. Chromatogr.*, *30*, 502 (1967).
173. H. Schmandke and G. Schmidt, *Int. Z. Vitaminforsch.*, *37*, 195 (1967).
174. J. Green, S. Marcinkiewicz, and P. R. Watt, *J. Sci. Food Agr.*, *6*, 274 (1955).
175. Analytical Methods Committee, *Analyst*, *84*, 356 (1959).
176. T. Aratani and Y. Asakawa, in *Densitometry in Thin Layer Chromatography—Practice and Applications* (J. C. Touchstone and J. Sherma, eds.), Wiley, New York, 1979, p. 695.

177. H. D. Stowe, *Arch. Biochem. Biophys.*, *103*, 42 (1963).
178. M. K. Govind Rao, S. Venkob Rao, and K. T. Achaya, *J. Sci. Food Agr.*, *16*, 1 (1965).
179. J. F. Pennock, F. W. Hemming, and J. D. Kerr, *Biochem. Biophys. Res. Commun*, *17*, 542 (1964).
180. K. J. Whittle and J. F. Pennock, *Analyst*, *92*, 423 (1967).
181. C. K. Chow, H. H. Draper, and A. Saari Csallany, *Anal. Biochem.*, *32*, 81 (1969).
182. K. J. Whittle, P. J. Dunphy, and J. F. Pennock, *Biochem. J.*, *100*, 138 (1966).
183. A. Jensen, *J. Sci. Food Agr.*, *20*, 449 (1969).
184. H. G. Lovelady, *J. Chromatogr.*, *78*, 449 (1973).
185. W. Müller-Mulot, *J. Amer. Oil Chem. Soc.*, *53*, 732 (1976).
186. J. G. Bieri and E. L. Prival, *Proc. Soc. Exp. Biol. Med.*, *120*, 554 (1965).
187. T. Tsugo, K. Yamauchi, and C. Kanno, *J. Agr. Chem. Soc. (Japan)*, *42*, 367 (1968); *CA*, *69*, 58406c (1968).
188. H. E. Schmidt, *Fette, Seifen, Anstrichm.*, *69*, 913 (1967).
189. J. Blattná, J. Manoušková, and J. Davídek, *Z. Lebensm. Unters. Forsch.*, *134*, 242 (1967).
190. P. G. Roughan, *J. Chromatogr.*, *29*, 293 (1967).
191. H. Schmandke, *J. Chromatogr.*, *14*, 123 (1964).
192. G. Biernoth, *Fette, Seifen, Anstrichm.*, *69*. 635 (1967).
193. W. Piorr, L. Tóth, and N. Novakovic, *Z. Lebensm. Unters. Forsch*, *138*, 11 (1968).
194. J. Glavind and G. Holmer, *J. Amer. Oil Chem. Soc.*, *44*, 539 (1967).
195. H. R. Bolliger and A. König, in *Thin-layer Chromatography*, 2nd ed. (E. Stahl, ed.), Springer, New York, 1967, p. 286.
196. W. A. Skinner, R. M. Parkhurst, and P. Alaupovic, *J. Chromatogr.*, *13*, 240 (1964).
197. P. A. Sturm, R. M. Parkhurst, and W. A. Skinner, *Anal. Chem.*, *38*, 1244 (1966).
198. A. Seher, *Microchim. Acta*, *2*, 308 (1961).
199. G. F. M. Ball and P. W. Ratcliff, *J. Food Technol.*, *13*, 433 (1978).
200. M. Malaiyandi, J. P. Barrette, and M. Lanouette, *J. Chromatogr.*, *101*, 155 (1974).
201. H. J. Kayden, C.-K. Chow, and L. K. Bjornson, *J. Lipid Res.*, *14*, 533 (1973).
202. S. D. Stamatov, N. K. Nikolov, A. V. Schischkov, and S. A. Ivanov, *Talanta*, *22*, 686 (1975).
203. T. Aratani, K. Mita, and F. Mizui, *J. Chromatogr.*, *79*, 179 (1973).
204. M. Koleva, M. Joneidi, and O. Budewski, *Pharmazie*, *30*, 168 (1975).

205. J. L. Hess, M. A. Pallansch, K. Harich, and G. E. Bunce, *Anal. Biochem.*, *83*, 401 (1977).
206. S. Dayton, S. Hashimoto, D. Rosenblum, and M. L. Pearce, *J. Lab. Clin. Med.*, *65*, 739 (1965).
207. L. C. Argrett and C. R. Seward, *J. Ass. Offic. Anal. Chem.*, *51*, 1236 (1968).
208. M. Jáky, *Fette, Seifen, Anstrichm.*, *69*, 507 (1967).
209. W. Müller-Mulot, *Z. Lebensm. Unters. Forsch.*, *148*, 201 (1972).
210. P. G. Roughan, *Anal. Biochem.*, *19*, 461 (1967).
211. E. J. DaSilva and A. Jensen, *Biochim. Biophys. Acta*, *239*, 345 (1971).
212. E. J. DaSilva and A. Jensen, *J. Sci. Food Agr.*, *22*, 308 (1971).
213. R. R. Crawford, D. C. Naramore, and O. K. Esmerian, *J. Pharm. Sci.*, *57*, 1716 (1968).
214. A. L. Livingston, J. W. Nelson, and G. O. Kohler, *J. Agr. Food Chem.*, *16*, 492 (1968).
215. R. F. Bayfield, R. H. Falk, and J. D. Barrett, *J. Chromatogr.*, *36*, 54 (1968).
216. R. F. Bayfield and L. F. Romalis, *Anal. Biochem.*, *97*, 264 (1979).

4

VITAMIN K: PHYLLOQUINONE AND MENAQUINONES

MARC F. L. LEFEVERE and ALBERT E. CLAEYS

Laboratorium voor Analytische Chemie, Rijksuniversiteit Gent, Gent, Belgium

ANDRÉ P. DE LEENHEER

Laboratoria voor Medische Biochemie en vor Klinische Analyse, Rijksuniversiteit Gent, Gent, Belgium

1. INTRODUCTION

1.1 Vitamin K: Structure, Occurrence, and Chemical Properties

Vitamin K is the generic name for a series of 2-methyl-1,4-naphthoquinone derivatives with a common physiological activity (prevention of hemorrhagic diseases). The natural and synthetic forms of the vitamin are treated separately in view of their greatly different chromatographic properties.

1.1.1 Natural K Vitamins

In nature, vitamin K is formed by both plants and bacteria. Depending on the source, different homologs and analogs were found [1-8] (Fig. 4.1).

Vitamin K_1, or phylloquinone, is from plant origin and is the principal form of vitamin K in the human diet [3]. We henceforth refer to it as K_1. The natural form possesses a 2',3'-trans configuration, whereas synthetic preparations always contain some cis isomer, too. The latter, however, has little or no physiological activity [9].

Vitamin K_2, on the other hand, was isolated from putrified fish meal [5-7]. It was shown later to consist of a number of menaquinone homologs (MK-n), which are characterized by a polyunsaturated side chain with n isoprene units. The menaquinones MK-6 to MK-13 are products of bacterial biosynthesis [3,8], but, according to some authors, MK-4 can be formed by biotransformation of other K vitamins in animals and birds [9a].

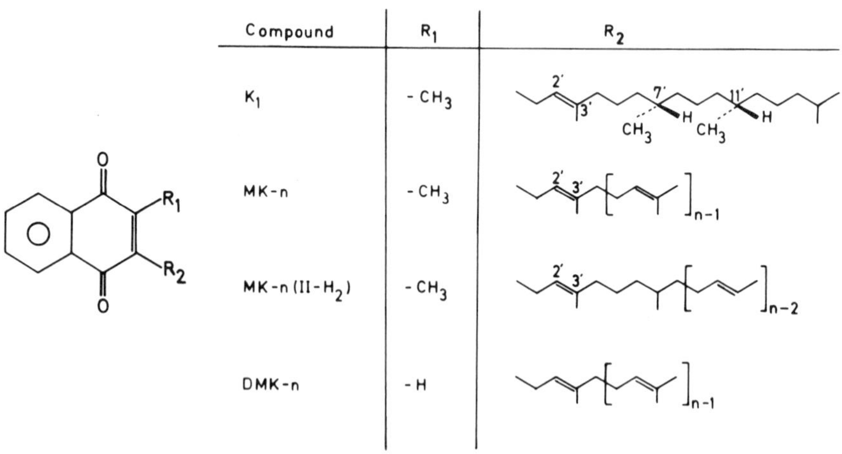

Figure 4.1 Natural K vitamins.

The liver of omnivorous animals, including humans, was shown to contain both phylloquinone (K_1) and some menaquinone homologs. The first clearly is from dietary origin, and the menaquinones are believed to be formed by the intestinal flora [3,4,10].

Furthermore, the existence of menaquinones with a partly saturated side chain has been demonstrated [4,8,10]. These so-called hydromenaquinones, MK-n(H_x), are also of bacterial origin.

Finally, a number of demethylmenaquinones (DMK-n) and a demethyl K_1 analog are also known. They are characterized by the absence of the methyl group in position 2 on the naphthoquinone nucleus.

All these natural compounds have a high molecular weight (from 444 for MK-4 to 1056 for MK-13) and a strong lipophilic nature in common. They are readily soluble in lipophilic solvents, such as acetone, chloroform, petroleum ether, and hexane, but, depending on their molecular weight, sparingly soluble or even totally insoluble in polar organic solvents (alcohol and acetonitrile). They are sensitive to light degradation, alkali hydroxides, and reducing agents.

Both K_1 and MK-n show a characteristic UV absorption spectrum with maxima at 242, 248, 260, 269, and 325 nm (hexane). Their molar absorption at 248 nm is of the order of 20,000.

1.1.2 Synthetic Derivatives

The discovery and characterization of the natural K vitamins initiated a search for synthetic compounds with an equal physiological activity, but enhanced water solubility and/or stability [3]. It was soon discovered that a methyl group in position 3 and an unsubstituted benzene ring were essential for vitamin K activity [3]. The vitamin K activity of unsubstituted 2-methyl-1,4 naphthoquinone was demonstrated by Ansbacher and Fernholtz [11]. The current belief is that this vitamin K_3 or menadione is a provitamin (MK-0), which is metabolized in the liver to MK-4 [3]. From this parent molecule, several other synthetic derivatives have been derived (Fig. 4.2).

Menadione (structure I) itself is even more sensitive to light than K_1 or MK-n. Several esters of the corresponding hydroquinone, such as menadiol diacetate, however, possess vitamin K activity but are insensitive to light. Others, like the diphosphate (Synkavite, Roche), form water-soluble sodium salts useful for parenteral administration. A collective term for these esters is vitamin K_4 (structure II). Other water-soluble derivatives are menadione sodium bisulfite (structure III), Synkamin or vitamin K_5 (structure IV) and 2-methyl-1,4-naphthalenediamine or vitamin K_6 (structure V). Unfortunately, all these synthetic derivatives were shown to be more or less toxic, especially in newborns. Both for this reason and for their slower response, they are no longer used in human medicine. Menadione and menadione bisulfite, however, are still in use as feed supplements.

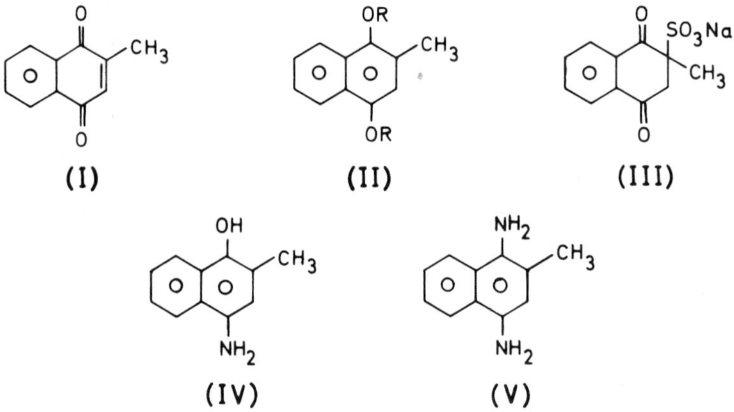

Figure 4.2 Synthetic K derivatives.

1.1.3 *Metabolites*

Phylloquinone-2,3-epoxide (Fig. 4.3) was detected as a metabolite in 1970 [12]. It was shown to accumulate in plasma and liver after the administration of [^{3}H]K_1 to warfarin-treated animals. Later, it was demonstrated that it originates from an enzymatic interconversion between the quinone, hydroquinone, and epoxide form of the vitamin, which is blocked by coumarin anticoagulants (vitamin K cycle) [12a]. Hence, its determination in plasma or tissue homogenates can be used to measure the extent of the enzymatic block in prothrombin synthesis.

The catabolism of [^{3}H]phylloquinone in humans has been reviewed by Shearer et al. [13]. It was shown to involve the formation of more polar metabolites, which are excreted into the urine and via the bile into the feces. Part of the urinary radioactivity was attributed to glucuronide conjugates of acidic aglycons (C 5-7 side chain); the fecal metabolites have a more lipophilic character (esters?) [13,14].

1.2 Biological Significance of Vitamin K

In mammals and birds vitamin K deficiency causes bleeding due to a coagulation defect of the blood. This is the result of an inadequate synthesis of biologically active prothrombin and other vitamin K-dependent clotting factors (factors VII, IX and X). The role of vitamin K in this synthesis has recently been shown to be one of a cofactor in the posttranslational carboxylation of glutamic acid residues to γ-carboxyglutamic acid (GLA). The GLA residues provide the clotting factors with the calcium-binding properties that are essential for the characteristic Ca-phospholipid-clotting factor interaction [15].

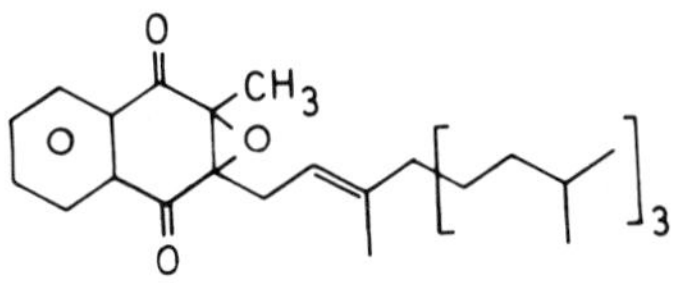

Figure 4.3 Phylloquinone 2,3-oxide (K_1-epoxide).

After this discovery, GLA residues have also been demonstrated in a number of proteins and tissues that are not associated with the hemostatic process [15-24]. This would imply that vitamin K most probably plays an important role outside the hemostatic system, too. There is strong experimental evidence that at least some of these vitamin K-dependent proteins are involved in the regulation of calcium metabolism, and would play an important role both in normal and pathological mineralization [16,25]. This view is supported, for instance, by the teratogenic effect of warfarin and by the beneficial effect of vitamin K administration in osteoporosis patients [16,17].

Finally, it should be mentioned that, according to certain authors, vitamin K depletion would exert an antimetastatic effect [26,27]. In plants and microorganisms, vitamin K homologs play an equally important role in electron transport, oxidative phosphorylation, and photosynthetic processes [28-33].

1.3 Review of Early Methodology

1.3.1 Biological Methods

Until the early 1970s much use was made of the chick bioassay [34, 35]. It was more selective and sensitive than most of the existing physicochemical methods, and allowed a direct comparison of the biological activity of various preparations. From an analytical standpoint, however, the assay has major disadvantages, such as variable results, high labor requirements, and a total lack of specificity [36].

The determination of the blood clotting or prothrombin time, on the other hand, has long remained the only way to evaluate physiological vitamin K status. Before a direct chromatographic vitamin K assay became available, it was the only tool to diagnose vitamin K deficiency [37].

1.3.2 Physicochemical Methods

From the early days of vitamin K discovery, colorimetric and spectrophotometric methods have been used for identification (Dam-Karrer test and Irreverre reaction [38,39]), as well as for quantitative purposes [40,41]. Synthetic vitamin K analogs, which lack a side chain in position 3, form intense colored reaction products with dinitrophenylhydrazine, piperidine, and malononitrile. Spectrophotometric assays based on these relatively specific color reactions have been applied to pharmaceutical [42-44] and feed [45] preparations. Natural vitamin K homologs do not react due to steric hindrance of the side chain. Alternatively, those compounds can be determined spectrophotometrically by taking advantage of their strong ultraviolet (UV) absorption [46, 47].

More recently, procedures based on the formation of fluorescent derivatives have been reported for menadione and water-soluble vitamin K analogs [48], as well as for K_1 [49]. Other physicochemical methods include titration with cerium sulfate [50], polarographic determinations [51-55], microcoulometry [56], and phosphorimetry [56a].

In most instances, however, these methods require an extensive and time-consuming sample cleanup, and lack the specificity required to differentiate various homologs. From the beginning of vitamin K research, chromatography has been used to separate the K vitamins from accompanying compounds and from each other.

Preliminary fractionation of biological extracts was mostly carried out by adsorption chromatography on neutral alumina [57] or silica [58] columns. Other adsorbents used include silicic acid, Decalso, Florisil, Permatite, Darco, and magnesium and zinc carbonate [32,59]. Basic alumina should not be used as it causes isomerization [36]. In some cases, this first separation yielded a vitamin fraction sufficiently pure to allow direct UV spectrophotometric quantitation [60]. Mostly, however, it had to be followed by further fractionation of compounds by partition and/or size exclusion chromatography (SEC) [32]. This often led to very elaborate and time-consuming chromatographic procedures, as exemplified by the determination of vitamin K in liver [4,10,61].

Early partition methods include reversed-phase paper and column chromatography. The first has long been replaced by the faster and more efficient thin-layer chromatography (TLC) on hydrocarbon-impregnated plates. Reversed-phase column chromatography was first described in 1967 by Matschiner and Taggart [62]. In contrast to paper chromatography, columns filled with a mixture of hydrophilic Celite and polyethylene could handle large lipid loads (13 mg to 6.5 g), which made them useful for preparative as well as for analytical work [58,62]. Monitoring the large columns for many dilute fractions of K isoprenologs, however, was a very tedious process.

Applications of SEC include the separation of menaquinone homologs [63], the analysis of mixtures of vitamins A, E, and K [64], the separation of naphthoquinones and lipophilic vitamins [65], and the demonstration of alkyl naphthoquinones in cigarette smoke condensate [66]. Attempts were made to calculate the molecular weights of the separated compounds from their elution volumes [65]. It appears, however, that the mechanism involved in the separation is a combination of partition chromatography and size exclusion processes [32]. Besides identification purposes, gel chromatography has also been reported to be suitable for the removal of such contaminants as high-molecular-weight di- and triglycerides [65]. Sephadex LH-20 columns have been used for the preliminary cleanup of food samples prior to liquid chromatographic (LC) analysis [67].

At present, column chromatographic methods have largely been replaced by TLC and modern LC (high-performance liquid chromatography, HPLC) methods. They do, however, remain of some importance for preparative scale separations, as well as for sample cleanup prior to quantitation with TLC [57,68], gas chromatography (GC) [69,70], or LC [67,71].

2. NATURAL VITAMIN K HOMOLOGS AND THEIR METABOLITES

2.1 Thin-Layer Chromatography

Some years ago, Dunphy and Brodie [32], Sommer and Kofler [72], and Bolliger and Konig [73] wrote excellent reviews on the thin-layer chromatography of vitamin K. Since then there has been considerable improvement in quantitation techniques, but the chromatographic procedures have essentially remained the same. They can be divided into three main types: adsorption chromatography, argentation chromatography (Ag^+ TLC), and partition chromatography, the latter being practiced most in the reversed-phase mode (RP-TLC).

2.1.1 Adsorption TLC

Thin-layer chromatography with silicic acid as an adsorbent has proved to be extremely useful for the purification and identification of various vitamin K-related naphthoquinones [32]. Separation is based mainly on polarity differences, which make the procedure the method of choice for the isolation of naphthoquinones from other lipid classes. Adsorption TLC was used, for instance, as a purification step in the preparative isolation of K_1 and MK-n from liver extracts [4,10] and cell cultures [74]. The required degree of selectivity can be adjusted by an appropriate choice of solvent systems, as is discussed more extensively in Sec. 2.3.2. Some interesting examples of this solvent selectivity can be found in Table 4.1. Quinones will either be separated as a single band from polar and nonpolar lipids, such as phospholipids and carotenoids [75], or will be resolved into naphthoquinones (K_1 and MK-n) and benzoquinones (ubiquinones) or even further into compounds with a fully saturated (K_1) and a nonsaturated (MK-n) side chain. Adsorption TLC, however, will not, or only very poorly, distinguish between isoprenologs (MK-n). On the other hand, it easily

TABLE 4.1 Thin-Layer Chromatographic Properties of Various Naphthoquinones on Silica Gel G ($R_f \times 100$)

Compound	Diisopropyl ether-petroleum ether (1:9)	Hexane-benzene (1:1)	Butyl ether-hexane (1:9)	Methyl ethyl ketone-hexane[a] (1:6)	Dimethyl formamide-water[b] (49:1)	Acetone-water[b] (19:1)
cis-Vitamin K_1	35	36	51	65	63	53
trans-Vitamin K_1	32	32	45	64	63	53
cis-MK-9(II-H_2)	37	37	51	39	25	25
trans-MK-9(II-H_2)	35	35	43	38	26	25
MK-9	33	33	38	9	35	30
MK-10	—	—	—	7	23	25

[a]Silver nitrate-impregnated silica gel G.
[b]Plates impregnated with paraffin prior to chromatography.
Source: Ref. 32.

resolves geometric isomers, such as the 2',3'-phytyl cis and trans isomers of K_1 (Table 4.1). In this type of separation, which cannot be achieved either by Ag^+ TLC or by RP-TLC, solvent selectivity plays an important role. According to Dunphy and Brodie, the best results were obtained with di-*n*-butyl ether-modified systems [32].

Adsorption TLC is also useful for the separation of the aglycons of urinary metabolites [76,77] and for the resolution of photochemical degradation products [98,99], some of which, however, were further degraded during the TLC procedure [30,78]. The important epoxide metabolites, on the contrary, were better separated from their parent compounds (K_1 and MK-n) by reversed-phase chromatography [79,80] or Ag^+ TLC [80] than by adsorption TLC.

2.1.2 Argentation TLC

Argentation chromatography is based on the property of olefins to form complexes with silver ions. In Ag^+ TLC, $AgNO_3$ is incorporated into the silica gel layer in concentrations from 0.5 to 10% w/w [32]. As a result, lipids containing unconjugated double bonds in their side chain display higher retention than their saturated counterparts.

Ag^+ TLC is complementary both to adsorption and reversed phase TLC. As can be seen in Table 4.1, the resolution between isomeric forms is almost completely lost, but in return, a very good separation between saturated (K_1), partly saturated [MK-n(H_x)], and fully unsaturated homologs (MK-n) became possible. Some of these can be separated by RP-TLC as well, but Ag^+ TLC displays a greater selectivity toward the partly saturated hydromenaquinones. The difference in R_f value for each saturated double bond, e.g., between MK-9 and MK-9(H_2), is much larger in Ag^+ TLC than in RP-TLC (approximately equivalent to a difference of two isoprene units in the fully unsaturated series of MK) [32].

Not only the R_f increment for each saturated double bond, but also its direction, is different. Hydromenaquinones migrate faster than their corresponding MK-n homolog on the Ag^+ TLC plate, but they are more retained by the reversed-phase system (Tables 4.2 and 4.3). In Ag^+ TLC the accent is on the degree of unsaturation, whereas R_m values, log (1/Rf - 1), in RP-TLC can be directly correlated with the length of the side chain [75]. In consequence, argentation in conjunction with reversed-phase TLC is a simple and useful tool to determine both the length and degree of unsaturation of menaquinone side chains [32].

From a practical standpoint, Ag^+-TLC is advantageous over RP-TLC since the loading capacity of the Ag^+ plates is much greater and nondestructive visualization is easy [32]. Silver ions have no destructive effect on naphthoquinones, which can be collected for further identification and quantitation. As a matter of fact, the collection of individual samples from Ag^+ plates was reported to be easier than from

TABLE 4.2 R_f Values (× 100) for a Number of Naphthoquinones on Silica Gel G Impregnated with $AgNO_3$

Compound	15% MEK-hexane	5% Methanol-benzene
cis/trans-K_1	63	80
Demethyl-K_1	60	79
MK-2	54	75
MK-3	48	70
MK-4	43	65
MK-5	36	59
MK-6	30	53
MK-7	25	46
Chlorobiumquinone	16	46
MK-8	21	38
Demethyl-MK-9	12	30
MK-9	15	33
MK-9(II-H_2)	26	46
MK-10	11	27
MK-9(H_8)	62	75
MK-9(H_6)	54	69
MK-9(H_4)	37	55
MK-9(H_2)	26	45

Source: From Ref. 32.

TABLE 4.3 R_f Value (× 100) of Naphthoquinones on Reversed-Phase Thin-Layer Chromatography[a]

Compound	93% A/W[b]	97% DMF-water[c]
cis/trans-K_1	61	56
Demethyl-K_1	65	61
MK-2	91	87
MK-3	85	83
MK-4	78	77
MK-5	69	66
MK-6	58	54
MK-7	44	42
MK-8	33	28
Demethyl-MK-9	30	23
MK-9	21	16
MK-9(II-H_2)	17	11
MK-10	16	10
MK-9(H_8)	6	2
MK-9(H_6)	8	3
MK-9(H_4)	14	7
MK-9(H_2)	17	11

[a]Plates were impregnated by dipping them in a 5% paraffin-petroleum ether solution prior to chromatography, and the solvent systems employed were saturated with liquid paraffin.
[b]Acetone-water.
[c]Dimethylformamide-water.
Source: From Ref. 32.

(hydrocarbon-impregnated) RP-TLC plates [32]. However, high-molecular-weight menaquinones might show irreversible adsorption on Ag^+ TLC [81].

2.1.3 Reversed-Phase TLC

Most reversed-phase separations have been performed on thin-layer plates impregnated with nonpolar, nonvolatile hydrocarbons, such as paraffin, silicones, or undecane. These plates (silica gel G, kiezelguhr, cellulose, or starch) can be prepared, either by dipping them into a solution of the hydrocarbon in a volatile solvent, followed by evaporation of this solvent [32], or alternatively by ascending chromatography [75]. The type and purity of the stationary phase are critical, both in the separation process itself, and in the subsequent recovery of purified fractions. Hammond and White [75] recommend the use of highly purified hexadecane. Recently, thin-layer plates with chemically bonded phases (C_2, C_8, and C_{18}) have become available. They are fairly expensive, but in return offer good sample capacity, efficiency, and reproducibility. Their use also circumvents detection problems associated with the potential contaminating effects of physically coated hydrocarbon stationary phases [32,75]. They may be particularly advantageous for quantitative work with densitometric scanning. Collins et al. [82] described the use of such RP-18 TLC plates for the separation of natural mixtures of bacterial menaquinones.

Retention data for K vitamins on various RP-TLC systems were given by Dunphy and Brodie (Table 4.3) and Hammond and White [75]. The resolution between MK-n homologs is about the same as in Ag^+ TLC. Unlike Ag^+ TLC, however, RP-TLC, which is essentially governed by hydrophobic interactions, is capable of separating demethylmenaquinones from their methylated counterparts (MK-n). There may also be overlapping pairs in Ag^+ TLC, such as MK-7 and MK-9(H_2) that can be separated by RP-TLC, and vice versa. Together, Ag^+ TLC and RP-TLC will fully resolve all vitamin K isoprenologs with the exception of geometric isomers.

2.1.4 Other Systems

Although less common, straight-phase partition systems have also been applied successfully to the separation of vitamin K and other fat-soluble vitamins (FSV). Perisic-Janjic [82a] described the use of talc thin-layer plates. Two different solvent systems were used, one, consisting of water-dioxane-acetone-formaldehyde (85:20:15:25), to resolve the more polar synthetic vitamin K analogs K_3, K_4, and K_5, and the other, acetone-concentrated acetic acid (30:20), to differentiate K_1 and other FSV. Other stationary phases described include polyethylene glycol 200 [65,83] and β,β'-oxydipropionitrile [83], impregnated on a silica or alumina support. The use of polyamide as a stationary phase was first reported by Campbell and Bentley [84]. The versatility of this material was demonstrated by Chiang and Jan [85], who separated water and fat-soluble vitamins together in one run on a mixed polyamide-silica gel layer.

2.1.5 Detection and Quantification

Quinones of vitamin K type can be visualized in a nondestructive way by fluorescence quenching of dichlorofluorescein or rhodamine B (G). All UV-absorbing lipids will be detected, and the method is not very sensitive (detection limit of 9.5 $\mu g/cm^2$) [32]. The method, however, is suited to localize the zones of interest after a preparative fractionation step.

A more specific and sensitive detection method for quinones on paper and thin-layer chromatograms was described by Lester and Ramasarma [86]. It is based on the reduction of quinones with sodium borohydride and visualization of the resulting quinols with Emmerie-Engel reagent [87]. Plates are sprayed first with a solution of $NaBH_4$ in absolute ethanol (up to 1%), followed by 1% ethanolic hydrochloric acid (to decompose the excess borohydride), and finally the Emmerie-Engel reagent ($FeCl_3$-α,α-dipyridyl). The method is specific for quinones and quinols, which form pink spots (detection limit around 1 $\mu g/cm^2$). It is effective on adsorption and RP-TLC plates, but not useful for Ag^+-impregnated plates [32]. The original Emmerie-Engel reagent can be replaced by neotetrazolium (0.2% aqueous solution) [75].

A number of other spray reagents used to visualize K vitamins are summarized in Table 4.4. Most of them are common lipid-staining reagents and not specific for quinones. An exception is the leukomethylene blue method, which allows one to differentiate vitamin K and other lipoquinones [36]. Leukomethylene blue fails to react with K_1 and MK-n, but other lipoquinones, e.g., ubiquinones, become visible as brilliant blue spots [91]. As can be seen in Table 4.4, these color reactions have lower detection limits than UV quenching, but they do destroy the quinones [32]. When the purified fractions are to be recovered, part of the TLC plate should be covered while the zones of interest are localized [68]. Specific color reactions for synthetic vitamin K analogs are discussed in Sec. 3.1.

In modern quantitative TLC, densitometric scanning has completely superseded the elution of separated bands followed by UV spectrophotometry. The method is faster, less tedious, more sensitive, and more precise. Even with earlier equipment, such as the Zeiss PMQ II scanner, detection limits of as low as 0.01 μg were obtained [36]. A typical example of such a densitometric scan is shown in Fig. 4.4. For this type of work, the use of standardized, precoated plates is recommended. Variation coefficients for replicate determinations of K_1 in infant formulas and elemental diets were of the order of 4-14% [36,57, 68]. With modern scanning equipment, high performance TLC plates

TABLE 4.4 Spray Reagents for Staining Quinones and Lipids

Reagent	Development	Specificity	Sensitivity (MDQ)	Color produced	Remarks	Reference
$NaBH_4$/HCl/Emmerie-Engel	Cold	Quinones Quinols	1 μg/cm^2	Reddish-pink	Not for Ag^+-impregnated plates	32,86
$NaBH_4$/(HCl)/ Neotetrazolium	Cold	Quinones Quinols	4 μg	Red	Not for Ag^+-impregnated plates	75,88
Phosphomolybdic acid, 10% in ethanol	100°C for 2-4 min	All lipids		Blue		32,59
Iodine vapor	Cold	Unsaturated lipids		Yellow, brown		32,74,76
Anisaldehyde and sulfuric acid	100°C for 2-3 min	Most lipids		Gray-green		32,89
Concentrated sulfuric acid	110°C for 30-45 min	?				59
Perchloric acid (70%)	105°C for 5-10 min	?	1-2 μg	Violet		65,90
$SbCl_5$ in chloroform	Cold	K_1 and other fat-soluble vitamins		Yellow; blue; orange-brown	No reaction with K_3, K_4, and K_5; color specific for A, D, E, or K	82a
Leukomethylene blue	Cold	Ubiquinones + other lipoquinones		Blue	No reaction with K_1 or MK-n	36,91

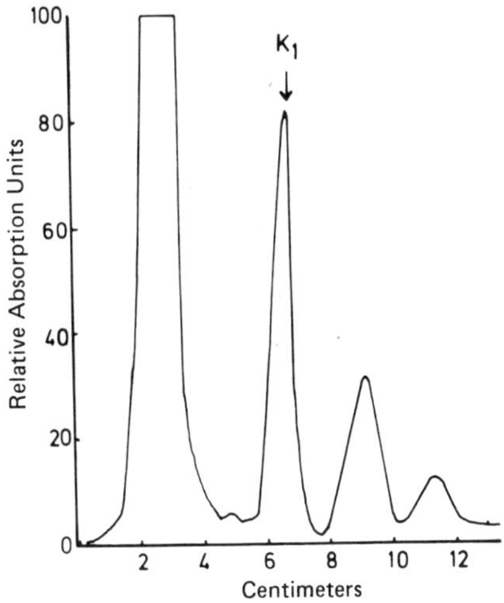

Figure 4.4 Reflectance densitometric scan (270 nm) of elemental diet. (Reproduced from Ref. 68 with permission of the copyright owner.)

and the use of an appropriate internal standard, sensitivity and precision can be further improved [92].

Other detection techniques include autoradiography [75], radiochromatoscanning [93], and liquid scintillation counting [75-77].

2.1.6 Applications

Table 4.5 summarizes some typical applications of TLC for the analysis of natural vitamin K homologs and their metabolites. They can be divided into three categories: quantitative methods, isolation and identification procedures, and process monitoring.

2.1.6.1 Quantitative Methods. Analytical TLC has the advantage over GC and LC that several samples can be processed simultaneously. With modern HPTLC, this number may be as high as 24 (linear development) to 40 (circular instrumental HPTLC) [100]. This high sample capacity makes the procedure particularly well suited to processing large numbers of samples in a relatively short time (process and food control and drug screening). The use of data pairing may contribute to an increased assay precision [101,102]. The choice between HPTLC or conventional TLC mainly depends on the number of samples to be analyzed, the sensitivity required, and the available instrumentation.

Rittich et al. [94] determined vitamin K_1, K_3, and K_4 in concentrates and pharmaceutical preparations with adsorption TLC on precoated alufoil plates. For a complete resolution of the vitamins, elution

TABLE 4.5 Applications of TLC to Vitamin K Analysis[a]

Application	Sample pretreatment	Plate material	Eluent	R_f; compounds	Detection	Comments	Reference
Quantitation methods							
K_1 in infant formulas	Deproteinization (EtOH); extraction (PE/Et_2O); alumina column chromatography	Silica (Adsorbosil 5 Prekote)	CCl_4; benzene		Densitometry (275 nm); MDQ = 10 ng	Precision, 4-14%; range, 50-400 ng	36,57
K_1 in elemental diets	Extraction (PE); silica column chromatography	Silica (activated)	Toluene	0.40 (K_1)	Densitometry (270 nm)	Precision, 7%; range, 50-1000 ng	68
K_1, K_3, and K_4 in vitamin concentrates and pharmaceutical preparations		Silica	Benzene B-MeCOEt	K_1, K_3, K_4 oxidation products	Elution spectrophotometry (270 nm)	Separation from oxidation products	94
K_1, α-TC, and β-carotene in spinach	Extraction	Silica (HPTLC)	Benzene; PE-B (6:1)	K_1, α-TC; β-carotene	Densitometry	Precision, 0.5-2.2%	92
A, D_2, E, and K in fish, crustaceans	Reflux KOH; extraction (Et_2O)	Silica	Hex-Et_2O (70:30)		Spray 10% phosphomolybdate		95
Preparative isolation and identification							
MK-n from actinomycetes and corynebacteria	Extraction ($CHCl_3$-MeOH)	Silica	PE-Et_2O (85:15)	0.7 (MK-n)	UV spectrophotometry; mass spectrometry	MK-n elute as 1 band, separated from UQ	8
MK-n from bacteria	Extraction ($CHCl_3$-MeOH)	Silica; hexadecane-impregnated silica	$CHCl_3$-*i*-Oct (2:1); A-H_2O (95:5)	0.6 (MK-n)	UV quenching; $NaBH_4$-neotetrazolium; UV spectrophotometry	Preparative isolation MK-n fraction; resolution isoprenologs (MK-n, DMK-n)	75
MK-n from bacteria	Extraction (Et_2O-EtOH); silica column chromatography	Silica	Hex-Et_2O-HAc (70:5:1)		I_2 vapor	MK-n separated as a whole from other lipids	74

Vitamin K from liver (human, bovine, chick, horse)	Extraction (EtOH-A); partition (A-Hex-H_2O); column chromatography, silica; column chromatography, RP	Silica;	Hex-Et_2O (98:2) or Hept-B (2:8);		UV quenching;	Preparative isolation K fraction; separation and identification of K_1 and MK-n	4,10,61, 96,97
		silica-$AgNO_3$;	benzene;	0.94 (K1); 0.33 (MK-4); 0.12 (MK-10)	H_2SO_4 spray;		
		silica-paraffin	A-H_2O (96:4)	0.85 (MK-4); 0.71 (K_1); 0.32 (MK-10)	UV spectrophotometry; mass spectrometry		
Vitamin K from Liver	Extraction ($CHCl_3$-MeOH); Precipitation phospholipids (cold acetone); column chromatography	Silica	PE-Et_2O (95:5)	0.66 (K_1); 0.53 (MK-4);		Separation between	88
			or benzene	0.69 (K_1); 0.63 (MK-4)			
K_1 epoxide		Silica-paraffin	A-H_2O (92:8)				79
Separation K_1; lipids (serum-fecal)	Extraction ($CHCl_3$-MeOH)	Silica activated; silica-$AgNO_3$ 1%	PE-DIPE (95:5); Hept-B (2:8)	0.72 (K_1); 0.25 (MK-4); 0.09 (MK-10)	I_2 vapor; liquid scintillation counting	Separation K_1, MK-n	76
Separation metabolites (serum-bile-urine-feces)	Extraction ($CHCl_3$-MeOH)	Silica (activated)	cHex-Et_2O-EtCH (16:4:3)	0.65 (K_1); 0.35 (γ-lactone)	Liquid scintillation counting	cHex = cyclohexane	77
K_1, MK-n, and metabolites in liver, heart muscle	Cell fractionation: extraction (A-$CHCl_3$)	Silica; silica-$AgNO_3$; silica-paraffin	B-$CHCl_3$ (1:1): Hept-B (4:1); A-H_2O (95:5)		Radiochromatography scanning		93
Process monitoring							
Photochemical degradation	UV irradiation	Silica (activated)	$CHCl_3$	0.63 (K1)	Visual inspection (color; fluorescence)	Separation of K_1, dihydro-K_1, phthiocol, other degradation products	98
Photochemical degradation	UV irradiation	Silica (activated)	$CHCl_3$-B (1:1)	0.59 (K_1)	Color/UV quenching; H_2SO_4 spray; fluorescence scan		99

[a]Abbreviations: UQ = Ubiquinone; α-TC = α-tocopherol; PE = petroleum ether; Hex = hexane; Hept = heptane; iOct = iso-octane; Et_2O = diethyl ether; DIPE = diisopropyl ether; EtOH = ethanol; MeOH = methanol; A = acetone; B = benzene; MeCOEt = methyl ethyl ketone; HAc - acetic acid.

with benzene was recommended; peroxides and other oxidation products could be separated with a mixture of benzene and methyl ethyl ketone. K_1 was determined spectrophotometrically at 270 nm after elution with isopropanol.

Perisic-Janjic [82a] and Chiang and Jan [85] used partition rather than adsorption chromatography to separate K vitamins in vitamin premixes. The former workers employed paraffin-impregnated cellulose and starch, as well as unimpregnated talc, to resolve K_1 and MK-n from a number of other fat-soluble vitamins. The eluent was acetone-concentrated acetic acid (30:20). Spraying with $SbCl_5$ permitted further differentiation of the vitamins, on the basis of the specific colors formed. Chiang and Jan succeeded in the separation of both fat- and water-soluble vitamins together on mixed polyamide-silica gel layers using a mixture of ethanol-diethyl ether-cyclohexane-actic acid (23.5:6:21:2) as an eluent. No details were given on the quantitation procedures or the precision obtained in the above assays.

The analysis of K_1 in infant formulas [57] and elemental diets [68] was generally based on the same analytical scheme: total lipid extraction with petroleum ether or petroleum ether-diethyl ether (1:1), removal of polar and neutral lipids by column chromatography on alumina [57] or silica columns [68], analytical adsorption TLC to separate K_1 from remaining food excipients, and spectrophotodensitometric scanning. An example of such a densitometric scan is shown in Fig. 4.4. Horn [68] analyzed each sample and standard in duplicate on alternate tracks (data pairing) and obtained an average relative standard deviation (RSD) of 7.0% for samples containing 0.09-0.15 μg K_1 per gram. All operations were conducted in subdued light, and citric acid was used to prevent oxidation of K_1 during the evaporation step. We personally would also strongly recommend the use of an internal standard to compensate for losses during the extraction and sample workup procedures.

Other applications in food chemistry include the reflectance spectrophotometric determination of α-tocopherol, K_1, and β-carotene in spinach [92] and the determination of vitamins A, D_2, E, and K in fish and crustaceans [95].

2.1.6.2 Isolation and Identification Procedures. During the last years, LC methods have superseded TLC procedures for trace analysis of K homologs in biological matrices. The resolving power of LC is superior, the system can be fully automated (e.g., on-line sample clean-up), substances are no longer exposed to air and light during their separation, and the method provides excellent sensitivity and selectivity.

Nevertheless, TLC may be advantageous for screening purposes. It allows a parallel examination of the presence of K_1, menaquinones, partly saturated menaquinones, demethyl analogs, and their geometric isomers by complementary techniques, such as adsorption, argentation, and reversed-phase chromatography, without a need for supplementary equipment. When sufficient material is present, e.g., in bacterial extracts, the identity of the separated bands can be confirmed by spraying with specific color reagents (Emmerie-Engel or leukomethylene blue), densitometric scanning, and direct-inlet mass spectrometry [4,10,36,82]. The original and tedious cleanup procedures [4,10,61] might be replaced by a single preparative LC step with on-line sample cleanup (Sec. 2.3.6). Care should be taken, however, to prevent isomerization and degradation of these naphthoquinones, especially when adsorbed onto an active surface [104,105]. We therefore would recommend all operations be performed in subdued light and an inert atmosphere (N_2 stream).

Unlike HPLC, TLC also allows a fast screening for compounds of unknown polarity. This is illustrated, for instance, by the characterization of urinary, fecal, and biliary metabolites of K_1 [13,76,77]. Once the component(s) of interest has (have) been identified or characterized, quantitative LC procedures can be worked out. In our opinion TLC and LC thus should be looked upon as complementary.

2.1.6.3 Process Monitoring. TLC forms a fast and easy way to monitor the synthesis of vitamin K homologs and analogs [106] and the formation of (photochemical) degradation products [98,99], some of which, however, may be further degraded during the TLC procedure [30,78].

2.2 Gas-Liquid Chromatography

2.2.1 General Considerations

Since Nair and Turner [107] first indicated the possibility of using gas-liquid chromatography (GLC) for the analysis of vitamin K, several papers have reported the gas chromatographic analysis of both natural (K_1 and MK-n) and synthetic (K_3, K_4, and K_5) vitamin K homologs and analogs on a variety of stationary phases. An excellent review in this area has been given by Sheppard et al. [108]. GLC, however, never became as popular for the analysis of natural K vitamins, as it did for many other compounds of biomedical interest. Most papers deal with the separation and quantitation of standard mixtures of K_1 and MK-n. By 1971, only three attempts to measure K_1 and/or MK in a biological matrix had been reported [107-109], and this situation has not since changed much. Two of the most obvious reasons for the relative unpopularity of GLC are the high retention times and possible on-column degradation of vitamin K-related compounds (see Sec. 2.2.2). However, gas chromatography does display some particularly interesting features for vitamin K analysis as well. It can be used in conjunction with specific and highly sensitive detectors, such as electron capture (GC-ECD) [110], mass spectrometry (GC-MS), and selected ion monitoring (GC-SIM).

2.2.2 Specific Analytical Problems

Natural K vitamins have a very low volatility. In most GC systems they elute well after the other fat-soluble vitamins, or do not elute at all (higher menaquinone homologs). Retention times of the order of 50 min for K_1 and over 2 hr for MK-7, are not uncommon (Table 4.6). This makes the procedure rather slow and has an adverse effect on detectability.

Moreover, on-column degradation was reported by several authors [111-114]. It seems quite inevitable on columns with a low liquid phase loading. Vetter et al., however, observed a similar phenomenon on columns with a 5-20% stationary phase load [114]. They attributed this to a thermal conversion of K vitamins into their respective chromenol and chromanol isomers (Fig. 4.5). A similar observation was made for the structurally related ubiquinones [114,119]. It was concluded that direct GC of the natural K vitamins was impossible. Therefore, the formation of dihydro-trimethylsilyl (TMS) ether derivatives was recommended (Sec. 2.2.4). Other authors, however, strongly opposed this method [104,109]. They point out that the observed "thermodegradation" products may have been present in the sample itself [109]. There is indeed a close resemblance between the direct gas chromatograms shown by Vetter et al. [114] and those of an irradiated K_1 solution [116,117]. The possibility of thermodegradation, however, should always be taken into account.

2.2.3 GC on Packed Columns Without Derivatization

Most GC determinations have been performed on packed columns without derivatization (Table 4.6). Nair and Turner used a binary stationary phase (SE 52-XE 60) to separate α-, β-, γ-, and δ-tocopherol, α-tocopherylquinone, α-tocopherylhydroquinone, and vitamin K_1 [107]. Ninomiya and coworkers [111], Caroll and Herting [113], and Libby et al. [112] used both polar (Craig's polyester succinate, FS-1265, XE 60, and QF-1) and apolar (SE 30) stationary phases. Later, the use of a 3% OV-17 column was reported for the determination of K_1, K_1 epoxide, and MK-4 in plasma samples [110].

All other applications were made on apolar phases (Apiezon L, SE-30, UCW-982, Dexsil 300, and OV-1). Libby et al. selected SE 30 for the majority of their research, primarily because K_1 could be chromatographed without the interference of tocopherols [112]. In our own experience, apolar columns usually yielded the best peak symmetry, and they afford a good separation between K_1, K_1-epoxide, and lower molecular weight menaquinones [109,115]. Compounds with a saturated side chain, e.g., K_1, display considerably lower retention times compared with their unsaturated counterparts (MK-4) [109,115]. For menaquinones (MK-n), a linear relationship between the number of isoprene units (n) and the log of the retention time (log T_R) was observed.

TABLE 4.6 GC Experimental Conditions

Stationary phase	Column temperature (°C)	Retention time	Application	Comments	References
Packed columns[a]					
2.5% SE 52 + 2.5% XE 60 on Gas Chrom PAW-DMCS	228	K_1, 53.2 min	Separation tocopherol isomers, tocopheryl quinone, hydroquinone and K_1	DMCS-treated support reduces bleeding	107
1% SE 30	200-220		Standards	Signs of breakdown (leading)	111
1% SE 30 on Gas Chrom P, silanized	250	K_1, 10 min	Standards	Characteristic shoulder on K_1 peak; multiple peaks after cooling column; no satisfactory results MK-n	112
3% QF-1; 3% SE 30 on Gas Chrom P	195 205	K_1, 27.8 min; K_1, 41.7 min, 50.3 min	K_1 standard; K_1 standard	Single peak; 2 peaks (ratio 4:1)	113
5% Apiezon L; 10% UCW-98; 3% UCW-98 on Gas Chrom Q	275 250 280	 K_1, 30 min MK-9, 53 min	K_1, MK-7, and MK-9 as their TMS derivatives	Resolution of cis and trans isomers; thermodegradation without derivatization	114
10% SE 30 on Celite 545 (1.2 m)	225	K_1, 90 min	Homologous series of K_1, MK-n (1-4) and UQ		115
3.8% UCW 982 on Diatoport S (1.2 m)	260 295	K_1, 11 min; MK-4, 18 min MK-7, 140 min	Quantitation of K_1, MK-4, and α-tocopherol without derivatization (liver)	No thermodegradation observed	109; 122

TABLE 4.6 (*Continued*)

Stationary phase	Column temperature (°C)	Retention time	Application	Comments	References
3% OV-1 on Chromosorb W AW-DMCS	240; 270	29.7 min; 9.0 min	Determination K_1 and photodegradation products	Small peak just before K_1 peak, slight tailing (IS n-triacontane)	116,117
2.5% Dexsil 300 on Chromosorb G AW-DMCS	290	26 min	K_1 in vegetables	No sign of thermal breakdown of K_1 or of column deterioration; IS: Dotriacontane	104
3% OV-17 on Anachrom Q	302	K-ep 4 min; K_1 4.5 min	Quantitative determination K_1 and K_1-2,3-epoxide plasma	No degradation observed; IS: MK-4	110
Capillary columns					
Silar 10C (100 m) (cyanoethylsilicone); CP-TM-WAX (20 m) (PEG)	195; 250	*cis*-K_1, 140 min; *trans*-K_1, 150 min	Separation diastereomers of K_1 in the form of their dihydrodimethylethers; separation cis trans isomers	Dimethylethers more stable than TMS derivatives; Grob injector	118

[a]All columns were 1.8-2 m in length except when stated otherwise.

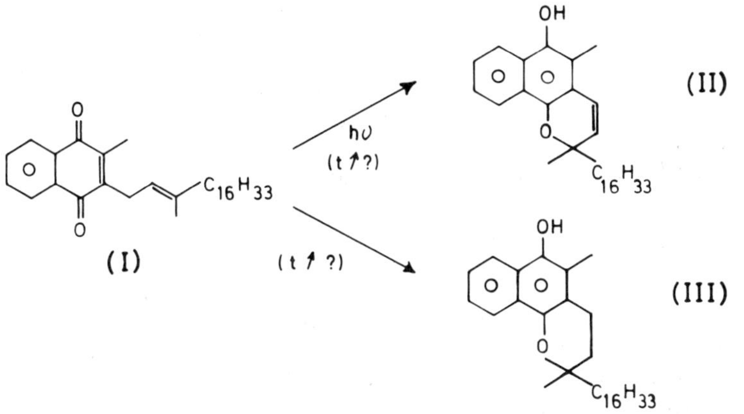

Figure 4.5 Conversion of K_1 (structure I) into its chromenol (structure II) and chromanol (structure III) isomers.

Likewise, log T_R was found to decrease linearly with increasing column temperature [115]. Chromatography on OV-1 separates K_1 from its photodegradation products [116,117], and a chromatographic separation of the free quinone, hydroquinone, chromenol, and chromanol forms has been achieved on UCW-982 and on Apiezon L. Positional isomers, e.g., *cis*- and *trans*-K_1, on the other hand, were not resolved unless they were first derivatized to their dihydro-TMS ethers [114].

To avoid very long retention times, some authors used columns with very low liquid phase loadings (1-2%) and/or high column temperatures (295°C) (Table 4.6). However, the use of low liquid phase loadings, together with bleeding at high column temperatures, holds an increased risk for column deterioration and artefacts. Libby et al. observed that once these columns had been allowed to cool after use or were stored, they became unreliable for further K_1 analysis. Reproducibility was lost, and multiple peaks were observed [112]. Even with new columns, signs of on-column degradation (leading and tailing) were observed for small amounts of K_1, and chromatography of mena quinones proved totally unsatisfactory [111-113]. In our opinion, this must be attributed to the formation of active sites on the support, which have a catalytic effect in the degradation process.

Dialameh and Olson [109], on the other hand, did not encounter any problems of thermal breakdown or multiple peak formation when analyzing fresh samples on a 3.8% (SGR) UCW-982 column at 260°C. Linear calibration curves were obtained for underivatized K_1, MK-4, α-tocopherol, and phytylubiquinone. K_1 and MK-4 were recovered unchanged after chromatography on this column, and no traces of chromenol or chromanol forms were observed. Other authors did not report on-column degradation either and recommended direct GC as the method of choice for K_1 and MK-n analysis [104,110,115-117]. We would advise keeping the stationary phase load above 3%. The resulting need for higher column temperatures (260-300°C) is largely outweighed by the reduced chance for adsorptive degradation.

Special attention should also be paid to the supports used. Nair and Turner demonstrated the advantages of a dimethyldichlorosilane (DMCS) treatment [107]. This procedure not only eliminates remaining active sites, but also forms methoxy groups on the silanized support, which are believed to take part in a chemical bonding of the stationary phase. Substitution of DMCS by trimethylchlorosilane (TMCS) resulted in excessive bleeding and a concomitant loss in performance [107]. High-quality supports, such as Gas Chrom Q and Chromosorb W AW-DMCS, were used for all the more recent work on vitamin K. They are DMCS treated by the manufacturer, but should be treated very carefully to avoid active site formation during column packing and other manipulations. One author used chromosorb G AW-DMCS, a support especially recommended for its mechanical strength and chemical inertness [120].

Finally, the injection temperature has to be kept as low as possible [121] and an all-glass system should be used to avoid adsorptive losses in metal transfer lines (Lefevere et al., 1978, unpublished results).

2.2.4 *Derivatization Procedures*

According to Vetter et al., the dihydro-TMS derivatives of K_1 and MK are more thermostable and improve the assay's precision, while also affording the separation of the K_1 geometrical isomers and MK up to MK-9. They are usually prepared by catalytic reduction with hydrogen gas and subsequent reaction of the hydroquinone with *N*-trimethylsilylacetamide [114]. As an alternative, reduction with $NaHSO_3$ [109,114] has been recommended for menaquinones to avoid hydrogenation of the unsaturated side chain. The major criticism about these derivatives is that they are very unstable. In spite of the encouraging results reported by Vetter et al. [114], Dialameh and Olson always found a peak of unreacted phylloquinone (K_1), even after TLC purification of the TMS-K_1 derivative [109]. These conflicting observations perhaps can be explained by considering the amount of starting material used. Whereas Dialameh and Olson [109] aimed at a microtechnique, suitable for the determination of K_1 and/or MK-n in submicrogram amounts, Vetter et al. [114] used 100 to 250 mg pure vitamin K as starting material.

More recently, Vecchi et al. recommended the formation of dihydrodimethyl ethers [118]. They are more stable than the TMS derivatives, and were found useful in the separation of the diastereomers of *cis*- and *trans*-K_1. Reduction with sodium dithionite and a reaction of the resulting hydroquinone with dimethylsulfate and methanolic sodium hydroxide was shown to yield 90% of the desired dihydrodimethyl ether, but its usefulness as a microtechnique still remains to be established [118].

Other derivatization attempts, such as reductive acylation and methoxime formation, failed or resulted in the formation of multiple derivatives (Lefevere et al., 1978, unpublished results).

As a general rule, we fully support the view of Dialameh and Olson [109], that no derivatization should be performed unless it has clearly been proven to be profitable under the given analytical conditions.

2.2.5 *GC on Capillary Columns*

The only report on the use of capillary GC for the analysis of K vitamins dramatically illustrates the high resolving power of this technique. On a 100 m long polar Silar 10 C (cyanoethylsilicone) column, Vecchi et al. [118] resolved the four diastereomers of synthetic K_1: 7'(*R*),11'(*R*)*cis*-, 7'(*R,S*),11'(*R,S*)*cis*-, 7'(*R*),11'(*R*)*trans*-, and 7'(*R,S*),11'(*R,S*)*trans*-, K_1 in the form of their dihydrodimethyl ethers. This separation, which had never been achieved before by TLC, GLC, or LC, required an analysis time of 150 min. Routine analyses, such as the determination of the ratio of *cis*- and *trans*-K_1, were carried out much faster on a 20 m glass capillary coated with CP-TM-WAX (Chrompack, The Netherlands) at an oven temperature of 250°C [118].

Our hope is that, besides affording this excellent selectivity, capillary GC will also limit degradation problems. In the course of preliminary experiments with cold on-column injection (40-70°C, modified Grob injector, Carlo Erba Instrumatione) on a 18 m × 0.32 mm OV-1 column, we were able to separate underivatized MK-2, $K_{1(15)}$, MK-3, $K_{1(20)}$ epoxide, $K_{1(20)}$, MK-4, and $K_{1(25)}$ within 20 min (temperature program from 150 to 280°C) but higher MK-n homologs (MK-8, 9, and 10) did not elute.

2.2.6 *Detection and Quantification*

Most of the work on vitamin K has been done with the flame ionization detector (FID). Dialameh and Olson [109] showed its response to be linear over a range 20 ng to 1 μg, but no doubt this linear range extends further into the microgram range, as evidenced from the work of other authors [104,117,121]. Sheppard et al. [108] claimed a detection limit of as low as 1-2 ng K_1 on packed columns. For complex, e.g., biological samples, however, a practical quantitation limit of about 50 ng for K_1 seems more realistic [104,122]. Moreover, an elaborate purification procedure was needed to compensate for the lack of selectivity of this detector [109]. The use of a more specific and sensitive detector is therefore indicated. Other applications, e.g., the study of the photodegradation of K vitamins [116,117], on the other hand, require a detection system as universal as possible. There, FID is a good choice as it allows the detection of all photodegradation products, even those subject to important structural changes.

Although the usefulness of ECD (electron-capture detection) for the determination of vitamin K and ubiquinones was reported in 1964 [123], it took until 1979 for the first practical application of GC-ECD to the analysis of natural K vitamins to appear [110]. A detection limit as low as 240 pg for K_1 and K_1 epoxide was obtained, and the selectivity was such that 0.1 ml plasma extracts could be injected without any preliminary cleanup. No endogenous material present in the plasma of untreated subjects was found to interfere with the ECD signal of either substance [110].

2.2.7 *Applications*

GC has been used for the determination of natural K vitamins as part of stability studies [116,117] and in food chemistry [104] and biological analysis [110,122].

Nakata and coworkers studied the separation of K_1 from photodegradation products, formed under aerobic and anaerobic conditions, on SE-30, OV-1, and OV-17 columns [116,117]. A 3% OV-1 column was found to give the best separation within a reasonable time. The column temperature was kept below 270°C to avoid on-column degradation. Under these circumstances, and with *n*-triacontane as an internal standard (IS), the determination of the residual vitamin K content was found to be reproducible both before [coefficient of variation (CV) of 0.7%] and after irradiation (CV of 1.1%). Under anaerobic conditions two peaks of photoproducts were found and the principal one identified as the naphthochromenol. More volatile photoproducts, formed under both aerobic and anaerobic conditions, could be separated at a column temperature of 200°C.

Vecchi et al. described the chromatographic analysis of the diastereomers of K_1 in conjunction with optical rotation measurements as a useful approach for the characterization of the origin of K_1 samples [118].

In food chemistry, GC was used to determine K_1 in vegetables [104]. After extraction of the freeze-dried sample with *n*-hexane and column chromatography on alumina, K_1 was quantified on a 2.1 m × 2 mm ID Dexsil 300 GC column at 290°C. It took about 45 min to elute all FID-responsive peaks (Fig. 4.6). Some 10% of standard K_1 added to spinach was lost during the cleanup step, and a fairly high CV (10.5%) was obtained for foods containing 0.76 ± 0.08 mg K_1 per 100 g. Obviously, this performance might be improved by using an analog as internal standard and a more specific and sensitive detector (ECD). Moreover, the contact of K_1 with alumina or silica gel apparently resulted in a double peak in the final chromatogram. This change of K_1

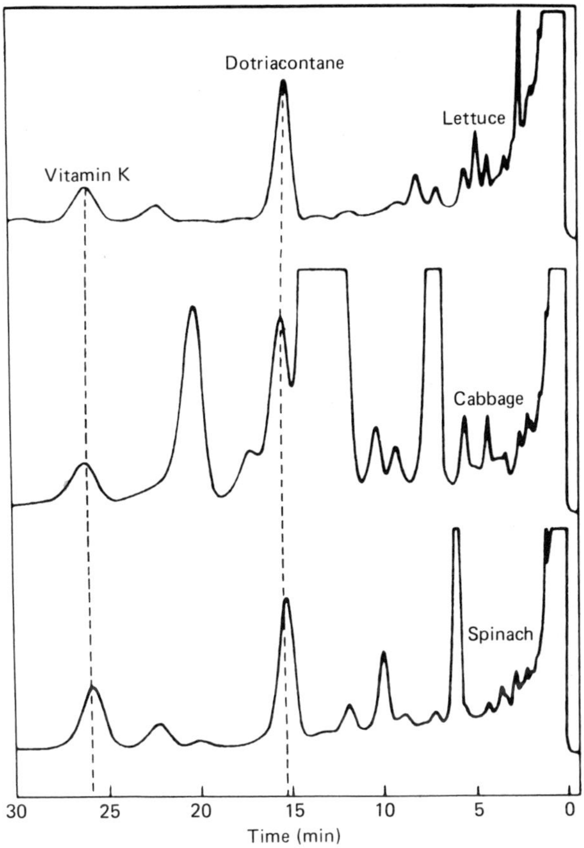

Figure 4.6 GC-FID chromatogram of K_1 in vegetable extracts. (Reprinted from Ref. 104. Copyright 1979 American Chemical Society.)

upon column chromatography shows close resemblance with the chromenol formation reported by Green et al. [105] as an artefact of alumina chromatography. As many other authors, however, did not report such degradation during adsorption chromatography, the possibility exists that these isomerizations would be due to environmental effects (photodegradation of K_1 adsorbed on an active surface).

The first successful GC determination of vitamin K in animal tissue was reported by Dialameh et al. [122]. They used the technique for the identification of MK-4 in chick liver extracts. An elaborate purification procedure was needed to concentrate the very small amounts of MK-4 in a sufficiently pure form for GC with FID: (1) extraction of desiccated tissue with diethyl ether; (2) column chromatography on silicic acid; (3) reversed-phase column chromatography according to Matschiner and Taggart [62]; (4) TLC on silica gel; and (5) TLC on silica gel impregnated with $AgNO_3$ [109].

With ECD, on the other hand, Bechthold and Jahnchen [110] could determine K_1 and K_1 epoxide in plasma extracts without any preliminary cleanup step (Fig. 4.7). The method was sufficiently sensitive to determine plasma levels above 3 ng/ml, such as those that occur following oral K_1 administration and phenprocoumon inhibition. A combination of GC and ECD with a single LC cleanup step might allow a more efficient use of the method's inherent sensitivity (detection limit of 240 pg), by allowing the injection of larger amounts of serum or plasma extract.

2.3 Modern Liquid Chromatography

2.3.1 General Considerations

The use of HPLC for the analysis of vitamin K was first reported in 1972 by Williams et al. [124]. By now, the technique has become the method of choice for most types of vitamin K determinations and the last few years have shown a real proliferation of papers on the subject. Several factors underlie this success.

Modern LC eliminates the risk of thermodegradation and protects the compounds from light and other external influences during the chromatographic run. The highly lipophilic, neutral character of natural K vitamins renders them compatible with a multitude of LC systems, and their strong UV-absorbing characteristics guarantee sensitive on-line detection. However, the greatest advantage of modern LC probably lies in the combination of a highly sensitive and selective technique with a unique accuracy and reproducibility even when very small vitamin K quantities are involved [125]. As is discussed in detail later in this chapter, both the sample workup and the chromatographic procedure itself are usually much simpler and faster than with TLC or GLC. Most often a single cleanup step suffices to remove crude lipid material.

In a recent article, Donnahey et al. [126] compared the respective advantages and disadvantages of TLC and HPLC for the analysis of vitamin K. They found that the separation speed was the same for both techniques but that LC is more sensitive and has a higher resolving power. Thus, it became possible to perform separations in a single LC run, which previously required two or more TLC steps [81,103,126, 127]. Furthermore, LC is very convenient when vitamin K fractions are to be recovered for further chromatographic or spectrometric investigation. For cleanup purposes and/or an increased selectivity, online column switching in LC even obviates the need for collection or any further sample manipulation (Sec. 2.3.4 and 2.3.6).

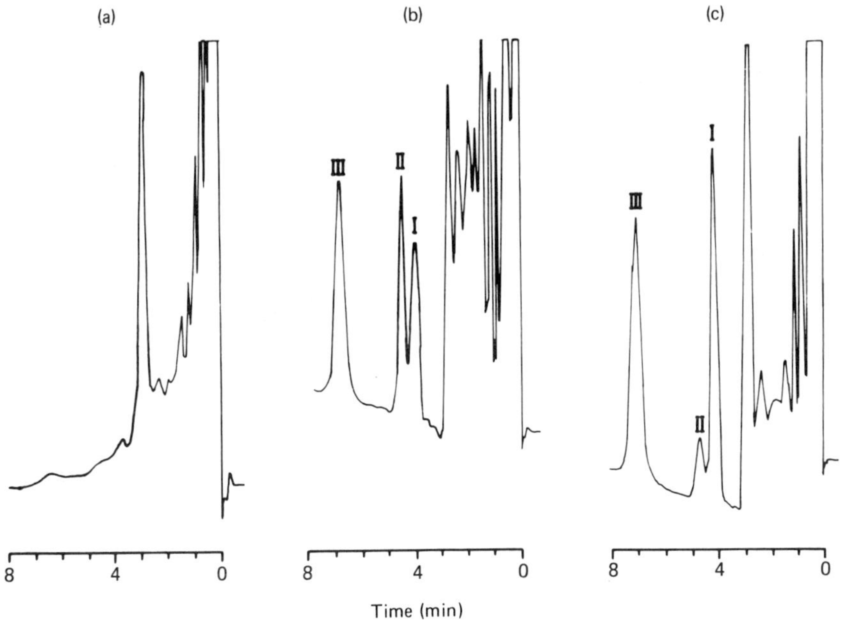

Figure 4.7 GC-ECD chromatograms of (a) blank plasma extract (0.1 ml); (b) extract of plasma, spiked with 0.1 μg/ml K_1-epoxide (structure 4.1) and K_1 (structure 4.2) with MK-4 (0.17 μg/ml) as internal standard (structure 4.3); and (c) plasma extract of patient treated with phenprocoumon and K_1. (From Ref. 110.)

2.3.2 Normal Phase Separations

2.3.2.1 Underivatized Silica. As the interactions between the highly lipophilic K vitamins and the polar silanol groups on the silica surface are very weak, eluents of very low eluotropic strength (ε_0) have to be used. The basic solvent is always a saturated hydrocarbon, e.g., *n*-hexane, which is moderated (modified) by the addition of a solvent of greater eluotropic strength. Typical moderator concentrations are less than 0.05% for alcohols, in the range 0.1-0.5% for acetonitrile and dioxane, and around 2-3% for ethers. With medium polarity solvents, such as chlorinated hydrocarbons, concentrations required to obtain the same eluotropic strength are considerably higher (10-20%). The nature (selectivity class) of this modifier plays an important role in determining the selectivity of the system [128,129].

Another important point in the choice of a proper solvent system pertains to adsorbent deactivation. This is necessary since the silanol groups on the silica surface differ in strength. So-called strong adsorption sites have to be deactivated first to (1) reduce peak tailing; (2) improve column efficiency; (3) increase sample capacity (more linear adsorption isotherms); (4) prevent chemical modification of solutes; and (5) reduce retention drift [130]. In modern LC, two different approaches are possible: the use of water as a moderator, and the use of polar organic solvents.

Water is a very strong moderator and strongly influences retention behavior. With medium-polarity solvents, e.g., mixtures of dichloromethane and *n*-hexane, different elution patterns may be obtained depending on the water content of the mobile phase [103,125,131]. The problem with nonpolar solvent mixtures, however, is that they are very difficult to saturate with water, and hence it takes a very long time before sufficient eluent is passed to deactivate all strong sties. Saunders [132] and Engelhardt [131] therefore recommend the use of polar organic moderators, such as acetonitrile or alcohols. In our experience, however, the residual water content of the mobile phase continues to have some influence on the retention, even when a polar modifier such as acetonitrile is present. For applications requiring highly reproducible retention times (quantitative work and peak identification), one should try to stabilize the water content of different batches of *n*-hexane and moderator or at least make regular calibrations. Attempts to make different eluent mixtures isohydric [133,134] by adding water only to the polar modifier failed. The reason for this is that the amount of modifier needed to obtain the correct eluotropic strength was so low that the release or uptake of water by the *n*-hexane could no longer be neglected (Lefevere et al., 1981, unpublished results).

Of the four modifiers tested in our laboratory, i.e., acetonitrile, isopropanol, dichloromethane, and diisopropyl ether, the last one gave the best day-to-day reproducibility in retention characteristics, together with excellent selectivity. In contrast to what one might expect on the basis of the observations of Saunders and Engelhardt [131, 132], it was not necessary to saturate the eluent with water to obtain good column efficiencies. With dichloromethane, on the other hand, very disappointing results were obtained. Whether it was dry or (partly) saturated with water, we found it very difficult to equilibrate the columns, and strong peak tailing impaired the separations. Moreover, concentrations needed to obtain similar capacity ratios k' varied by as much as 10-25%. Yet, Shearer's group [71,81,103] obtained excellent results with this modifier. These quite conflicting observations might be rationalized by considering the presence of different stabilizers in the dichloromethane from various suppliers. Anyhow, the optimal choice of a moderator also depends on the particular separation envisaged. Whereas diisopropyl ether, e.g., yielded an excellent selectivity for the separation of *cis*-$K_{1(25)}$, *cis*-$K_{1(20)}$, *trans*-$K_{1(25)}$, *trans*-$K_{1(20)}$; and *trans*-$K_{1(15)}$ (Fig. 4.8), menaquinones are much

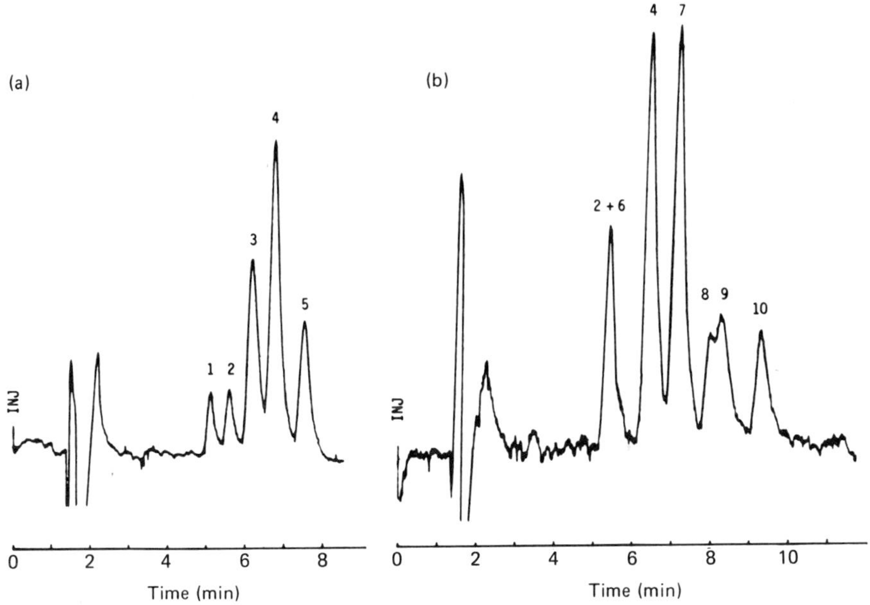

Figure 4.8 Adsorption LC with 2% diisopropyl ether as a modifier. (a) Separation of geometric isomers of $K_{1(20)}$ and its synthetic homologs $K_{1(15)}$ and $K_{1(25)}$. (b) Separation of dihydro-$K_{1(20)}$, *cis*- and *trans*-$K_{1(20)}$, $K_{1(20)}$- epoxide, MK-4, MK-8, and MK-10. Peaks: (1) *cis*-$K_{1(25)}$; (2) *cis*-$K_{1(20)}$; (3) *trans*-$K_{1(25)}$; (4) *trans*-$K_{1(20)}$; (5) *trans*-$K_{1(15)}$; (6) dihydro-$K_{1(20)}$; (7) ν $K_{1(20)}$-epoxide; (8) MK-10; (9) MK-8; and (10) MK-4.

better resolved from K_1 epoxide and *trans*-$K_{1(15)}$ using acetonitrile as a modifier (Fig. 4.9). This is due to a reversal in the elution order of low- and high molecular weight menaquinones, a typical example of solvent selectivity. The experimental conditions used by different authors are summarized in Table 4.7.

Generally speaking, adsorption LC will (1) resolve K_1 and menaquinones from other fat-soluble vitamins and coextracted lipids of different polarities; (2) separate geometric isomers; and (3) separate K_1 and K_1 epoxide.

It is interesting to note that homologs of the K_1 series [$K_{1(15)}$, $K_{1(20)}$, or phylloquinone, and $K_{1(25)}$] can be resolved from each other, while menaquinones that differ in as much as three isoprene units coelute (Fig. 4.8). Compounds with a saturated side chain (K_1 series) elute in front of those with an unsaturated one (menaquinones). In our view

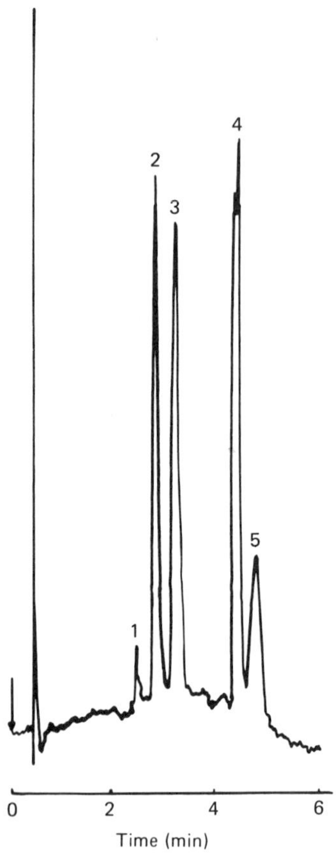

Figure 4.9 Adsorption LC with 0.2% acetonitrile as a modifier. Peaks: (1) *cis*-$K_{1(20)}$; (2) *trans*-$K_{1(20)}$; (3) $K_{1(20)}$-epoxide; (4) MK-2+4; (5) MK-9.

the interactions of the double bonds in the side chain with the adsorbent predominate lipophilicity differences. Haroon et al., on the other hand, could resolve hydromenaquinones with a difference of at least two saturated isoprene units on a spherical packing (Spherisorb-5, Phase Separations), using 25% dichloromethane (50% H_2O saturated) as a mobile phase [81]. Shearer observed that adsorbent deactivation with water had a much more pronounced effect in reducing the retention of menaquinones than on the retention of the K_1 isomers [125]. They normally elute as two peaks, one for the lower homologs (MK-2, 3, and 4) and one for the higher homologs (MK-8, 9, and 10), but

TABLE 4.7 Experimental Conditions Used in Adsorption LC[a]

Stationary phase	Eluent(s)	Application	Comments	References
Li Chrosorb Si 60 (Merck)	0.1-1.25% DIOX in *n*-Hex	Separation of β-carotene, K_1, PQ-9, α-TC, UQ-9/10, phytol, and αTQ	Column switching as an alternative for gradient elution	Lichtenthaler and Prinzel [135]
Li Chrosorb Si 60 (Merck)	0.03% IPA in *n*-Hex	Separation of natural $K_{1(20)}$, synthetic $K_{1(15)}$, and $K_{1(25)}$; MK-n	MK-n elute as 1 peak; system used later for fractionation of extracts (vegetables, milk)	Thompson et al. [136]
Partisil-5,10 (Whatman)	20% MC in *n*-Hex (50% H_2O saturated)	Separation of *cis*-K_1, *trans*-K_1, MK-10, and MK-4	Same conditions as used for fractionation of milk and serum extracts	Shearer et al. [71]
Rsil 5 μm (Alltech Europe)	0.2% CH_3CN in *n*-Hex; 10% MC in *n*-Hex	Separation of *cis*-K_1, *trans*-K_1, K_1-epoxide, MK-2, MK-4, and MK-9	Same conditions as used for fractionation of serum extracts	Lefevere et al. [127]
Zorbax SIL (Du Pont)	2% DIPE in *n*-Hex (replaced later by DNBE)	Separation of cis and trans isomers of K_1 and MK-4		Yamano et al. [144,144a]
Partisil-5 (Whatman); Spherisorb-5 (Phase Separations)	0.5% CH_3CN in *n*-Hex; 25% MC in *n*-Hex (50% H_2O saturated); 25% MC in *n*-Hex (dry)	Separation of cis and trans isomers of K_1, K_1-epoxide, and 2-chloro-K_1 (vitamin K antagonist)	Separation of K_1 epoxide isomers only on Spherisorb-5 column with eluent 3	Haroon et al. [103]
Partisil-5 (irregular); Spherisorb-5 (spherical)	25% MC in *n*-Hex (50% H_2O saturated); 0.5% CH_3CN in *n*-Hex	Separation of *cis*-K_1, *trans*-K_1, MK-10, and MK-4; separation λ = hydromenaquinones with same chain length and their geometric isomers	Isoprenologs intermediate length partially resolved; dry MC causes peak tailing	Haroon et al [81]

[a]Abbreviations: Diox = dioxane; IPA = *i*-propanol; MC = methylene chloride; DIPE = diisopropyl ether; DNBE = di-*n*-butyl ether; PQ-9 = plastoquinone-9; α-TC = α-tocopherol; α-TQ = α-tocopherylquinone; UQ-9/10 = ubiquinones 9 and 10; K_1 = natural form $K_{1(20)}$.

Thompson found that all menaquinones coeluted when using 0.03% isopropanol as a moderator [136].

For the separation of geometric isomers of the K_1 series excellent selectivity was obtained with diisopropyl ether or isopropanol as a modifier. The selectivity of ether-modified systems for this type of separation corresponds well with similar observations in TLC [32,125]. The separation of *cis*- and *trans*-K_1 epoxide, however, is more difficult, probably due to steric hindrance of the 2',3' double bound by the epoxide ring on the quinoid nucleus [125]. This separation was achieved by Haroon et al. on a Spherisorb-5 column, using 25% dichloromethane (dry) in hexane as the eluent. In this case it was found that a dry eluent greatly improves the resolution of all six isomeric forms of K_1, K_1 epoxide, and chloro-K_1 [103].

Our group was the first to demonstrate that K_1 and K_1 epoxide could be separated by adsorption LC [127]. The selectivity again was shown to depend on the moderator used. Although the epoxide eluted after *trans*-K_1 with 0.2% acetonitrile in *n*-hexane (as could be expected from its slightly greater polarity) this resolution was lost using dichloromethane as a moderator. A similar observation was made by Haroon et al. [103] using another irregular packing material (Partisil 5, Whatman). On the highly efficient, spherical Spherisorb-5 material, however, both were separated with dichloromethane in *n*-hexane, but in reversed order (K_1 epoxide in front of K_1). A similar observation was made later by us using 0.05% isopropanol in *n*-hexane as the eluent, and diisopropyl ether again showed the elution order K_1, K_1 epoxide.

Finally, it should be noted that adsorption LC is very sensitive to temperature change. A raise in column temperature of a few degrees results in a marked increase in retention time, especially when a polar modifier such as acetonitrile is used. This negative temperature coefficient of the retention volume (V'_R) would indicate a dispersive type of interaction in which the moderator plays a dominant role [137-139]. It therefore is necessary to thermostabilize the columns.

2.3.2.2 Chemically Bonded Phases (CN, NH_2, and Diol). Interactions with polar bonded phases are even weaker than with the silanol groups of underivatized silica. Moderator concentrations have to be reduced accordingly, from 2% diisopropyl ether for the NH_2 phase to pure hexane for the diol column. Shearer [125] attributes this phenomenon to the lower proportion of available adsorption sites on the bonded phases. The order of elution of K homologs essentially was the same as for underivatized silica. In spite of such advantages as a more rapid equilibration and a more uniform surface, only a very limited number of applications of this type of phase to vitamin K analysis have been described [81,127].

2.3.3 *Reversed-Phase Separations*

2.3.3.1 Nonaqueous Reversed-Phase Chromatography. Although the true mechanisms underlying reversed-phase separations are still subject to continuous investigation [140-143], hydrophobic interactions are generally believed to be predominant. Compounds will be eluted in reverse order of their polarity. The usefulness of this chromatographic mode for the resolution of homologs and metabolites of vitamin K has been recognized ever since the time of paper chromatography. The difference with the older RP approaches, however, is that in RPLC chemically bonded rather than physically coated stationary phases are used. This greatly improves the reproducibility and efficiency of separations and permits larger sample loads and an easier recovery of separated fractions.

All initial separations [124,144,145] were carried out on pellicular packings (Permaphase ODS or Vydac RP) using semiaqueous eluents (in analogy to RP-TLC). By now, however, these pellicular packings have been completely replaced by the more efficient totally porous microparticulate materials (5-10 μm). The higher hydrocarbon load ensures higher sample capacity and accounts for sufficient retention in conjunction with nonaqueous mobile phases [81,103,127]. This so-called nonaqueous reversed-phase liquid chromatography (NARP) has the advantage of an enhanced solubility of nonpolar samples in the mobile phase. This is particularly important for biological extracts and long-chain menaquinones [125]. The latter are indeed so lipophilic that they are even insoluble in pure methanol. Another advantage cited by Parris [146], and confirmed by Shearer [125], as well as by our own experiences, are the long-term stability and excellent efficiencies of columns eluted in the NARP mode.

Differences in selectivity and chromatographic performance among stationary phases used may be explained by [125]: (1) particle characteristics (shape and porosity), (2) the chemical nature of bonded groups (octyl, octadecyl, or phenyl), (3) the number of organic groups bound (low versus high carbon load) and their surface configuration, and (4) the concentration of accessible unreacted silanol groups (capped versus not fully capped packings). Two stationary phases of particular interest for this work are Zorbax ODS (du Pont), which owing to its very high carbon load (22%) is ideally suited for the NARP of nonpolar compounds [81,103,146], and the low-load Rsil C18LL 5 μm packing (Alltech Europe), which has proven to yield particularly good column efficiencies and an interesting selectivity for the determination of K_1 in serum extracts [127].

2.3.3.2 Separations on C_8 and C_{18} Phases. RPLC on various octyl (C_8) and octadecyl (C_{18}) stationary phases has been successfully applied as an alternative to RP-TLC for the determination of epoxide metabolites and their respective parent vitamins in biochemical [145, 147] and pharmacological [148] studies. Eluents employed were semi-aqueous [126,145,147], pure alcohols [126,127,148], or based on mixtures with other organic solvents [103,127]. As an example of solvent selectivity we found that the resolution between K_1 epoxide and MK-4 markedly increased by the addition of acetonitrile to the eluent [127]. Hiroshima et al. [148] employed a mixture of methanol and ethanol (7:3, v/v) to separate K_1, MK-4, their epoxide metabolites and a number of other FSV on Nucleosil C_{18}. The resolution between K_1 epoxide and MK-4, however, was unsatisfactory. Donnahey et al. [126] found it necessary to devise two separate RPLC systems to separate K_1 and K_1 epoxide on the one hand (μBondapak C_{18}, 100% methanol) and MK-3, MK-4 and their epoxy derivatives on the other (μBondapak C_{18}; acetonitrile-water, 85:15, v/v).

RPLC is also ideally suited for the resolution of the various homologs of the K_1 and MK-n series. As expected for a homologous series, a linear relationship between log k' and the carbon number of the prenyl side chain of menaquinones was observed under isocratic conditions [81,127,144]. Similar observations were made for $K_{1(20)}$ and its synthetic analogs $K_{1(15)}$ and $K_{1(25)}$ (Lefevere et al., 1981, unpublished results). Saturated homologs are more hydrophobic than the corresponding derivatives with an unsaturated side chain and hence will be longer retained. Thus, the progressive saturation of the double bonds of the side chain of hydromenaquinones will separate them from their unsaturated counterparts [81], and K_1 will be well resolved from MK-4 as well as from dihydro K_1 (personal experience). Finally, RPLC will also differentiate demethylmenaquinones [126] and decomposition products of the menaquinones [144].

Thompson [136] resolved a mixture of six menaquinones (MK-1, 2, 4, 6, 8, and 10), natural $K_{1(20)}$, and its synthetic homologs $K_{1(15)}$ and $K_{1(25)}$, on a LiChrosorb RP column using a 30 min gradient of 15% water in methanol to pure methanol. Although the type of RP phase was not specified, it probably was a C8 column, given the low retention of MK-10. Donnahey et al. [126] used acetonitrile-water (66:33, v/v) to separate menadione (MK-0), MK-3, MK-4, MK-4 epoxide, and 2-demethyl-MK-4, on a Partisil 10 ODS column. Almost simultaneously, and independently, both Shearer's group [71] and we ourselves [127] introduced the use of nonaqueous eluents for the separation of K_1, K_1 epoxide, and menaquinones MK-2 through 10. We compared the retention of K_1, K_1 epoxide, and the menaquinones MK-2, 4, and 9 on four different phases: one with a short alkyl chain bonded (Li-Chrosorb RP8 5 μm, Merck), one heavy load C18 phase (Rsil C18 HL 10 μm, Alltech Europe), and two low-load octadecyl reversed phases (Rsil C18 LL) with 5 and 10 μm particles. The best results in view of applicability to serum extracts were obtained on the not fully capped (free silanol groups) Rsil C18LL phase using pure methanol or acetonitrile-dichloromethane (92:8, v/v) as an eluent. Shearer, on the other hand, recommended gradient elution [71,125].

His work finally led to the development of a system that was capable of resolving MK-4, MK-5, K_1, MK-6, MK-7, MK-8, MK-9, MK-8(H_4), MK-10, MK-9(H_6), and MK-9(H_8). To this end a gradient of from 20 to 50% dichloromethane in methanol on a Zorbax ODS column (25 cm × 4.6 mm inner diameter, ID) was used [81]. Baseline separation of the 12 different K vitamins was obtained in a single run within 30 min (Fig. 4.10). A methanol-based eluent was used, since mixtures of acetonitrile and dichloromethane or tetrahydrofuran abolished the separation between K_1 and MK-6, in spite of an increased column efficiency. Acetronitrile, on the other hand, would improve the selectivity for the pair K_1 epoxide and MK-4 [127], as well as for the pair K_1 and chloro-K_1 [81,103].

RPLC permits separation of almost all the natural K vitamins (with the exception of geometric isomers) in a single run. Certain samples,

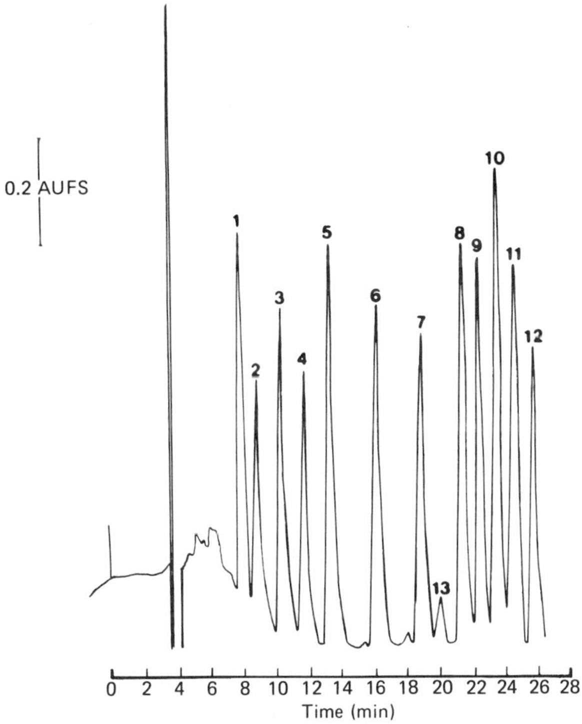

Figure 4.10 Separation of vitamin K compounds by reversed-phase HPLC on Zorbax-ODS using gradient elution. Peaks: (1) MK-4; (2) K_1-epoxide; (3) MK-5; (4) K_1; (5-8) MK-6 through 9; (9) MK-8(H_4); (10) MK-10; (11, 12) MK-9(H_6) and MK-9(H_8); (13) unknown. (From Ref. 81.)

however, such as mixtures of chloro-K_1, K_1, K_1-epoxide, and MK-6, will require separate runs using different eluents. In this particular case adsorption LC might offer a solution.

2.3.3.3 Argentation Systems. The combination of reversed-phase and argentation chromatography as reported recently by Mack [149] shows great promise. Unlike Ag^+ TLC, silver ions in argentation LC are added to the mobile phase, and the compounds are separated on a reversed-phase type of packing. As a result, the retention of menaquinones will be shortened proportionally to the number of unsaturated bonds in the side chain. Various concentrations of silver ions (from 5 to 50 mM) were tested for the resolution of K_1 and MK-0 through 10, with methanol or methanol-dichloromethane (85:15, v/v) as the mobile phase. Apart from forming an alternative to gradient elution [125,149], this technique also shows great potential for the identification of menaquinones in biological extracts. The linear relationship between log k' and the number of isoprenoid units in the side chain existing in classic RPLC was maintained, and the k' shift upon successive additions of Ag^+ ions to the eluent provides information about the unsaturated character of the compounds [149]. This extra means of selectivity may also be helpful to improve the separation between hydrogenated menaquinone forms, as well as their separation from K_1 and other lipids.

2.3.3.4 Other Chemically Bonded Phases (CN, Phenyl, and NH_2). Apart from the classic octyl and octadecyl materials, several other stationary phases have been tested. Although no retention in the reversed-phase mode was obtained on cyanopropyl (CN) and aminopropyl (NH_2) columns, K vitamins did chromatograph on a phenyl phase (Rsil Phenyl 5 μm, Alltech Europe) (Lefevere et al., 1981, unpublished results. As expected on the basis of literature data [150], the retention, however, was much lower than on C8 and C18 phases (k' of 0.6 for MK-9 in pure methanol). At least 10% water had to be added to obtain a k' of 3.2 for K_1. This in turn led to tailing peaks and very poor efficiency, especially for the menaquinones. Eluents investigated were mixtures of methanol, acetonitrile, water, and ethyl acetate. A total incompatibility with acetonitrile was observed. It is not clear whether this is a peculiarity of the Rsil Phenyl packing material, or whether it also applies to phenyl phases from other manufacturers. Barnett and Frick [150] used coupled μBondapak Phenyl-μBondapak C_{18} columns (Waters) in combination with solvent and flow programming, to determine vitamin A acetate, vitamin D_2, and vitamin E acetate in tablets. Our present data for vitamin K homologs, however, do not seem very promising with respect to this type of separation. In our opinion, they rather indicate that a phenyl phase could be useful as packing material for guard columns in an on-line cleanup system.

2.3.4 Column-Switching and Multidimensional Systems

Whenever a single chromatographic system does not provide the selectivity needed to resolve all the compounds of interest, or when compounds with strongly different retention characteristics must be analyzed, switching between different LC columns or systems may be a solution [151-157]. Of the variety of possibilities described, those of interest for this work can be divided into three different types.

1. Switching between systems of similar selectivity but with a different stationary phase ratio, as an alternative to gradient elution [151].
2. Switching between a guard column and an analytical column as a means for on-line cleanup of complex extracts. Compounds that would contaminate the analytical column or cause late-eluting peaks are retained on the guard column and (back) flushed afterward [155,156].
3. Multidimensional LC is a special form of column swithcing useful when part of the chromatogram requires an extremely high degree of resolution [153] or when it contains traces of a compound masked by an excess of contaminating material [157]. The area of interest is "heart-cut" and rechromatographed on a second chromatographic system with different selectivity.

For a better understanding of column switching and multidimensional LC, it should be realized that resolution in LC is governed by selective interactions of the solutes with both the stationary and the mobile phase. Figure 4.11 graphically illustrates stationary and mobile phase selectivity of some systems investigated in our laboratory. A practical example of solvent and selectivity is shown in Figure 4.12 in which a serum extract was chromatographed twice on the same silica column, once with 0.2% acetonitrile and once with 3% diisopropyl ether in *n*-hexane as the eluent. As to stationary phase selectivity, the greatest selectivity differences occur between systems in which the separation is based on entirely different mechanisms, e.g., straight- and reversed-phase columns (Fig. 4.11). Switching between such systems is called multidimensional chromatography [152,154].

As a general rule, such systems not only consist of different stationary phases, but are also eluted with different mobile phases. This puts high demands on the equipment (need for two pumps, trapping column or loops, etc.). Prevention of band broadening is crticial to maintain the effect of the first separation. Therefore, a prerequisite

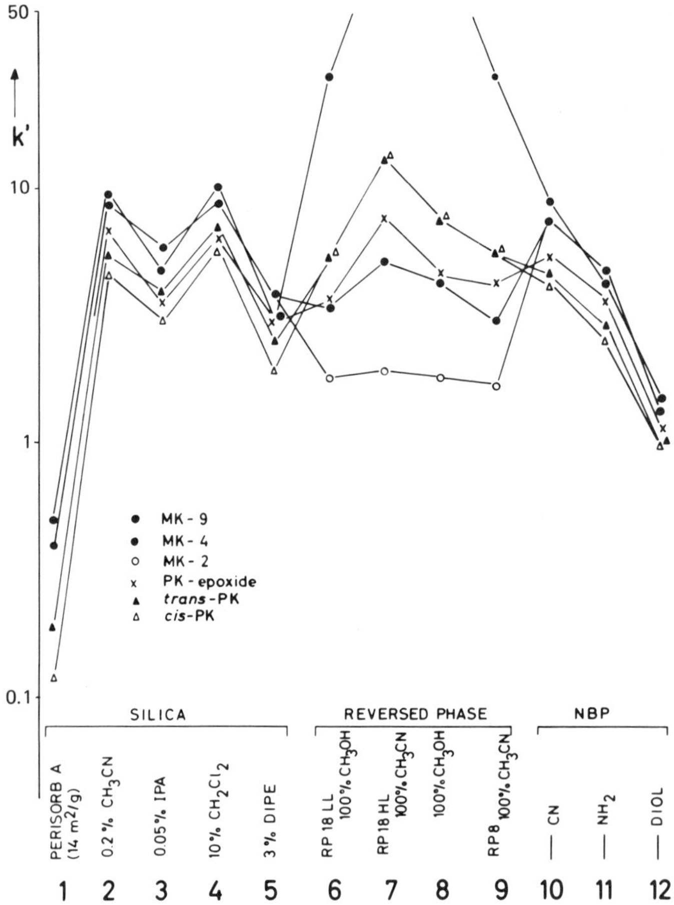

Figure 4.11 Capacity ratios of K vitamins as a function of mobile and stationary phase selectivity. Systems 2-5: underivatized silica; 6-9: reversed-phase columns of different chain length and organic carbon content; and 10-12: bonded-phase columns eluted in the straight-phase mode (NBP). System 1 was a silica column with a very low specific surface (14 m^2/g) compared with 550 m^2/g for the other silica columns. IPA is isopropanol and DIPE is diisopropyl ether.

for on-line multidimensional chromatography is a step to preconcentrate the separated fractions from the first system on the second column [152]. When the respective solvent systems are incompatible an evaporation step is to be included. Until now this has always been performed off-line [71,127,136]. A small fraction containing the compound(s) of

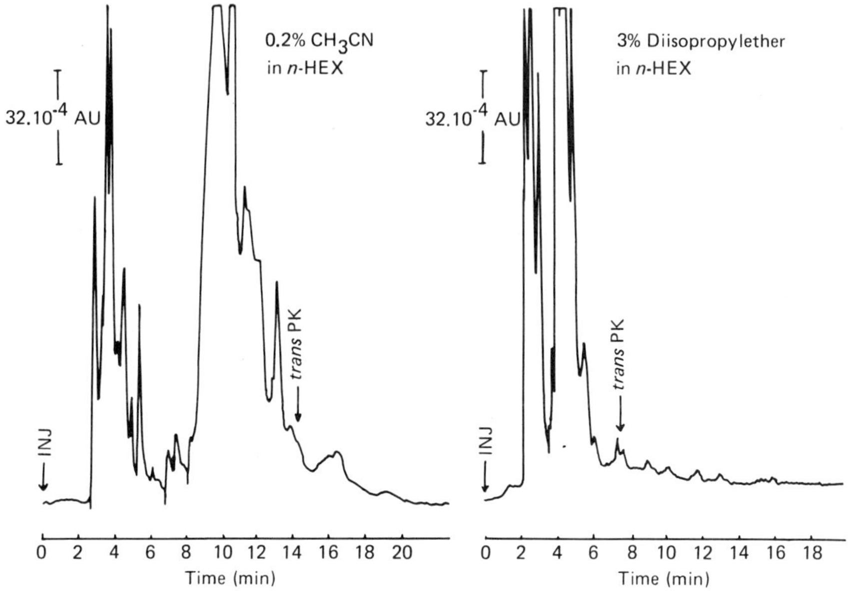

Figure 4.12 Solvent selectivity for serum extracts on an adsorption column (20 cm × 7.0 mm ID Rsil 5 μm) eluted with (a) 0.2% acetonitrile and (b) 3% diisopropyl ether in *n*-hexane.

interest is collected from the first system, evaporated under a nitrogen stream, redissolved in a small amount of the second eluent, and reinjected into system 2.

A typical problem associated with such multidimensional systems is the choice of an appropriate internal standard. Apart from satisfying the classic requirements (structural analog, not naturally present in the sample, similar stability, and extraction recovery as the compound of interest), it should also coelute with the compound(s) of interest during the first chromatographic step while being separated from them during the second one.

One or more examples of each of these types of column switching is discussed in Sec. 2.3.6.

2.3.5 Detection

Quantitative trace analysis in LC is usually carried out by peak height measurements, because this approach permits the highest accuracy and adequate precision [158]. With the exception of electrochemical detection, all the detectors used for the quantification of vitamin K are concentration dependent. This means that the peak height is directly proportional to C_{max}, the solute's concentration in the eluent at the peak maximum, and that the best sensitivity will be obtained by using short narrow-bore columns filled with 5-6 μm particles [158,159]. We found a 15 cm × 3.2 mm ID column to offer a good compromise between absolute sensitivity and sample capacity [127]. Furthermore, the capacity ratio k' should be kept as small as is compatible with a good separation. Gradient elution may seem a good solution to this problem, but unfortunately many detectors will show considerable drift and increased noise at the high sensitivity levels required.

2.3.5.1 UV Detection. The variable wavelength UV detector is by far the most used in modern LC. In contrast to the original ones, recent types are almost as sensitive as the fixed wavelength 254 nm detectors (0.005-0.0025 a.u.f.s.)[1] and allow quantification of components at their UV absorption maximum. Advantages include the ease of operation, excellent reproducibility, which eliminates the need for daily calibrations, and compatibility with most chromatographic systems. The sensitivity and selectivity are functions of the compounds' UV characteristics. In the case of vitamin K, the sensitivity is good (large molecular extinction coefficient), except for the epoxide metabolites. The selectivity, however, is much less favorable, especially for trace analysis in a complex matrix. This often will necessitate the use of an elaborate (multidimensional) separation scheme. Unlike Shearer [125], we did not observe any substantial increase in selectivity when the K vitamins were detected at their quinoid absorption maxima (262 and 270 nm) instead of at 248 nm (benzenoid absorption band) [127]. This is probably due to the different nature of biological samples investigated: vegetables and milk [125] versus serum extracts [127]. The sensitivity was best at 248 nm with a detection limit of the order of 400-500 pg for K_1 (less than 1 pmol) [127,156]. Apart from its use for the quantification of vitamin K, the variable wavelength UV detector is also useful to check peak identity and purity [71,127]. This is particularly true for vitamin K, which has a very distinctive UV-absorption spectrum [125].

An interesting development is the recent marketing of a fixed wavelength detector with a cadmium lamp (LDC, Riviera Beach, Fla.). This very stable light source permits detection at 318 nm, very near the weak but selective vitamin K absorption maximum of 325 nm, with an unrivalled sensitivity of 0.001 a.u.f.s. It remains to be established, however, how substantial this increase in selectivity will be, compared to the use of specific detectors.

2.3.5.2 Specific Detectors. Apart from its rather poor selectivity (at 248-270 nm), UV detection has the disadvantage that one must always work at or near the limit of instrumental performance to detect vitamin K traces in, e.g., body fluids or tissues. This makes the

[1] Absorption units full scale.

technique very vulnerable to all kinds of disturbances and, among others, to interferences from solvent impurities [127] and sample contamination. In order to reduce sample pretreatment to a minimum, as well as to improve the assay's selectivity and sensitivity, several more specific detection techniques have been worked out [148,160-164]. Some typical aspects of vitamin K analysis with these detectors are discussed below.

Some years ago, Ikenoya et al. [161] published a method for the amperometric detection of K_1 after reversed-phase LC. The applied potential was -0.30 V versus $Ag^+/AgCl$. In spite of the promising results shown (sufficient selectivity for a direct determination of K_1 in serum extracts by a single RPLC step and a detection limit of 100 pg), reductive mode amperometric detection in a NARP eluent in our experience can be very problematic. In spite of a continuous helium degassing of the eluent, very high background currents were observed (up to a 1000 times the actual reduction current to be measured). As a consequence, extreme precautions had to be taken to keep the system stable. The slightest variation in eluent flow resulted in a peak equivalent to 1 ng K_1. The whole setup had to be thermostated, mounted in a Faradic cage, and left in operation day and night (very long equilibration time. Under these circumstances we did indeed obtain a detection limit of 150 pg, although the result strongly depended on the type of amperometric detector used and the condition of the working and auxiliary electrodes (surface contamination). We came to the conclusion that amperometric detection is indeed more selective than UV detection at 248 or 270 nm and that, under very stringent conditions, very low detection limits may be obtained. However, both the instrumentation and the technique still await further development before routine use in the NARP-reductive mode at the sensitivity limits required for the determination of subnanogram amounts of vitamin K would become possible.[1] On the other hand, we had no problems in working at medium sensitivities such as those required for the quantification of pharmacological doses of K_1 in serum [161].

Although originally devised for use with RPLC systems, Hiroshima et al. [162] later adapted this amperometric method to straight-phase chromatography by a postcolumn addition of electrolyte solution. Differential amperometric detection, on the other hand, may overcome the problem of the high background currents and further increase the selectivity, but until now it has only been applied to the analysis of K_3 in aqueous eluents [163].

In our laboratory much more encouraging results have been obtained with fluorimetric reaction detection of the quinol forms of K_1 and MK-n. Although these compounds exhibit no native fluorescence, fluorophores with λ_{exc} = 320 nm and λ_{em} = 425 nm can be formed either by a chemical

---

[1]Note added in proof: An interesting new development is the combination of coulometric reduction with amperometric reoxidation or fluorimetric detection as described by Langenberg et al. [163a].

reduction with $NaBH_4$ [160] or $NaHSO_3$-HCl and $NaBH_3CN$ [148], or by a photochemical reaction in the presence of ascorbic acid [164]. Detection limits of the chemical reduction methods were originally reported to be of the order of 400 pg [160] to 3 ng [148]. With a well-optimized air (N_2)-segmented postcolumn reaction system, however, a detection limit of 150 pg could be obtained for K_1, with both chemical and photochemical reduction [164]. This means that both methods are about four times as sensitive as UV detection. With a suitable internal standard, i.e., one with the same reaction kinetics as the compounds to be analyzed, a within-run precision of the order of 1-4% was obtained. The gain in selectivity and specificity was such that serum samples could be analyzed without any preliminary cleanup with $K_{1(15)}$ as internal standard [164]. Literature data indicate that the photochemical reaction method could be adapted to straight-phase eluents [165]. Epoxides, on the other hand, were detected only after reduction with $NaHSO_3$-HCl and $NaBH_3CN$ [148].

Finally, detection of K_1, K_1 epoxide, and MK-n with the prototype electron-capture LC detector developed by Willmott and Dolphin at Pye Unicam Laboratories, Cambridge, England [166], has been investigated, too [166a]. It was shown that, after a modification on the transfer tube (vaporizer) and outlet of the ECD, high-boiling compounds such as K_1 and MK-n could be detected with an equal sensitivity, and K_1 epoxide with an even better sensitivity than with UV detection (248 nm). The use of this detector is strictly limited to straight-phase systems with very low moderator concentrations. Moreover, the *n*-hexane used should be carefully purified by refluxing with metallic sodium, and after distillation further dried over silica and/or molecular sieve. In spite of these limitations, the technique looks quite promising for identification purposes and for the detection of the epoxide metabolites [166a], but unfortunately we do not know of any such detector currently available on the market.

2.3.6 Applications

A general overview of the use of LC in vitamin K analysis is given in Table 4.8. Typical applications are as follows.

1. The determination of K_1 in mixtures with other FSV and related lipids [135,167].
2. Control of the stability of (pharmaceutical) formulations and monitoring for decomposition products [168].
3. Determination of K_1 in vegetables, dairy products, and infant formulas [67,71,167,169].
4. Assay for K_1, MK-n, and their metabolites in animal tissues and fluids [71,127,145,147,148,160,162,171].

The first application requires the separation of compounds of considerably different polarity and lipophilicity. As the retention times for β-carotene, *cis*- and *trans*-K_1, plastoquinone-9, α-tocoquinone, and phytol were very different, Lichtenthaler and Prinzel [135]

TABLE 4.8 Applications of LC to the Analysis of Natural K Vitamins

References	Chromatographic mode	Application(s)
Williams et al., 1972 [124]	RP	Separation and quantitation of FSV A, D, E, K;
Elliot et al., 1976 [145]	TP	Separation K_1 and K_1 epoxide added to liver homogenates
Lichtenthaler and Prinzel, 1977 [135]	Adsorption; column switching	LC of natural prenylquinones
Yamano et al., 1977-78 [144,168]	Adsorption; RP;	Separation of homologs, geometrical isomers, and decomposition products of K_1, MK-4, and UQ-10
Bjornsson et al., 1978 [147]	RP	Preparative separation of $[^3H]K_1$ and K_1-epoxide in plasma
Thompson et al., 1978-1979 [136,169]	Adsorption; RP; multidimensional	Separation of $K_{1(15)}$, $K_{1(20)}$. $K_{1(25)}$; and MK-n; determination of K_1 in vegetables, milk
Donnahey et al., 1979 [126]	RP	Comparison of LC and TLC for separation of MK-0, MK-3, MK-4, epoxy-MK-4, demethyl-MK-4, K_1, and epoxy-K_1
Lefevere et al., 1979 [127]	Adsorption/NARP/NBP; multidimensional	Separation of K_1, epoxy-K_1, MK-2, 4, and 9; determination of K_1 in serum
Shearer et al., 1980 [71]	Adsorption; NARP; multidimensional	Separation of K_1, chloro-K_1, epoxy-K_1, MK-n; determination of K_1 in vegetables, milk, serum (treated)
Abe et al., 1979 [148, 160]	NARP	Fluorimetric reaction detection of K_1, MK-4, and their epoxides in serum (treated)
Haroon et al., 1980 [103]	Adsorption; NARP	Separation of K_1, chloro-K_1, epoxy-K_1, isomers

TABLE 4.8 (*Continued*)

References	Chromatographic mode	Application(s)
Hiroshima et al., 1979, 1981 [161,162]	Adsorption; RP	Electrochemical (amperometric) determination of K_1 and MK-4 in serum (treated)
Mack, 1980 [149]	Argentation, RP	Separation of MK 0-10 and K_1 by Ag^+ chromatography
Barnett et al., 1980 [167]	NARP	Simultaneous determination of A, D_2, D_3, E, and K_1 in foods
Haroon et al., 1981 [170]	Adsorption; NARP; NBP	Separating of MK 4-10, K_1, and hydromenaquinones
Lefevere et al., 1981 [155, 156]	Multidimensional	Determination of physiological K_1 levels in serum; on-line cleanup with back-flushing
Chiu et al., 1981 [171]	Multidimensional	Determination of K_1 in plasma (untreated) (no *cis*-K_1 or MK-n reported)
Lefevere et al., 1982 [161]	NARP	Photochemical reaction detection of K_1 and MK-n

proposed switching between a short and a long adsorption column as an alternative to gradient elution. α-Tocoquinone, phytol, and other "polar" lipids were separated on a 12.5 cm Lichrosorb SI 60 column with 1.25% dioxane in *n*-hexane and this fraction fed directly to the detector; the apolar fraction (β-carotene, K_1, and plastoquinone-9) was further separated on the second, 25 cm long SI 60 column. An even better solution would have been to switch between columns with a low and a high specific surface (e.g., SI 1000 and SI 60) [151]. Barnett et al., on the other hand achieved the separation of vitamin A alcohol, δ-tocopherol, D_2 or D_3, α-tocopherol, α-tocopheryl acetate, K_1, and A-palmitate on two Zorbax ODS 25 cm columns connected in series, using a methanol-ethyl acetate (86:14) acetonitrile gradient and flow programming [167]. Cholesterol phenylacetate was used as the internal standard.

Monitoring of vitamin K synthesis involves an accurate determination of the ratio between the nonactive cis and the biologically active trans isomer. This separation can only be accomplished by straight-phase chromatography. Control of the photostability of K_1 and MK-4 in injection and infusion solutions, on the other hand, was done by RPLC using α-heptadecanylnaphthoate and α-pentadecanylnaphthoate as the respective internal standards [168].

Foods and other biological samples often contain a number of unknown contaminants that are difficult to resolve from the compounds of interest. Typical problems are [125]:

1. The low physiological concentrations of K vitamins in animal tissues and fluids
2. The multiplicity of molecular forms of K vitamins
3. The instability of the quinone forms to alkali, which rules out saponification as an aid to purification
4. The presence of contaminants that persistently resist attempts to remove them by conventional chromatographic techniques

With LC, sample pretreatment usually can be limited to the removal of bulk lipids by a single chromatographic step on Sephadex LH-20 [67], hydroxyalkoxypropyl Sephadex (HAPS) [169], or Silica [71]. An additional advantage is that this cleanup step may be integrated online into the LC system [155,156]. As an alternative to the chromatographic removal of lipids, Barnett et al. [167] suggested an enzymatic hydrolysis step with *Candida cyclindraccae* lipase. By adjusting the pH of the resulting hydrolysate, fatty acids could be precipitated and subsequent solvent extraction yielded an organic phase containing primarily the fat-soluble vitamins, trienols, and other sterols. The method, which is considerably faster and less drastic than the extraction with Claisen's alkali [88], was applied successfully to the determination of vitamins A, D_2, D_3, E, and K_1 in infant formulas, milk, margarines, laboratory rodent chows, cattle supplement, and a variety of liquid and powdered nutritional specialty products [167].

Sample pretreatment and the chromatographic procedure itself depend not only on the nature of the sample but also on the concentrations to be measured. A single RP step has been sufficient for the determination of K_1 in green leafy vegetables and milk powder [67], as well as for the assay of the K_1-2,3-epoxide-reducing system in liver homogenates [145], but more elaborate and sophisticated techniques have been found necessary for human milk (1-2 ng K_1 per milliliter) [71] and serum (1-8 ng K_1 per milliliter) [155,156]. Here, LC opened the way for determinations that were practically impossible with other chromatographic techniques. It is interesting to note that, independently of each other, similar multidimensional procedures have been worked out by Thompson [169], Shearer et al. [71], and Lefevere et al. [127]. In all three cases a fractionation step on a silica column, followed by quantification on a RP system, was used for the determination of K_1 in vegetables and milk [71,169] and human serum [127], respectively.

As a sample cleanup and to prevent adsorption of polar lipids at the top of the first (silica) HPLC column, Thompson and Shearer used an off-line pre-purification on HAPS [169] or silica [71]. In our laboratory, a low specific surface guard column is used for the on-line cleanup of serum extracts (Fig. 4.13). Upon the injection of an extract, eluent 1 (3% diisopropyl ether in *n*-hexane) is first passed through the guard column (Perisorb A, Merck, specific surface 14 m^2/g) until the compounds of interest (K_1 and MK-n), together with other equally lipophilic serum constituents, elute onto the fractionation column (porous silica, specific surface 550 m^2/g). As soon as they arrive there, the valve is switched, and the guard column back-flushed with 0.5% methanol in *n*-hexane (to remove more strongly retained compounds), while a first separation of the lipophilic serum constituents takes place on the fractionation column (AC), which continues to be eluted with eluent 1. As can be seen in Fig. 4.14a, the K_1 peak at this point is still masked by interferences. Therefore a fraction of the eluent corresponding to the elution time of K_1 is collected and, after evaporation and redissolution in methanol, rechromatographed on a reversed-phase column that, owing to its totally different selectivity, separates K_1 from the remaining UV-absorbing constituents (Fig. 4.14b). The collection is facilitated by the use of *trans*-$K_{1(25)}$ and *trans*-$K_{1(15)}$, which elute just in front of and just behind the *trans*-K_1 peak on the straight-phase column, as markers (Fig. 4.14a).

It is important to note that, in order to avoid interfering peaks during the final RP step, the volume of the fraction(s) collected from the adsorption system should be kept as small as possible [125] and that although the loading capacities of 4.6 mm ID microparticulate silica columns are generally sufficient [71,127], an improved resolution may

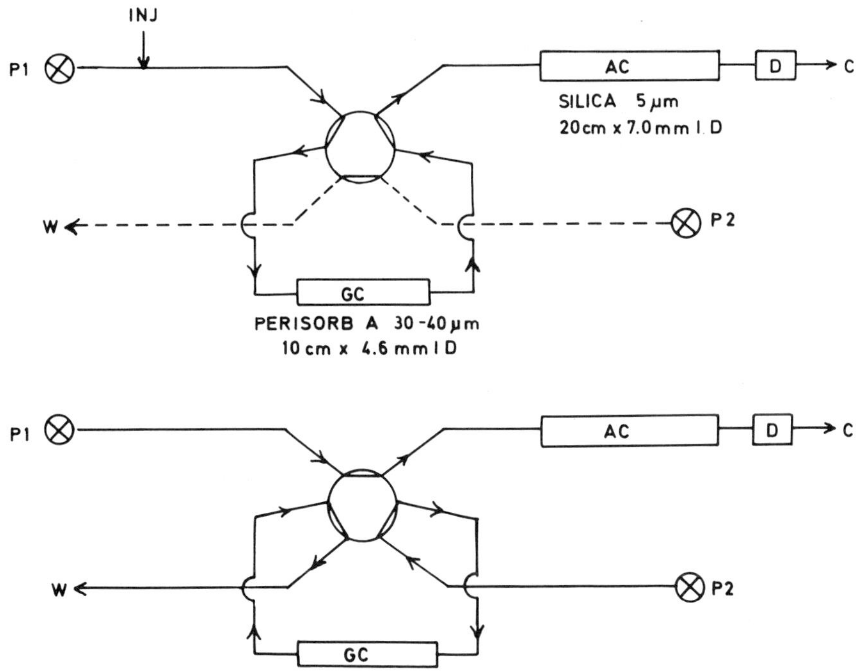

Figure 4.13 Scheme of first step in multidimensional assay for K_1 (PK) in serum.

be obtained using semipreparative (7.0 mm ID) columns for the first step [156,169]. For the final resolution of K_1 from remaining UV-absorbing contaminants on the RP system, Thomson used a solvent gradient [169], but Shearer et al. [71] and we [127,156] used isocratic elution with nonaqueous eluents. According to Shearer [125], methanol-based eluents tend to give a better resolution of K_1 in food samples, but no such selectivity difference was observed for serum samples [127,156].

Problems associated with the choice of an appropriate internal standard for multidimensional systems were mentioned in Sec. 2.3.4. Shearer [71] proposed the use of K_1 epoxide as an internal standard for K_1. Both substances indeed coelute in his Partisil 5 dichloromethane-n-hexane (20:80, v/v) straight-phase system, while being separated later during the RP step on Zorbax ODS. Although the stability of the epoxide is not exactly the same as for the native K_1, the results shown by Shearer et al. indicate that is a suitable internal standard for the determination of K_1 in vegetables, milk, and normal serum samples, in which natural K_1 epoxide levels are so low that they do not interfere.

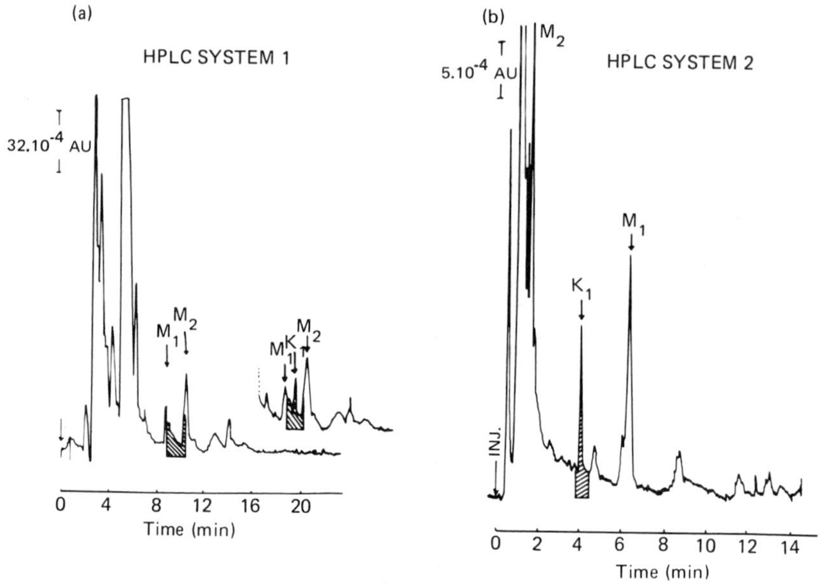

Figure 4.14 (a) LC system 1. Fractionation of serum extract by adsorption chromatography on silica. Hatching = fraction containing K_1, collected for rechromatography on LC system 2. M_1 and M_2 are the marker compounds [*trans*-$K_{1(25)}$ and $K_{1(15)}$] used to identify the K_1 fraction. Upper trace: serum extract spiked with K_1. (b) LC system 2. Quantification of K_1 fraction by reversed-phase chromatography. Hatching = K_1 peak collected for liquid scintillation counting of IS. Injected fraction corresponds to 1 ml serum. Experimental conditions: 15 cm × 3.2 mm ID Rsil C18 LL 5 μm column, eluted with 100% methanol, 1 ml/min. UV detection, 248 mm.

Serum samples from coumarin-treated subjects or persons with a genetic vitamin K epoxide reductase deficiency, however, may contain considerable amounts of this K_1 epoxide. Moreover, K_1 and its epoxide did not coelute in the straight-phase systems used in our multidimensional assay [127,156]. We therefore use radiolabeled $[1',1'-{}^3H_2]K_1$ as internal standard.

These assays have been validated by examining the identity and purity of the chromatographic peaks measured. Peak purity was examined by tests of chromatographic homogeneity in different LC systems [125,127] and by verification of the characteristic UV-absorption spectrum [71,127]. Further confirmation of the peak identity was obtained by collecting the putative peak of K_1 as it eluted from the RP system and subjecting the pooled eluates to analysis by mass spectrometry

[125,156] and GC-ECD [156]. As to the precision of the assay, Shearer found a CV of 1-2% for the three chromatographic steps and an overall precision of 5-7% for vegetable samples [125]. The greatest source of error lay in the difficulty to obtain uniform samples and in the extraction. Using the $[^3H]K_1$ IS method, we obtained a within-day precision of 5.3% for serum samples [155,156].

The merits of electrochemical [161-163] and fluorimetric reaction-detection [148,160,164] for the determination of K_1 and MK-n in serum extracts have already been pointed out in Sec. 2.3.5. Owing to their great selectivity they not only allow determination of these compounds by a single chromatographic step (Fig. 4.15), but they are less sensitive to interferences from column contaminants as well [164]. This increased selectivity is also likely to apply to biological samples other than serum.

Finally, that LC has enabled both the first detailed survey of dietary sources of vitamin K available to the newborn infant [170] and the first systematic study on physiological K_1 levels in human blood [155,156] clearly illustrates its importance in the analysis of the natural vitamin K content of biological materials. The published work today is largely limited to the specific assay of K_1, probably the most important individual vitamin K source to humans [125]. Attempts to determine the natural menaquinone content of human serum have failed so far [156,171].

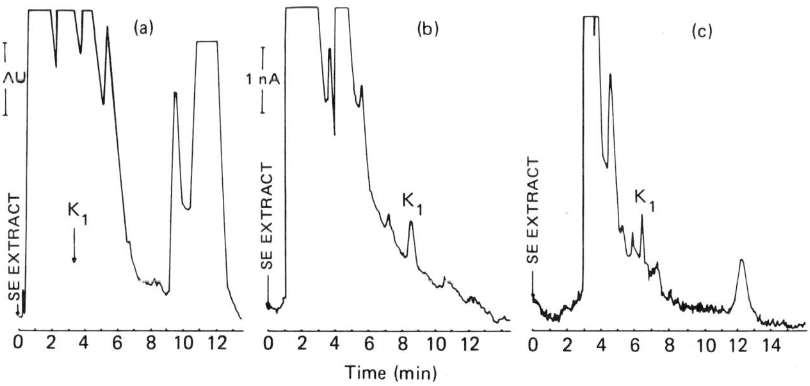

Figure 4.15 Direct determination of K_1 in serum extracts: a comparison between (a) UV (248 nm); (b) amperometric (-0.3 V versus Ag/AgCl); and (c) fluorimetric (λ_{ex} = 320 nm; λ_{em} = 420 nm) reaction detection. Injected amount corresponds to 0.4-0.5 ml serum extracted. Experimental conditions: 15 cm × 3.2 mm ID Rsil C18 LL 5 μm column, eluted with methanol-water (95:5, v/v) for a and c, and methanol-water-$HClO_4$ (95:5:0.1, v/v) with 0.05 M $LiClO_4$ as carrier electrolyte for b.

3. SYNTHETIC VITAMIN K ANALOGS

The chromatographic behavior of menadione (M), menadione sodium bisulfite (MSB), and other synthetic analogs is quite different from that of the natural vitamin K homologs. As a result of the absence of a side chain in position 3, they are much more volatile and less lipophilic. In fact, some of them are even water soluble or form water-soluble salts. With no side chain in position 3, isomerization to the chromenol form, such as occurs during the anaerobic photolysis and/or gas chromatographic analysis of K_1, is obviously excluded. However, under anaerobic conditions, a photoreduction to the corresponding hydroquinone has been reported [54]; and under aerobic conditions photoepoxidation and a very fast decomposition of menadione were reported [172]. This increased susceptibility to air and light degradation should be taken into account when planning sample preparation and TLC procedures.

Unlike natural K vitamins, there is very little demand for assays of synthetic vitamin K analogs in biological materials. Rather, they are determined in different types of vitamin formulations and feed premixes (e.g., poultry and swine feeds). Sufficiently large amounts of sample are usually available, and detection is not a problem. The focus is on the separation of the compound(s) of interest from synthetic by-products and degradation products as part of quality control of the commercialized materials. To this end, TLC as well as GC and LC methods are used.

3.1 Thin-Layer Chromatography

The separation of menadione (vitamin K_3), menadione sodium bisulfite, menadiol diphosphate (vitamin K_4), and a number of related quinones by adsorption, reversed-phase, and normal-phase partition TLC was studied by Dummler and Dorfling [173]. Menadione could be separated from 1,2-naphthoquinone, 2-naphthol, 1-naphthol, juglone (5-hydroxy-1,4-naphthoquinone), and 1,4-naphthoquinone by adsorption chromatography on mixed silica gel and kieselgur (1:1) plates. With the two solvent systems investigated, petroleum ether-methyl ethyl ketone (MEK) (85:15) and benzene-ethyl acetate (97:3), menadione sodium bisulfite, and menadiol diphosphate (MDP) remained at the origin (R_f = 0). Natural K_1, MK-7, and plastoquinone-9, on the other hand, migrated very fast (R_f values, 0.79-0.82). Individual separation of the last three and of menadione was accomplished by RP-TLC on paraffin-impregnated plate with acetone-water (90:10) as an eluent [$R_{f(M)}$ = 0.9]. MSB and MDP were determined on the normal-phase partition system silica gel G and benzene-methanol (60:40).

Later reports include the separation of menadione and menadione diacetate on silica gel and silica gel-impregnated with PEG 200 [65], and of vitamins K_3, K_4, and K_5 on an unimpregnated talc thin layer with water-dioxane-acetone-formaldehyde (85:20:15:25) as the eluent [82a]. In each of these reports, retention data for other FSV were given as well.

Rittich et al. [94] determined K_1, K_3, and K_4 in concentrates and pharmaceutical formulations by adsorption chromatography on silufoil. With benzene as an eluent, these authors obtained R_f values of 0.61, 0.30, and 0.13 for K_1 (phylloquinone), K_3 (menadione), and K_4 (menadiol diacetate), respectively. Addition of ethyl acetate or methyl ethyl ketone to the mobile phase impaired the separation of K_3 and K_4, but improved their separation from decomposition products [65,94]. After elution of the spots, the vitamin K content was determined spectrophotometrically at 252 nm (K_3) and 286 nm (K_4).

Spots of synthetic K analogs can be visualized by spraying the plates with perchloric acid (detection limit 0.4-1 μg) [65], rhodamine B [173], sulfochromic acid [173], aromatic amines (detection limit 0.8-2 μg) [174], 4-acetophenazone [175], and 4-aminophenazone [176]. Some of these reagents react only with synthetic vitamin K analogs, others with both synthetic forms and natural homologs.

3.2 Gas-Liquid Chromatography

3.2.1 Menadione

Owing to its higher volatility, GC analysis of menadione is much less problematic than for the natural K homologs. No signs of on-column degradation were reported, expect by Ninomiya and coworkers [111] (Table 4.9). In view of the well-known air and light susceptibility of the compound, the two distinct peaks they observed on SE 30, Craig's polyester succinate, and FS-1265 stationary phases, however, may very well have been due to sample degradation.

Libby and Sheppard [177] found a 6 ft × 4 mm ID column, packed with 1% NPGS stationary phase on Gas Chrom P, ideal in terms of peak configuration and sensitivity. This same column was used later for the routine GLC assay of the menadione and MSB content of pharmaceutical formulations [178,179]. The method is very simple: tablets are ground and extracted with diethyl ether, and these extracts diluted to a concentration of approximately 1.0 mg menadione per milliliter. After the addition of BHA (butylated hydroxyanisole) as an internal standard, 4 μl of the sample and standard solutions are injected in duplicate.

A more polar stationary phase, 5% Silar-5CP, was used by Mee et al. [172] to monitor the photoepoxidation of menadione. At a column temperature of 180°C, menadione, menadione epoxide, and a second degradation product were well separated. Proof as to the high stability of K_3 during GC analysis was obtained by comparison of direct-inlet and GC/MS mass spectra. The epoxide remained stable during GC analysis as well [172].

TABLE 4.9 Experimental GLC Conditions for Synthetic K Analogs

Stat. phase	Temperature (°C)	T_R (min)	Application	Comments	References
Menadione					
1% SE 30; 1% Craig's PES; 1% FS-1265	90-130 (K_3)		Selection of optimal GLC conditions for K_1, K_3, and K_4-diacetate	K_3 gives rise to 2 distinct peaks on all 3 phases	111
5% DOW silicone	180	1	M. in extracts of *F. ferooxidans*	ECD	119
1% NPGS	145	6	M. content of tablets	BHA internal standard	177-179
5% Silar-5 CP	180	4 (M); 6 (ep)	Separation M and M.epoxide in photodegradation solutions		172
5% XE 60	190	13	Determination of purity of menadione	Separation of synthesis by-products	180
Esters and water-soluble compounds					
1% SE 30; 1% Craig's PES; 1% FS-1265	150-180		Menadiol diacetate	Direct GLC	111
3% OV-1	175	2.4 (K_5-TMS) 8.5 (K_3) 20.8 (K_5)	K_5 determination in foods	TMS derivatization	181
1.5% SE 30	130	12	K_4 disodium disulfate	Hydrolysis, oxidation → M	182
3% Dexsil 300; 2% OV-17	140		MSB in feed premixes; IS = diethylphthalate	On-column pyrolysis	183, 184
1% NPGS	145	6	MSB in pharmaceuticals; IS = butylated hydroxyanisole (BHA)	Determination M. after alkaline hydrolysis	177-179

Castello et al. [180] used a stationary phase of intermediate polarity (5% XE 60) to determine the purity of menadione (K_3). Apart from traces of methylnaphthalene (precursor in synthesis of menadione) and of 1,4-naphthoquinone, the presence of low concentrations of a previously unrecognized synthetic by-product was also demonstrated. This compound eluted close to the menadione peak and was tentatively identified as the 6-methyl-1,4-naphthoquinone isomer. A 2.5% SE 30 column did not separate this by-product, but on 1% NPGA and 20% Craig polyester columns, a tailing peak was observed. For the quantitative analysis of commercial menadione, three different internal standards were considered: *n*-eicosane (C_{20}), 1,4-naphthoquinone, and diethylphthalate [180].

As to the FID response for menadione, Libby and Sheppard reported a detection limit of 4 ng without an established upper limit of linearity [177].

3.2.2 Esters and Water-Soluble Compounds

Water-soluble forms, such as menadione disodium disulfate and menadione sodium bisulfite are commonly analyzed as menadione following hydrolysis and oxidation of the parent compounds [108]. Alkaline hydrolysis of the bisulfite addition product, however, tends to yield variable results and low recoveries due to problems with the extraction of menadione from alkaline solutions [185]. Therefore, a direct GLC method for MSB in feed premixes was developed by Winkler and Yoder [183]. This method utilizes on-column pyrolysis of MSB and subsequent GLC of the resulting menadione. Differential thermal analysis and GC-MS data showed a quantitative pyrolysis at an injector temperature of 250°C. In a subsequent collaborative study, Dexsil 300 (1-3% on Gas Chrom Q) was shown to be the column material of choice, but OV-17 also gave acceptable results provided the support had been silylated with BSA-TMCS and not with MSTFA-TMCS [184]. With diethyl phthalate as an internal standard, coefficients of variation of the order of 5.5-6.5% were obtained by eight different laboratories.

Ninomiya et al. [111] reported the direct GLC analysis of menadione diacetate on SE 30, Craig's polyester succinate, and FS-1265 columns. In view of the use of K_5 as a food preservative and the potential hydrolysis of K_5 to K_3, Cornelius and Yang [181] developed a GLC method for the simultaneous determination of K_5 (as its TMS ether) and K_3 in dried carrots. On a 3% OV-1 column at 175°C, K_5-TMS; K_5; and K_3, eluted with a retention time of 2.5, 20.8, and 8.5 min, respectively.

3.3 Modern Liquid Chromatography

Although only a limited number of papers have been published on the subject, modern LC certainly is of value equal to GLC for the determination of synthetic K vitamins.

The first report on the subject was that of Williams et al. [124], who used methanol-water (20:80, v/v) to separate menadione (K_3) from other FSV on a Permaphase ODS column. In this paper a number of interesting suggestions for the isolation of fat-soluble vitamins from capsules, multilayered tablets, and oil matrices were given.

Ranfft and Ruckemann [186] described a method for the rapid determination of K_3 in vitamin premixes and mineral supplements. Menadione was extracted as its bisulfite adduct with ethanol-water (40:60, v/v) and, after tannin precipitation and conversion into menadione with Na_2CO_3, taken up into *n*-hexane. A 20 μl aliquot of this *n*-hexane phase was injected directly on a C_{18} column (Kieselgel ODS, Latek, Heidelberg, West Germany) with methanol-1% ammonium carbonate (95:5, v/v) as the eluent. Although the injection of *n*-hexane in a RP system is rather uncommon and no internal standard was used, a coefficient of variation of 3.9% was reported for feeds containing an average amount of 29.8 g MSB per kilogram.

Elton-Bott and Stacey [187], used an alkylnitrile bonded phase (MicroPak-CN, 10 μm, Varian Associates) to separate the fat-soluble vitamins A, D, E, and K_3 in formulated feeds and cod-liver oil without prior saponification. A straight-phase system was selected to prevent contamination of the column by triglycerides, as would have occurred on RP columns. In the case of cod-liver oil, samples simply were dissolved in the mobile phase (final concentration, 10-20 mg/ml). Powdered feed samples were first extracted with *n*-hexane-chloroform-ethanol (6:3.4:0.5, v/v). Without any further cleanup the samples were chromatographed with a mobile phase of methylene chloride-chloroform-*n*-hexane (3:2:15, v/v) using either thymol, phenol, or benzyl alcohol as the internal standard. Menadione eluted with a retention time of 4.6 min between all-*trans*-retinyl acetate and 13-*cis*-retinyl acetate. Detection was performed at 254 nm. Stopped-flow scanning was used for the identification of the isolated vitamins. Samples containing MSB were first subjected to an alkaline hydrolysis of MSB into menadione. For the extraction of FSV from gelatin-protected beadlets, dispersion of the sample in warm dimethylsulfoxide was recommended.

Very recently, a comparison of the dinitrophenylhydrazine colorimetric method and HPLC for the determination of K_3 in mixed feeds was published. LC compared favorably from a standpoint of reproducibility and precision. The method was used for stability studies of various synthetic analogs during pelleting and storage of the feeds [188].

Finally, Brunt et al. [163] reported the use of a differential amperometric detector for the LC determination of menadione. His system consisted of a Nucleosil RP-8, 5 μm column (15 cm × 4.6 mm ID) eluted with methanol-water-1.0 M aqueous acetic acid (60:30:10, v/v) adjusted to pH 5.5 with sodium hydroxide. A linear dynamic range of 0.5-200 ng was observed for menadione, which according to Brunt (Ph.D. thesis), compares favorably with other techniques.

4. CONCLUSION

Liquid chromatography seems to us the method of choice for the analysis of both natural and synthetic K vitamins. It provides an excellent selectivity and sensitivity, and sample preparation, e.g., cleanup procedures, are by far the simplest with this technique. Moreover, chances for sample degradation during the chromatographic procedure itself are minimal.

TLC, on the other hand, may be advantageous when large numbers of samples must be analyzed in parallel, as well as for fast screening of different chromatographic systems. Examination of a sample by adsorption, argentation, and reversed-phase HPLC would indeed immobilize at least two liquid chromatographs for a considerable period of time. One must take account, however, of the labile nature of the natural K vitamins and menadione. They should be protected from air and light, in particular when adsorbed onto the active surface of the TLC plate.

Finally, GC in our opinion will never become a routine method for the analysis of natural K vitamins. Future applications will rather be restricted to the area of specific detection and separation problems (determination of diastereomers, GC-MS, and GC-ECD). GC analysis of synthetic K analogs, on the other hand, is less problematic. The technique is still in use in many routine laboratories, but nevertheless is gradually being replaced by LC methods.

ACKNOWLEDGMENTS

The authors wish to thank Dr. H. Nelis for his stimulating advice and C. De Groote for her very valuable technical assistance in preparing this manuscript.

REFERENCES

1. E. A. Doisy, D. W. MacCorquodale, S. A. Thayer, S. B. Binkley, and R. W. McKee, *Science, 90*, 407 (1939).
2. H. Dam, A. Greiger, J. Glavind, P. Karrer, W. Karrer, E. Rothschild, and H. Salomon, *Helv. Chim. Acta, 22*, 310 (1939).
3. R. E. Olson, Vitamin K, in *Human Nutrition—A comprehensive Treatise*, Vol. 3B (R. Alfin-Shtarand and D. Katchovski, eds.), Plenum Publishing, New York, 1980, p. 267.
4. T. J. Duello and J. T. Matschiner, *J. Nutr., 102*, 331 (1972).
5. R. W. McKee, S. B. Binkley, S. A. Thayer, D. W. MacCorquodale, and E. A. Doisy, *J. Biol. Chem., 131*, 327 (1939).
6. S. B. Binkley, R. W. McKee, S. A. Thayer, and E. A. Doisy, *J. Biol. Chem., 133*, 721 (1940).
7. O. Isler, R. Ruegg, L. H. Chopard-dit-Jean, A. Winterstein, and O. Wiss, *Helv. Chim. Acta, 41*, 786 (1958).

8. M. D. Collins, T. Pirouz, and M. Goodfellow, *J. Gen. Microbiol.*, *100*, 221 (1977).
9. J. T. Matschiner and R. G. Bell, *J. Nutr.*, *102*, 625 (1972).
9a. M. Billeter, W. Bolliger, and C. Martius, *Biochem. Z.*, *340*, 290 (1964).
10. P. Rietz, U. Gloor, and O. Wiss, *Int. Z. Vitam. Forsch.*, *40*, 351 (1970).
11. S. Ansbacher and E. Fernholz, *J. Amer. Chem. Soc.*, *61*, 1924 (1939).
12. J. T. Matschiner, R. G. Bell, J. M. Anelotti, and T. E. Knauer, *Biochim. Biophys. Acta*, *201*, 309 (1970).
12a. J. W. Suttie, *Fed. Proc.*, *39*, 2830 (1980).
13. M. J. Shearer, A. McBurney, and P. Barkhan, *Vitam. Horm.*, *32*, 513 (1974).
14. M. J. Shearer and P. Barkhan, *Biochim. Biophys. Acta*, *297*, 300 (1973).
15. C. M. Jackson and J. W. Suttie, *Prog. Haematol.*, *10*, 333 (1977).
16. P. M. Gallop, J. S. Lian, and P. V. Hauschka, *N. Engl. J. Med.*, *302*, 1460 (1980).
17. J. B. Lian, P. V. Hauschka, and P. M. Gallop, *Fed. Proc.*, *37*, 2615 (1978).
18. A. E. Griep and P. A. Friedman, Purification of a protein containing 3-carboxyglutamic acid from bovine kidney, in *Vitamin K Metabolism and Vitamin K-Dependent Proteins* (J. W. Suttie, ed.), University Park Press, Baltimore, 1980, p. 307.
19. R. J. Levy, J. B. Lian, and P. M. Gallop, 3-Carboxyglutamic acid and arteriosclerotic plaque, in *Vitamin K Metabolism and Vitamin K-Dependent Proteins* (J. W. Suttie, ed.), University Park Press, Baltimore, 1980, p. 269.
20. J. Scheinbuhs, The presence of γ-carboxyglutamic acid in nascent polypeptides and ribosomal proteins from rat liver, in *Vitamin K Metabolism and Vitamin K-Dependent Proteins* (J. W. Suttie, ed.), University Park Press, Baltimore, 1980, p. 279.
21. R. G. Bell, Vitamin K dependent carboxylation in lung microsomes, in *Vitamin K Metabolism and Vitamin K-Dependent Proteins* (J. W. Suttie, ed.), University Park Press, Baltimore, 1980, p. 286.
22. S. D. Buchtal and R. G. Bell, Vitamin K-dependent carboxylation in spleen and kidney, in *Vitamin K Metabolism and Vitamin K-Dependent Proteins* (J. W. Suttie, ed.), University Park Press, Baltimore, 1980, p. 299.
23. H. P. Traverso, P. V. Hauschka, and P. M. Gallop, Vitamin K-dependent carboxylation in microsomal preparations from cultured kidney cells, chick embryo fibroblasts and pancreas, in *Vitamin K Metabolism and Vitamin K-Dependent Proteins* (J. W. Suttie, ed.), University Park Press, Baltimore, 1980, p. 311.
24. R. Delanoy and P. N. Gray, *Fed. Proc.*, *38*, 710 (1979).
25. R. E. Olson and J. W. Suttie, *Vitam. Horm.*, *35*, 59 (1977).
26. P. Hilgard, The use of oral anticoagulants in tumor therapy, in

Malignancy and the Hemostatic System (M. R. Donnati et al., eds.), Raven Press, New York, 1981, p. 103.
27. H. P. Traverso, P. V. Hauschka, and P. M. Gallop, *Calcif. Tissue Int.*, *30*, 73 (1980).
28. M. Shoyab and G. J. Todaro, *J. Biol. Chem.*, *255*, 8735 (1980).
29. G. Leary and C. Porter, *J. Chem. Soc.*, *A*, 2273 (1970).
30. C. D. Snyder and H. Rapoport, *J. Amer. Chem. Soc.*, *91*, 731 (1969).
31. H. Taber, Function of vitamin K_2 in microorganisms, in *Vitamin K Metabolism and Vitamin K-Dependent Proteins* (J. W. Suttie, ed.), University Park Press, Baltimore, 1980, p. 177.
32. P. J. Dunphy and A. F. Brodie, *Methods Enzymol.*, *18C*, 407 (1971).
33. J. F. Pennock, *Vitam. Horm.*, *24*, 307 (1966).
34. H. Dam and J. Glavind, *Biochem. J.*, *32*, 485 (1938).
35. H. J. Almquist, *Biol. Symp.*, *12*, 508 (1947).
36. J. D. Manes, Jr., D. A. Cook, and D. L. Schneider, Vitamin K, in *Densitometry in Thin Layer Chromatography, Practice and Applications* (J. C. Touchstone and J. Sherma, eds.), J. Wiley and Sons, New York, 1979, p. 729.
37. R. M. Bertina, W. Van der Marel-Van Nieuwkoop, J. Dubbeldam, R. J. Boekhout-Mussert, and J. J. Veltkamp, *Clin. Chim. Acta*, *105*, 93 (1980).
38. H. Dam, A. Greiger, J. Glavind, P. Karrer, W. Karrer, E. Rotschild, and H. Salomon, *Helv. Chim. Acta*, *22*, 310 (1939).
39. F. Irreverre and M. X. Sullivan, *Science*, *94*, 497 (1941).
40. J. V. Scudi and R. P. Buhs, *J. Biol. Chem.*, *141*, 451 (1941).
41. P. J. Dunphy and A. F. Brodie, *Methods Enzymol.*, *18C*, 427 (1971).
42. J. S. M. Hassan, *Methods Enzymol.*, *67*, 125 (1980).
43. J. S. M. Hassan, *J. Ass. Offic. Anal. Chem.*, *64*, 611 (1981).
44. L. B. Burns, R. J. Stedman, and M. M. Tuckerman, *J. Pharm. Sci.*, *67*, 258 (1978).
45. A. Gaudiano, G. Bellomont, E. Sanzini, G. Gilardi, and S. Civalleri, *Riv. Soc. Ital. Sci. Aliment.*, *6*, 193 (1977).
46. V. Dadak and L. Krivankova, *Methods Enzymol.*, *67*, 128 (1980).
47. J. C. Viré, G. J. Patriarche, R. J. Nowak, and H. B. Mark, Jr., *Anal. Chem.*, *49*, 1343 (1977).
48. J. Van Damme and R. De Nève, *Anal. Lett.*, *B11*, 501 (1978).
49. J. J. Aaron, *Methods Enzymol.*, *67*, 139 (1980).
50. G. J. Patriarche and Lingane, *Anal. Chim. Acta*, *49*, 241 (1970).
51. I. T. Takahashi, *Al. Dent. Akad. Wiss. Berlin Klin. Med.*, *4*, 589 (1966).
52. K. Takamura and F. Watanabe, *Anal. Biochem.*, *74* (1976).
53. K. Takamura, M. Sakamoto, and Y. Hayakawa, *Anal. Chim. Acta*, *106*, 261 (1979).

54. J. C. Viré, G. J. Patriarche, and G. D. Christian, *Anal. Chem.*, *51*, 752 (1979).
55. J. C. Viré, G. J. Patriarche, and G. D. Christian, *Pharmazie*, *35*, 209 (1980).
56. J. C. Viré, G. J. Patriarche, and G. D. Christian, *Mikrochim. Acta*, *1*, 17 (1977).
56a. J. J. Aaron and J. D. Winefordner, *Anal. Chem.*, *44*, 2122 (1972).
57. J. D. Manes, H. B. Fluckiger, and D. L. Schneider, *J. Agr. Food Chem.*, *20*, 1130 (1972).
58. J. T. Matschiner, W. V. Taggart, and J. M. Amelotti, *Biochemistry*, *6*, 1243 (1967).
59. H. Mayer and O. Isler, *Methods Enzymol.*, *18C*, 469 (1971).
60. D. H. Bishop and H. L. King, *Biochem. J.*, *85*, 550 (1962).
61. T. J. Duello and J. T. Matschiner, *Arch. Biochem. Biophys.*, *138*, 640 (1970).
62. J. T. Matschiner and W. V. Taggart, *Anal. Biochem.*, *18*, 88 (1967).
63. E. Nystrom and J. Sjovall, *J. Chromatogr.*, *24*, 212 (1966).
64. M. N. Merzlyak, I. I. Ivanov, and B. N. Taruzov, *Stud. Biophys.*, *30*, 29 (1972).
65. E. Rittich, M. Simek, and J. Coupek, *J. Chromatogr.*, *133*, 345 (1977).
66. M. E. Snook, P. J. Fortson, and O. T. Chortyk, *Tob. Int.*, *182*, 78 (1980).
67. C. T. Leth, Determination of vitamin K_1 by HPLC, with the use of menaquinone-4 as internal standard, in Proceedings of the Vth FSV Symposium, Leeds, U.K., 1981.
68. G. B. Horn, *J. Pharm. Sci.*, *67*, 834 (1978).
69. G. H. Dialameh and R. E. Olson, *Anal. Biochem.*, *32*, 263 (1969).
70. R. M. Seifert, *J. Agr. Food Chem.*, *27*, 1301 (1979).
71. M. J. Shearer, V. Allan, Y. Haroon, and P. Barkhan, Nutritional aspects of vitamin K in the human, in *Vitamin K Metabolism and Vitamin K-Dependent Proteins* (J. W. Suttie, ed.), University Park Press, Baltimore, 1980, p. 317.
72. P. Sommer and M. Kofler, *Vitam. Horm.*, *24*, 349 (1966).
73. H. R. Bolliger and A. Konig, Vitamin K Gruppe und verwandte Chinone, in *Dunnschicht-Chromatographie* (E. Stahl, ed.), Springer-Verlag, Berlin, 1967, p. 280.
74. C. Lopes Silva and T. Ioneda, *Rev. Microbiol. (S. Paulo)*, *8*, 39 (1977).
75. R. K. Hammond and D. C. White, *J. Chromatogr.*, *45*, 446 (1969).
76. M. J. Shearer, P. Barkhan, and G. R. Webster, *Br. J. Haematol.*, *18*, 297 (1970).
77. M. J. Shearer, C. N. Mallison, G. R. Webster, and P. Barkhan, *Br. J. Haematol.*, *22*, 579 (1972).
78. R. M. Wilson, T. F. Walsh, and S. K. Gee, *Tetrahedron Lett.*, *21*, 3459 (1980).

79. J. T. Matschiner, R. G. Bell, J. M. Amelotti, and T. E. Knauer, *Biochim. Biophys. Acta, 201*, 309 (1970).
80. P. L. Donnahay, V. T. Burt, H. H. Rees, and J. F. Pennock, *J. Chromatogr., 170*, 272 (1979).
81. Y. Haroon, M. J. Shearer, and P. Barkhan, *J. Chromatogr., 206*, 333 (1981).
82. M. D. Collins, H. N. Shah, and D. E. Minnikin, *J. Appl. Bacteriol., 48*, 277 (1980).
82a. N. Perisic-Janjic, S. Petrovic, and P. Hadric, *Chromatographia, 9*, 130 (1976).
83. B. Rittich, M. Krska, M. Simek, and J. Coupek, *Nauk. Zootech., 5*, 33 (1978).
84. I. M. Campbell and R. Bentley, *Biochemistry, 8*, 4651 (1969).
85. H. C. Chiang and C. M. Jan, *T'ai-Wan Yao Hsuch Tsa Chich, 23*, 37 (1971).
86. R. L. Lester and T. Ramasarma, *J. Biol., 234*, 672 (1969).
87. A. Emmerie and C. Engel, *Rec. Trav. Chim., 57*, 1351 (1938).
88. J. W. Hamilton and R. D. Dallam, *Arch. Biochem. Biophys., 123*, 514 (1968).
89. G. P. McSweeney, *J. Chromatogr., 17*, 183 (1965).
90. M. H. Hashmi, F. R. Chughtai, and M. I. D. Chughtai, *Microchim. Acta, 53*, 1969).
91. Y. J. Abul-Hajj, *J. Biol. Chem., 253*, 2356 (1978).
92. L. Ersov and R. Duden, *Lebensm. Wiss. Technol., 13*, 198 (1980).
93. T. Konishi and S. Baba, *Chem. Pharm. Bull., 71*, 1906 (1973).
94. B. Rittich, M. Simek, and M. Krska, *Cesk. Farm., 25*, 64 (1976).
95. I. V. Stepanova and M. B. Aman, *Rybn. Khor. (Moscow), 1*, 75 (1977).
96. J. T. Matschiner and J. M. Amelotti, *J. Lipid Res., 9*, 176 (1968).
97. T. T. Duello and J. T. Matschiner, *Arch. Biochem. Biophys., 144*, 330 (1971).
98. G. Katsui and M. Ohmae, *J. Vitaminol., 12*, 117 (1966).
99. O. Jansson, *Acta Chem. Scand., 24*, 2839 (1970).
100. J. Ripphahn and H. Halpaap, Application of a new high performance layer in quantitative thin-layer chromatography, in *High Performance Thin-Layer Chromatography* (A. Zlatkis and R. E. Kaiser, eds.), Elsevier, Amsterdam, 1977, p. 189.
101. U. B. Hezel, Potential and experience in quantitative HPTLC, in *High Performance Thin-Layer Chromatography* (A. Zlatkis and R. E. Kaiser, eds.), Elsevier, Amsterdam, 1977, p. 147.
102. H. Bethke, W. Santi, and R. W. Frei, *J. Chromatogr. Sci., 12*, 392 (1974).
103. Y. Haroon, M. J. Shearer, and P. Barkhan, *J. Chromatogr., 200*, 293 (1980).
104. R. M. Seifert, *J. Agr. Food Chem., 27*, 1301 (1979).
105. J. Green, E. Edwin, A. Diplock, and D. McHale, *Biochem. Biophys. Res. Commun, 2*, 269 (1960).

106. D. Misiti, H. W. Moore, and K. Folkers, *Biochemistry, 4*, 1156 (1965).
107. P. Nair and D. Turner, *J. Amer. Oil Chem. Soc.*, *40*, 353 (1963).
108. A. J. Sheppard, A. R. Prosser, and W. D. Hubbard, *J. Amer. Oil Chem. Soc.*, *49*, 619 (1972).
109. G. H. Dialameh and R. E. Olson, *Anal. Biochem.*, *32*, 263 (1969).
110. H. Bechthold and E. Jahnchen, *J. Chromatogr.*, *164*, 85 (1979).
111. T. Ninomiya, K. Kodokoro, M. Horiguchi, and N. Higosaki, *Vitamins (Kyoto), 27*, 349 (1963).
112. D. A. Libby, A. R. Prosser, and A. J. Sheppard, *J. Ass. Offic. Anal. Chem.*, *50*, 806 (1967).
113. K. K. Carroll and D. C. Herting, *J. Amer. Oil Chem. Soc.*, *41*, 473 (1964).
114. W. Vetter, M. Vecchi, H. Gutmann, R. Ruëgg, W. Walther, and P. Meyer, *Helv. Chim. Acta, 50*, 1866 (1967).
115. M. T. Yanotovskii, E. I. Korlov, E. A. Obol'nikova, O. I. Volkova, and G. I. Samokhvalov, *J. Gas Chromatogr.*, *6*, 520 (1968).
116. Y. Nakata, Y. Mita, and S. Khono, *Yakugaku Zasshi, 96*, 53 (1976).
117. Y. Nakata and E. Tsuchida, *Methods Enzymol.*, *67*, 150 (1981).
118. M. Vecchi, M. Schmid, W. Walther, and F. Gerber, *J. High Resolut. Chromatogr. Chromatogr. Commun.*, *4*, 257 (1981).
119. H. Morimoto, I. Imada, T. Amano, M. Toyoda, and Y. Ashida, *Biochem. Med.*, *7*, 169 (1973).
120. J. H. M. Van Den Berg, in *Gaschromatografie*, course manual, T. H. Eindhoven, Netherlands, May 1976, p. 50.
121. A. J. Sheppard, *Methods Enzymol.*, *18C*, 461 (1971).
122. G. H. Dialameh, T. S. Najafi, and R. E. Olson, *Fed. Proc.*, *27*, 436 (1968).
123. P. Dugan and D. Lundgren, *Anal. Biochem.*, *8*, 312 (1964).
124. R. C. Williams, J. A. Schmit, and R. A. Henry, *J. Chromatogr. Sci.*, *10*, 494 (1974).
125. M. J. Shearer, High-Performance Liquid Chromatography of K Vitamins and Their Antagonists, in *Advances in Chromatography, Vol. 24* (J. Calvin Giddings, E. Grushka, J. Cazes and P. R. Brown, eds), Marcel Dekker, New York, 1983, p. 243.
126. P. L. Donnahey, V. T. Burt, H. H. Rees, and J. F. Pennock, *J. Chromatogr.*, *170*, 272 (1979).
127. M. F. Lefevere, A. P. De Leenheer, and A. E. Claeys, *J. Chromatogr.*, *186*, 749 (1979).
128. L. R. Snyder and J. J. Kirkland, *Introduction to Modern Liquid Chromatography*, 2nd ed., J. Wiley and Sons, New York, 1979, p. 246.
129. L. R. Snyder, *J. Chromatogr. Sci, 16*, 223 (1978).
130. D. L. Saunders, *Anal. Chem.*, *46*, 470 (1974).

131. H. Engelhardt, *J. Chromatogr. Sci.*, *15*, 380 (1977).
132. D. L. Saunders, *J. Chromatogr.*, *125*, 163 (1976).
133. J. P. Thomas, A. Brun, and J. P. Bounine, *J. Chromatogr.*, *139*, 21 (1977).
134. J. P. Thomas, A. Brun, and J. P. Bounine, *Analusis*, *7*, 221 (1979).
135. H. K. Lichtenthaler and U. Prinzel, *J. Chromatogr.*, *135*, 493 (1977).
136. J. N. Thompson, G. Hatina, and W. B. Maxwell, in *Trace Organic Analysis: A New Frontier in Analytical Chemistry*, Special Publication 519 of National Bureau of Standards, Gaithersburg, Maryland, 1979, p. 279.
137. R. J. Maggs, *J. Chromatogr. Sci.*, *7*, 145 (1969).
138. W. R. Sisco and R. K. Gilpin, *J. Chromatogr. Sci.*, *18*, 41 (1980).
139. R. P. W. Scott, *J. Chromatogr. Sci.*, *18*, 297 (1980).
140. R. M. McCormick and B. L. Karger, *Anal. Chem.*, *52*, 2249 (1980).
141. W. Melander, J. Stoveken, and C. Horvath, *J. Chromatogr.*, *199*, 35 (1980).
142. A. Nahum and C. Horvath, *J. Chromatogr.*, *203*, 53 (1981).
143. K. E. Bij, C. Horvath, W. R. Melander, and A. Nahum, *J. Chromatogr.*, *203*, 65 (1981).
144. Y. Yamano, S. Ikenoya, T. Tsuda, M. Ohmae, and K. Kawabe, *Yakugaku Zasshi*, *97*, 486 (1977).
144a. Y. Yamano, S. Ikenoya, M. Ohmae, and K. Kawabe, *Yakugaku Zasshi*, *99*, 1102 (1979).
145. G. R. Elliot, F. M. Odam, and M. S. Townsend, *Biochem. Soc. Trans.*, *4*, 615 (1976).
146. N. A. Parris, *J. Chromatogr.*, *15*, 161 (1978).
147. T. D. Bjornsson, S. E. Swezey, P. J. Meffin, and T. F. Blaschke, *Thromb. Haemost.*, *30*, 466 (1978).
148. O. Hiroshima, K. Abe, S. Ikenoya, M. Ohmae, and K. Kawabe, *Yakugaku Zasshi*, *99*, 1007 (1979).
149. D. O. Mack, *J. Liquid Chromatogr.*, *3*, 1005 (1980).
150. S. A. Barnett and L. W. Frick, *Anal. Chem.*, *51*, 641 (1979).
151. J. F. K. Huber and F. Eisenbeiss, *J. Chromatogr.*, *149*, 127 (1978).
152. F. Erni and R. W. Frei, *J. Chromatogr.*, *149*, 561 (1978).
153. R. L. Miller, K. Ogan, and A. F. Poile, *Int. Lab.*, *11*(6), 60 (1981).
154. D. H. Freeman, *Anal. Chem.*, *53*, 1 (1981).
155. M. F. Lefevere, A. P. De Leenheer, and A. E. Claeys, *Int. J. Vitam. Nutr. Res.*, *51*, 179 (1981).
156. M. F. Lefevere, A. P. De Leenheer, A. E. Claeys, I. V. Claeys, and H. Steyaert, *J. Lipid Res.*, *23*, 1088 (1982).

157. H. Hulpke and U. Werthmann, *Chromatographia, 13*, 305 (1980).
158. L. R. Snyder and J. J. Kirkland, *Introduction to Modern Liquid Chromatography*, 2nd ed., J. Wiley and Sons, New York, 1979, p. 560.
159. B. L. Karger, M. Martin, and G. Guiochon, *Anal. Chem. 46*, 1640 (1974).
160. K. Abe, O. Hiroshima, K. Ishibashi, M. Ohmae, K. Kawabe, and G. Katsui, *Yakugku Zasshi, 99*, 192 (1979).
161. S. Ikenoya, K. Abe, T. Tsuda, Y. Yamano, O. Hiroshima, M. Ohmae, and K. Kawabe, *Chem. Pharm. Bull. 27*, 1237 (1979).
162. O. Hiroshima, S. Ikenoya, M. Ohmae, and K. Kawabe, *Chem. Pharm. Bull, 29*, 451 (1981).
163. K. Brunt, C. H. P. Bruins, and D. A. Doornbos, *Anal. Chim. Acta, 125*, 85 (1981).
163a. J. P. Langenberg and U. R. Tjaden, *J. Chromatogr., 305*, 61 (1984).
164. M. F. Lefevere, R. W. Frei, A. H. Scholten, and U. A. Brinkman, *Chromatographia, 15*, 459 (1982).
165. J. J. Aaron, J. E. Villafranca, V. R. White, and J. M. Fitz-gerald, *Appl. Spectr., 30*, 159 (1976).
166. F. W. Willmott and R. J. Dolphin, *J. Chromatogr. Sci., 12*, 695 (1974).
166a. M. F. Lefevere, R. W. Frei, U. A. Brinkman, A. P. De Leenheer, and A. E. Claeys, A critical evaluation of electrochemical electron-capture and fluorimetric reaction-detection for the selective quantification of vitamin K in LC effluents. Paper presented at the Detection in HPLC Symposium, Amsterdam, January 1982.
167. S. A. Barnett, L. W. Frick, and H. M. Baine, *Anal. Chem., 52*, 610 (1980).
168. Y. Yamano, S. Ikenoya, M. Anze, M. Ohmae, and K. Kawabe, *Yakugaku Zasshi, 98*, 774 (1978).
169. J. N. Thompson, Application of high pressure liquid chromatographic methods for determination of fat soluble vitamins A, D, E and K in foods and pharmaceuticals, in *Proceedings 1978 Symposium, Association of Vitamin Chemists*, Chicago, 1978, p. 69.
170. Y. Haroon, M. J. Shearer, G., McEnery, V. E. Allan and P. Barkhan, *Proc. Nutr. Soc., 39*, 48A (1980).
171. Y. J. D. Chiu, R. K. Zee-Cheng, and R. E. Olson, *Fed. Proc., 40*, 873 (1981).
172. J. M. L. Mee, C. C. Brooks, and K. H. Yanagihara, *J. Chromatogr., 110*, 178 (1975).
173. W. Dummler and P. Dorfling, *J. Chromatogr., 30*, 263 (1967).
174. H. Thielemann, *Sci. Pharm., 45*, 242 (1977).
175. H. Thielemann, *Mikrochim. Acta, 2*, 227 (1972).
176. H. Thielemann, *Sci. Pharm., 40*, 40 (1972).
177. D. A. Libby and A. J. Sheppard, *J. Ass. Offic. Anal. Chem., 48*, 973 (1965).
178. A. J. Sheppard and W. D. Hubbard, *J. Ass. Offic. Anal. Chem., 53*, 1093 (1970).

179. A. J. Sheppard, W. D. Hubbard, and A. D. Prosser, *J. Ass. Offic. Anal. Chem.*, *55*, 660 (1972).
180. G. Castello, E. Bruschi, and G. Ghelli, *J. Chromatogr.*, *139*, 195 (1977).
181. J. Cornelius and H. Y. Yang, *J. Gas Chromatogr.*, *5*, 327 (1967).
182. S. Silvestri and J. Staibano, *Farm. Ed. Prat.*, *23*, 613 (1968).
183. V. W. Winkler and J. M. Yoder, *J. Ass. Offic. Anal. Chem.*, *55*, 1219 (1972).
184. V. W. Winkler, *J. Ass. Offic. Anal. Chem.*, *56*, 1277 (1973).
185. W. D. Hubbard and A. J. Sheppard, *J. Ass. Offic. Anal. Chem.*, *51*, 15 (1968).
186. K. Ranfft and H. Ruckemann, *Z. Lebensm. Unters. Forsch.*, *167*, 150 (1978).
187. R. R. Elton-Bott and C. I. Stacey, *Anal. Chim. Acta*, *127*, 213 (1981).
188. W. Mueller-Mulot and H. Lindner, *Kraftfutter*, *64*, 356 (1981).

5

ASCORBIC ACID AND RELATED COMPOUNDS

MAI H. BUI-NGUYÊN

Swiss Federal Dairy Research Institute, Swiss Institute of Vitamins, Lausanne, Switzerland

1. INTRODUCTION

1.1 Physical and Chemical Aspects

1.1.1 Structure

Ascorbic acid, also referred to as L-ascorbic acid or vitamin C, is a carbohydrate derivative. Carbohydrates are formed in green plants as the result of photosynthesis, which is the chemical combination or fixation of carbon dioxide and water by utilization of energy gained through the absorption of visible light. Although it is called an acid, this substance is actually not a carboxylic acid but a lactone and owes its acidic properties and ease of oxidation to the presence of an enediol grouping. It belongs to the L series by the glyceraldehyde convention (Fig. 5.1). Erythorbic acid, also known as D-ascorbic acid and isoascorbic acid (Fig. 5.1), is an epimer of L-ascorbic acid.

1.1.2 Physical and Chemical Aspects

Dry crystals of ascorbic acid are stable to air and daylight at ordinary room temperature. It is soluble in water, methyl alcohol, ether, oil, fats, and fat solvents. Ascorbic acid is less soluble in methyl alcohol than its isomer, erythorbic acid. Absorption maxima are at 245 nm (log ε = 3.88) in acid solution and 265 nm (log ε = 3.85) in neutral solution. Aqueous solutions are rapidly oxidized on exposure to air and light. The reaction is accelerated by alkalis, iron, and copper. To inhibit the oxidation of ascorbic acid reductants, i.e., hydrogen sulfide, homocysteine, 1,2-ethanedithiol, 1,3-propanedithiol, thiourea, and sodium metabisulfite, vitamine K_5 and erythorbic acid have been used. Acid solutions, i.e., metaphosphoric acid, orthophosphoric acid, perchloric acid, trichloroacetic acid and acetic acid, and ethylenediaminetetraacetic acid (EDTA) as ligands to form coordination compounds with metal ions (e.g., iron and copper), have also been employed. Stabilizers for ascorbic acid are added to some fruit juices [1] and in chemical assays, which will be further described.

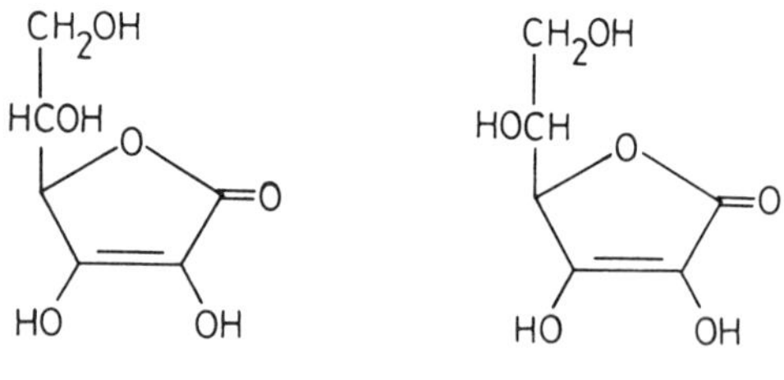

L-Ascorbic acid (vitamin C)

Erythorbic acid (isoascorbic acid)

Figure 5.1 Ascorbic acid isomer structure.

As is evident from the structural formula, ascorbic acid possesses a relatively high reducing power. It is easily and reversibly oxidized to dehydroascorbic acid, which is still physiologically active, though much less stable. Further conversion beyond the dehydroascorbic acid stage results in the irreversible formation of physiologically inactive diketogulonic acid (Fig. 5.2). Like ascorbic acid, erythorbic acid is a strong reducing agent. Erythorbic acid has been used in foods as an antioxidant but has 5% the vitamin C activity of L-ascorbic acid.

Much recent work has been done to clarify the chemical nature of compounds formed by the reversible oxidoreduction of ascorbic acid. Using in situ radiolysis electron spin resonance, Laroff et al. [2] have shown that the electron in the radical intermediates produced by oxidation of ascorbic acid is spread over a highly conjugated tricarbonyl system (Fig. 5.2). The same radical is produced from pH 1 to 13. The radical is relatively unreactive and reacts slowly with oxygen. Other workers—Kirino and Kwan [3], McAlpine et al. [4], Steeken and Ollrich [5]—confirm these results. Pfeilsticker et al. [6] have demonstrated on the basis of gas-liquid chromatographic (GLC), ultraviolet (UV), infrared (IR), nuclear magnetic resonance (NMR), and mass spectrometric (MS) data that the dehydroascorbic acid obtained by the oxidation of ascorbic acid in aqueous solution mainly exists as a bicyclic hydrated lactone. Hvoslef [7], with Nordenson [8,9], reached the same conclusion on the basis of x-ray diffraction of the crystal and NMR spectra. The commonly written structure of dehydroascorbic acid containing a tricarbonyl unit is an unlikely formula due to its relative instabilities. A fused-ring system seems to be kinetically more stable than a tricarbonyl. The more appropriate structure is shown in Fig. 5.2.

1.2 Biochemical Aspects

1.2.1 Functions of Ascorbic Acid

Ascorbic acid was first known to prevent scurvy [10]. This disease is a rare clinical finding today. Yet, interest in ascorbic acid persists. Pauling [11] and Cameron et al. [12] have demonstrated the fundamental role of this vitamin in optimizing health and resistance to infectious diseases. Wilson [13] has investigated the role of ascorbic acid in the prophylaxis of the common cold and hay fever. Ascorbic acid is largely used in therapy as an anti-infectious factor. It is essential to the normal functioning of cells. In a study on levels of total ascorbic acid and histamine in blood of women during pregnancy, the function of ascorbic acid as a modulator to control the synthesis and/or release of histamine in the body have been suggested by Scharma et al. [14]. Ascorbic acid is used in chronic alcoholic patients after short-term intravenous therapy to normalize their blood level in ascorbic acid [15]. Vitamin C reduced the formation of nitrites and N-nitroso compounds in achlor-

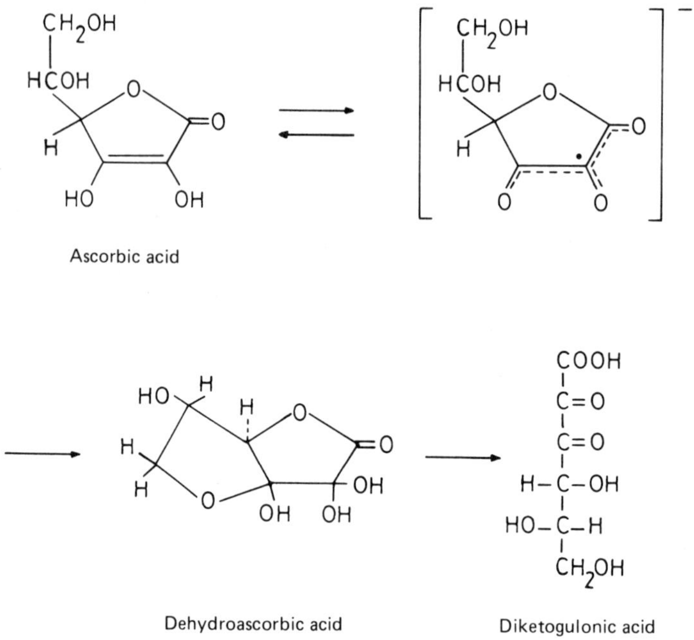

Figure 5.2 Ascorbic acid oxidation and reduction.

hydric patients and may protect susceptible individuals from the risk of gastric cancer [16]. Many other possible functions of ascorbic acid have been suggested (Table 5.1).

1.2.2 Biological Activity

Although much progress has been made in ascorbic acid biochemistry and physiology, its exact mode of action is not fully understood in chemical terms. Ascorbic acid is known as a cofactor and activator for prolyl hydroxylase in collagen synthesis [17]. From studies of the inactivation of viruses by ascorbic acid, it was concluded by Banič [35] that free radicals produced as intermediates during oxidation are responsible for the inactivating effect on the phage. Ascorbic acid is a potent reducing agent. The oxidized form, dehydroascorbic acid, is thought to play an important role in the oxidation of phenylalanine and tyrosine, but since only a small proportion of the vitamin is present as dehydroascorbic acid in the body, this is probably not its main activity. Clegg and MacDonald [36] found in a common cold survey that ascorbic acid appears to reduce the effect of the common cold virus, but not necessarily by the antiscorbutic metabolic pathway. Solanki et al. [37] have reported that a metabolite of ascorbate, such as oxalate,

TABLE 5.1 Possible Functions of Ascorbic Acid in Humans

Function	References
Collagen synthesis	17-19
Drug metabolism	20,21
Folate metabolism	22
Leukocyte function	23-25
Cyclic nucleotide metabolism	26
Interferon production	27,28
Amine metabolism	29,30
Cholesterol catabolism	31,32
Iron metabolism	33
Carnitine synthesis	34

could be responsible for the inhibition of gluconeogenesis in rat liver hepatocytes. Yourga et al. [38] and Ikeuchi [39] have shown that erythorbic acid contained little or no antiscorbutic activity in guinea pigs or in ascorbic acid-deficient humans.

1.3 Analytical Systems

Increased interest in ascorbic acid research and pharmaceutical and food controls has led to a proliferation of methodology for ascorbic acid analyses.

1.3.1 Biological Methods

Bioassay methods for the determination of ascorbic acid are time consuming and leave much to be desired insofar as precision is concerned. However, bioassays have the advantage of measuring the summation of chemical entities that possess vitamin C activity. Among the various bioassay methods, those based on a study of the odontoblasts of the incisor teeth appear to have the highest degree of specificity [40,41].

Microbiological methods are less time consuming. A qualitative microbiological assay using *Myrothecium verrucacia* was reported by Meyer et al. [42] for the estimation of ascorbic acid and erythorbic acid. Marchesini et al. [43], Posadka and Macholan [44], and Liu et al. [45] have described an enzymatic method that uses ascorbate oxidase for the assay of ascorbic acid in spinach, fruit juices, and serum or plasma, respectively. Enzymatic determination was performed using an amperometer (Clark's electrode) [43,44] or a spectrophotometer [45].

1.3.2 Chemical Methods

Chemical methods generally based on the reducing power of ascorbic acid are less specific than bioassay and microbiological methods. But, they are much faster to perform and also more reproducible. Chemical methods may be divided into two procedures, the determination of the reduced form of ascorbic acid and that of "total" vitamin C, represented by ascorbic acid, dehydroascorbic acid, and 2,3-diketogulonic acid.

Analyses of the reduced form of ascorbic acid are usually based on its reducing potential and subsequently allowed to determine ascorbic acid. Tillmans [46] introduced 2,6-dichlorophenol indophenol as the oxidizing agent for the reaction. Other oxidimetric reagents used are 0.1 N iodine solution of 0.1 N chloramine-T solution [47]. The reduction of ferric iron by ascorbic acid followed by formation of a complex of the ferrous iron products and α,α'-dipyridyl has been used to determine ascorbic acid [48,49]. Jagota and Dani [50] reported a colorimetric technique for the assay of ascorbic acid using folin phenol reagent.

Analyses of total vitamin C are based on the oxidation of ascorbic acid to diketogulonic acid and the subsequent formation of an osazone between diketogulonic acid and 2,4-dinitrophenylhydrazine, as introduced by Roe and Oesterling [51]; ascorbic acid, dehydroascorbic acid, and 2,3-diketogulonic acid are determined simultaneously or differentially [52]. Several workers [53-66] have tried to increase the specificity of these reactions.

1.3.3 Electrochemical Methods

Electrochemical techniques for the determination of ascorbic acid are more simple than the chemical method using the 2,4-dinitrophenylhydrazine reagent described above. A polarographic assay for ascorbic acid was reported by Lindquist and Ferroha [67]. Alternating current polarography [68] and isotachophoresis [69] have been developed for ascorbic acid and erythorbic acid determination. Stopped-flow methods for ascorbic acid have been described by Hiromi et al. [70] and applied to the determination of ascorbic acid in orange juic by Obata et al. [71].

2. MODERN LIQUID CHROMATOGRAPHY

2.1 Sample Handling

2.1.1 Sample Preparation

Ascorbic acid analysis presents no special problems in sample preparation for fruit juices and pharmaceutical preparations. Samples have only to be diluted with an extracting solution to obtain a final assay in the range of the working curve. In fruit products a solution of methanol and 6% metaphosphoric acid has been suggested by Shaw and Wilson [72] for extraction.

In protein-containing samples a treatment for removing protein is a major requirement. Protein precipitation may be done by using strong acids, a high concentration of salt, or organic reagents. However, samples containing strong acids and a high concentration of salt could not be chromatographed directly. Dilute perchloric acid (50 mmol/liter) has been suggested by Pachla and Kissinger [73] for the deproteinization of milk products. Another conventional technique of protein precipitation entails mixing a small volume of serum or plasma with two volumes of methanol or acetonitrile. The mixture is then centrifuged and an aliquot injected directly into the chromatograph. Protein precipitation with organic reagents is relatively easy to perform and minimizes the amount of protein injected onto the column. However, for assays in which ascorbic acid is present at low concentration this technique results in dilution of the samples. In these cases ultrafiltration of the plasma seems to be a more convenient method for removing plasma protein [74,75].

2.1.2 Stability of Ascorbic Acid

As mentioned before (Sec. 1.1.2), many stabilizers for ascorbic acid have been used. Metaphosphoric acid and oxalic acid have been found by Ponting [76] to be the most suitable for extraction of ascorbic acid. Plasma ascorbic acid demonstrated maximum stability when stored at -70°C with or without trichloroacetic acid or metaphosphoric acid [77]. For modern liquid chromatography, classic extracting solution, which affords excellent stability for ascorbic acid, could not be chromatographed directly because of its high ionic strength and strong acidity.

Generally, aqueous metaphosphoric acid alone or with acetic acid used as extractant solution affords excellent stability for ascorbic acid. Cold dilute perchloric acid, proposed by Pachla and Kissinger [73,78], Veazey and Nieman [79], also seems to be a good medium to inhibit the oxidation of ascorbic acid for ion-exchange chromatography (IEC) and ion-pair chromatography (IPC). Similarly, 4% metaphosphoric acid with 0.1% 2,3-dimercapto-1-propanol (BAL) was suggested by Floridi et al. [80] to avoid the oxidation of ascorbic acid. A solution containing 2 mg/ml sodium metabisulfite and 10% orthophosphoric acid saturated with EDTA has been used by Mason et al. [81] as a "preservative" solution for an ascorbic acid assay on amino reversed-phase chromatography (RPC). The stability of aqueous solutions of ascorbic acid for the analysis by liquid chromatography is improved by the addition of EDTA [82]. A 3% citric acid medium proposed by Wimalasini and Wills [83] also appears to have a stabilizing effect on ascorbic acid and shows no interference when measuring the absorbance at 214 nm used for dehydroascorbic acid analysis.

Acetic acid (1%) is also found to be a good medium to inhibit the oxidation of ascorbic acid. Ascorbic acid standards in 1% acetic acid solution stored at 4°C and N_2 are found fairly stable over a day. Moreover, in biological fluids (e.g., urine and serum), the stability of ascorbic acid is particularly poor, so that processing should be carried out in the shortest possible time and in an inert atmosphere to avoid significant changes of ascorbic acid prior to analysis.

2.2 Liquid-Liquid Chromatography

Classic liquid chromatography is quite specific but time consuming. This is due to poor column efficiency. In the past several years significant theoretical and technological advances have been made in the equipment, columns, and column packings; modern liquid chromatography (LC) becomes more attractive for the analysts. Excellent separation by LC can be achieved within a few minutes or less. Many applications of LC to the determination of ascorbic acid in different compounds (foods, pharmaceutical products, and biological fluids) have been reported. Ascorbic acid analysis by liquid chromatography has been performed based on three different types of mechanisms: ion-exchange chromatography, partition chromatography, and ion-pair chromatography.

2.2.1 Ion-Exchange Chromatography

Ion-exchange chromatography is the preferred method of separation for the analysis of strongly ionic compounds. The mobile phase is always an aqueous solution. The chromatographic separation may be optimized by altering the pH or ionic strength of the mobile phase. Nelson [84] introduced strong anion-exchange chromatography on a Zipax SAX column with UV detection for the separation of ascorbic acid and other food additives. Williams et al. [85] used the same technique for the analysis of water-soluble vitamins in food and pharmaceutical products. Electrochemical detection has been proposed by Thrivikraman et al. [86] for the analysis of ascorbic acid in brain tissue using IEC on a Bondapak AX/Corasil column. Kissinger et al. [87] have reported a similar technique on a Zipax SAX column for uric acid and ascorbic acid in urine, and Pachla and Kissinger [73] have applied it to the determination of ascorbic acid in foodstuffs, pharmaceuticals, and body fluids. Liquid chromatographic electrochemical results and titrimetric assay were compared, and highly consistent values were obtained; different stationary phases (Vydac SAX, Zipax SAX, and Bondapak AX/Corasil) were evaluated, and a significant variation in the retention of ascorbic acid has been observed with these packings. Brunt and

Bruins [88] have compared the response of the amperometric detector with that of an UV detector in the analysis of ascorbic acid in combination with IEC on a Nucleosil SB-10 column. The response of the amperometric detector with that of an UV detector in the analysis of ascorbic acid in combination with IEC on a Nucleosil SB-10 column. The response of the amperometric detector was about five times more sensitive than that of the UV detector. Ion-exchange chromatography with an amperometric detector has also been sued by the same author for ascorbic acid determination in a pharmaceutical preparation and in urine. The analysis of ascorbic acid and riboflavine in citrus juice has been proposed by Rouseff [89] using IEC on a Partisil-10 PAC column with UV and fluorometric detection. Pachla and Kissinger [78] have described the determination of ascorbic acid in biological samples, food, and pharmaceutical preparations using IEC and ion-pair chromatography with amperometric detection. Ion-exchange chromatography on Partisil-10 SAX column separation of ascorbic acid and other reducing agents has been proposed by Vaezey and Nieman [79] using chemiluminescence as the detector. Obata et al. [90] have reported on the determination of ascorbic acid and other reductones in orange juice and in skim milk by IEC on a SAX-801 Yanaco column. Carr and Neff [91] have described a similar technique on a Partisil-10 SAX column with amperometric detection for ascorbic acid assay in the tissues of marine animals. The ascorbic acid content in plasma, serum, and leukocytes was determined by Tsao and Salimi [92] using IEC with electrochemical detection. Using a similar technique, Saeki et al. [93] have determined simultaneously ascorbic acid and flavine adenine dinucleotide (FAD) in intravenous fluids with UV and fluorometric detection. Liebes et al. [94] assayed ascorbic acid in extracts of human lymphocytes using IEC with UV detection. Floridi et al. [80] reported on the use of an Aminex A-14, Cl^- form anion-exchange chromatography with UV detection for the determination of ascorbic acid in food and compared LC results with other methods: the thin-layer chromatography (TLC) of Vuillemier [95], the classic liquid chromatography of Bourgeois and Mainguy [96], and the chemical method of Deutsch and Weeks [55] showed excellent agreement. Ascorbic acid has been determined by Grün and Loewus [97] in algae by LC on a strong cation-exchange resin, Aminex HPX-87, with electrochemical detection.

2.2.2 *Partition Chromatography*

Partition chromatography may be subdivided into two types: normal phase and reversed phase. For ascorbic acid assay in the osazone derivative form, both normal and reversed-phase chromatography have been employed. Generally, ascorbic acid was assayed in its ionic form. The reversed-phase mode has been used on nonpolar (C_{18} and C_8) and polar (NH_2, CN) chemically bonded phases.

2.2.2.1 Normal-Phase Chromatography. Normal-phase chromatography on a Si-100 column has been used by Manso [98] to determine the total ascorbic acid in wheat flour; TLC, LC, and gas-liquid chromatographic methods have been compared by the same author, and LC was found to be more rapid. A similar LC technique has been reported by Sauberlich et al. [99] to determine ascorbic acid and dehydroascorbic acid in biological tissues. The best results were obtained on a Porasil column.

2.2.2.2 Reversed-Phase Chromatography. Vitamins B_2, B_{12}, C, and K_3 have been separated by Stillman and Ma [100] using reversed-phase chromatography on Corasil C_{18} with polarographic detection. Wills et al. [101] have resolved ascorbic acid and the other seven water-soluble vitamins in a multivitamin preparation using RPC on Bondapak C_{18} and Bondapak NH_2 with UV detection; the effect of solvent polarity, solvent pH, and the addition of various salts to the solvent on vitamin resolution was examined. The use of RPC on Bondapak C_{18} with UV detection for the determination of ascorbic acid in urine has been described by Wagner et al. [102]. Using a similar technique, Tsushida, with Fukazaw [103] and Takeo [104], has evaluated ascorbic acid and combined ascorbic acid and dehydroascorbic acid in tea after reduction of the latter to ascorbic acid with H_2S. Rueckemann [105] reported ascorbic acid assay in milk formulas by RPC on C_{18} with an electrochemical detector. Ascorbic acid in tomato products was determined by Miki [106] using RPC on Fine Sil C_{18} with UV detection and the indophenol method (AOAC); the values obtained by the former method were lower compared with those obtained by applying the latter one. Using RPC, Garcia-Castineiras et al. [107] have evaluated the osazone derivative form of ascorbic acid in human aqueous humor. Watada [108] achieved good recoveries of ascorbic acid in fresh fruits and vegetables using RPC on a C_{18} cartridge in a radial compression module. A similar technique with a radial compression module and a UV detector has been used by Shaw and Wilson [72] to determine ascorbic acid content in tropical fruits; extraction with methanol-6% metaphosphoric acid has been suggested by the authors before LC determination. Van Boekel and Neewissen [109] have determined the hydrazone derivative of ascorbic acid using RPC on C_{18} with UV detection; this method was applied to the analysis of ascorbic acid in beer.

Ascorbic acid and erythorbic acid are well resolved by RPC on polar chemically bonded silica. This technique on a Lichrosorb-NH_2 column with a UV detector has been described by Bui-Nguyên [110] to separate ascorbic acid and its isomer; the procedure was applied to fruit juices (Fig. 5.3). Carnevale [111] has reported a simulatneous estimation of ascorbic, sorbic, and benzoic acids in citrus juices by LC on a Bondapak CN column with UV detection. Mason et al. [81] have used an amino reversed-phase chromatography for the determination of

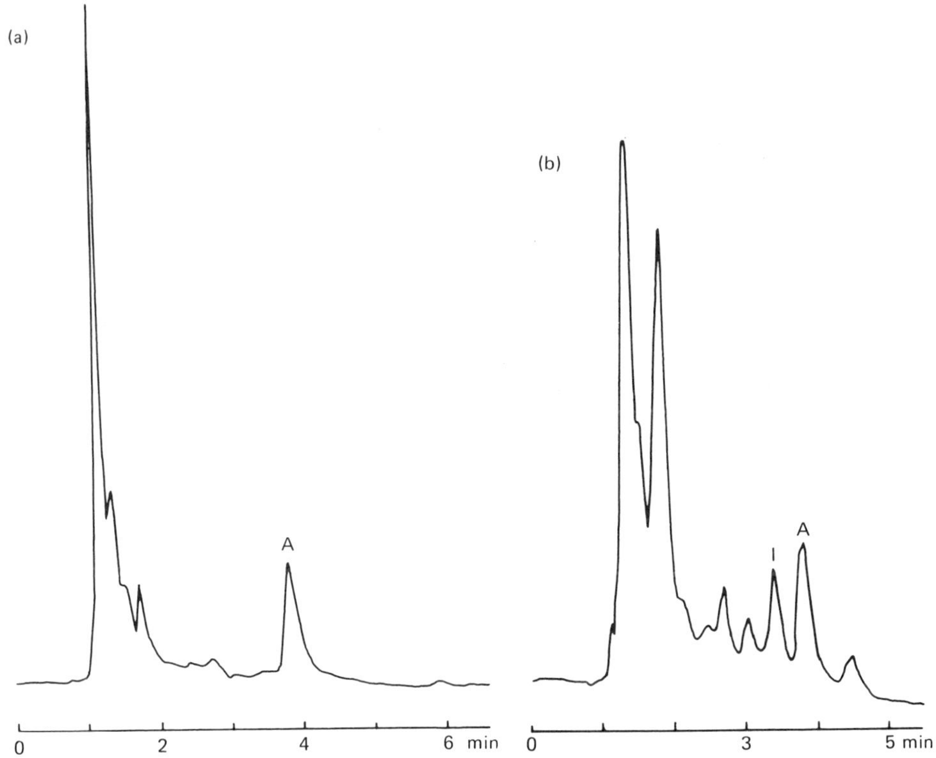

Figure 5.3 Fruit juice extract. (a) Fruit juice containing ascorbic acid (A). (b) Fruit juice containing both ascorbic acid (A) and isoascorbic acid (I). Conditions: 250 × 4.6 mm (ID) Brownlee Lichrosorb NH_2, acetonitrile-water with 5 mM/liter KH_2PO_4 (75:25), 3 ml/min, UV at 268 nm. (From Ref. 130.)

ascorbic and erythorbic acids in citrus juices and in meat products. Ascorbic, erythorbic, and nicotinic acids in curing adjuncts have been determined by Hoffmann [112] using a similar technique with UV detection. Geigert et al. [113] have used a carbohydrate analysis column to differentiate ascorbic acid and its isomer and measured UV absorbance of the components in fruit juices. Reversed-phase chromatography on Lichrosorb-NH_2 for separation of the two isomers has been described by Arakawa et al. [114] and applied to the determination of these compounds in animal tissues by Otsuka et al. [115]. A UV detector has been employed by these workers. Lichrosorb NH_2 column has also been used by Rose and Nahrwold [116] for ascorbic and dehydroascorbic acid analysis in biological samples, foods, and pharmaceutical products with UV detection, and for ascorbic acid in plasma and animal tissues with UV and scintillographic detection for radioactive material [117]. Dennison et al. described RPC on a Bondapak NH_2 column with UV detection [118] for ascorbic acid and combined ascorbic acid and dehydroascorbic acid (DHA) determination in beverages; "total" ascorbic acid was obtained by reducing DHA to AA with homocysteine. Ascorbic acid, erythorbic acid, dehydroascorbic acid, dehydroerythorbic acid, diketogulonic acid and diketogluconic acid have been resolved by RPC on Zorbax NH_2 using UV detection for ascorbic and erythorbic acids and refractive index detection for all six compounds [119]. The authors have evaluated these components in orange juice and urine with the same procedure. A technique similar to that on a Bondapak carbohydrate column with a UV detector has been reported by Lookhardt et al. [120] to determine ascorbic acid in wheat flours, bread dough conditioners, and commercial vitamin C tablets. A simultaneous assay of ascorbic acid and dehydroascorbic acid in fruits and vegetables has been described by Wimalasini and Wills [83] using RPC on Bondapak NH_2 with UV detection at 254 nm for ascorbic acid and 210 nm for dehydroascorbic acid (Fig. 5.4).

2.2.3 *Ion-Pair Chromatography*

Ion-pair chromatography, also called extraction chromatography, ion-pair extraction (or partition) chromatography, chromatography with a liquid ion exchanger, or soap chromatography, is a technique in which ion exchange and partition chromatography are combined [121-123]. The chromatographic separation may be optimized by altering the ion pair reagent, pH, and ionic strength of the mobile phase. Quaternary or primary ammonium salts have usually been employed as ion-pairing reagents for ascorbic acid assay.

Ion-pair chromatography with UV detection has been used by Wills et al. [101] for water-soluble vitamin separation and by Sood et al. [124] for ascorbic acid determination in foods and multivitamin products. Ascorbic acid and uric acid in plasma prepared by ultrafiltration has been determined by Green and Perlman [75] using IPC on a Bondapak C_{18} column with an amperometric detector. Augustin et al. [125] reported a technique similar to that above for ascorbic acid determination in potatoes. Finley and Duang [126] have resolved ascorbic, dehydroascorbic, and diketogulonic acids by IPC on two Bondapak C_{18} columns with UV detection and applied this method to orange juice. Ion-pair chromatography on Lichrosorb C_{18} or C_8 for the separation of ascorbic acid from its isomer in food has been described by Coustard and Sudraud [127]. Parolari [128] has determined ascorbic acid in meat products using IPC on a Bondapak C_{18} column; the same author has also resolved L-ascorbic acid and D-ascorbic acid using Bondapak C_{18} coupled with Bondapak CN columns. The determination of ascorbic

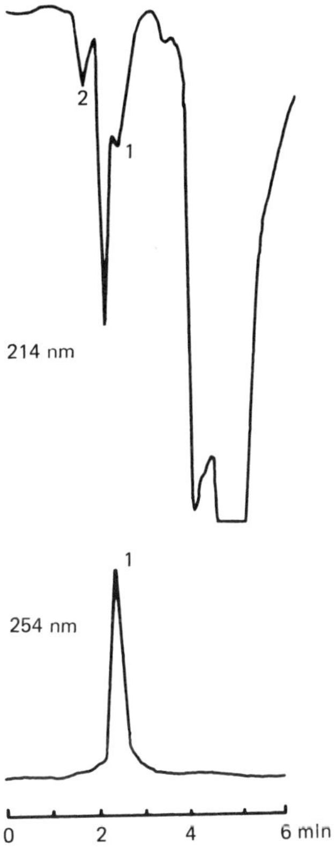

Figure 5.4 Brussels sprout extract in 3% citric acid after passage through a C_{18} Sep-pak. Conditions: 30 cm × 4 mm (ID) Bondapak/carbohydrate, acetonitrile-water (70:30, v/v) with 10 mM $NH_4H_2PO_4$, pH 4.3, 2 ml/min, UV at 254 and 214 nm. Peak 1, ascorbic acid; 2, dehydroascorbic acid. (From Ref. 143.)

acid and dehydroascorbic acid in orange juice was achieved by Keating and Haddad [129] using IPC with UV detection, following precolumn reaction of dehydroascorbic acid with 1,2-phenylenediamine. Tsao and Salimi [130] differentiated ascorbic acid and erythorbic acid by IPC and electrochemical detection and applied it to animal tissues (Fig. 5.5). The measurement of the ascorbic acid content of leukocytes using IPC with electrochemical detection has been determined by Lee et al. [131].

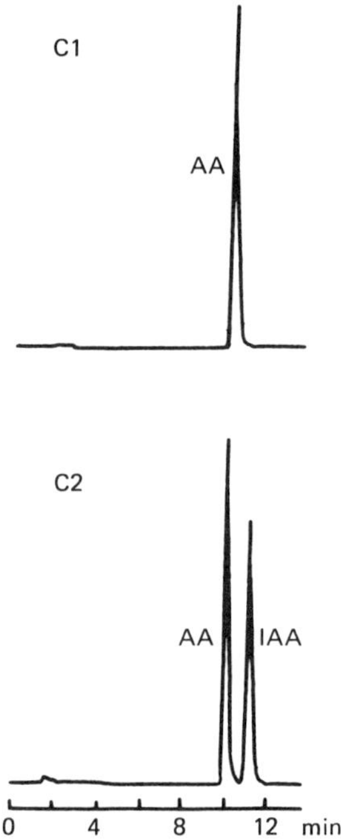

Figure 5.5 Tissue extract sample of mouse brain. (C1) Brain extract containing ascorbic acid (AA). (C2) Brain extract sample with 7 ng isoascorbic acid (IAA) standard addition. Conditions: 25 cm × 4.6 mm ID Altex C_{18}, 40 mM acetate buffer + 1 mM decylamine in methanol-water (15:85), 0.93 ml/min. Applied potential: 0.7 V versus Ag/AgCl. (From Ref. 153.)

2.3 Optimization of Chromatographic Conditions

2.3.1 Solvent Polarity and Temperature

As mentioned in partition chromatography (Sec. 2.2.2) ascorbic acid (AA) and erythorbic acid (EA) are well resolved by RPC on polar chemically bonded phase. Table 5.2 shows a comparison of resolution of AA and EA between a completely new column (Lichrosorb-NH_2), 250 × 4.6 mm inner diameter (ID), 10 μm, and the same column after 50-100 AA assays. The table also summarizes the effects of solvent polarity and column temperature on the capacity factors k' of AA and EA. The main features are as follows.

On a completely new column, AA and EA show an excellent resolution at 35°C column temperature.

TABLE 5.2 Effects of Solvent Polarity and Column Temperature on the Capacity Factor k' of Ascorbic Acid and Erythorbic Acid on an Amino Chemically Bonded Phase

Mobile phase (salt, 5 mmol/liter; flow, 3 ml/min)	Column temperature (°C)	k'_{EA}	k'_{AA}	Resolution (R_s)
New column				
KH_2PO_4-MeCN	35	3.0	3.7	1.8
"Old" column				
KH_2PO_4-MeOH (25:75, v/v)[a]	40	3.0	3.2	0.5
KH_2PO_4-MeOH (15:85)	40	4.6	4.9	0.7
KH_2PO_4-MeOH (25:75)	60	2.1	2.6	1.7

[a]"Old" column: see Sec. 2.3.1.

With the same columns after 50-100 AA analyses, there is a strong decrease in the resolution of AA and EA.

A slight decrease in polarity (methanol) varying from 75 to 85% in phosphate solution does not significantly affect the resolution of the two isomers.

With increasing column temperature (from 35 to 60°C) there is a strong increase of resolution of the two compounds.

With the reversed-phase chromatography on an amino-bonded phase under the conditions described in Fig. 5.6, ascorbic acid and EA eluted as two sharp, symmetrical, and well-resolved peaks. The separation efficiency of ascorbic acid and EA was greatly influenced by changing thermal conditions. Column temperature appeared to have more effect on the k' value of ascorbic acid than that of EA. Over the course of a day, coefficients of variation of k'_{AA} and k'_{EE} were found to be 3 and 1%, respectively.

2.3.2 *Salt Effects and pH*

Reversed-phase chromatography on nonpolar bonded-phase and ion-pair chromatography were used with varied pH on two kinds of commercial packings with a particle size of 7 μm diameter and an irregular shape (Lichrosorb RP-18), and a particle size of 5 μm diameter and a

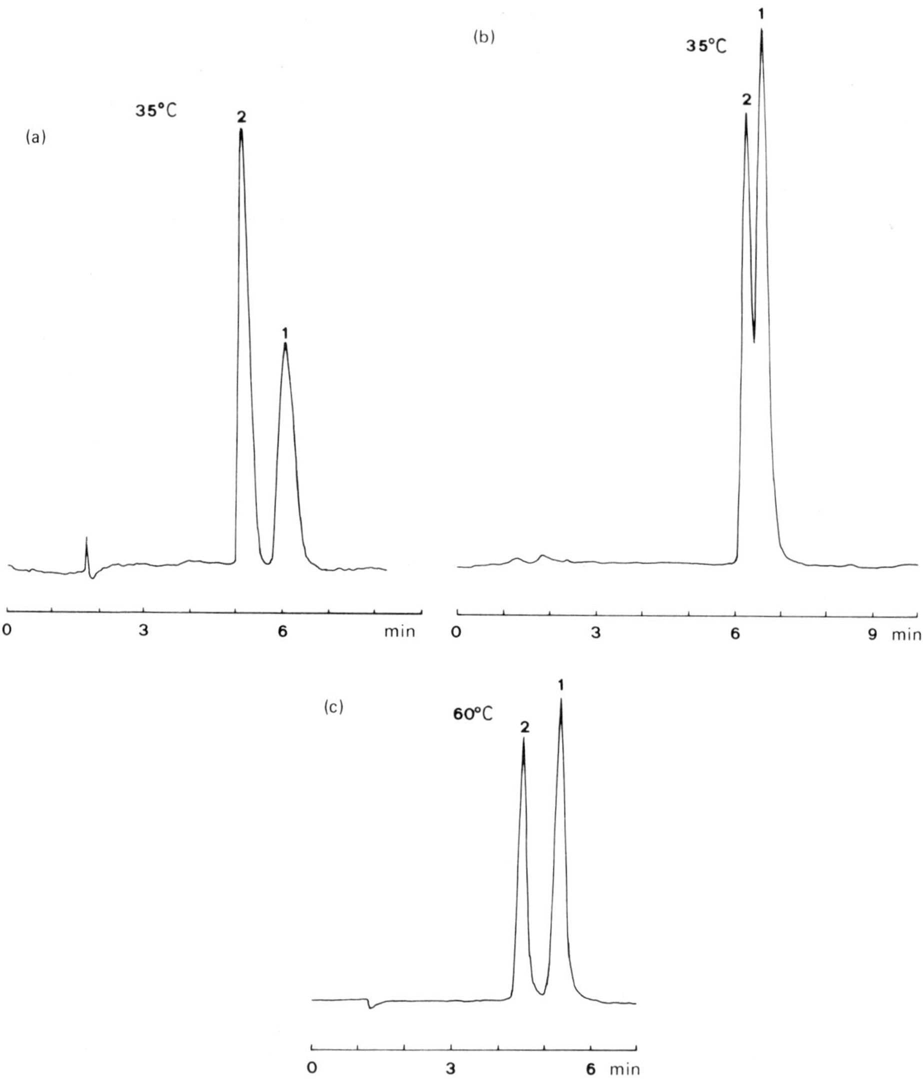

Figure 5.6 Separation of ascorbic acid (1) and erythorbic acid (2) on a Lichrosorb NH_2, 250 × 4.6 mm (ID). Conditions: methanol or acetonitrile-water with 5 mM KH_2PO_4 (75:25), 3 ml/min, UV at 265 nm. (a) new column, t_c = 35°C. (v) "old" column, t_c = 35°C. (c) "old" column, t_c = 60°C. See text.

spherical shape (Zorbax ODS). The results of the effects of salt and pH on the capacity factors of AA and EA may be summarized as follows (Table 5.3).

In a RPC with orthophosphoric acid and potassium hydrogen phosphate as mobile phase, k' values are relatively low (k' = 0.02-0.5) for both columns. An increase in pH entails a decrease in k' value.

Using cetyltrimethylammonium bromide (CTMA) as counterion, AA is strongly retained on the column. With this same counterion in phosphate buffer and methanol (9:1, v/v) as the mobile phase, Coustard and Sudraud [127] have resolved AA and EA. Using tetrabutylammonium hydroxide (TBA) in phosphate solution there is also a good separation of AA and EA: large band broadening at a pH lower than or equal to 5 and sharp, symmetrical peaks with increasing pH to about 6 (Fig. 5.7a). This effect of pH on the retention of AA has also been observed by Finley and Duang [126]. Using TBA without phosphate solution, AA and EA are well resolved but there is an inverse variation of the k' value with pH, with sharp and symmetrical peaks at pH values as low as 3 (Fig. 5.7b) and large band broadening with increasing pH to about 5. With a very small pH variation, the k' values of AA and EA with TBA vary greatly. An increase in k' value does not much affect the resolution of AA and EA. The results were more or less the same on both columns.

The separation efficiency of AA and EA is greatly influenced by changing pH in ion-pair chromatography, especially without buffer solution. The optimal pH with TBA in a phosphate buffer is of the order of 6, and without phosphate with TBA at pH 3. AA and particularly dehydroascorbic acid have been found to be fairly stable below pH 4.0. Above pH 4.0, dehydroascorbic acid readily undergoes an irreversible rearrangement to biologically inactive material. For this reason TBA solution was especially selected for AA analyses. With extracts from urine and serum, the optimal pH for the separation of AA and EA from interfering peaks was found to be near 3 with this solution. Over the course of a day, coefficients of variation of k'_{AA} and k'_{EA} were 0.2 and 3%, respectively.

The following remarks can be made when comparing Tables 5.2 and 5.3. The more polar substances, like AA, elute first on RP-18 (Table 5.3), whereas less polar substances, like EA, elute first on RP-NH_2 (Table 5.2). Therefore, AA is more polar than EA. Such a difference in polarity for the two compounds is related to the fact that AA is less soluble in methyl alcohol than its isomer, EA.

2.3.3 Summary

For ascorbic acid assay, ion-exchange chromatography would be a powerful tool. However, IEC has still not been applied for the resolution

TABLE 5.3 Effects of Salt and pH on the Capacity Factors k' of Ascorbic Acid and Erythorbic Acid in a Nonpolar Bonded Phase

		Lichrosorb RP-18			Zorbax ODS		
Mobile phase flow rate	pH	k'_{AA}	k'_{EA}	R_s	k'_{AA}	k'_{EA}	R_s
H_3PO_4, 0.2%, 2 ml/min	2.3	0.3	0.4				
KH_2PO_4, 5 mmol/liter, 1.5 ml/min	2.4	0.5	0.5		0.4	0.5	
	2.7	0.2	0.3				
	2.9				0.3	0.4	
	3.8				0.02	0.02	
$CH_3(CH_2)_{15}N(CH_3)_3Br$, 2 mmol/liter, 3 ml/min	4.6	10.3					
KH_2PO_4, 5 mmol/liter; $[CH_3(CH_2)_3]_4NOH$, 2 mmol/liter, 2 ml/min	4.0	4.8	5.6	2.0			
	4.3				5.3	6.6	1.7
	5.2	4.3	5.0	2.0			
	5.4				4.8	5.9	1.7
	6.4	1.8	2.2	1.3			
	6.8				1.6	2.0	1.4
$[CH_3(CH_2)_3]_4NOH$, 2 mmol/liter, 2 ml/min	2.7	1.2	1.4	1.2			
	2.9	2.2	3.2	1.7	3.2	4.2	1.5
	3.4	3.7	4.7	1.5			
	5.4	5.9	6.7	1.9	6.0	7.0	1.5

of ascorbic acid and related compounds. Similarly, partition chromatograpy on nonpolar chemically bonded phase does not resolve ascorbic acid from erythorbic acid and can therefore be used only for the determination of ascorbic acid at low pH (about 3) (Table 5.2).

Both ion-pair chromatography and partition chromatography on an amino-bonded phase lead to a good resolution of ascorbic acid and related compounds: dehydroascorbic, diketogulonic, and erythorbic acids. However, when RPC is used on an amino phase column, the temperature strongly affects the retention (Table 5.2) [132,133]; there is little effect of temperature on solute retention in IPC. Moreover, the column used in the case of IPC can be used for a long period. Over 500 extracts have been injected into the system with no discernible change in the chromatographic characteristics. In contrast, after about 100 injections of extracts, a decrease became apparent in the resolution

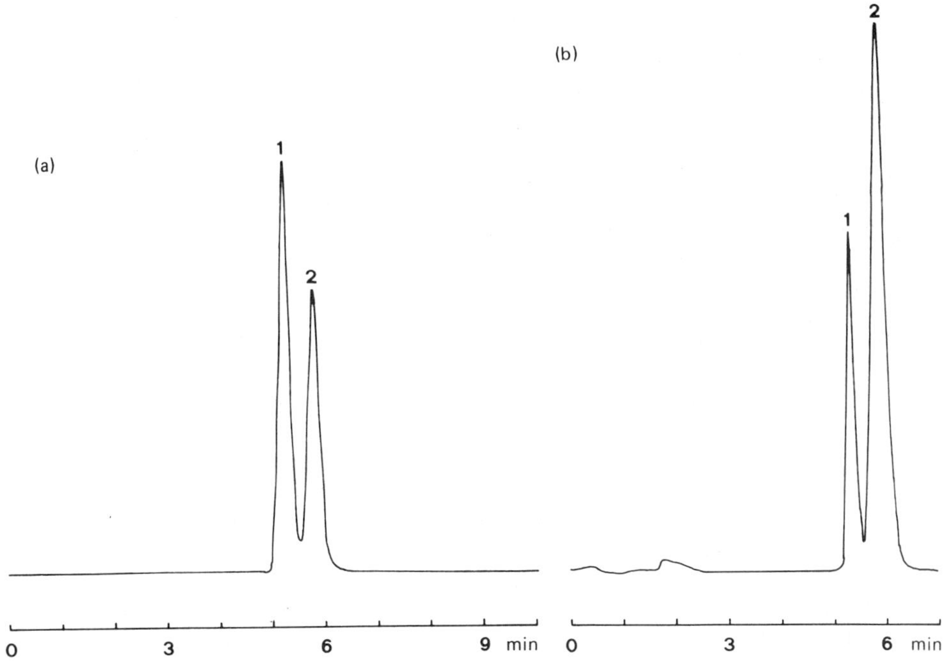

Figure 5.7 Separation of ascorbic acid (1) and erythorbic acid (2). Conditions: 250 × 4 mm ID, Knauer Lichrosorb C_{18} 10 μm, 2 ml/min, UV at 245 nm, injected volume 10 μl. (a) tetrabutylammonium hydroxyde, 2 mM, in KH_2PO_4, 5 mM, pH 6.4. (b) tetrabutylammonium hydroyde, 2 mM, pH 2.9.

of ascorbic acid and erythorbic acid in the column used in RPC on amino phase. This deterioration may be explained by the loss of an amino group, probably due to Schiff base formation [134]. Many attempts have been made to increase amino column [135,136]. Rose and Nahrwold [116] have proposed the use of a precolumn to protect the analytical column.

2.4 Detectors

For all the liquid chromatographic techniques described above, chemiluminescence, electrochemical, radiochemical, refractive index, and/or UV detection have been employed. The electrochemical and the UV detectors were by far the most widely used.

2.4.1 Electrochemical and UV Detectors

The electrochemical detector is quite selective and highly sensitive. A comparison of the two detectors made by Brunt and Bruins [88] indicated that for the detection of ascorbic acid the electrochemical detector is about five times more sensitive than the UV detector. On the other hand, the UV detector is relatively insensitive to flow and temperature changes and has a high sensitivity for a large number of compounds. Ultraviolet detectors are ideal for use with gradient elution. They are more commonly found in laboratories than the detector required for the electrochemical technique. Moreover, the use of microbore columns is known to enhance the sensitivity of the analyzed compounds, but also requires pump delivering microflow and low flow cell volume.

Electrochemical and UV detectors at 254 nm are not directly suitable for dehydroascorbic acid determination. Assay of the latter can be achieved by reduction to ascorbic acid prior to analysis. Using the UV detector, dehydroascorbic acid can also be detected after treatment with 2,4-dinitrophenylhydrazine or 1,2-phenylenediamine, or in some cases directly at 210 nm absorbance without precolumn derivatization (Fig. 5.4). Either detector can be used for erythorbic acid assay. Both detectors have been employed for analyses of ascorbic acid and related compounds, as well as for food, pharmaceutical products, biological fluids, and tissue samples.

The two detectors are not competitive but complementary. The nature of the compounds and the requirement of the analyses are the main factors that will determine the choice between the two types of detectors.

2.4.2 Peak Identification

Peak identification was mainly based on retention time and the stopped-flow scanning method. Under the conditions described in Table 5.4, the retention times of AA and EA were highly reproducible. Coefficients of variation of the retention times of AA and EA were found to be 0.2 and 0.3%, respectively.

Figure 5.8 shows the chromatogram of a human urine extract containing 78 mg/liter of AA, to which a known amount of internal standard, EA, has been added. Such conditions of analysis are shown in Table 5.4. AA eluted as a sharp and symmetrical peak. It was identified on the basis of its retention time relative to the internal standard. This identification was confirmed by stopped-flow UV spectra (210-350 nm). The comparison of the UV spectra obtained by the stopped-flow scanning method are shown in Fig. 5.9. Close agreement between the spectra of the ascorbic acid reference compound and the chromatographic peak in the urine indicates the identity of the peak under study.

TABLE 5.4 Conditions of Ascorbic Acid Analysis

Column	250 × 4.6 mm ID and 30 × 4.6 mm guard column, Lichrosorb C_{18} 10 μm
Mobile phase	Tetrabutylammonium hydroxyde, 2 mM, pH 2.9
Flow rate	2.0 ml/min
Pressure	1000 psi
Column temperature	35°C
Wavelength	245 nm
AU	32×10^{-4} cm^{-1}
Internal standard	Erythorbic acid (2)
Sample	Concentration: 20-300 ng ascorbic acid Quantity injected: 30-10 μl
Paper speed	1 cm/min

2.5 Applications

LC can be applied to determine ascorbic acid and related compounds in food and pharmaceutical products, as well as in biological fluids.

2.5.1 Biological Samples

Some results of ascorbic acid assay in biological fluids are given in Table 5.5 and Figs. 5.8, 5.10 and 5.11.

Powdered milk samples were weighed and diluted with 1% AcOH solution to yield a nominal 10 mg/liter solution. The solution was then shaken, about 10 ml was transferred into a centrifuge tube, overlayered with N_2, and centrifuged at 2800 × *g* for 20 min. The aqueous layer was filtered and used for the chromatographic assay (Fig. 5.10a).

Human milk was analyzed by adding to 3 ml samples 1.0 ml internal standard, EA (80 mg/liter), in 1% acetic acid in a centrifuge tube. The solution was then worked up further in the same manner as the milk powder (Fig. 5.10b).

Human plasma or serum was deproteinized with methanol or acetonitrile. In 50 μl plasma were added 100 μl EA (60 mg/liter) dissolved in methanol. The mixture was then shaken and centrifuged for 5 min at 2000 × *g*. A total of 10 μl of the resulting deproteinized supernatant was used for the chromatographic assay. A human plasma extract spiked with AA is shown in Fig. 5.11.

Human urine was diluted twofold or more if required with 1% acetic acid solution to give a final concentration range about 10 mg/liter. To

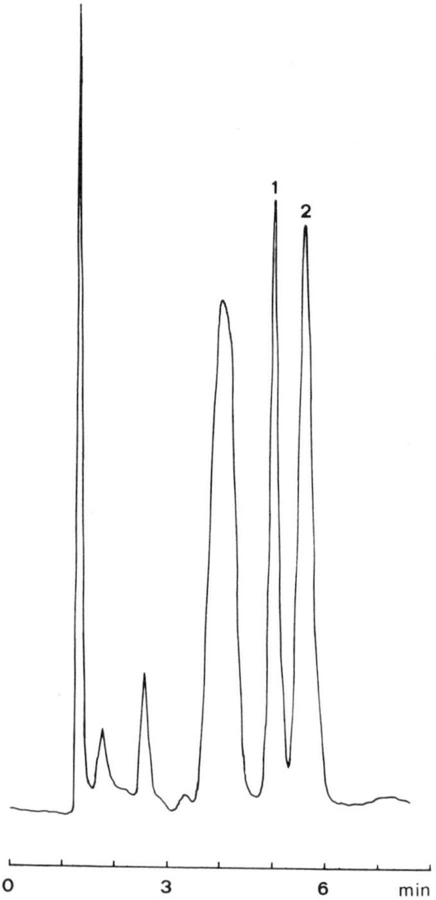

Figure 5.8 Human urine extract. Conditions same as in Fig. 5.7b. Peak 1, ascorbic acid; 2, erythorbic acid as internal standard.

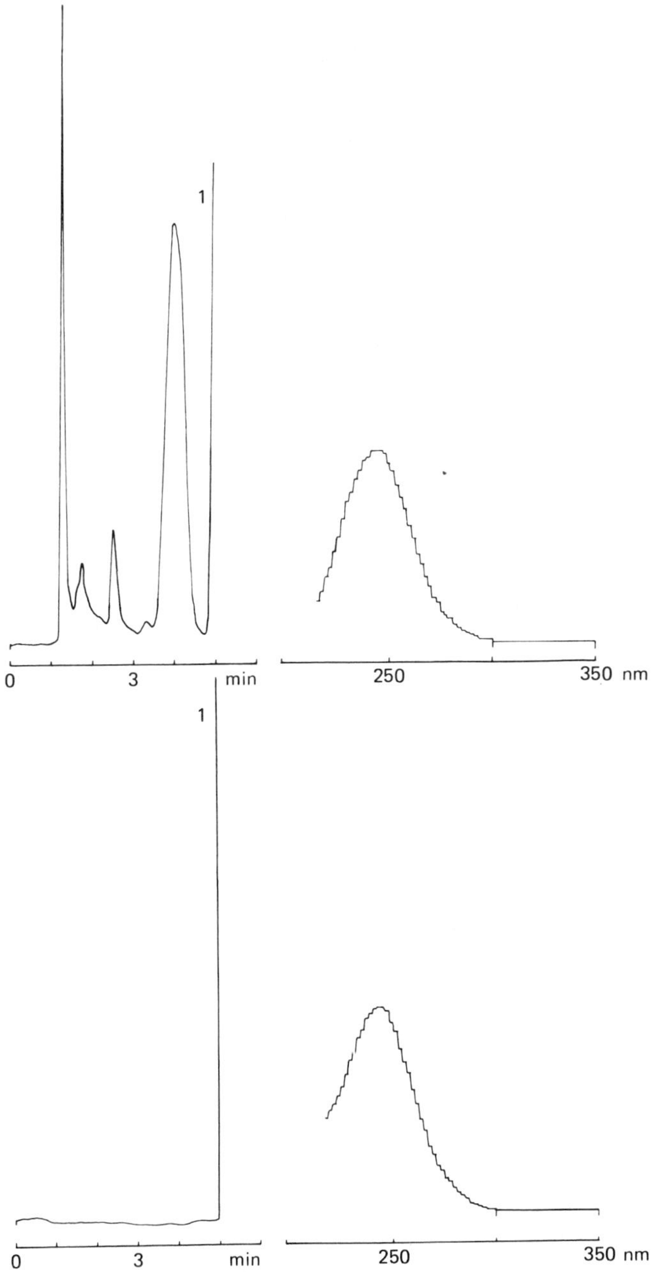

Figure 5.9 UV scan spectra of peak 1 of Fig. 5.8 in human urine and that of ascorbic acid standard.

TABLE 5.5 Some Results on Ascorbic Acid in Biological Fluids

Samples	Ascorbic acid (mg/l) (n = 3)
Human milk collected on the second day after delivery	80
Same human milk stored at -20°C 1 month after delivery	57
Human urine, sample 1	32
Human urine, sample 2	78
Human urine, sample 3	22
Infant human urine, collected 2 hr after ingestion of multivitamin syrup	173
Same urine stored at 4°C one night after ingestion	Traces

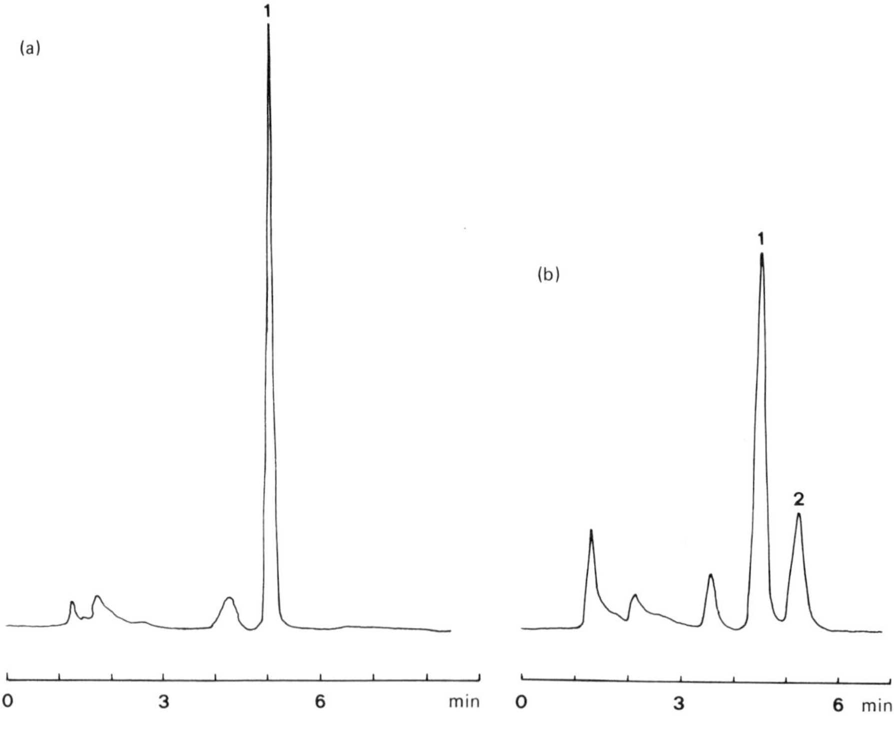

Figure 5.10 (a) Infant formula extract. (b) Human milk extract. Conditions same as in Fig. 5.7b. Peak 1, ascorbic acid; 2, erythorbic acid as internal standard.

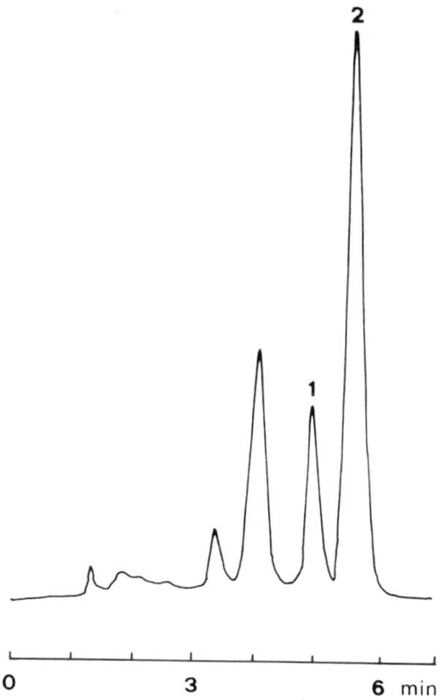

Figure 5.11 Human plasma extract spiked with 80 ng AA. Conditions same as in Fig. 5.7b. Peak 1, ascorbic acid; 2, erythorbic acid as internal standard.

100 μl human urine aliquots, 100 μl internal standard, EA (80 mg/liter), in 1% acetic acid was added. The mixture was then shaken and centrifuged for 2 min at 2000 × *g*. A 10 μl portion of the supernatant liquid was injected (Fig. 5.8).

2.5.2 LC and Conventional Chemical Results

Table 5.6 illustrates a comparison between the IPC method (Table 5.4) and conventional chemical methods: indophenol method (AOAC) and/or hydrazine assay (Roe method). The three methods give approximately the same results on ascorbic acid contents, i.e., neither erythorbic acid nor appreciable oxidation products can be detected in the compounds analyzed.

For LC ascorbic acid assays, if appreciable amounts of dehydroascorbic acid are present in the sample, there may be errors in the determination of ascorbic acid activity if DHA is not taken into account. In these cases, DHA can be assayed directly, after derivatization with hydrazine or after reduction to ascorbic acid with hydrogen sulfide. Hydrogen sulfide reduces dehydroascorbic acid and leaves diketogulonic acid unchanged in pH ranges from 4.7 to 1.2 [52]. An example of the

TABLE 5.6 Comparative Analysis of Ascorbic Acid in Selected Foods and Pharmaceutical Products

Samples	Chromatographic (n = 3)	Visual titration (n = 3)	Hydrazone method (n = 3)
Tablet 10 vitamins	48.2	49.7	
Powder 10 vitamins	74.3	75.8	
Powder 6 vitamins	31.8	32.6	
Drops 10 vitamins	59.0	60.9	
Apple juice	20.1	21.5	
Banana juice	23.5	24.1	
Tomato juice	15.4	16.1	
Apricot juice	23.7	23.8	
Apple juice	78.2		81.4
Strawberry juice	46.0		47.8
Peach juice	35.7	37.9	
Cherry juice	56.7	58.1	
Peach juice	38.9	39.7	41.4
Infant formula	63.7	64.1	65.6
Infant formula	50.8	51.5	53.0
Table 10 vitamins	84.9	85.6	87.3

determination of ascorbic acid in peach juice extract by LC (Table 5.4), conventional chemical methods, and LC after treatment with hydrogen sulfide is illustrated in Table 5.7 and Fig. 5.12a.

The mean value of LC ascorbic acid assays after treatment with H_2S is close to that obtained with 2,4-dinitrophenylhydrazine but clearly higher than the values obtained without hydrogen sulfide treatment and visual titration. Such values may be explained by the presence of dehydroascorbic acid and the absence of diketogulonic acid in the sample.

3. OTHER CHROMATOGRAPHIC METHODS

3.1 Paper Chromatography

Paper chromatography is limited by the volume of extracts but is quite rapid and specific. Ulmann [137] reported a paper chromatography for the separation of ascorbic acid and dehydroascorbic acid in plant extracts. Mitchell and Patterson [138] differentiated ascorbic acid and its isomer; Chen et al. [139] and Kadin and Osadca [140] performed quantitative estimation of these two compounds in biological fluids using

TABLE 5.7 Determination of Ascorbic Acid in Peach Juice by Different Methods

Methods	Ascorbic acid (mg/l) (n = 3)
Extraction and LC as described in Table 5.4	389
Visual titration	397
2,4-Dinitrophenylhydrazine	414
Extraction saturated with H_2S for 15 min and LC as described in Table 5.4	423

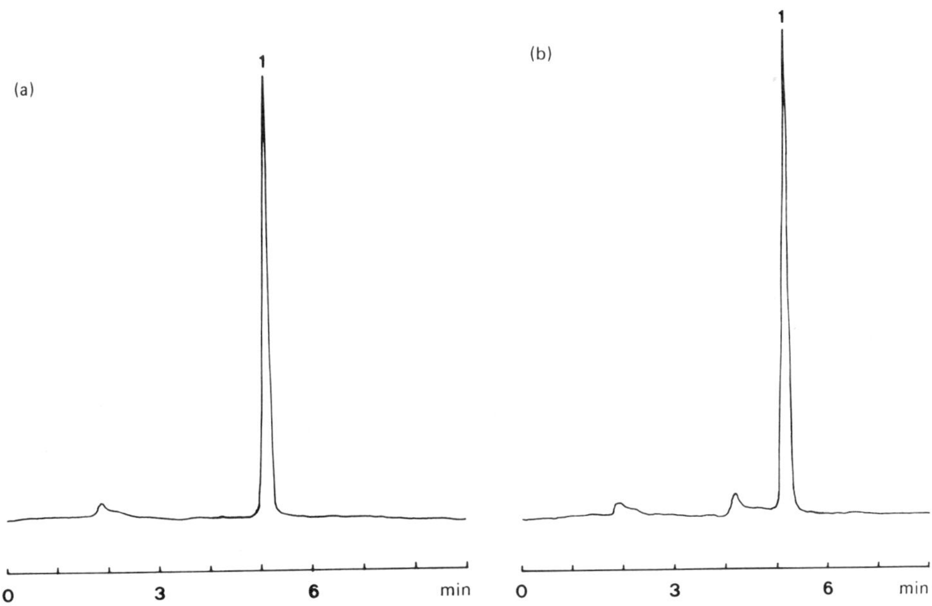

Figure 5.12 (a) Peach juice extract. (b) Peach juice extract after saturation with H_2S. Conditions same as in Fig. 5.7b. Peak 1, ascorbic acid.

paper chromatography and 2,5-dichlorophenol indophenol as the visualization reagent. The use of impregnated papers increases paper chromatography applicability to the separation of ascorbic acid and related compounds. Weeks and Deutsch [141] achieved good recoveries of ascorbic acid and its isomer in fruit beverages using silicic acid-impregnated glass fiber chromatography followed by a microfluorometric assay. A silica gel-impregnated glass-fiber paper chromatographic method was reported by Hornig [142] to separate ascorbic acid, dehydroascorbic acid, 2,3-diketogulonic acid, 2-ketogulonic acid, ascorbylpalmitate, isoascorbic acid, and 3-*O*-methylascorbic acid.

3.2 Thin-Layer Chromatography

Thin-layer chromatography has become one of the most widely used of modern separation methods. TLC, like paper chromatography, is inexpensive. It is a rapid, yet versatile technique. Many applications for ascorbic acid assays have been developed using this method.

Strodecker and Pies [143], Vuillemier and Nobile [144], Vuillemier [95], and Matthiessen [145] have evaluated ascorbic acid in food with TLC after purification of the ascorbic acid osazone derivative by column chromatography. Strodecker [146] has described a similar technique for ascorbyl palmitate in fats. Brenner et al. [147] have resolved ascorbic acid and erythorbic acid by TLC; zone position was detected by 2,6-dichlorophenol indophenol or iodine. Vuillemier [95] separated ascorbic and erythorbic hydrazone derivatives by using TLC. Pelletier and Godin [148] reported a quantitative determination of the two isomers in urine and organs of guinea pigs using TLC with molybdophosphoric acid as spray reagent. Zloch et al. [149] described a radiochemical evaluation of ascorbic acid after isolation of the oxidized form by TLC. Bunton et al. [150] measured the oxidized form of ascorbic acid and erythorbic acid in meat using TLC and fluorometry. Chatterjee and Banerjee [151] have used TLC with spectrophotometry to separate the hydrazone of ascorbic and dehydroascorbic acids in blood and plasma, and to identify those compounds. A technique similar to that described above has been used by Blesova and Zahradnicek [152] to determine ascorbic acid in pharmaceutical products. TLC separation of ascorbic acid and its oxidation products has been proposed by Castelli et al. [153] and by Koen and Yankova [154]. Lyle and Tehrani [155] have resolved vitamins B_5, B_6, and C using TLC, evaluated by a pyrolysis-gas chromatographic technique.

3.3 Classic Liquid Chromatography

In contrast to paper chromatography and TLC, large amounts of extract can be separated by liquid chromatography. Crossland [156] described a method of separation of ascorbic acid from interfering substances in natural products and foodstuffs by chromatography on a cellulose column. Ascorbic acid was determined in the eluate by means of diazotized nitroanisidine.

The separation of ascorbic, dehydroascorbic, and diketogulonic acids has been reported by Hegenauer and Saltman [157] using anion-exchange chromatography. A colorimetric 2,4-dinitrophenylhydrazine method was used to assay the chromatographically separated acids. Vitamin C (ascorbic acid and dehydroascorbic acid) has been determined by Bourgeois and Mainguy [96] in foods and feeds using an anionic Sephadex column and 4-nitro-1,2-phenylenediamine as colorimetric reagent.

3.4 Gas-Liquid Chromatography

Gas liquid chromatography is suitable for microanalytical work. Sweeley et al. [158], Roberts et al. [159], Dalgliesh et al. [160], Horning et al. [161], and Vecchi and Kaiser [162] described a gas chromatographic determination of ascorbic acid in the form of a trimethylsilyl ether derivative with a hydrogen flame ionization detector. Allison and Stewart [163] have applied this method to the quantitative analysis of ascorbic acid in tissues. Zaprometova et al. [164] differentiated ascorbic acid with the hydrate of diisopropylidene-2-keto-L-gulonic acid and with 2-keto-L-gulonic acid using GC and a catherometer as detector. A similar technique has been used by Watanabe and Marota [165] to determine ascorbic acid after treatment with phenylborate.

ACKNOWLEDGMENTS

The author wishes to thank Prof. Dr. H. Isliker, director of the Department of Biochemistry and Swiss Institute of Vitamins, University of Lausanne, for the reading of the final draft of this paper.

REFERENCES

1. J. B. Domingues, *Anais Fac. Farm. Odontol. Univ. Sao Paulo, 18*, 55 (1961).
2. G. P. Laroff, R. W. Fessenden, and R. H. Schuler, *J. Amer. Chem. Soc., 94*, 9062 (1972).
3. Y. Kirino and T. Kwan, *Chem. Pharm. Bull., 202*, 651 (1972).
4. R. D. McAlpine, M. Cocevira, and H. Chen, *Can. J. Chem., 51*, 1682 (1973).
5. S. Steenken and G. Ollrich, *Photochem. Photobiol., 18*, 43 (1973).
6. K. Pfeilsticker, F. Marx, and M. Bochish, *Carbohyd. Res., 45*, 269 (1975).
7. J. Hvoslef, *Acta Crystallogr., B28*, 916 (1972).
8. J. Hvoslef and S. Nordenson, *Acta Crystallogr., B32*, 448 (1976).

9. J. Hvoslef and S. Nordenson, *Acta Crystallogr.*, *B32*, 1665 (1976).
10. A. Szent-Györgi, *Biochem. J.*, *22*, 1387 (1928).
11. L. Pauling, *Proc. Nat. Acad. Sci. U.S.*, *67*, 1643 (1970).
12. E. Cameron, L. Pauling, and B. Leibovitz, *Cancer Res.*, *39*, 663 (1979).
13. C. W. M. Wilson, *Acta Vitaminol. Enzymol.*, *2*(3,4), 120 (1980).
14. S. C. Sharma, A. Molloy, M. Walzman, and L. Bonnar, *Int. J. Vitam. Nutr. Res.*, *5*, 266 (1981).
15. S. K. Majumdar, S. Patel, G. K. Shaw, P. O'Gorman, and A. D. Thomson, *Int. J. Vitam. Nutr. Res.*, *5*, 274 (1981).
16. P. I. Reed, K. Summers, P. L. R. Smith, C. L. Walters, B. A. Bartholomew, and M. J. Hill, *Gut*, *24*(5), A 492-A 493 (1983).
17. M. J. Barnes, *Ann. N.Y. Acad. Sci.*, *258*, 264 (1975).
18. S. R. Pinnel, S. Murad, K. A. Lindberg, and S. Tajima, *J. Invest. Dermatol.*, *76*(4), 355 (1981).
19. B. A. Booth and J. Uitto, *Biochim. Biophys. Acta*, *G675*(1), 117 (1981).
20. A. D. Beattie and S. Sherlock, *Gut*, *17*, 571 (1976).
21. V. G. Zannoni and P. H. Sato, *Ann. N.Y. Acad. Sci.*, *258*, 119 (1975).
22. P. L. Stokes, V. Melikian, R. L. Leeming, H. Portman-Graham, J. A. Blair, and W. T. Cooke, *Am. J. Clin. Nutr.*, *28*, 126 (1975).
23. R. Anderson and A. Theron, *S. Afr. Med. J.*, *56*, 429 (1979).
24. C. J. Schorah, *Int. J. Vitam. Nutr. Res.*, *Supplement 19*, 167 (1979).
25. A. Rebora, F. Dallegri, and F. Patrone, *Br. J. Dermatol.*, *102*, 49 (1980).
26. S. Lewin, *Vitamin C: Its Molecular Biology and Medical Potential*, Academic Press, London, 1976.
27. B. V. Siegel, *Infect. Immun.*, *10*, 409 (1974).
28. L. C. Tolbert, T. N. Thomas, L. D. Middaugh, and J. W. Zemp, *Life Sci.*, *25*, 2189 (1979).
29. W. Losert, H. Vetter, and H. Wendt, *Arzneim. Forsch.*, *30-1*, 21 (1980).
30. P. Sandhya and U. N. Das, *IRCS Clin. Pharmacol. Ther.*, *9*(7,8), 618 (1981).
31 E. Ginter, *Lancet*, *2*, 958 (1979).
32. A. Hanck and H. Weiser, *Int. J. Vitam. Nutr. Res.*, *Supplement 19*, 83 (1979).
33. J. S. Prasad, *Clin. Chim. Acta*, *59*, 101 (1975).
34. J. D. Hulse, S. R. Ellis, and L. M. Henderson, *J. Biol. Chem.*, *253*, 1654 (1978).
35. S. Banič, *Int. J. Vitam. Nutr. Res.*, *Supplement 16*, 235 (1977).
36. K. M. Clegg and J. M. MacDonald, *Proc. Nutr. Soc.*, *34*(1), 7A (1975).

37. K. Solanki, U. Moser, and P. Walter, *Int. J. Vitam. Nutr. Res.*, *51*, 186 (1981).
38. F. J. Yourga, W. B. Esselen, Jr., and C. R. Fellers, *Food Res.*, *9*, 188 (1944).
39. M. Ikeuchi, *Vitaminology*, *1*, 193 (1955).
40. M. E. Reid, *Estimation of Vitamins, Biological Symposia XII*, Jacques Cattell Press, Lancaster, Penn., 1947, p. 373
41. F. W. Crampton, *J. Nutr.*, *33*, 49 (1947).
42. B. J. Meyer, H. Lagally, and M. L. Dodds, *Anal. Biochem.*, *4*, 88 (1965).
43. A. Marchesini, F. Montuori, D. Muffato, and D. Maestri, *J. Food Sci.*, *39*, 568 (1974).
44. P. Posadka and L. Macholan, *Collect. Czech. Chem. Commun.*, *44*, 3395 (1979).
45. T. Z. Liu, N. Chin, M. D. Kiser, and W. N. Bigler, *Clin. Chem.*, *28*(11), 2225 (1982).
46. J. Tillmans, *Z. Unters. Lebensm.*, *54*, 33 (1927).
47. H. Leonhardt and W. Moeser, *Z. Anal. Chem.*, *122*, 3 (1941).
48. M. X. Sullivan and H. C. N. Clarke, *J. Ass. Offic. Agr. Chem.*, *38*, 514 (1955).
49. R. P. Maickel, *Anal. Biochem.*, *1*, 498 (1960).
50. S. K. Jagota and H. M. Dani, *Anal. Biochem.*, *127*(1), 178 (1982).
51. J. H. Roe and M. J. Oesterling, *J. Biol. Chem.*, *152*, 511 (1944).
52. J. H. Roe, M. B. Mills, M. J. Oesterling, and C. M. Dameron, *J. Biol. Chem.*, *174*, 201 (1948).
53. M. Schmall, C. W. Pifer, and E. G. Wollish, *Anal. Chem.*, *25*, 1486 (1953).
54. M. Schmall, C. W. Pifer, E. G. Wollish, R. Duschinsky, and H. Gainer, *Anal. Chem.*, *26*, 1521 (1954).
55. M. J. Deutsch and C. E. Weeks, *J. Ass. Offic. Agr. Chem.*, *48*, 1248 (1965).
56. M. J. Deutsch, *J. Ass. Offic. Agr. Chem.*, *50*, 798 (1967).
57. S. S. Wilson and R. A. Guillan, *Clin. Chem.*, *15*(4), 282 (1969).
58. O. Pelletier, *Can. J. Biochem.*, *47*, 499 (1969).
59. V. Zannoni, M. Lynch, S. Goldstein, and P. Sato, *Biochem. Med.*, *11*, 41 (1974).
60. D. J. Cameron, *Proc. Nutr. Soc.*, *34*(1), 7A (1975).
61. J. W. Ralls, *Agr. Food Chem.*, *23*, 609 (1975).
62. S. T. Omaye, J. D. Turnbull, and H. E. Sauberlich, *Methods Enzymol.*, *62*, 3 (1979).
63. D. L. Dunmire, J. D. Reese, R. Bryan, and M. Seegers, *J. Ass. Offic. Anal. Chem.*, *62*, 648 (1979).
64. U. Muralikrishna, Y. V. S. K. Sesha Sayi, and I. V. Raghu Ram, *J. Indian Chem. Soc.*, *56*, 794 (1979).
65. P. Vasileva-Aleksandrova and A. Neicheva, *Mikrochim. Acta*, *II*, 337 (1979).

66. H. O. Beutler and G. Beinstingi, *Dtsch. Lebensm. Rundsch.*, *76*, 69 (1980).
67. J. Lindquist and S. M. Ferroha, *Analyst (London)*, *100*, 377 (1975).
68. C. Ratzkowski and J. Korol, *J. Inst. Can. Sci. Technol. Aliment.*, *10*, 215 (1977).
69. A. Baldesten, G. Hjalmarsson, and G. Newmann, *Z. Anal. Chem.*, *290*, 148 (1978).
70. K. Hiromi, C. Kuwamoto, and M. Ohnishi, *Anal. Biochem.*, *101*, 421 (1980).
71. H. Obata, T. Tokuyama, Y. Nitta, M. Takagi, and K. Hiromi, *Agr. Biol. Chem.*, *43*, 2191 (1979).
72. P. E. Shaw and C. W. Wilson, III, *J. Agr. Food Chem.*, *30*, 396 (1982).
73. L. A. Pachla and P. T. Kissinger, *Anal. Chem.*, *364* (1976).
74. R. A. Hartwick, D. Van Haverbeke, M. McKeag, and P. R. Brown, *J. Liquid Chromatogr.*, *2*, 725 (1979).
75. D. J. Green and R. L. Perlman, *Clin. Chem.*, *26*, 796 (1980).
76. J. D. Ponting, *Ind. Eng. Chem. Anal. Ed.*, *15*, 389 (1943).
77. D. W. Bradley, G. Emery, and J. E. Maynard, *Clin. Chim. Acta*, *44*, 47 (1973).
78. L. A. Pachla and P. T. Kissinger, *Methods Enzymol.*, *62*, 15 (1979).
79. R. L. Veazey and T. A. Nieman, *J. Chromatogr.*, *200*, 153 (1980).
80. A. Floridi, R. Coli, A. A. Fidanza, C. F. Bourgeois, and R. A. Wiggins, *Int. J. Vitam. Nutr. Res.*, *52*, 193 (1982).
81. W. D. Mason, E. N. Amick, and W. Heft, *Anal. Lett.*, *13*, 813 (1980).
82. A. W. Archer, V. R. Higgins, and D. L. Perryman, *J. Ass. Publ. Anal.*, *18*, 99 (1980).
83. P. Wimalasini and R. B. H. Wills, *J. Chromatogr.*, *256*, 368 (1983).
84. J. J. Nelson, *J. Chromatogr. Sci.*, *11*, 28 (1973).
85. R. C. Williams, D. R. Baker, and J. A. Schmit, *J. Chromatogr. Sci.*, *14*, 618 (1973).
86. K. V. Thrivikraman, C. Refshauge, and R. N. Adams, *Life Sci.*, *15*, 1335 (1974).
87. P. T. Kissinger, L. J. Felice, R. M. Riggin, L. A. Pachla, and D. C. Wenke, *Clin. Chem.*, *20*, 992 (1974).
88. K. Brunt and C. H. P. Bruins, *J. Chromatogr.*, *172*, 37 (1979).
89. R. Rouseff, in *Liq. Chromatogr. Anal. Food Beverages (Proceedings of a Symposium on the Analysis of Foods and Beverages by HPLC)*, Vol. 1 (G. Charalambous, ed.), Academic Press, New York, 1979, p. 161.
90. H. Obata, W. Tsuchihashi, and T. Tokuyama, *Agr. Biol. Chem.*, *44*, 1435 (1980).
91. R. S. Carr and J. M. Neff, *Anal. Chem.*, *52*, 2428 (1980).
92. C. S. Tsao and S. L. Salimi, *J. Chromatogr.*, *224*, 477 (1981).
93. T. Saeki, Y. Katagiri, S. Nagasako, K. Narita, and M. Hayashibara, *Tekugaku Zasshi*, *101*, 836 (1981).

94. L. F. Liebes, S. Kuo, R. Krigel, E. Pelle, and R. Silber, *Anal. Biochem.*, *118*, 53 (1981).
95. J. P. Vuillemier, *Int. Z. Vitaminforsch.*, *37*, 504 (1967).
96. C. F. Bourgeois and P. R. Mainguy, *Int. J. Nutr. Res.*, *45*, 70 (1975).
97. M. Grün and F. A. Loewus, *Anal. Biochem.*, *130*, 191 (1983).
98. L. Manso, *EFCE Publ. Ser.* (Proc.-Congr. Nac. Quim., 3rd 1980, Vol. 2), *12*, 137 (1980).
99. H. E. Sauberlich, M. D. Green, and S. T. Omaye, *Adv. Chem.*, *200*, 199 (1982).
100. R. Stillman and T. S. Ma, *Mikrochim. Acta*, 641 (1974).
101. R. B. H. Wills, C. C. Shaw, and W. R. Day, *J. Chromatogr. Sci.*, *15*, 1335 (1974).
102. E. S. Wagner, B. Lindley, and R. D. Coffin, *J. Chromatogr.*, *163*, 225 (1979).
103. T. Tsushida and M. Fukazawa, *Chagyo Gijutsu Kenkyu*, *58*, 29 (1980).
104. T. Tsushida and T. Takeo, *Chagyo Gijutsu Kenkyu*, *58*, 34 (1980).
105. H. Rueckemann, *Z. Lebensm. Unter. Forsch.*, *171*, 446 (1980).
106. N. Miki, *Nippon Shokuhin Kogyo Gakkaishi*, *28*, 264 (1981).
107. S. Garcia-Castineiras, V. D. Bonnet, R. Figueroa, and M. Miranda, *J. Liquid Chromatogr.*, *4*, 1619 (1981).
108. A. E. Watada, *Hortscience*, *17*, 334 (1982).
109. M. A. J. S. Van Boekel and C. A. J. M. Neewissen, *J. Chromatogr.*, *261*, 176 (1983).
110. M. H. Bui-Nguyên, *J. Chromatogr.*, *196*, 163 (1980).
111. J. Carnevale, *Food Technol. Aust.*, *32*, 302 (1980).
112. B. Hoffmann, *Lebensm. Gericht. Chem.*, *34*, 114 (1980).
113. J. Geigert, D. S. Hirano, and S. L. Neidleman, *J. Chromatogr.*, *206*, 396 (1981).
114. N. Arakawa, M. Otsuka, T. Kurata, and C. Inagaki, *J. Nutr. Sci. Vitaminol.*, *27*, 1 (1981).
115. M. Otsuka, T. Kurata, E. Suzuki, N. Arakawa, and C. Inagaki, *J. Nutr. Sci. Vitaminol.*, *27*, 9 (1981).
116. R. C. Rose and D. L. Nahrwold, *Anal. Biochem.*, *114*, 140 (1981).
117. R. C. Rose and D. L. Nahrwold, *Anal. Biochem.*, *123*, 389 (1981).
118. D. B. Dennison, T. G. Brawley, and G. L. K. Hunter, *J. Agr. Food Chem.*, *29*, 927 (1981).
119. L. W. Doner and K. B. Hicks, *Anal. Biochem.*, *115*, 225 (1981).
120. G. L. Lookhart, S. B. Hall, and K. F. Finney, *Cereal Chem.*, *59*, 69 (1982).
121. C. Horvath, W. Melander, and I. Molnar, *Anal. Chem.*, *49*, 142 (1977).
122. P. T. Kissinger, *Anal. Chem.*, *49*, 883 (1977).
123. E. Tomlinson, C. M. Riley, and T. M. Jefferies, *J. Chromatogr.*, *173*, 89 (1979).

124. S. P. Sood, L. E. Sartori, D. P. Wittmer, and W. G. Haney, *Anal. Biochem.*, *48*, 796 (1976).
125. J. Augustin, C. Beck, and G. I. Marousek, *J. Food Sci.*, *46*, 322 (1981).
126. J. W. Finley and E. Duang, *J. Chromatogr.*, *207*, 449 (1981).
127. J. M. Coustard and G. Sudraud, *J. Chromatogr.*, *219*, 338 (1981).
128. G. Parolari, *Ind. Conserve*, *57*, 19 (1982).
129. R. W. Keating and P. R. Haddad, *J. Chromatogr.*, *245*, 249 (1982).
130. C. S. Tsao and S. L. Salimi, *J. Chromatogr.*, *245*, 355 (1982).
131. W. Lee, P. Hamernyik, M. Hutchinson, V. A. Raisys, and R. F. Labbe, *Clin. Chem.*, *28*, 2165 (1982).
132. R. Schwartzenbach, *Biol. Biomed. Appl. Liquid Chromatogr. II*(12), 193-222 (1979).
133. R. K. Gilpin and W. R. Sisco, *J. Chromatogr.*, *194*, 285 (1980).
134. H. Woidich, W. Pfannahuser, and G. Blaicher, *Lebensm. Chem. Gericht. Chem.*, *32*, 74 (1978).
135. K. Aïtzemüller, *J. Chromatogr.*, *156*, 354 (1978).
136. K. Aïtzemüller, *Chromatographia*, *13*, 432 (1980).
137. M. Ulmann, *Parmazie*, *9*(6), 523 (1954).
138. L. C. Mitchell and W. I. Patterson, *J. Ass. Offic. Agr. Chem.*, *36*, 1127 (1953).
139. Y. T. Chen, F. A. Isherwood, and L. W. Mapson, *Biochem. J.*, *55*, 821 (1953).
140. H. Kadin and M. Osadca, *Agr. Food Chem.*, *7*, 358 (1959).
141. C. E. Weeks and M. J. Deutsch, *J. Ass. Offic. Agr. Chem.*, *50*, 793 (1967).
142. D. Hornig, *J. Chromatogr.*, *71*, 168 (1972).
143. R. Strodecker and H. Pies, *Z. Lebensm. Untersuch. Forsch.*, *118*, 394 (1962).
144. J. P. Vuillemier and S. Nobile, Assay of Vitamin C in Compound Feeds, *Proceedings World's Poultry Congress*, 238-261 (1962).
145. E. Matthiessen, *Dtsch. Lebensm. Rundsch.*, *74*, 403 (1978).
146. R. Strodecker, *Fette Seifen Anstrichm.*, *66*, 787 (1964).
147. G. S. Brenner, D. F. Hinkley, L. M. Perkins, and S. Weber, *J. Org. Chem.*, *29*, 2389 (1964).
148. O. Pelletier and C. Godin, *Can. J. Physiol. Pharmacol.*, *47*, 985 (1969).
149. Z. Zloch, J. Cerven, and E. Ginter, *Anal. Biochem.*, *43*, 99 (1971).
150. N. G. Bunton, N. Jennings, and N. T. Crosby, *J. Ass. Publ. Anal.*, *17*, 105 (1979).
151. I. B. Chatterjee and A. Banerjee, *Anal. Biochem.*, *98*, 368 (1979).
152. M. Blesova and M. Zahradnicek, *Csek. Farm.*, *28*, 374 (1979).

153. A. Castelli, S. Lippa, A. M. Frosca, E. Meucci, R. Zenobi, and G. E. Martorana, *Acta Vitaminol. Enzymol.*, *3*, 111 (1981).
154. V. Koen and M. Yankova, *Tr. Nauchnoizsled Khim. Farm. Inst.*, *11*, 185 (1981).
155. S. J. Lyle and M. S. Tehrani, *J. Chromatogr.*, *236*, 31 (1982).
156. I. Crossland, *Acta Chem. Scand.*, *4*, 805 (1960).
157. J. Hegenauer and P. Saltman, *J. Chromatogr.*, *74*, 133 (1972).
158. C. C. Sweeley, R. Bentley, M. Makita, and W. W. Wells, *J. Amer. Chem. Soc.*, *85*, 2497 (1963).
159. R. N. Roberts, J. A. Johnston, and B. W. Fuhr, *Anal. Biochem.*, *10*, 792 (1966).
160. C. E. Dalgliesh, E. C. Horning, M. G. Horning, K. L. Knox, and K. Yarger, *Biochem. J.*, *101*, 792 (1966).
161. M. G. Horning, E. A. Boucher, and A. M. Moss, *J. Gas Chromatogr.*, 5, 297 (1967).
162. M. Vecchi and K. Kaiser, *J. Chromatogr.*, *26*, 22 (1967).
163. J. M. Allison and M. A. Stewart, *Anal. Biochem.*, *43*, 401 (1971).
164. L. M. Zaprometova, V. M. Svetlaeva, M. T. Yanatovskii, N. N. Egorshina, L. A. Tobaeva, and I. D. Frizen, *Khim. Farm. Zh.*, *15*, 120 (1981).
165. T. Watanabe and S. Murota, *Kobe Joshi Daigaku Kiyo*, *12*, 87 (1981).

6

CHROMATOGRAPHIC STUDIES OF FOLIC ACID AND RELATED COMPOUNDS

JOHN J. McCORMACK and ROBERT A. NEWMAN

Vermont Regional Cancer Center and Department of Pharmacology, College of Medicine, University of Vermont, Burlington, Vermont

Derivatives of folic acid (pteroylglutamic acid) play an extremely important role in a wide range of biological processes, and inhibition of aspects of folate metabolism has been shown to produce clinically useful effects in neoplastic, protozoal, and bacterial diseases. The biochemical and pharmacological importance of folate derivatives and analogs makes it essential that biomedical scientists have at their disposal a variety of analytical methods, of appropriate sensitivity and selectivity, to examine mechanisms involved in the biological action of such compounds. It is our intent, in this chapter, to describe and evaluate a variety of chromatographic approaches that have been used in the solution of analytical problems encountered in studies of the biological actions of folate coenzymes and antagonists. We have decided to focus on developments in the period 1976-1981, and many of the examples that we choose to illustrate analytical approaches are taken from recent communications.

1. CHEMICAL AND BIOCHEMICAL PROPERTIES OF FOLATES

1.1 Structure

The structure of folic acid (Fig. 6.1) can be considered to be composed of three units: a 2-amino-4-oxopteridine (pterin) moiety, a *p*-aminobenzoic acid moiety, and a glutamic acid residue. Folic acid, per se, is not an active coenzyme and requires enzymatic reduction, in two stages, to 7,8-dihydrofolate and 5,6,7,8-tetrahydrofolate (Fig. 6.1) in order to be "activated" for subsequent participation in biochemical reactions. The reduction of folic acid can be carried out chemically not only by catalytic hydrogenation but also, and more conveniently, by use of sodium dithionite or borohydride [1]. Removal of the glutamic acid portion of the folate molecule produces pteroic acid; we will comment later about compounds related to pteroic acid in our discussion of analytical methods suitable for the study of the 4-amino derivatives of folate, such as methotrexate. Attachment of additional glutamic acid residues, in peptide linkage, to the glutamate portion of folate can be accomplished both chemically and biochemically. Such polyglutamate derivatives of folic acid represent the predominant forms of folate in a variety of foodstuffs [2,3], as well as in most animal tissues [4]. The

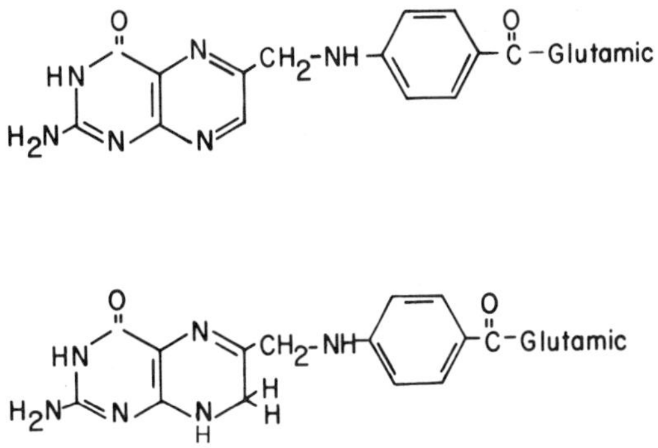

Figure 6.1 Chemical structures of folic acid and 7,8-dihydrofolic acid.

presence of the additional anionic residues in the polyglutamates, as well as the increase in molecular size, presents both problems and opportunities for analytical exploitation for those who investigate aspects of folate metabolism.

1.2 Biochemical Function

Folate coenzymes participate in several one-carbon transfer reactions that are prominent in the cellular economy of a variety of biological species. Tetrahydrofolate (THF) serves as the acceptor for one-carbon units that are bound at positions 5 and/or 10 of the molecule. 5-Methyl-THF is the major "transport form" of folate in plasma; this derivative of THF is required for the formation of methylcobalamin, a vitamin B_{12} coenzyme that plays a role in the conversion of homocysteine to methionine [5]. Tetrahydrofolate coenzymes also donate or accept one-carbon units in enzymatic processes that subserve the conversion of serine to glycine, the synthesis of thymidylic acid and purines for use in the synthesis of nucleic acid precursors, and the catabolism of histidine, and THF derivatives may be of clinical value in ameliorating the toxicity associated with methanol ingestion [6]. The structures of several important THF coenzymes, together with a designation of the reactions in which they play a role, are illustrated in Fig. 6.2.

1.3 Spectroscopic Properties

Analytical studies of pteridine derivatives, including folate coenzymes, are facilitated greatly by the absorption characteristics of these compounds in the ultraviolet and visible ranges. In addition, many pteridine derivatives exhibit high native fluorescence or are capable of being converted, by facile oxidation reactions, to fluorescent compounds.

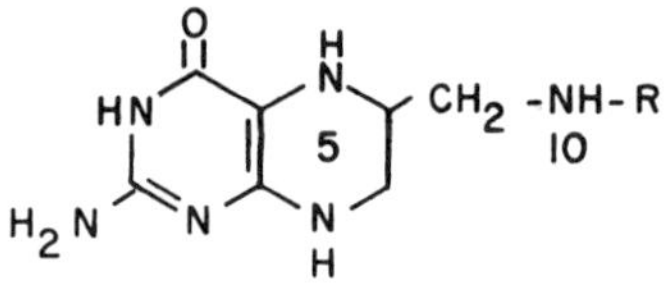

R = p-aminobenzoyl-glutamate

Compound	Representative Biochemical Role
5-Methyl	Methionine formation
5-Formyl (folinic acid; citrovorum factor; leucovorin)	"Reversal" of MTX toxicity
10-Formyl	Purine synthesis
5,10-Methylene	Thymidylate synthesis
5,10-Methenyl	Purine synthesis
5-Formimino	Histidine catabolism

Figure 6.2 Chemical structures of 5,6,7,8-tetrahydrofolic acid, its coenzyme derivatives, and examples of biochemical roles of tetrahydrofolate coenzymes.

The important monograph by Blakley [7] contains an excellent description and discussion of earlier studies of pteridine derivatives and should be consulted for a summary of much valuable information, not only on aspects of the spectroscopic behavior of pteridines but also on other facets of the chemistry and biology of these compounds. 2-Amino-4-oxopteridines (pterins) generally show three major absorption maxima in the ultraviolet range at neutral pH values; these maxima appear, for 2-amino-4-oxopteridine itself, at 339, 270, and 233 nm, respectively. The extinction coefficients for the latter two maxima are considerably higher than that for the peak at the longest wavelength. That compounds of this type contain ionizable functions, both acidic and basic, permits an additional approach to detection and determination of such compounds, namely, examination of the effects of acidification and/or alkalinization on the ultraviolet absorption spectrum. For example, with 2-amino-4-oxopteridine, the lowest energy maximum of the neutral species undergoes a hypsochromic shift of 20 nm in acidic solution and a quantitatively similar bathochromic shift in alkaline solution. The ultraviolet absorption spectra of folate derivatives differ from those of corresponding simpler pteridines in that the folate derivatives exhibit greatly enhanced absorption in the 250-290 region of the spectrum due to the presence of the additional aromatic nucleus (*p*-aminobenzoic acid [8]). Reduction of folate derivatives produces pronounced alterations in spectroscopic properties; 7,8-dihydrofolate (DHF), at neutral pH, does not have the high wavelength maximum observed for folate at approximately 340 nm but the maximum at 280 nm is retained. Further reduction of DHF to THF results in an ultraviolet absorption pattern characterized by peaks of similar intensity at 270 and 290 nm. Chromatographic studies of folate derivatives discussed later will, in many instances, involve ultraviolet spectroscopic detection of compounds of analytical interest. Knowedge that many pteridines are capable of emitting fluorescence when activated at suitable wavelengths affords an additional, more sensitive approach that is illustrated by a recent description of the analysis of pteridine derivatives in rat brain [9] using liquid chromatography (LC) with fluorescence detection according to methodology developed by Fukushima and Nixon [10]. The fluorescence characteristics of THF are sufficiently different from those of DHF to permit fluorimetric assay of the enzymatic reduction of DHF by monitoring the increased fluorescence in 360 nm (excitation at 310 nm) attributable to formation of the fully reduced derivative [11]. Phosphorescence, observed at room temperature, recently has been reported for several pteridines; phosphorescence emission maxima were observed in the 500 nm range when samples were activated in the range 350-380 nm [12]. This technique may offer promise for use in a number of analytical studies of pteridine derivatives, including folates.

1.4 Chemical Stability

2-Amino-4-oxopteridines and related compounds are stable substances at moderate temperatures and neutral pH when protected from light. Reduction of such compounds, on the other hand, leads to products that are readily oxidized, in solution, under conventional laboratory conditions [1] and also are susceptible to oxidation in the solid state. Tetrahydropteridines, such as tetrahydrobiopterin and THF, are particularly susceptible to oxidation, under relatively mild conditions, to the corresponding 7,8-dihydro derivatives. Oxidative transformations of reduced pteridines that bear substituents in position 6 can involve not only dehydrogenation but also cleavage of the side chain, a reaction that may lead to formation of partially reduced 2-amino-4-oxo-6-pteridine aldehyde (pterin aldehyde [13]). The stability of folate under conditions in which reduced derivatives are degraded extensively is well exemplified in studies of the anaerobic oxidation of reduced folates, with ferricyanide as oxidant [14]; these studies by Chippel and Scrimgeour also illustrate the effective use of a variety of chromatographic techniques to investigate the chemical properties of folate derivatives. It should be mentioned that recent studies by Whitely and Russell have further documented the complexity of oxidative transformations of reduced folates that includes rearrangement reactions observed for peroxide-mediated oxidation of 5-methyl-THF [15]. The instability under oxidizing conditions of reduced pteridine derivatives has presented significant problems to investigators studying the chromatographic behavior of such compounds; Bailey and Ayling, in disucssing their contribution of LC of pteridine derivatives at the Fifth International Pteridine Symposium, reported that the use of "carefully degassed solvents" permitted analysis of tetrahydropteridine derivatives, in neutral solvents, with no requirement for inclusion of reducing agents in the solvents [16]. Pretreatment of eluent buffers with nitrogen also has been reported to facilitate chromatography of reduced folates with minimal degradation in what is essentially an anaerobic system [17]. Light-induced transformation can play a major role in the instability of pteridine derivatives. Baur and his colleagues have studied several photochemical degradations of pteridines and found evidence for anaerobic "dealkylation" of biopterin and neopterin at alkaline pH resulting in formation of 5,8-dihydropterin-6-aldehyde with subsequent oxidation of the dihydro compound to pterinaldehyde when the reaction mixture is oxygenated [18].

1.5 Biological Assays

Our chapter is concerned primarily with chromatographic techniques that are useful in elucidating aspects of the chemistry, biochemistry, and pharmacology of folates and related compounds. Nevertheless, to provide appropriate perspective, we consider it worthwhile to discuss, briefly, examples of biological (or biochemical) approaches that have contributed valuable information to our store of knowledge about these interesting and important compounds. Analysis of levels of 2,4-diamino pteridine analogs of folate, such as methotrexate (MTX), in biological samples is accomplished frequently by taking advantage of the powerful "stoichiometric" (tight-binding) inhibition of the enzyme, dihydrofolate reductase (DHFR [19]) produced by compounds such as MTX. This approach involves determination of enzyme-mediated changes in ultraviolet absorption at 340 nm; the presence of MTX, or compounds of equivalent biochemical potency [20], in serum and other biological samples that are added to the assay mixture results in a concentration-dependent diminution of the characteristic absorbance change. The DHFR inhibition assay gives pharmacokinetic results comparable to those obtained by radioimmunoassay and related methods [21]. The DHFR inhibition approach to determining MTX and other potent inhibitors of DHFR, though extremely convenient and quite sensitive, has some significant limitations. For example, this method detects metabolites that retain the ability to inhibit DHFR at low concentrations and, on the other hand, fails to detect possible prominent metabolites that are only weakly active against DHFR. The use of LC to detect and quantitate MTX, considered in detail later in this chapter, overcomes these limitations of the enzyme inhibition assay. Competitive ligand binding assays for MTX and related compounds have been developed [22] and are useful for some pharmacokinetic studies; such assays are based on the displacement of radiolabeled MTX from DHFR by nonlabeled MTX (or equivalent drug) present in a biological fluid or tissue extract. Ligand binding assays for MTX can be carried out with kits that are obtainable from commercial sources. It is also possible to measure concentrations of methotrexate in biological samples by estimating the inhibition of the growth of microorganisms caused by MTX present in such samples; such bioassays are relatively slow, owing to the requirement of incubation of microbial cultures for sufficient time periods to permit adequate colony growth. Mehta and Hutchinson have described this type of microbiological assay for MTX, using a strain of *Streptococcus faecalis* [23]. These authors have developed clinically useful microbiological assays, using appropriate microbial strains (e.g., *Pediococcus cerevisiae* and *Lactobacillus casei*), to analyze levels of 5-formyl-THF (citrovorum factor or Leucovorin) and other folate derivatives. The use of radioimmunoassay for folate derivatives is exemplified by a recent report [24] of successful application of this approach to the analysis of the investigational anticancer agent, 5-methyltetrahydrohomofolate, in plasma and tissue samples. This analysis involved displacement, by unlabeled drug, of a 5-methyltetrahydrohomofolate derivative (labeled with either ^{3}H or ^{125}I) from a complex with an antibody raised in rabbits to a 5-methyltetrahydrohomofolate-bovine serum albumin conjugate. As indicated above in the brief description of studies by Mehta and Hutchinson, folate and certain of its derivatives can be determined microbiologically by serveral methods; one such method involves measuring the growth stimulation of *L. casei* produced by folates, and is exemplified by a protocol that represents a modification [25] of a standard method described by Herbert and his colleagues [26]. *Lactobacillus casei* is a particularly suitable test organism for folate analysis since it responds to 5-methyltetrahydrofolate, the major form of "circulating" folate in mammals. *Lactobacillus casei* does not respond to a variety of folate polyglutamates, but this problem is overcome readily by enzymatic hydrolysis of polyglutamates to active folate derivatives. Microbiological assays have been employed extensively to study unconjugated pteridines, such as biopterin. The most commonly used organism for these studies in the flagellated protozoan, *Crithidia fasciculata* [27]. *Crithidia fasciculata* bioassays provide sensitive indices of the concentration of biopterin and other compounds capable of serving as *Crithidia* growth factors, but are very time consuming. This bioassay has been used extensively to estimate levels of biopterin and equivalents in serum and urine; a recent example of use of this approach is afforded by the communication by Leeming and Blair on the relationship of biopterin levels to different physiological and pathological states [28]. An enzymatic assay [29] for reduced biopterin equivalents, based on the cofactor activity of such compounds for phenylalanine hydroxylase, can be used to estimate levels of biopterin derivatives in tissues, but this approach, like the *Crithidia* assay, suffers from lack of specificity.Nagatsu and his colleagues have recently reported a highly sensitive and selective radioimmunoassay for biopterin [30] that can be used to assay biopterin equivalents in blood, urine, cerebrospinal fluid, and tissue extracts. This type of assay procedure, coupled with LC approaches, will undoubtedly contribute much to the resolution of analytical problems that have hampered progress in studies of pteridine metabolism, especially in mammals.

2. CHROMATOGRAPHY OF FOLATES

As discussed in Sec. 1 of this chapter, folate does not exist as a single chemical entity, but can occur as a variety of structurally related compounds differing in (1) the state of oxidation of the pteridine ring, (2) the number of glutamate residues, and (3) the presence or absence of a single carbon substitution on the 5- and 10-nitrogen in the pteridine ring. Most of these compounds are too closely related to allow for separate identification and accurate quantitation by standard chemical methods. Therefore, conventional chromatographic procedures for separating individual folates of interest were developed. These have included ion-exchange (IEC) and thin-layer (TLC) chromatographic techniques, which have traditionally been associated with an ultraviolet (UV)-spectrophotometric or a microbiological assay technique for specific product identification.

Within the past few years, traditional chromatographic procedures such as TLC and IEC have given way to modern liquid chromatographic methods, which, for the most part, have progressed in step with the development of high-performance liquid chromatographic (HPLC) instrumentation. This portion of the review of the chromatography of folates will focus primarily on recent analytical methods for separation and identification of pteroylmonoglutamate and pteroylpolyglutamate standards, as well as the application of these methods for the detection of naturally occurring folate derivatives in biological fluids and pharmaceutical and food preparations.

2.1 Thin-Layer Chromatography

The separation and identification of the naturally occurring pteroylmonoglutamates by TLC has recently been reviewed by Scott [31]. Although simple and relatively inexpensive, TLC is not now commonly used for routine chromatography of pteroylmonoglutamates, having been largely replaced by LC methods. As pointed out by Scott [31], although numerous solvent systems have been described to separate folate derivatives, a single solvent system (3% w/v NH_4Cl) containing 0.5% (v/v) mercaptoethanol (MET) as antioxidant gives good separation of most of the important naturally occurring pteroylmonoglutamates (Table 6.1). The separated folates can be detected directly on cellulose-coated TLC plates containing a fluorescent indicator by absorption, blue color, fluorescence, quench, or yellow color when exposed to UV light at 254 or 366 nm. Alternatively, where biological samples have been separated by this method, low levels of folate present are best determined by removal of powder in sections from the plate followed by desorption and microbiological assay using *L. casei* [32].

A specific TLC procedure for the purification of folic acid standards has been reported by Reif et al. [33]. These authors have demonstrated that commercial preparations of folic acid contain fluorescent

TABLE 6.1 TLC of Naturally Occurring Folates (Pteroylmonoglutamates) and Pteroate on MN 300 UV_{254} Powder

Compound[a]	NH_4Cl[b]	Sodium phosphate[c] A	Sodium phosphate[c] B	Detection[d]
Pteroate	0	0	0	Q/A
DHF	10	26	24	BF/BF
FA	24	56	40	Q/A
THF	56	70	59	BF/BF
5-Formyl-THF	72	85	76	Q/YBF
5,10-Methenyl-THF	32	79	45	Q-BF/BF
5,10-Methylene-THF	75	—	82	Q
5-Methyl-THF	80	86	79	Q
10-Formyl-THF	82	81	82	Q
5-Methyl-DHF	87	58	84	Q
10-Formyl-FA	70	58	—	Q
10-Formyl-DHF	73	72	—	Q/BF
Methotrexate	67	67	74	Q/A
p-ABG	91	94	97	Q
p-ABA	56	85	73	Q

[a]All tetrahydro compounds are racemic mixtures.
[b]3.0% (w/v) NH_4Cl containing 0.5% (v/v) 2-mercaptoethanol as antioxidant (pH 6.2).
[c](A) 0.1 M sodium phosphate buffer, pH 7.0, and (B) 0.1 M sodium phosphate buffer, pH 6.0, both containing 0.5% (v/v) 2-mercaptoethanol.
[d]Exposure to UV light, 254, 360 nm. Abbreviations: A = absorption, B = blue, F = fluorescence, Q = quench, and Y = yellow.
[e]Mean R_f values × 100 of 2-11 determinations.
Source: Table taken, in entirety from Ref. 31, with the kind permission of Dr. John Scott. Compound nomenclature has been adjusted for consistency with that used in this review.

pteridines, diazotizable amines, and compounds other than folic acid that can interfere with standard colorimetric assays. The reported method permits separation of folic acid from the following contaminants: pterin-6-carboxylic acid, *N*-(*p*-aminobenzoyl)-1-glutamic acid, and pterin. This method utilizes a 250 μm layer of silica gel containing a fluorescent indicator. Plates are predeveloped and air dried. Chromatograms are developed using a 1.48 N ammonium hydroxide-1-propanol (1:3, v/v) solvent system. This method appears impractical for routinely resolving a wide variety of reduced folate derivatives from folic acid, but it does permit determination of the relative purity of folic acid standards that may contribute to color formation in standard spectrophotometric assays.

2.2 Liquid Chromatography

As discussed by Kas and Cerna [34,35], several methods of column chromatography have been developed for the separation of folates. None of the currently existing methods, however, allows for the separation of all the naturally occurring folates by means of one separation technique. Thus, the choice of any one particular method often depends principally on which specific pteroylmonoglutamates an investigator wishes to separate.

Column chromatographic materials that have been found useful for the separation of folates include different kinds of ion exchanger (e.g., Dowex 1, DEAE-, and TEAE-cellulose and QAE-Sephadex A-25) and gel chromatographic materials (e.g., Sephadex G-10, G-15, and G-25). Detection of folates eluted from these chromatographic materials are typically achieved by either UV spectrophotometry or by the use of specific bacterial cultures (i.e., *L. casei*, ATCC 7469, which detects all eluting folate activity; *S. faecalis*, ATCC 8043, which does not respond to methyltetrahydrofolate derivatives; and/or *P. cerevisiae*, ATCC 8081, which does not respond to either methyltetrahydrofolate or unreduced folate derivatives). The qualitative responses of each of these assay microorganisms to different pteroylmono-, di-, and triglutamate derivatives have recently been summarized by Kas and Cerna [35].

Very small differences in molecular weight exist between individual folate forms, particularly among the reduced derivatives, making separation of these compounds by molecular weight alone difficult. Nevertheless, Kas and Cerna [34] have reported Sephadex G-10 chromatography to be useful for separation 5-methyl-THF, 5-formyl-THF, THF, 10-formylfolic acid, DHF, 10-formyl-DHF, and folid acid (FA). For example, in their system, 5-methyl-THF appears at an elution volume of 65 ml, DHF appears at 95 ml, and FA appears at 155 ml. Although other investigators [36,37] have also reported separation of folate derivatives by exclusion chromatography on Sephadex G-15 and G-25, there can be problems associated with the use of gel materials. For example, Shin et al. [38] have reported that pteroylpolyglutamates may become adsorbed to the gel matrix. Whether this adsorption involves an ion-exchange interaction is unclear. A further problem in separating pterolylglutamates by gel-permeation techniques is that the separation is usually affected by the number of γ-glutamyl residues. This type of chromatography appears impractical for the resolution of compounds with the same number of γ-glutamyl derivatives but differing in oxidation state.

Rao and Noronha [39] have reported an ion-exchange method for the separation of folates that, unlike prior IEC methods, provides information not only on the number of γ-glutamyl residues but also on the state of reduction of the pteridine ring and the nature of the attached C1 moiety. The method employed by these investigators is a modification of that previously described by Silverman et al. [40]. Rao and Noronha [39] report a correlation between the log of the buffer phosphate concentration and the number of glutamate residues in successive pteroyl-γ-glutamyl derivatives eluting from a DEAE-cellulose column. The eluted fractions were assayed microbiologically before and after treatment of the eluted fractions with a folyl-poly-γ-glutamylcarboxypeptidase preparation. The separation and identification of 5-methyl, 5-formyl, and 10-formyl derivatives of folic acid and several polyglutamates are presented.

Another approach to the separation and identification of folate derivatives by ion-exchange chromatography involves an oxidative or reductive cleavage of the C9-N10 bond using either potassium permanganate [41] or a zinc catalyst [42], which coverts natural pteroylmono- and pteroylpolyglutamates to a reportedly homologous series of *p*-aminobenzoyl-γ-glutamates (Fig. 6.3). These then can be separated by DEAE- and TEAE-cellulose [43-46] or QAE-Sephadex [36,37] and Dowex [47] IEC chromatography. Individually eluted folate derivatives are measured quantitatively by colorimetric estimation of the primary aromatic amine or by radiometric determination. Reed and Scott [48] have used this approach to examine the pteroylmonoglutamate content in various cell types. To allow unambiguous separation of folate polyglutamates, the C9-N10 bond connecting the pteridine ring to the *p*-aminobenzoylglutamate is first cleaved and the resulting *p*-aminobenzoylpolyglutamates are then separated. Prior to analysis on DEAE-cellulose 52, the C9-N10 bond is cleaved either oxidatively or reductively (see Fig. 6.3). The procedure as described by Reed and Scott [48] does present some technical problems, however. As the authors themselves suggest, the preparation of folate polyglutamate markers may be difficult for many groups, and these are not usually commercially available. The applications of selective cleavage of the C9-N10 bond of folic acid and its derivatives include analyses of pharmaceuticals [49], determination of the distribution of radioactive label in ^{3}H-labeled folates [16], and estimation of polyglutamate chain lengths in natural folate derivatives [41,42].

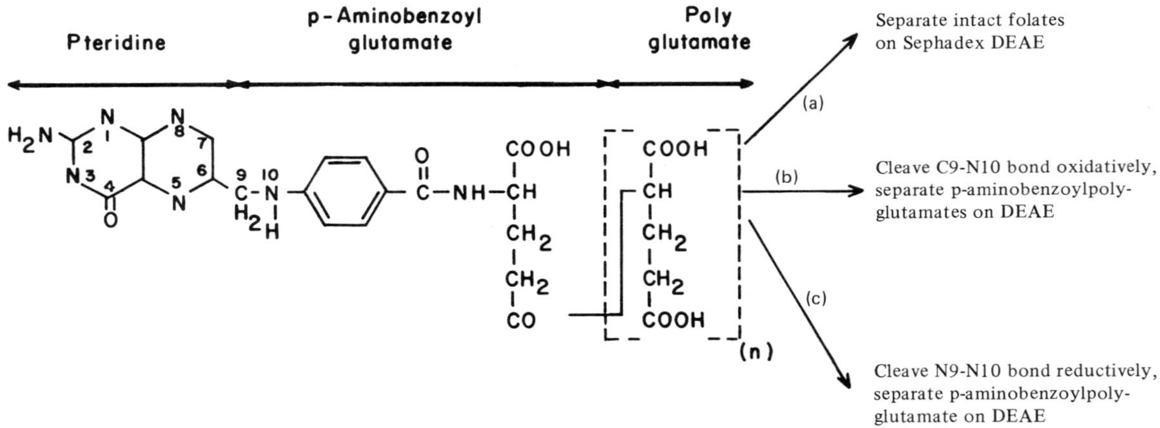

Figure 6.3 Chemical structure of polyglutamates of folic acid and approaches to their chromatographic identification. (From Ref. 48.)

Maruyama et al. [50] and Lewis and Rowe [51] have both presented critical analyses of the oxidative and reductive folate cleavage techniques. Maruyama et al. [50] report that 5-methyl-THF is resistant to attempted oxidative cleavage, and *p*-Aminobenzoylglutamic acid (*p*-ABG) is susceptible to oxidation. Lewis and Rowe [51] have confirmed these findings and, in addition, have shown that the reductive technique is as ineffective as the oxidative method for cleaving the C9-N10 bond of the predominant mammalian tissue folate derivative, 5-methyl-THF. These authors therefore suggest that the failure of these cleavage techniques to produce a homologous series of *p*-ABG derivatives differing only in the length of the γ-polyglutamate chain may lead to serious errors of interpretation. More recently, important contributions by Foo and his colleagues [52] and Eto and Krumdieck [53] have led to useful modified cleavage reactions for the analysis of folate derivatives.

As recently discussed by McMartin et al. [54], analytical approaches, such as those discussed in the previous sections of this chapter, are currently available to accurately measure the total level of folates in fluids and tissues. However, such approaches do not permit the rapid determinations of individual folate forms with the reliability that is now considered necessary to evaluate changes in tissue folic acid metabolism and interconversions of folic acid derivatives. In order to address these shortcomings in analytical methodology, numerous reports have recently been published describing the application of LC techniques to particular folate derivative separations. Most of these methods employ anion-exchange [55-57] or reversed-phase paired-ion techniques [54, 65,58-62]. However, as with the more traditional methods, each of the reported LC methods has been developed for the separation of specific folate derivatives. No single LC method has yet been devised that is applicable for the determination of all known naturally occurring pteroylmono- and pteroylpolyglutamates.

One of the first applications of LC for the separation of certain water-soluble vitamins was that reported by Williams et al. in 1973 [63]. These investigators demonstrated the separation of nicotinic acid, riboflavine monophosphate, and folic acid using an anion-exchange column eluted with a concave exponential gradient of phosphate buffer. One of the first useful methods that used high-pressure anion-exchange chromatography for the separation of several different naturally occurring pteroylmonoglutamates was that reported by Reed and Archer [57]. AL-Pellionex-WAX, a pellicular, weak anion-exchange material, consisting of an aliphatic matrix with amino functional groups, was specifically developed to extend the classic weak anion-exchange chromatographic separations, such as DEAE-cellulose, to modern LC. Using a 4-1/2 ft long column with an isocratic 0.025 M sodium dihydrogen phosphate (pH 4.8) solvent system, these authors achieved a baseline separation of THF, DHF, and FA. Use of a weaker NaH_2PO_4 buffer (0.006 M, pH 4.0) solvent resulted in separation of 5-methyl-THF and

5-formyl-THF within 20 min. Mercaptoethanol, the antioxidant used in both separations, eluted at the solvent front along with 5,10-CH-THF. *p*-Aminobenzoylglutamate was detected in both separation schemes and was believed to be due to a small amount of contamination or decomposition of the standard compounds. Ultraviolet-absorbance measurements at 254 nm permitted detection of an little as 0.02 μg THF and folic acid and 0.1 μg DHF.

AL-pellionex-WAX has also been found to be a useful material for the separation of products resulting from the interaction of folates, particularly those occurring in food, with nitrous acid. One of the products of this reaction, 10-nitrosofolic acid, has been reported to be a carcinogen in mice [64]. Archer and Reed [17] described the separation by LC of 2-hydroxy-10-nitrosofolic acid, 2-hydroxyfolic aicd, 10-nitrosofolic acid, and FA and suggest that chromatographic separations of this type may be useful in studying the chemical kinetics of such reaction as well as the environmental distribution of these folate derivatives. In the system employed by Archer and Reed, FA and its 10-nitroso derivative have retention times of approximately 3 and 2 min, respectively.

Strong anion-exchange (SAX) LC columns are also useful for the separation of particular pteroylmonoglutamate derivatives. Clifford and Clifford [55c] have established conditions for the separation of THF, 5-methyl-THF, DHF, and FA using a pellicular strong anion-exchange resin and a gradient of potassium chloride in a phosphate buffer. As the authors point out, this is not a method for the separation of all naturally occurring pteroylmonoglutamates. For example, 5-formyl-THF cochromatographed with 5-methyl-THF. Furthermore, pteroylpolyglutamates derived from 5-methyl-THF cochromatographed with 5-methyl-THF. These findings demonstrate that the separation of the various folates was dependent on the substitutions of the pteridine portion of the molecule and was independent of the number of glutamate residues in excess of one.

A new liquid chromatographic technique, paired-ion chromatography (IPC), has recently been developed to solve the analytical problems traditionally associated with samples containing ionic species. Ion-pair chromatography employs a large organic counterion in the mobile phase that enters into an equilibrium complex with an ionic species. The resulting species behaves as an electrically neutral and nonpolar (lipophilic) compound that can then be easily chromatographed by a reversed-phase liquid chromatographic technique. This new approach has been used within the past few years by a number of laboratories [56,61] to achieve useful separations of specific folate derivates.

Branfman and McComish [58] described one of the first uses of the IPC technique for the separation of folic acid derivatives. Their method utilizes a μBondapack (Waters) reversed-phase column with a simple isocratic methanol-water (35:65, v/v) solvent system (containing 0.005

M tetrabutylammonium phosphate). They report the separation of standard solutions of FA, DHF, and THF, as well as *p*-ABG. Chapman et al. [56] have also examined the utility of IPC in the separation of specific pterolymonoglutamates. These investigators used an isocratic system that was very similar to that of Branfman and McComish [58]. Although a distinct separation of DHF from 5-formyl-THF was achieved, 5-methyl-THF could not be separated from pteroylglutamic acid by this technique (although they could be differentiated by microbiological assays). Because determination of FA, 5-formyl-THF, and 5-methyl-THF is considered of possible clinical importance, our laboratory [60] extended the findings of previous investigators who attempted to separate 5-substituted pteroylmonoglutamates from folic acid. Using a reversed-phase column, the paired-ion technique was combined with a specific methanol-water solvent gradient program. Although relatively complex, this gradient pattern is easily achieved by modern microprocessor-controlled LC equipment. The method results in distinct separations of 5-formyl-THF, DHF, and *p*-ABG within 20 min. Quantitation of individual folate peaks was performed by UV spectrophotometry at 285 nm. Under these conditions, the minimum detectable levels were FA (25 ng), DHF (15 ng), 5-methyl-THF (25 ng), and 5-formyl-THF (50 ng). We also found recently (unpublished observations) that this method is applicable to the separation and quantitation of 5-methyl-tetrahydrohomofolate (NSC-139490) an investigational anticancer agent that is currently undergoing phase I clinical trails. The use of IEC (DEAE-cellulose) to purify tetrahydrohomofolate for subsequent use in the preparation of a methenyl derivative as well as a TLC system for establishing the homogeneity of the methenyl compound have been reported recently [61a], as well as the application of LC (reversed-phase) to the analysis of tetrahydrohomofolates in biological systems [61b,61c]. Horne and coworkers [62] have utilized reversed-phase IPC to achieve separation of naturally occurring monoglutamate derivatives of folic acid, as well as several oxidation and decomposition products of reduced folates. This method is one of the most useful LC approaches reported to date for the separation of a wide variety of folates, such as *p*-ABG, THF, 5-formyl-THF, DHF, 5-methyl-THF, and folate itself; the approximate elution times for several folate derivatives are summarized in Table 6.2. This method differs from the method developed in our laboratory [60] primarily with respect to the composition of the mobile phase. Horne et al. [62] have employed a solvent gradient consisting of (1) the ion-pairing reagent, tetrabutylammonium phosphate (10 mM) and (2) a 1:1 mixture of water and 95% ethanol. The composition of (2) was changed from 15 to 25% over 50 min by a programmed concave gradient. It is important to note, as Horne et al. have stressed, that current LC assay methods using UV-absorbance detection are not sensitive enough to determine naturally occurring levels of folates in serum (2-10 ng/ml) without injecting unduly large volumes of serum or concentration of the sample prior to analysis.

TABLE 6.2 Chromatographic Behavior of Folate Derivatives[a]

Compound	Approximate retention time (min)
p-ABG	12
10-Formyl-THF	21
THF	25
5-Formyl-THF	32
5-Methyl-THF	48
FA	55

[a] μBondapak C_{18} column; elution with a nonlinear ethanol gradient; detection at 280 nm.
Source: Data from Ref. 62.

The most useful approaches toward analyses of pteroylpolyglutamates involve the high-performance liquid chromatographic technique. One of the initial successful techniques described by Stout and coworkers [55b] is a method suitable for the separation of various substituted pteroylmonoglutamates and pteroylpoly-γ-L-glutamates. The technique involves the use of a Partisil-10 SAX column with a linear gradient solvent system composed of NaCl and $NaNH_4PO_4$ (pH 6.5). The authors observed that solute retention is largely determined by the interaction between the folate-carboxylic groups and the quaternary ammonium groups attached to the organic moiety, which is covalently bound to the silica support. The report by Stout et al. [55b] included a chromatogram depicting the separation (in order of elution) of (1) 5-formyl-THF and 5-methyl-THF, which coelute, (2) 10-formyl-THF, (3) DHF, (4) FA, (5) 5-methyl-THF diglutamate, (6) pteroyltriglutamate, (7) 5-methyl-THF tetraglutamate, (8) pteroylpentaglutamate, (9) 5-methyl-THF hexaglutamate, and (10) pteroylheptaglutamate. Retention times for selected folates and γ-oligoglutamates are listed in Table 6.3.

Cashmore et al. [65] have recently reported two modifications of the method of Stout et al. [55b] for the separation of pteroylpolyglutamates. The first method uses a strong anion-exchange column and gradient elution with aqueous phosphate buffer in order to separate oligoglutamates in the order of increasing number of glutamyl residues. In the second method, a C_{18} ODS (octadecylsilica) reversed-phase column is used with gradient elution at increasing concentrations of acetonitrile in the aqueous elutent. Under these conditions, the order of elution of pteroylpolyglutamates is the opposite of that described in the first method. The authors report that the concomitant use of the two

TABLE 6.3 Chromatographic Behavior of Folates and γ-Glutamate Derivatives[a]

Compound	Approximate retention time (min)
5-Formyl-THF	8
5-Methyl-THF	8
FA (pteroylglutamate)	10
Pteroyltriglutamate	15
5-Methyltetrahydropteroyl pentaglutamate	19
Pteroyl pentaglutamate	21
Pteroyl heptaglutamate	26

[a]Partisil-10 SAX column; NaCl gradient elution at pH 6.5. Detection at 254 nm.
Source: Data from Ref. 55b.

analytical methods facilitates the identification of the individual peaks on the basis of different quantitative structure-retention relationships. Although the combined analyses of samples with both columns appears to be a useful method for the separation of standard preparations of the oligo-γ-L-glutamates of folic acid, the application of this method for the determination of naturally occurring pteroylpolyglutamates must be done with caution, as has been pointed out by Lewis and Rowe [51].

Each of these LC methods is adequate for the separation of a particular series of standard folate derivatives that differ with regard to the numbers of glutamic acid moieties attached via γ-glutamyl linkages, but they still appear inadequate for resolving folates, which may differ in the oxidation state of the pteridine ring and in the one-carbon constituents on nitrogen position 5 or 10, as well as the number of glutamate residues. MacMartin et al. [54] have approached this problem by hydrolyzing tissue polyglutamate forms to pteroylmonoglutamates prior to separation by paired-ion, reversed-phase LC technique and analysis of eluted peaks with microbiological assay. The pteroylmonoglutamates THF, 5-formyl-THF, 5,10-methenyl-THF, 10-formyl-THF, 5-10-methylene-THF, and DHF were separated under these conditions. Folic acid and 5-methyl-THF coeluted but were distinguishable with microbiological analysis. Levels of specific pteroylpolyglutamates were not determined; however, the authors assayed rat liver folate levels prior and subsequent to incubation with hog kidney polyglutamate hydrolase, thereby obtaining pteroylmonoglutamate versus total folate quantitation. The increment in individual pteroylmonoglutamate levels after hydrolase treatment of the samples presumably represents the contribution from the respective pteroylpolyglutamate forms.

Folic acid and related analogs, including several chemotherapeutic agents, have recently been analyzed by a novel mass spectrometry technique [66]. Various folate compounds, including FA, DHF, THF, and 5-formyl-THF, were dissolved in a solution of dimethylsulfoxide (DMSO). Sodium methylsulfinylmethide in DMSO was added, followed by methyl iodide. After extraction of the permethylated products, the derivatized samples were introduced into a mass spectrometer, where electron impact ionization of the permethylated derivatives provided unique spectra. The applicability of this particular new detection technique to the routine analyses of folate derivatives in clinical, food, and other samples, however, remains to be demonstrated.

2.3 Selected Applications of Folate Chromatography

Two studies have been published within the past few years that were designed to determine folate derivatives in selected foods by high-performance liquid chromatography. Clifford and Clifford [55c] have used an ion-exchange method to determine 5-methyl-THF, THF, DHF, and FA content in raw peanut extracts, lemon juice, lemonade, grapefruit juice, pineapple-orange juice, apple juice, almonds, wheat germ, and raw peanuts. It should be noted that a tabulation of the folate levels in the various foods is not presented. A more recent analytical HPLC technique for the determination of pterin-6-carboxylic acid, DHF, FA, and 5- and/or 10-formyl-THF has been reported by Day and Gregory [67]. These authors have used coupled octadecylsilica and phenylsilica columns with an isocratic, acidic-phosphate-acetonitrile mobile phase system. Detection was performed either by monitoring UV absorption at 280 nm or by a postcolumn fluorogenic oxidative derivatization method. This HPLC procedure was applied to the determination of total (i.e., γ-glutamylcarboxypeptidase-treated) folate content in samples of cornflakes, an infant formula preparation, and fresh beef liver. Representative chromatograms from the analyses of each of these samples are presented. The minimal detectable levels by ultraviolet absorption at 280 nm were THF (2 ng), 5-methyl-THF (3 ng), DHF (5 ng), 5-formyl-THF (10 ng), and FA (3 ng). The use of fluorometric detection after postcolumn derivatization improved the detection sensitivity for THF (0.3 ng), DHF (0.9 ng), and FA (0.2 ng), which are comparable to minimum levels detectable with the microbiological assay.

Various liquid chromatographic procedures for determining water-soluble vitamins have been developed during the past few years. Methods developed for analyzing pharmaceutical preparations have typically made use of C_{18} reversed-phase columns with appropriate ion-pair reagents added to their mobile phase [68]. Recently, Vandemark and Schmidt [69a] have developed a rapid LC procedure using a C_8 bonded-phase column for separating simultaneously the major water-soluble vitamins. These include, in the order of elution, ascorbic acid, niacin, niacinamide, pyridoxine, folic acid, riboflavine, and thiamine. The mobile phase consisted of various concentrations of methanol and acetonitrile in a three-step gradient run over a 17 min period. Compounds were detected by UV absorption at their individual optimal wavelength by use of a stop-flow UV scanning technique. This method appears well suited to the separation of a large number of the major water-soluble vitamins in tablet and capsule preparations. It is not adequately sensitive, however, to determine the low levels of two other water-soluble vitamins, pantothenic acid and vitamin B_{12}, and individual determinations of these two vitamins appear necessary. A comparable procedure for the assay of folate in pharmaceutical preparations containing several vitamins as well as mineral supplements has been described by Tafolla et al. [69b]. This procedure allows excellent separations of folate from possible degradation products.

A number of conditions (pregnancy, old age, and vitamin B_{12} deficiency) and chemicals (anticonvulsant and chemotherapeutic drugs, alcohol, and oral contraceptives) are thought to alter folate metabolism and/or the requirement for folate [70]. A number of laboratories have, therefore, sought to explore the utility of measuring serum folate levels. Chapman et al. [56] have used both ion-exchange and ion-pair partition chromatographic techniques to assay serum folate levels in human subjects pretreated with pteroylglutamic acid. Allen and Newman [60] also developed a LC technique that was specifically designed to assess changes in serum folate derivative levels in humans and rats receiving antifolate chemotherapy. Both of the previously mentioned methods, however, really lack the sensitivity required for routine determination of the relatively low content of serum folates in humans. Since the endogenous plasma concentration of 5-methyl-THF, the major circulating folate present in humans, lies between 10^{-9} and 10^{-8} M, the determination of this particular folate demands high efficiency of a chromatographic system. Lankelma and Van Der Kleijn [71] have approached this problem of detection sensitivity using on-column sample concentration and electrochemical detection techniques. Their method, which effectively determined only 5-methyl-THF and not other folate derivatives, such as FA or 5-formyl-THF, has a detection limit of 2×10^{-9} M (0.9 ng/ml) for both plasma and spinal fluid. This method appears directly applicable to the determination of the behavior of 5-methyl-THF in the body after the administration of Leucovorin during anticancer therapy. The utility of electrochemical detection in other situations in which changes in serum folate levels are suspected, however, is not certain.

Recently, a simple biochemical assay for Leucovorin has been reported [72] that is based on the formylation of bacterial initiator methionyl-tRNA (met-tRNA) in a cell-free extract from *Escherichia coli*. Product formation in this novel assay is dependent on added folate derivatives. This assay is not only MTX independent but is also able to distinguish between Leucovorin and 5-methyl-THF, the major folate derivative in human serum. Hamel et al. [72] have used this assay in a preliminary study of the pharmacokinetics of Leucovorin disappearance and 5-methyl-THF formation after an intravenous dose of Leucovorin to several patients. The assay is less sensitive than other MTX-independent radioimmune and microbiologic techniques, although it can detect approximately 5×10^{-8} M *l*-Leucovorin and 5×10^{-7} M 5-methyl-THF. In addition to the analysis of serum, this biochemical method of folate quantitation may be useful when applied to the chromatographic separation and determination of clinically important folate derivatives.

3. CHROMATOGRAPHY OF 4-AMINO DERIVATIVES OF FOLATE (METHOTREXATE AND RELATED COMPOUNDS)

3.1 Thin-Layer Chromatography

Valerino and his colleagues [73] have shown that MTX can be separated reasonably cleanly from its degradation product DAMPA (2,4-diamino-*N*-10-methyl pteroic acid) using DEAE-cellulose plates and development with 0.1 M ammonium bicarbonate (pH 8.8); both MTX and DAMPA had higher mobilities in several paper chromatographic (PC) systems than were observed with ion-exchange TLC. In principle, the solvents used for these PC studies should be suitable for TLC studies using cellulose plates ("regular" and microcrystalline cellulose). Methotrexate, DAMPA, and folate can be separated efficiently on cellulose plates using aqueous solutions containing mercaptoethanol; the mercaptoethanol is included to reduce the rate of oxidation of folate derivatives [74].

3.2 Liquid Chromatography

Column chromatography, using DEAE-cellulose, has been used frequently to achieve separation of a number of folate derivatives, and this approach was applied successfully in studies of the metabolism of MTX by intestinal flora [73]. Such studies demonstrated the presence of a major microbial metabolite of MTX, DAMPA, that was eluted considerably more rapidly than MTX, from a DEAE-cellulose column, using a linear gradient of ammonium bicarbonate (0.01-0.25 M). A similar IEC system is useful for the separation of MTX from its polyglutamate derivatives [74,75]; with such a system MTX is eluted from the column first followed by successively substituted polyglutamates. This technique has been applied to studies that demonstrate substantial inhibition by bile salts of intracellular formation of MTX polyglutamates [76]. Polyglutamate derivatives of MTX may play a critical role in producing biochemical lesions responsible for the cellular toxicity associated with MTX administration, and consequently analytical methods for detection and determination of these compounds are of great importance to research workers. Separation of MTX from polyglutamates by use of size-exclusion chromatography (SEC) on Sephadex G-15 columns can be accomplished by use of dilute basic aqueous solutions [77]; application of this technique is exemplified in a recent communication by Galivan [78] on the formation of polyglutamate derivatives of MTX in cultured hepatoma cells. Methotrexate is frequently purified by IEC prior to use in biological experiments, and failure to do this may seriously complicate interpretation of experimental data, especially when isotopic methods are employed [79]. Purification of MTX and isolation of impurities by IEC have resulted in the finding that one impurity, N^{10}-methylfolic acid, is responsible for the inhibition of folate binding to high-affinity binding proteins, observed with unpurified samples of MTX [80].

The first report of liquid chromatographic analysis for MTX appears to be that by Tong et al. [81], who described in in 1975, the separation of MTX from several ultraviolet-absorbing impruities using an octadecylsilane (ODS; μBondapak) column and elution with either ammonium acetate-acetonitrile (85:15) or ammonium formate-acetonitrile (85:15). The elution time for MTX with the latter solvent (at pH 3.5) was approximately twice that observed when the former solvent (at pH 5.0) was used. This report was intended to demonstrate the application of LC to the monitoring of levels of impurities in MTX and the advantages of this approach over those previously used [82] in terms of speed and accuray. The successful separation of MTX, by LC, from closely related compounds [83], of course, clearly suggested the possible value of LC methodology for the detection and determination of MTX in biological samples.

Nelson and his colleagues, in 1977, described an LC method for the analysis of MTX in human plasma using a fluorescence detection system based on permanganate oxidation of MTX to 2,4-diaminopteridine-6-carboxylic acid [84]. These investigators used an ODS (C_{18}) reversed-phase column with either of two solvent systems, 0.1 M tris-phosphate (pH 6.7), or this solvent supplemented with methanol (20%). The LC approach to the analysis of MTX was found to compare favorably to a standard enzyme inhibition method for MTX in plasma. 2-Hydroxyfolic acid was found by Nelson et al. to be a suitable internal standard for this LC method. Buice et al. described an LC procedure [85] in which MTX and 7-hydroxymethotrexate (7-OH-MTX) are separated from interfering substances and concentrated by absorption on reversed-phase (SEP-PAK) cartridges, washing with acetate buffer and dilute sodium hydroxide solution, and elution with methanol. The use of the SEP-PAK resulted in high recovery (90%) of MTX from serum and appears to be preferable to an earlier ion-pair extraction method for the analysis of low levels of MTX in biological materials. Buice and his coworkers found that the LC approach compares very well with three alternate methods (enzyme immunoassay, radioassay, and radioimmunoassay) for monitoring serum MTX levels over a 24 hr period; the LC method offers a clear advantage in that MTX and 7-OH-MTX can be measured independently during the chromatographic run. The potential importance of assessing levels of both MTX and its 7-hydroxy derivative has been stressed by Lankelma and his colleagues [86] and by Howell et al. [87]. These latter investigators have described a LC procedure that employs a Partisil PXS 10/25 ODS column and an eluting solvent containing 1-hexanesulfonic acid in aqueous methanol; this LC system is capable of separating MTX from a variety of metabolites (see Table 6.4); under the conditions used ordinarily this method is not suitable for analysis of polyglutamates in plasma because these compounds coelute with interfering endogenous substances in plasma samples. Liquid chromatography played a crucial role in the elucidation of the nature of a metabolic derivative of MTX that interferes with the radioimmunoassay of MTX. The interfering metabolite has been identified as DAMPA; in these studies, an ODS column was used with a mobile phase of methanol and 0.01 M acetate buffer (pH 5.5); plasma and urine samples were subjected to an ion-exchange procedure prior to analysis by LC; the ion-exchange procedure resulted in quite good (80-90%) recovery of the compounds of major analytical interest, namely, MTX and DAMPA [88]. Lankelma and Poppe [89] developed an elegant method for analysis of MTX in plasma samples in which the drug is concentrated on one column and transferred to a second column for subsequent chromatographic analysis. A hydrophobic concentration column was used

TABLE 6.4 LC of MTX and Related Compounds[a]

Compound	Retention time (min)
MTX triglutamate	6.3
MTX diglutamate	7.7
7-Hydroxy-MTX	9.1
MTX	10.5
DAMPA	28

[a]Mobile phase at pH 3.75; ultraviolet detection (280 or 305 nm).
Source: Data from Ref. 87.

prior to MTX analysis on a strong anion-exchange column (Partisil SAX). This procedure results in detection (at 306 nm) of plasma levels of MTX as low as 2×10^{-8} M. Lankelma and Poppe stated, in 1978, that their method could be used also for determination of MTX metabolites, a statement that has been confirmed experimentally [86].

Watson et al. described, in 1978 [90], a rapid method capable of determining MTX as well as 7-OH-MTX in human plasma. They used a 2,4-diaminoquinazoline analog of MTX as an internal standard for the procedure and employed a system that included an ion-exchange column (Partisil 10-SAX) and elution with phosphate buffer (0.025 m, pH 7.0). MTX and related compounds were detected at 315 nm, an analytical wavelength that gave more acceptable background absorbance values than those observed when detection was carried out at 254 nm. MTX and 7-OH-MTX were separated from plasma by use of an extraction method apparently dependent on ion-pair formation between the basic 2,4-diaminopteridine moiety of the drug (and its metabolite and the perchlorate counterion. This method yielded values for MTX concentrations in good agreement with those estimated by use of a competitive ligand binding method. Methotrexate is eluted from the Partisil column at approximately 7 min, about 2 min more rapidly than is 7-OH-MTX. Methotrexate can be detected at a sensitivity limit of 1×10^{-7} M with this approach, and the method can detect levels of 7-OH-MTX of the order of 1×10^{-6} M. Wisnicki et al., also in 1978, described a somewhat different approach to LC analysis of MTX and derived compounds in pharmaceutical preparations, and in urine of patients treated with high-dose MTX regimens [91]. This method represented a considerable improvement over that used earlier [81] to examine the purity of MTX preparations. The improvement was attributable to use of IPC on a μBondapak C_{18} column, with 0.005 M tetrabutylammonium phosphate (PIC reagent A) in aqueous acetonitrile. This system permitted detection of *21* distinct impurities, most of which were cleanly separated from MTX, which had a retention time, in this chromatographic system, of approximately 40 min. The separation of closely related compounds, such as aminopterin and DAMPA, from MTX, in this system is particularly efficient. Analysis of urine samples, after filtration through a 0.45 μm filter, using tetrabutylammonium phosphate as counterion in aqueous methanol (pH 7.5, 30% methanol), resulted in good separation of 7-OH-MTX (retention time 48 min) from MTX (retention time 35 min) when urine was obtained from pediatric patients treated with high-dose MTX. Wisnicki and his colleagues found that IPC gave better separation and resolution of peaks in the analysis of samples containing MTX and 7-OH-MTX than did LC using a strong anion-exchange column [91]. A recent communication by Tong et al. [92] describes a modification of the chromatographic approach outlined above for MTX analysis; the modification involves the use of a small precolumn (CO; Pel ODS) in conjunction with the μBondapak C_{18} analytical column. Inclusion of the precolumn allows for direct injection of plasma samples into the system. This modified approach is capable of determining levels of MTX, 7-OH-MTX, and dichloro-MTX in plasma samples obtained from rabbits and from humans and is rapid and sufficiently sensitive for many clinical applications. We agree with Tong et al. that their use of the precolumn, eliminating the need for tedious and often inefficient extraction procedures, not only facilitates the analysis of MTX and its close chemical relatives but also represents a general approach to overcoming extraction problems that complicate analytical studies of other anticancer drugs (and, indeed, drugs in general) in biological samples. Reversed-phase LC analysis of MTX and 7-OH-MTX serum can also be achieved by ion-pair extraction of the desired compounds into an ethyl acetate-isopropanol solution (10:1 [93]), chromatography at ambient temperature using a mobile phase consisting of pH 6.8 phosphate buffer-methanol (85:15), and use of an internal standard (*p*-aminoacetophenone). The relatively low efficiency of extraction observed in this study could be improved by the use of an approach based on precipitation of a MTX-silver complex, recently shown to be of value in the isotachophoretic determination of MTX in plasma [94]. The application of the reversed-phase method, with use of *p*-aminoacetophenone as the internal standard, to analysis of MTX levels in patients receiving both conventional and high doses of MTX has been reported by Chan et al. [95]. The internal standard used by these investigators is advantageous because it is readily available commercially and has a retention time, in a typical chromatographic run, of approximately 9 min whereas the retention times for MTX and 7-OH-MTX are 6.6 and 7.4 min, respectively. Their studies have demonstrated that MTX conversion to 7-OH-MTX occurs not only at high MTX doses but also at conventional doses. The finding of significant MTX metabolism in patients receiving conventional doses of the drug was also made independently by Lankelma et al. [86] and is of importance because it previously was assumed that MTX oxidation to 7-OH-MTX does not occur appreciably in patients receiving conventional doses. The failure to detect MTX oxidation at conventional doses in earlier studies seems to reflect methodological problems, such as inadequate sensitivity of detection systems and interference from endogenous substances. The method of Chan et al. [95] can be used also to detect both MTX and 7-OH-MTX in cerebrospinal fluid (CSF); CSF samples were introduced directly into the chromatographic system.

Another example of the application of LC to clinical analysis of MTX and 7-OH-MTX is provided by the report of Canfell and Sadee [96]. They used an extractive concentration method to improve the general sensitivity of their approach and performed LC analysis at room temperature using a μBondapak-phenyl reversed-phase column with sodium acetate (pH 4.6)-acetonitrile (89:1) as the mobile phase and ultraviolet detection (303 nm). Methotrexate concentrations were determined by comparing peak heights (or areas) with appropriate standards. This chromatographic procedure separates MTX (retention time, 6 min) from

Leucovorin (retention time, 2.8 min), 7-OH-MTX (retention time, 8.8 min) and the pteroic acid analog of MTX (retention time, 17.6 min). It deserves emphasis that the antibacterial and antiprotozoal agent, trimethoprim, is separated cleanly from MTX in this system, because trimethoprim, which may be administered to patients who also are receiving MTX, can interfere with some of the biochemically based assays for MTX [97]. Methotrexate and 7-OH-MTX can be measured in serum by a rapid reversed-phase technique, in which serum is treated with dilute percholoric acid containing a 2,4-diaminoquinazoline internal standard, and the supernatant solution, after centrifugation, injected directly onto a Hypersil-octadecylsilane column system (precolumn plus analytical column). The mobile phase for this method is 20% methanolic tris-phosphate buffer (pH 6.7 [98]); the procedure provides high recovery (80-100%) of MTX, 7-OH-MTX, and internal standard, and excellent separation of MTX and 7-OH-MTX from folate derivatives, as well as from DAMPA, is observed. This method is attractive because of speed and relative simplicity, but its sensitivity is not sufficiently high for accurate monitoring of MTX levels at times later than 48 hr after completion of drug administration. El Dareer and his colleagues have recently reported the use of LC to study the pharmacokinetics of MTX and 5-methyl-THF in rats [61b]; they used a Partisil ODS-2 column with sodium acetate (pH 3.6)-acetonitrile (15%) as mobile phase and were able to inject serum, bile, and urine samples directly into the chromatographic system.

Deen et al. have developed an assay procedure for MTX capable of detecting the drug at levels as low as 5×10^{-9} M [99]; this method involves dithionite reduction of MTX and subsequent degradation of the reduction product to a compound thought to be 2,4-diamino-6-methylpteridine. Chromatography was carried out using a C_{18} Corasil precolumn and an ODS-C_{18} analytical column with, for plasma samples, a mobile phase of 25% methanol containing tetrabutylammonium phosphate. The MTX assay is performed fluorimetrically (excitation, 367 nm; emission, 463 nm) with simultaneous analysis of the anticancer agent vincristine and of Leucovorin (to reduce MTX toxicity) by ultraviolet and fluorescence assay, respectively. Jolivet and Schilsky found recently that IPC, using a C_{18} μBondapak column eluted with tetrabutylammonium phosphate in aqueous acetonitrile, is capable of effecting excellent separation of MTX from polyglutamate forms in extracts prepared from cultured human breast cancer cells incubated with MTX [100]. Sample preparation included the use of a C_{18} cartridge as a preliminary "cleanup" procedure. The LC procedure permits much better resolution of MTX and polyglutamate derivatives formed by cultured cells than is achievable with such techniques as SEC, and it seems that use of the LC approach will contribute heavily to a clarification of the role of polyglutamates in the overall pharmacodynamic profile of MTX.

The use of LC for quantitative determination of MTX, in the presence of substances formed by hydrolytic or photolytic degradation processes, was described by Chatterji and Gallelli; two LC approaches were employed, one involving a single anion-exchange column and a second involving a bonded amine column in series with the anion-exchange column [101].

4. CHROMATOGRAPHY OF UNCONJUGATED PTERIDINES

Unconjugated pteridines, which lack a *p*-aminobenzoylglutamic acid residue, are of considerable interest not only because such compounds can be formed by oxidative transformation of reduced folate derivatives, but also because levels of compounds of this type in biological fluids have been reported recently to be elevated in certain pathological states [28,102]. A reduced pterin also plays a role, as a molybdenum complex, in the cofactor activity of several molybdoenzymes [103]. Therefore, we consider it appropriate to summarize some recent studies of approaches to the chromatographic analysis of these compounds.

4.1 Thin-Layer Chromatography

Two-dimensional TLC, using cellulose plates, has been employed by Descimon and his colleagues [104] to separate pteridines that are present in *Drosophila*. A modification of this approach has been reported [105] to give improved resolution and detection of fluorescent substances found in *Drosophila*; this two-dimensional TLC method involves initial development in 1:1 isopropanolol-2% ammonium acetate, subsequent development with 3% ammonium chloride, and fluorimetric quantitation in situ. The separation of several biologically important pteridine derivatives, using the TLC method, is illustrated in Ref. 105. It deserves emphasis that geometric isomers such as threo- and erythrobiopterin, differing in the relative positions of hydroxyl functions in the side chain, can be separated by the inclusion of boric acid in developing solvents [106]. Wilson and Jacobsen have used such solvents to detect L-erythrobiopterin in extracts prepared from *Drosophila* and detected a minor spot that may be the threo isomer; for example, chromatography of material identified as biopterin yields a single spot at R_f 0.06 in 3% ammonium chloride, but two spots (R_f 0.35 and 0.20) are observed when plates are developed with isopropanol-5% boric acid (4:1). We, in our laboratory at the University of Vermone [107], have made extensive use of TLC, primarily with cellulose plates, for studies of the oxidation of pteridine derivatives by xanthine oxidase and the related molybdoflavoprotein, aldehyde oxidase. We generally employed development solvents that were found earlier to be suitable for the separation of pteridines by paper chromatography. Kohashi has used TLC and LC to good advantage in a study of the content of unconjugated pteridine derivatives in soybean seeds [108]. R_f values for pterin-6-carboxylic acid, neopterin (erythro and threo), isoxanthopterin, 6-hydroxymethylpterin, pterin, 6-hydroxymethyllumazine, and lumazine are reported in five solvent systems for chromatography on cellulose plates with detection by fluorescence at 366 nm. Liquid chromatographic studies were accomplished with an ODS-B column with degassed 5% methanol as the mobile phase; pteridines were detected by fluorescence measured at 440 nm with excitation at 365 nm. The excellent LC separation of pterin-6-carboxylic acid (retention time, 3.1 min), isoxanthopterin (retention time, 11.1 min), and pterin (retention time, 16 min) are noteworthy features of Kohashi's report. The use of TLC and LC in studying the urinary excretion of pteridines in patients with cancer was described recently by Rokos and his colleagues [102c]; these investigators carried out two IEC "enrichment" steps and SEC with Sephadex G-10 prior to analytical chromatographic studies. Rokos et al. found elevation of levels of biopterin and neopterin in some patients with cancer, but it must be recognized that these workers have stated that "the number of cases examined is still too small to imply that altered pteridine excretion is a special feature of certain types of malignant disease" [102c].

4.2 Liquid Chromatography

Column chromatography on ECTEOLA-Sephadex and phospho-Sephadex provides a good approach for separating pteridines [109] and purifying reduced pteridines [9]. Phospho-Sephadex IEC had been used by Fukushima and Nixon to study iodine oxidation of 7,8-dihydrobiopterin; they found water to be a useful solvent for the separation of biopterin from pterin-6-carboxylic acid [110]. The use of IEC, together with other chromatographic techniques, to investigate the biosynthesis of pteridine derivatives in butterflies, was described by Descimon [111]. The importance of chromatographic techniques (IEC, TLC, and LC) in elucidating pathways for pteridine biosynthesis in mammalian tissues is well established [102,112]. High-voltage electrophoresis (HVE), in conjunction with TLC and PC provided information on levels of a dihydropteridine, identified as 7,8-dihydroxanthopterin, in the urine of patients with phenylketonuria [113]. The HVE was performed at pH 1.9 (formate-acetate buffer) for 30 min at 5000 V; pteridines were detected by fluorescence under long wavelength (365 nm) ultraviolet light and mobility expressed in relation to that of quinine sulfate. Similar use of HVE, on paper, in aqueous acidic solution, followed by PC, has been made to study urinary pterin excretion in an "atypical" phenylketonuric patient with 7,8-dihydropterin synthetase deficiency [114]. An interesting observation about the LC behavior of pteridines was made by Dewey and Kidder in 1967, who found that reduced pteridine derivatives can be separated from oxidized pteridines by chromatography on Sephadex G-25 and G-10 columns [115]; such separations clearly involve the operation of adsorption and/or ion-exchange phenomena in addition to SEC.

A recent example of the use of IEC in the investigation of chemical transformations of pteridine derivatives is provided by the report of Baumgarter and Bieri [116], who employed column chromatography with a DEAE-Sephadex matrix and elution with a linear NaCl gradient (in barbital buffer) to purify the monohydrate of xanthopterin-y-carboxylic acid. These investigators assessed the purity of 7-aminoxanthopterin monohydrate, prepared from xanthopterin-7-carboxylic acid by the use of cellulose TLC with 4¢ disodium hydrogen phosphate solution as a developing solvent; in this chromatographic system the carboxylic acid has a much higher mobility ($R_f = 0.63$) than did the amino derivative ($R_f = 0.17$).

Archer and Reed described recently a useful LC approach to the separation and detection of pteridines that involves the use of a strong cation-exchange column with citrate buffer (pH 6.0) as eluent [17]; resolution in this system is better when chromatography is carried out at an elevated temperature (65°C). Using this method, one can achieve good separation of closely related compounds, such as xanthopterin, dihydroxanthopterin, pterin, and pterin aldehyde. Tetrahydropteridines, such as tetrahydrobiopterin, play an important role in several biochemical reactions [117]. Kapatos and Kaufman [9] have demonstrated that high doses of tetrahydrobiopterin and 6-methyltetrahydropterin cross the blood-brain barrier after intraperitoneal administration to rats. Brain levels of pterins were determined using reversed-phase LC with fluorescence detection by a procedure essentially identical to that described by Fukushima and Nixon [10]. The approach to pteridine analysis devised by Fukushima and Nixon is extremely versatile and offers advantages in terms of speed, selectivity, and general applicability over previous analytical methods; their technique involves the use of an ODS column (Partisil or μBondapak) and aqueous methanol (5 or 20%) as the mobile phase. The relative mobilities of several biologically important pteridines, as determined by Fukushima and Nixon, are summarized in Table 6.5. The applicability of RPLC to problems encountered in the study of biological functions of pteridines is exemplified further in a recent communication of localization of tetrahydrobiopterin in brain striatal nerve terminals [118]. Reversed-phase LC, as well as LC using a cation-exchange column [119] and elution with a solvent containing phosphoric acid (0.003 M), methanol (7%), and acetonitrile (1%), was used to investigate biopterin synthesis in rat pineal glands in vivo and in vitro [120]. The detection and determination of eight pterin derivatives in blood and urine samples has been carried out with a LC system consisting of a cation-exchange column used in combination with a Dowex two column cleanup procedure [102a]. Milstein et al. [121] have described a similar approach that permits the detection of dihydropteridine reductase deficiency. Reports by Hausermann et al. [122] and Tanaka et al. [112a] demonstrate the utility of a variety

TABLE 6.5 LC of Pterins[a]

Compound	Relative elution volume[b]	
	5% Methanol	20% Methanol
Pterin	161	65
Biopterin	100	58
Sepiapterin	-	113
Deoxysepiapterin	-	169
Pterinaldehyde	151	65
Dihydropterinaldehyde	-	79
Xanthopterin	86	-
Isoanthopterin	115	59
Pterin-6-carboxylic acid	23	23
6-Hydroxymethylpterin	124	59
Dihydrobiopterin	123	-

[a]Modified from Ref. 10; Partisil 10 ODS column.
[b]Relative to that of biopterin (arbitrarily set at 100 using 5% methanol.

of chromatographic techniques (LC, IEC, and TLC) in studies of the enzymatic synthesis of pterins. Solvent systems for TLC on cellulose sheets are those described previously by Fukushima and Shiota for the assay of pterin levels in human urine [109a]. The reports of LC use summarized above clearly indicate the power of LC techniques in providing feasible solutions to difficult biological problems.

4.3 Gas Chromatography-Mass Spectrometry

The low volatility and rather poor solubility characteristics of pteridines present in biological materials have limited the use of gas chromatography-mass spectrometry (GC-MS) techniques for analytical studies of such pteridines. Nevertheless, important studies have been described on the application of GC-MS to pteridine analysis and these studies merit brief discussion. Lloyd and his colleagues, in 1971 [123], reported that the silylation of pteridines with *N,O*-bis-(trimethylsilyl)-acetamide in pyridine permits reasonably facile analysis by GC-MS. The silylation of all amino and hydroxyl functionalities attached to the pteridine ring and side chain results in very large differences in the mass spectral patterns between such compounds as pterin with no hydroxyl groups in the side chain and biopterin and related compounds. Rothler and Karobath [124] have demonstrated that GC-MS techniques can be used to analyze for pteridines in urine samples; their method required the use of 6,7-dimethylpterin as an internal standard and involves IEC sample preparation (Dowex columns), silylation, and GC on a 3% Dexsil 300 column. The GC-MS analysis of pterin levels in urine, though clearly feasible, appears to be less satisfactory for general application than the LC approaches briefly described above. Iwai and his colleagues have employed GC-MS (chemical ionization and electron impact) to confirm the identity of 6-hydroxymethylpterin isolated from spinach cloroplasts [125]. The isolation of this compound involved extensive use of LC approaches, including IEC on DEAE-Sephadex and DEAE-cellulose, SEC on Sephadex G-25, and preparative TLC on MN cellulose with propanol-ammonia as the developing solvent. A similar combination of GC-MS techniques with LC methodology (fluorometric detection) has been used in studies of 3'-hydroxysepiapterin excretion in relation to dihydrobiopterin deficiency [102a].

5. SAMPLE TREATMENT PRIOR TO ANALYSIS

We have mentioned briefly approaches to the preparation of samples containing folates, methotrexate (and metabolites), and unconjugated pteridine derivatives in biological samples, in the context of a description of several chromatographic procedures. In addition, however, it seemed appropriate to exemplify specifically approaches to sample preparation in order to provide a perspective for the practical aspects of the analyses that we have summarized.

5.1 Folates

Liquid chromatographic analysis of 5-methyl-THF in plasma can be performed after rather straightforward sample preparation; a recent study [126] describes a preparation procedure involving the addition of ascorbic acid to heparinized blood, which then was frozen (-20°C) until analysis. For analysis sodium carbonate solution was added to the blood sample and, after heating in boiling water and centrifugation, the sample (supernatant solution) was ready for injection into the chromatographic system.

Shane [127] and Eto and Krumdieck [128] have used derivatization of folate cleavage products to facilitate LC determination of folate polyglutamates from biological sources. Shane extracted intracellular folates from bacterial cells at high temperature (100°C) in phosphate buffer (pH 7.0) containing mercaptoethanol; rat liver extracts, containing folate derivatives, were prepared by homogenization in trichloroacetic acid (6.7%) with excess trichloracetic acid removed by treatment with diethyl ether. It is interesting to note that Shane's analytical procedure involves the separation of azo dye derivatives of folate cleavage products on polyacrylamide columns and reconversion of the azo dye to *p*-aminobenzoylpolyglutamate derivatives for LC analysis.

5.2 Methotrexate

Facile preparation of biological samples, such as serum or urine, for LC analysis of MTX and its metabolites is exemplified in a recent communication by Breithaupt et al. [129]. Blood was collected in heparinized tubes and centrifuged at 12,000X *g* for 3 min, and the serum was injected into a chromatographic system that consisted of a protective Perisorb RP 2 column coupled to a Spherisorb GDS column. Breithaupt and his colleagues simply diluted urine samples in distilled water (1:20-1:2000 dilutions) before LC analysis and they injected cerebrospinal fluid directly into the system (that includes the protective column). A recent communication by Fry et al. [130] describes a procedure useful for obtaining samples containing MTX and polyglutamate derivatives from Ehrlich ascites cells grown in culture. This procedure involves the extraction of a centrifuged cell pellet with trichloroacetic acid (10%), neutralization of the acid extracts with a mixture of KOH and K_2HPO_4, and injection of the neutralized extract directly onto an ODS column. Buice and Sidhu [131] prepare serum samples for the analysis of MTX and 7-OH-MTX by dilution of the serum sample in pH 9.0 buffer, application of the diluted sample to a C_{18} SEP-PAK, and, after several washes with aqueous solvents, elution of analytes with methanol.

5.3 Pteridines

Because neopterin excretion in human urine has been suggested as a useful marker of neoplasia we will emphasize this compound in our discussion of sample preparation for pteridine analysis. It is instructive to mention that the isolation of neopterin from human urine reported by Sakurai and Goto in 1967 [132] involved processing several hundred liters of urine by column chromatography on Florisil, adsorption on and elution from charcoal, and ion-exchange chromatography. Rokos and his colleagues [102c] treated urine samples for the analysis of neopterin and other pteridines by successive IEC using Dowex 50 (H^+) and Dowex 1 (OH^-) columns. The eluate after IEC is concentrated in vacuo and subjected to SEC on a Sephadex G-10 column. Samples "enriched" in pteridine content after the IEC and SEC procedures were then applied to SEP-PAK C_{18} cartridges and the pteridine eluted with dilute aqueous methanol prior to LC analysis. Hausen and his colleagues report a much less time-consuming approach to sample preparation that involves simple adsorption of urinary samples onto a SEP-PAK cartridge and elution with water [133].

REFERENCES

1. K. G. Scrimgeour, *Methods Enzymol.*, *66*, 517 (1980).
2. S. F. Zakrzewski, E. A. Evans, and R. F. Philips, *Anal. Biochem.*, *36*, 197 (1970).

3. T. Tamura and E. L. R. Stokstad, *Br. J. Haematol.*, *25*, 513 (1973).
4. C. M. Baugh and C. L. Krumdieck, *Ann. N.Y. Acad. Sci.*, *186*, 7 (1971).
5. K. Fujii, T. Nagasaki, and F. M. Huennekens, *J. Biol. Chem.*, *256*, 10329 (1981).
6. NIH Report, *J. Amer. Med. Ass.*, *242*, 1961 (1979).
7. R. L. Blakley, *The Biochemistry of Folic Acid and Related Pteridines*, Elsevier, New York, 1969.
8. S. W. Bailey and J. E. Ayling, *J. Biol. Chem.*, *253*, 1598 (1978).
9. G. Kapatos and S. Kaufman, *Science*, *212*, 955 (1981).
10. T. Fukushima and J. C. Nixon, *Anal. Biochem.*, *102*, 76 (1980).
11. C. A. Lindquist, E. C. Cadman, and J. R. Bertino, *Anal. Biochem.*, *83*, 20 (1977).
12. R. T. Parker, R. S. Freelander, E. M. Schulman, and R. B. Dunlap, *Anal. Chem.*, *51*, 1921 (1979).
13. B. L. Hillcoat, P. F. Nixon, and R. L. Blakley, *Anal. Biochem.*, *21*, 178 (1967).
14. D. Chippel and K. G. Scrimgeour, *Canad. J. Biochem.*, *48*, 999 (1970).
15. J. M. Whiteley and R. Russell, *Biochem. Biophys. Res. Commun.*, *101*, 1259 (1981).
16. S. W. Bailey and J. E. Ayling, in *Chemistry and Biology of Pteridines* (W. Pfleiderer, ed.), Walter de Gruyter, New York, 1975, p. 633.
17. M. C. Archer and L. S. Reed, *Methods Enzymol.*, *66*, 452 (1980).
18. R. Baur, M. Kappel, R. Mengel, and W. Pfleiderer, in *Chemistry and Biology of Pteridines* (R. Kisliuk and G. Brown, eds.), Elsevier-North Holland, New York, 1979, p. 13.

19a. J. R. Bertino and G. A. Fischer, *Methods Med. Res.*, *10*, 197 (1962).

19b. E. Scheufler, *Clin. Chem. Acta, III*, 113 (1981).

20. J. J. Heusner and J. J. McCormack, *J. Pharm. Sci.*, *70*, 827 (1981).
21. Y.-M. Wang, E. Lantin, and W. W. Sutow, *Clin. Chem.*, *22*, 1053 (1976).

22a. C. E. Meyers, H. M. Eliot, and B. A. Chabner, *Cancer Treat. Rep.*, *60*, 615 (1976).

22b. S. P. Rothenberg, M. Da Costa, and M. P. Iqbal, *Cancer Treat. Rep.*, *61*, 575 (1977).

23. B. P. Mehta and D. J. Hutchinson, *Cancer Treat. Rep.*, *61*, 597 (1977).
24. K.-L. Fung, D. H. Ho, R. S. Benjamin, F. Yang, J. Sickler, N. S. Brown, and G. P. Bodey, *J. Pharmacol. Exp. Ther.*, *218*, 344 (1981).
25. E. R. Eichner and V. L. Dickson, *Am. J. Clin. Pathol.*, *62*, 840 (1974).

26. V. Herbert, *J. Clin. Invest.*, *40*, 81 (1961).
27. H. Baker, O. Frank, A. Shapiro, and S. H. Hutner, *Methods Enzymol.*, *66*, 490 (1980).
28. R. G. Leeming and J. A. Blair, *Clin. Chim. Acta*, *108*, 103 (1980).
29. S. Kaufman, S. Berlow, G. K. Summer, S. Milstein, J. D. Schulman, S. Orloff, S. Spielberg, and S. Pueschel, *N. Engl. J. Med.*, *299*, 673 (1978).
30. T. Nagatsu, T. Yamaguchi, T. Katu, T. Sugimoto, S. Matsuura, M. Akino, S. Tsushima, N. Nakazawa, and H. Ogawa, *Anal. Biochem.*, *110*, 182 (1981).
31. J. M. Scott, *Methods Enzymol.*, *66*, 437 (1980).
32. J. M. Scott, V. Ghanta, and V. Herbert, *Am. J. Med. Technol.*, *40*, 125 (1974).
33. V. D. Reif, J. T. Reamer, and L. T. Grady, *J. Pharm. Sci.*, *66*, 1112 (1977).
34. J. Kas and J. Cerna, *J. Chromatogr.*, *124*, 53 (1976).
35. J. Kas and J. Cerna, *Methods Enzymol.*, *66*, 443 (1980).
36. K. V. Buehring, T. Tamura, and E. L. R. Stokstad, *J. Biol. Chem.*, *249*, 1081 (1973).
37. K. K. Batra, J. R. Wagner, and E. L. R. Stokstad, *Fed. Proc.*, *32*, 928 (1972).
38. Y. S. Shin, K. V. Buehring, and E. L. R. Stokstad, *J. Biol. Chem.*, *247*, 7266 (19).
39. K. N. Rao and J. M. Noronha, *Anal. Biochem.*, *88*, 128 (1978).
40. M. Silverman, L. W. Law, and B. Kaufman, *J. Biol. Chem.*, *236*. 2530 (1961).
41. C. M. Houlihan and J. M. Scott, *Biochem. Biophys. Res. Commun.*, *48*, 1675 (1972).
42. C. M. Baugh, E. Braverman, and M. G. Nair. *Biochemistry*, *13*, 4952 (9174).
43. D. G. Johns, I. H. Plenderleith, and B. A. Cooper, *Biochem. Pharmacol.*, *12*388 (1963).
44. M. E. Schertel, J. W. Boehne, and D. A. Libby, *J. Biol. Chem.*, *240*, 3154 (1965).
45. R. Corrocher, B. K. Bhunjan, and A. V. Hoffbrand, *Clin. Sci.*, *43*, 799 (1972).
46. B. T. Kaufman, K. O. Donaldson, and J. C. Keresztesy, *J. Biol. Chem.*, *238*, 1498 (1963).
47. M. R. Heinrich, V. C. Dewey, and G. W. Kidder, *J. Chromatogr.*, *2*, 296 (1959).
48. B. Reed and J. M. Scott, *Methods Enzymol.*, *66E*, 501 (1980).
49. M. Freed, in *Methods of Vitamin Assay*, Association of Vitamin Chemists, Ind., 3rd ed., Interscience, New York, 1966, p. 236.
50. T. Maruyama, T. Shiota, and C. L. Krumdieck, *Anal. Biochem.*, *84*, 277 (1978).
51. G. P. Lewis and P. B. Rowe, *Anal. Biochem.*, *93*, 91 (1979).

52. S. K. Foo, D. J. Cichowicz, and B. Shane, *Anal. Biochem.*, *107*, 109 (1980).
53. I. Eto and C. L. Krumdieck, *Anal. Biochem.*, *109*, 167 (1980).
54. K. E. McMartin, V. Virayotha, and T. R. Tephly, *Arch. Biochem. Biophys.*, *209*, 127 (1981).
55a. A. J. Clifford, *Adv. Chromatogr.*, *14*, 1 (1976).
55b. R. W. Stout, A. R. Cashmore, J. K. Coward, C. G. Horvath, and J. R. Bertino, *Anal. Biochem.*, *71*, 119 (1976).
55c. C. K. Clifford and A. J. Clifford, *J. Ass. Offic. Anal. Chem.*, *60*, 1248 (1977).
56. S. K. Chapman, B. C. Greene, and R. R. Streiff, *J. Chromatogr.*, *145*, 302 (1978).
57. L. S. Reed and M. C. Archer, *J. Chromatogr.*, *121*, 100 (1976).
58. A. R. Branfman and M. McComish, *J. Chromatogr.*, *151*, 87 (1978).
59. R. N. Reingold, M. F. Picciano, and E. G. Perkins, *J. Chromatogr.*, *190*, 237 (1980).
60. B. A. Allen and R. A. Newman, *J. Chromatogr.*, *190*, 241 (1980).
61a. C. A. Caparelli, P. Domanico, and S. J. Benkovic, *J. Med. Chem.*, *24*, 1086 (1981).
61b. S. M. El Dareer, K. F. Tillery, and D. L. Hill, *Cancer Treat. Rep.*, *65*, 101 (1981).
61c. K. J. Scanlon, A. R. Cashmore, B. A. Moroson, R. N. Dreyer, and J. R. Bertino, *Mol. Pharmacol.*, *19*, 481 (1981).
62. D. W. Horne, W. T. Briggs, and C. Wagner, *Anal. Biochem.*,*116*, 393 (1981).
63. R. C. Williams, D. R. Baker, and J. A. Schmit, *J. Chromatogr. Sci.*, *11*, 618 (1973).
64. G. N. Wogan, S. Paglialunga, M. C. Archer, and R. C. Tannenbaum, *Cancer Res.*, *32*, 1981 (1975).
65. A. R. Cashmore, R. N. Dreyer, C. Horvath, J. O. Knipe, J. K. Coward, and J. R. Bertino, *Methods Enzymol.*, *66*, 459 (1980).
66. R. G. Smith, J. C. Pegues, D. Farquhar, T. L. Loo, and Y.-M. Wang, *Biomed. Mass Spectrom.*, *8*, 144 (1981).
67. B. P. Day and J. F. Gregory, *J. Agr. Food Chem.*, *29*, 374 (1981).
68. S. P. Sood, D. P. Wittmer, S. A. Ismaiel, and W. G. Haney, *J. Pharm. Sci.*, *66*, 40 (1977).
69a. F. L. Vandemark and G. J. Schmidt, *J. Liquid Chromatogr.*, *4*, 1157 (1981).
69b. W. H. Tafolla, A. C. Sarapu, and F. R. Dukes, *J. Pharm. Sci.*, *70*, 1273 (1981).
70. K. E. McMartin, G. Martin-Amat, A. B. Makar, and T. R. Tephly, *J. Pharmacol. Exp. Ther.*, *201*, 564 (1977).
71. J. Lankelma and E. Van der Kleijn, *J. Chromatogr.*, *182*, 35 (1980).
72. E. Hamel, G. Johnson, and D. Glaubiger, *Cancer Treat. Rep.*, *65*545 (1981).

73. D. M. Valerino, D. G. Johns, D. S. Zaharko, and V. T. Oliverio, *Biochem. Pharmacol.*, *21*, 821 (1972).
74. J. P. Brown, G. E. Davidson, and J. M. Scott, *J. Chromatogr.*, *79*, 195 (1973).
75a. D. A. Gewirtz, J. C. White, J. K. Randolph, and I. D. Goldman, *Cancer Res.*, *39*, 2914 (1979).
75b. C. M. Baugh, C. L. Krumdieck, and M. G. Nair, *Biochem. Biophys. Res. Commun.*, *52*, 27 (1973).
76. D. A. Gewirtz, J. K. Randolph, and I. D. Goldman, *Cancer Res.*, *40*, 1852 (1980).
77a. V. M. Whitehead, M. Perrault, and S. Stelcner, *Cancer Res.*, *35*, 2985 (1975).
77b. V. M. Whitehead, *Cancer Res.*, *37*, 408 (1977).
78. J. Galivan, *Mol. Pharmacol.*, *17*, 105 (1980).
79. B. A. Kamen, A. R. Cashmore, R. N. Dreyer, B. A. Moroson, P. Hsieh, and J. R. Bertino, *J. Biol. Chem.*, *255*, 3254 (1980).
80. J. Holm, S. I. Hansen, and J. Lyngbue, *Biochem. Pharmacol.*, *29*, 3109 (1980).
81. W. P. Tong, J. Rosenberg, and D. B. Ludlum, *Lancet*, *2*, 71 (1975).
82. D. M. Valerino, *Lancet*, *2*, , 1025 (1972).
83. J. A. Montgomery, T. P. Johnston, H. J. Thomas, J. R. Piper, and C. Temple, Jr., *Adv. Chromatogr.*, *15*, 169 (1977).
84. S. A. Nelson, B. A. Harris, W. J. Decker, and D. Farquhar, *Cancer Res.*, *37*, 3970 (1977).
85. R. G. Buice, W. E. Evans, J. Karas, C. A. Nicholas, P. Sidhu, A. B. Straughn, M. C. Meyer, and C. R. Crom, *Clin. Chem.*, *26*, 1902 (1980).
86. J. Lankelma, E. Van der Klein, and P. Ramaekers, *Cancer Lett.*, *9*, 133 (1980).
87. S. K. Howell, Y.-M. Wang, R. Hosoya, and W. W. Sutow, *Clin. Chem.*, *26*, 734 (1980).
88. R. B. Donehower, K. R. Hande, J. C. Drake, and B. A. Chabner, *Clin. Pharmacol. Ther.*, *26*, 63 (1979).
89. J. Lankelma and H. Poppe, *J. Chromatogr.*, *149*, 587 (1978).
90. E. Watson, J. L. Cohen, and K. K. Chan, *Cancer Treat. Rep.*, *62*, 381 (1978).
91. J. L. Wisnicki, W. P. Tong, and D. B. Ludlum, *Cancer Treat. Rep.*, *62*529 (1978).
92. W. P. Tong, J. L. Wisnicki, J. Horton, and D. B. Ludlum, *Clin. Chim. Acta*, *107*, 67 (1980).
93. J. L. Cohen, G. H. Hisayasu, A. R. Barrientos, M. S. B. Nayar, and K. K. Chan, *J. Chromatogr.*, *181*, 478 (1980).
94. O. Driessen, H. Buekers, L. Belfroid, and A. Emonds, *J. Chromatogr.*, *181*, 441 (1980).
95. K. K. Chan, M. S. B. Nayar, J. L. Cohen, R. T. Chlebowski, H. Leibman, D. Stolinski, M. S. Mitchell, and D. S. Farquhar,

Res. Commun. Chem. Pathol.. Pharmacol., *28*, 551 (1980).
96. C. Canfell and W. Sadee, *Cancer Treat. Rep.*, *64*, 165 (1980).
97a. J. L. Bock and R. Pierce, *Clin. Chem.*, *26*, 1510 (1980).
97b. K. Hande, J. Gober, and R. Fletcher, *Clin. Chem.*, *26*, 1617 (1980).
98. G. J. Lawson, P. F. Dixon, and G. W. Aherne, *J. Chromatogr.*, *223*, 225 (1981).
99. W. M. Deen, P. Levy, J. Wei, and R. D. Partridge, *Anal. Biochem.*, *114*, 355 (1981).
100. J. Jolivet and R. L. Schilsky, *Biochem. Pharmacol.*, *30*, 1387 (1981).
101. D. C. Chatterji and J. F. Gallelli, *J. Pharm. Sci.*, *66*, 1219 (1977).
102a. A. Niederwiesis, A. Matasovic, H.-C. Curtius, W. Endres, and J. Schaub, *FEBS Lett.*, *118*, 299 (1980).
102b. R. J. Leeming, J. A. Blair, V. Melikian, and D. J. O'Gorman, *J. Clin. Pathol.*, *29*, 445 (1976).
102c. H. Rokos, K. Rokos, H. Frisius, and H. J. Kirstaedter, *Clin. Chim. Acta*, *105*, 275 (1980).
103. J. L. Johnson, B. E. Hainline, and K. V. Rajagopalan, *J. Biol. Chem.*, *255*, 1983 (1980).
104. H. Descimon and M. Barial, *J. Chromatogr.*, *25*, 391 (1966).
105. T. G. Wilson and K. B. Jacobsen, *Biochem. Genet.*, *15*, 307 (1977).
106a. G. W. Kidder and V. C. Dewey, *J. Biol. Chem.*, *243*, 826 (1968).
106b. R. J. Jackson and T. Shiota, *J. Biol. Chem.*, *246*, 7454 (1971).
107a. C. N. Hodnett, J. J. McCormack, and J. A. Sabean, *J. Pharm. Sci.*, *65*, 1150 (1976).
107b. J. O. Knipe and J. J. McCormack, *Biochem. Pharmacol.*, *26*, 2217 (1977).
108. M. Kohashi, *J. Biochem.*, *87*, 1581 (1980).
109a. T. Fukushima and T. Shiota, *J. Biol. Chem.*, *247*, 4549 (1972).
109b. T. Fukushima, K. I. Kobayashi, I. Eto, and T. Shiota, *Anal. Biochem.*, *89*, 71 (1978).
110. T. Fukushima and J. C. Nixon, in *Chemistry and Biology of Pteridines* (R. L. Kisliuk and G. Brown, eds.), Elsevier-North Holland, New York, 1978, p. 31.
111. H. Descimon, in *Chemistry and Biology of Pteridines* (R. L. Kisliuk and G. Brown, eds.), Elsevier-North Holland, New York, 1978, p. 93.
112a. K. Tanaka, M. Akino, Y. Hagi, M. Doi, and T. Shiota, *J. Bol. Chem.*, *256*, 2963 (1981).
112b. C. L. Lee, T. Fukushima, and J. C. Nixon, in *Chemistry and Biology of Pteridines* (R. L. Kisliuk and G. Brown, eds.), Elsevier-North Holland, New York, 1978, p. 125.

113. B. M. Watson, W. L. F. Armarego, P. Schlesinger, R. G. H. Cotton, and D. M. Danks, in *Chemistry and Biology of Pteridines* (R. L. Kisliuk and G. Brown, eds.), Elsevier-North Holland, New York, 1978, p. 159.
114. A. Niederwieser, H.-C. Curtius, O. Bettoni, J. Bieri, B. Schicks, and M. Viscontini, *Lancet, 1*, 131 (1979).
115. V. C. Dewey and G. Kidder, *J. Chromatogr., 31*, 326 (1967).
116. K. Baumgarten and J. H. Bieri, *Helv. Chim. Acta, 63*, 1805 (1980).
117a. S. Kaufman, in *Aromatic Amino Acids in the Brain* (G. E. W. Wolstenholme and D. W. Fitzsimms, eds.), Elsevier, New York, 1974, p. 310.
117b. J. L. Dhondt and J. P. Farriaux, *Pathol. Biol., 28*, 397 (1980).
118. R. A. Levine, L. P. Miller, and W. Lovenberg, *Science, 214*, 919 (1981).
119. B. Stea, R. M. Halpern, and R. A. Smith, *J. Chromatogr., 168*, 385 (1979).
120. G. Kapatos, S. Kaufman, J. L. Weller, and D. C. Klein, *Science, 213*, 1129 (1981).
121. S. Milstein, S. Kaufman, and G. K. Summer, *Pediatrics, 65*, 806 (1980).
122. M. Hausermann, S. Ghisla, A. Niederwieser, and H.-C. Curtius, *FEBS Lett., 131*, 275 (1981).
123. T. Lloyd, S. Markey, and N. Weiner, *Anal. Biochem., 42*, 108 (1971).
124. F. Rothler and M. Karobath, *Clin. Chim. Acta, 69*, 457 (1976).
125. K. Iwai, M. Kobashi, and T. Suzuki, *Methods Enzymol., 66*, 457 (1976).
126. P. Giulidori, M. Galli-Kienle, and G. Stramentinoli, *Clin. Chem., 27*, 2041 (1981).
127. B. Shane, *Amer. J. Clin. Nutr., 35*, 599 (1982).
128. I. Eto and C. Krumdieck, *Anal. Biochem., 120*, 323 (1982).
129. H. Breithaupt, E. Kuenzten, and G. Goebel, *Anal. Biochem., 121*, 103 (1982).
130. D. W. Fry, J. C. Yalowich, and I. D. Goldman, *J. Biol. Chem., 257*, 1890 (1982).
131. R. G. Buice and P. Sidhu, *J. Pharm. Sci., 71*, 74 (1982).
132. A. Sakurai and M. Gotu, *J. Biochem. (Tokyo), 61*, 142 (1967).
133. A. Hausen, D. Fuchs, K. Grunewald, H. Huber, K. Konig, H. Wachter, *Clin. Chim. Acta, 117*, 297 (1981).

7

NICOTINIC ACID AND NICOTINAMIDE

NIKOLAUS HENGEN

Human Pharmacology Institute, Ciba-Geigy GmbH, Tübingen, Federal Republic of Germany

JAN X. de VRIES

Medical Clinic, University of Heidelberg, Federal Republic of Germany

1. INTRODUCTION

1.1 History

Nicotinic acid (structure 7.1) and nicotinamide (structure 7.2) (Fig. 7.1) were identified as dietary factors whose absence caused the symptoms of the disease known as pellagra in humans and black tongue in dogs [1-3]. In the southern United States, pellagra became a serious problem following the American Civil War, when the social and economic upheaval of the war led to a large population living on maize as the dietary staple. The main work on the etiology of pellagra was therefore carried out in the United States. In 1913, the Federal Government sent a team of investigators to the southern states to attempt to discover the causes and cure of pellagra. The story of their discoveries is unfolded in a series of U.S. Government reports [4]. In 1913 there was still a considerable school that believed pellagra to be an infectious disease. It was the work by Goldberger's team that showed the disease could not be transmitted to healthy people by any of the conventional means, but could be contracted by otherwise healthy people living away from the region in which it was endemic if they ate a maize diet. In 1938, Spies et al. [5] showed that administration of nicotinic acid would cure pellagra, and it was gradually accepted that the disease was caused by a nicotinic acid or nicotinamide deficiency. This was supported by the work of Elvehjem et al. [6], who showed that nicotinamide was identical with the so-called pellagra-preventing factor isolated from protein-free yeast extracts by Goldberger's team. The administration of either of these substances or, as shown in Sec. 1.3, tryptophan made these symptoms disappear. Today primary dietary deficiencies are rare, at least in the industrialized countries, but secondary deficiencies may be caused by an inadequate intake as a result of esophago-

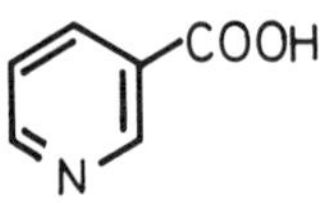

1

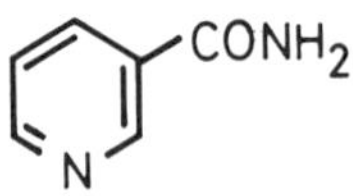

2

Figure 7.1 Structural formulas of nicotinic acid (structure 7.1) and nicotinamide (structure 7.2).

pharyngeal lesions, gastronintestinal disorders, alcoholism or carcinoid tumors [3,7].

Recently, Schoental [8] ascribed many of the reported outbreaks of pellagra not to the deficit of substances in the diet, but to the formation of toxic mycotoxins (trichothecenes), secondary metabolites produced by *Fusarium* species, which have good growing conditions in maize and in other cereals stored under high humidity and at low temperatures.

1.2 Chemical Properties

Nicotinic acid (3-pyridinecarboxylic acid, niacin, P.P. factor, pellagra-preventive factor, or antipellagra vitamin) and nicotinamide (nicotinic acid amide, niacinamide, 3-pyridinecarboxylic acid amide, or vitamin P.P.) are, as vitamins, equivalent [2,7,9]. They belong to the water-soluble vitamins. Nicotinic acid (molecular weight 123.11) forms colorless, nonhygroscopic needles and is stable in air. It sublimes without decomposition. It is amphoteric and forms salts with acids and bases; the pK_a is 4.9 and 2.07. The pH of the saturated aqueous solution is 2.7. A gram dissolves in 60 ml water. It is freely soluble in boiling water, boiling alcohol, alkali hydroxides, and carbonates. It is soluble in propylene glycol. It is insoluble in ether (nicotinamide is slightly soluble in ether and can be extracted from a water solution containing both the free acid and its amide).

Nicotinamide (molecular weight 122.12) has a melting point of 128-131°C. It can be recrystallized as needles from benzene. It distills at 150-160°C at 5×10^{-4} mm Hg. A gram dissolves in about 1 ml water, in about 1.5 ml alcohol, and in 10 ml glycerol. A 10% w/v solution in water is neutral to litmus. Nicotinamide is under normal conditions one of the most stable vitamins. A 10% solution in water can be autoclaved for 10 min at 120°C without decomposition. Nicotinamide possesses no acid properties and may be hydrolized with alkalis, acids, and enzymatically [10] to nicotinic acid. Both nicotinic acid and nicotinamide possess a strong ultraviolet (UV) absorption band at 261 nm whose intensity is pH dependent. They may be converted to methyl ester, ethyl ester, amides [11-13], and trimethylsilyl derivatives [14] useful for gas chromatographic separations.

A very often used reaction of nicotinic acid and nicotinamide is the formation of colored compounds with cyanogen bromide and amines, the so-called Koenig reaction [15], which has been extensively applied to the quantitative assay of both compounds and also to other pyridine derivatives [16]. The chemical reaction is as follows: an α-unsubstituted pyridine forms with cyanogen bromide a pyridinium salt, which reacts with an amine under ring opening to a derivative of glutaconaldehyde (see Scheme 7.1). With aromatic amines yellow polymethine dyes result, which can be measured spectrophotometrically. The

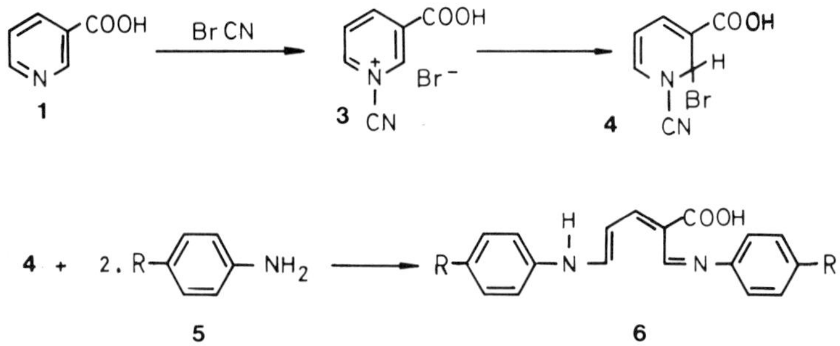

Scheme 7.1 Mechanism of the Koenig reaction. (1) nicotinic acid; (3) pyridinium salt; (4) postulated intermediate; (5) amine, e.g., tris-(hydroxymethyl)aminomethane [82]; and (6) glutacondialdehyde derivative. (From Ref. 2.)

colored products from nicotinamide are not as stable and intense as those from nicotinic acid. Therefore, prior to analysis, nicotinamide is frequently hydrolized to nicotinic acid [10,16].

1.3 Biochemistry

1.3.1 Metabolic Function of Nicotinamide Coenzymes

Nicotinic acid (Structure 7.1), nicotinamide (Structure 7.2), and tryptophan (Structure 7.14) are converted to the coenzymes nicotinamide adenine dinucleotide (NAD+; Structure 7.10) and nicotinamide adenine dinucleotide phosphate (NADP+; Structure 7.11). They combine with a wide variety of proteins and catalyze in that form a large number of oxidation-reduction reactions and substrates in living organisms (Scheme 7.2) [2,17]. The reduced pyridine nucleotides are dehydrogenated by flavine enzymes, which are oxidized by cytochromes, until at the end of this chain oxygen is attained. The oxidation of reduced nicotinamide adenine dinucleotide (NADH) may be coupled with phosphorylation [1,2,17]. The origin of the symptoms of the vitamin deficiency lies in the modification of the enzymatic systems coupled to NAD+ and NADP+. The normal daily human requirement of nicotinic acid or nicotinamide depends on the quantity of tryptophan intake in the form of proteins, with 60 mg tryptophan on average equivalent to 1 nicacin equivalent (1 mg nicotinic acid). The daily allowance recommended for adults is 6.6 niacin equivalents per 100 kcal and not less than 13 niacin equivalents at caloric intakes of less than 2000 kcal [18]. The recommended daily dietary allowance also varies with age and sex, from 6-19 niacin equivalents and with such circumstances as pregnancy and lactation, in which a supplement of 2 and 5 equivalents of niacin are indicated, respectively [2,3,7,18]. Other species have other

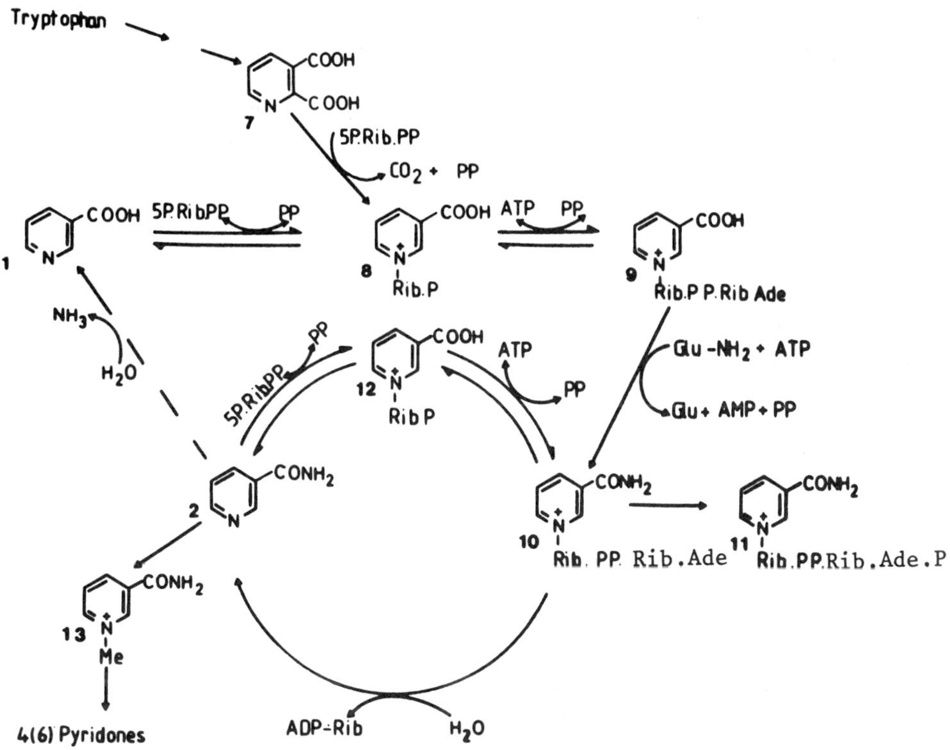

Scheme 7.2 Biosynthesis of nicotinamide adenine dinucleotide (NAD) in rat liver: (1) nicotinic acid; (7) quinolinic acid; (8) nicotinic acid mononucleotide (NaMN); (9) nicotinic acid adenine dinucleotide (NaAD); (10) nicotinamide adenine dinucleotide (NAD); (11) nicotinamide adenine dinucleotide phosphate (NADP); (12) nicotinamide mononucleotide (NMN); and (13) 1-methylnicotinamide (1-MN). Abbreviations: 5P Rib PP: 5-phosphoribosyl-1-pyrophosphate; ADP-Rib: adenosine diphosphate ribose. (From Refs. 23 and 2.)

requirements [2]. Nicotinamide is ordinarily the form of the vitamin in mammals [19] and is also found in human plasma (50 ng/ml) [20,21]. Nicotinic acid has not been found free in concentrations $\geq$100 ng/ml [22] in humans. Therefore, the reports on the occurrence of nicotinic acid in blood or tissues in higher concentrations may be due to derivatives that are converted to and expressed as total nicotinic acid during the analytical workup.

1.3.2 Biochemical Pathway from Tryptophan

The biosynthesis of nicotinic acid and derivatives in mammals [1,17] and other organisms takes place through the pathway (Scheme 7.3): tryptophan (structure 7.14), 3-hydroxyanthranilic acid, quinolic acid (Structure 7.7), nicotinic acid mononucleotide (NaMN+; Structure 7.8), nicotinic acid adenine dinucleotide (NaAD+; Structure 7.9), to nicotinamide adenine dinucleotide (NAD+; Structure 7.10) (see Scheme 7.2 and Refs. 1, 17, and 23). The pathway from tryptophan to NAD+ does not directly produce nicotinic acid or nicotinamide. NAD+ is transformed to nicotinamide, which is further N-methylated, reenters the pyridine nucleotide cycle, or may be deaminated to nicotinic acid, which is further converted to NaMN+ (Structure 7.8) (Scheme 7.2). In rat liver nicotinamide generated by NAD+-breakdown, is not deaminated but reutilized via nicotinamide mononucleotide. Nicotinic acid also is converted to nicotinamide for uptake into the NAD+ cycle. It may therefore be concluded that excretion of nicotinamide by the liver and subsequent deamidation in the intestinal tract is not important at normal nicotinamide concentrations [19,23]. Tryptophan may be metabolized by another route than the one represented, namely, through conversion to serotonin, indolacetic acid, indole alkaloids (in some plants), or further degraded to α-ketoadipic acid [17].

Nicotinic acid and nicotinamide are synthesized by most bacteria and higher plants by another pathway from aspartic acid and a three-carbon precursor [2,24]. The amount and pattern of the degradation products of nicotinic acid and nicotinamide are species and dose dependent. The basal excretion and the excretion after higher doses of nicotinic acid and nicotinamide in humans and other species is discussed in Sec. 5.

Although nicotinic acid and nicotinamide are equivalent as vitamins [1,7], they differ in their pharmacodynamic and pharmacokinetic behavior [2,25,26].

1.4 Pharmacological Properties

Nicotinic acid possesses vasodilating properties [27-29], in doses of 100-150 mg, probably due to the release of prostaglandin [29] and fibrinolytic properties when applied intravenously [28], and it lowers the concentration of plasma free fatty acids, cholesterol, triglycerides and the very low density (VLDL) and low-density (LDL) lipoproteins [26,30,31]. This latter property is the pharmacological basis for its therapeutic use in the management of the hyperlipoproteinemias, with doses of 2-9 g/day, divided into three to four portions [30]. Nico-

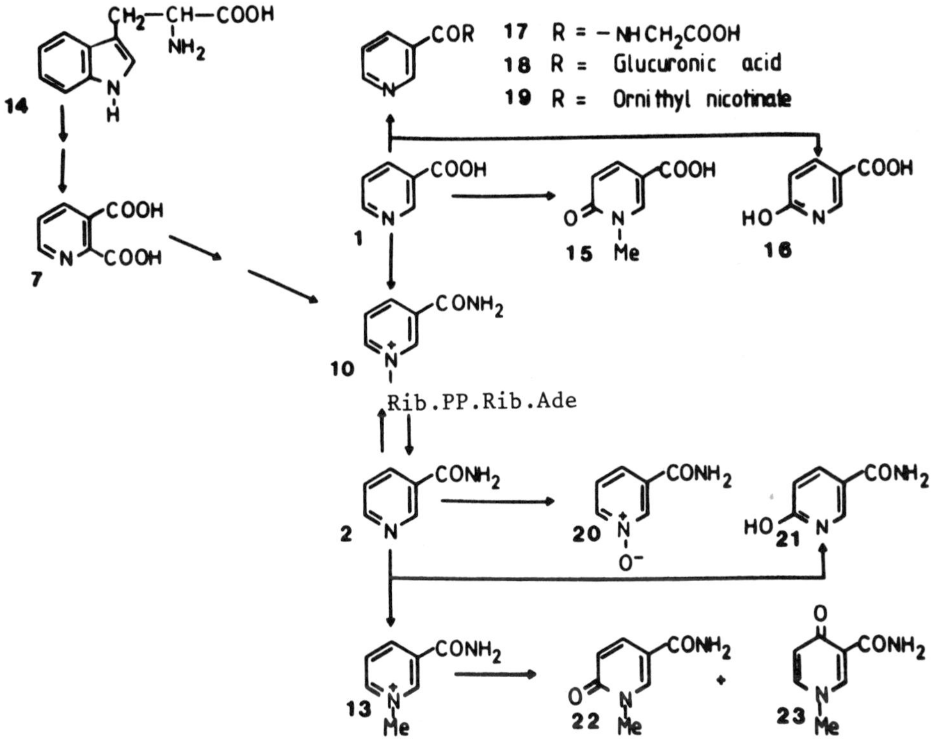

Scheme 7.3 Principal products from the metabolism of nicotinic acid and nicotinamide in humans and animals: (1) nicotinic acid; (2) nicotinamide; (7) quinolinic acid; (13) 1-methyl-nicotinamide; (14) tryptophan; (15) 1-methyl-6-pyridone-3-carboxylic acid; (16) 6-hydroxynicotinic acid; (17) nicotinuric acid; (18) β-nicotinoylglucuronide; (19) β-dinicotinoylornithine; (20) nicotinamide-1-oxide; (21) 6-hydroxynicotinamide; (22) 1-methyl-6-pyridone-3-carboxamide (6-pyridone); and (23) 1-methyl-4-pyridone-3-carboxamide (4-pyridone). (Adapted from Refs. 2, 25, 99, and 97.)

tinic acid has also been reported to be used in cases of peripheral vascular disorders, severe headache and migraine, leprosy, schizophrenia, senility, and vasculitis [7]. The pharmacokinetic behavior of nicotinic acid and nicotinamide are quite different and complex, depending on the dose and the species involved [2,25,26,32-35]. Both are readily absorbed.

Peak plasma levels of nicotinic acid were attained 30-60 min after an oral dose in humans [36-38] and in less than 30 min in rats [39-41]. In humans, approximately 88% of an oral dose was recovered in the urine as nicotinic acid and its metabolites, suggesting virtually complete adsorption. The fasting plasma levels of nicotinic acid are less than 100 ng/ml [22] in subjects who have not received nicotinic acid therapy; somewhat higher serum levels are found in lower animals. The flushing of the neck and face, a common side effect of nicotinic acid therapy, persisted as long as plasma nicotinic acid increased, but disappeared once a constant level had been attained [28].

The pharmacokinetics seem to be a dose dependent [36]. Nicotinic acid is rapidly eliminated from the plasma; in humans, the elimination half-life ($t_{1/2}$) ranged from 20 to 45 min [26,38]. In another report the shape of the elimination curve appeared to follow zero-order kinetics [41]. The $t_{1/2}$ of nicotinic acid was reported to be 1.1 hr in the rat and 1.4 hr in the rabbit [40].

Several nicotinic acid derivatives have been developed in an attempt to modify the bioavailability of nicotinic acid and to prevent the flushing associated with the rapid rise in circulating drug levels. The free acid is substituted by precursors [42] that may be esters of nicotinic acid, e.g., inositol nicotinate [43] or pentaerythritol tetranicotinate [44], or prodrugs, such as nicotinyl alcohol [45] or esters thereof (e.g., clofinol, the ester of *p*-chlorophenoxyisobutyric acid [46] which also acts lipolytic.

Nicotinamide has none of these properties [7] and does not possess the side effects of nicotinic acid. Apart from the prevention or treatment of dietary deficiencies it has been used occasionally in cases of migraine, vasomotor headache, and schizophrenia. For the treatment of pellagra a 50 mg dose given several times orally or 25 mg intravenously of nicotinic acid or nicotinamide per day is indicated [3].

The pharmaceutical formulations by which nicotinic acid is supplied are tablets, slow-release tablets, and solutions for injection. Nicotinamide is supplied as tablets and solutions for injection. If they are given as vitamin substitution, they may be associated with other vitamins, usually vitamin C or other vitamins from the B group, in pharmaceutical preparations or for the enrichment of food products.

Nicotinic acid and nicotinamide have a low toxicity. The lethal dose (LD_{50}) in rats is 5 g/kg for nicotinic acid and 1.6 g/kg for nictotinamide given subcutaneously.

2. PHARMACEUTICAL PREPARATIONS

2.1 Introduction

Previous methods for the analysis of nicotinic acid and nicotinamide may be classified as colorimetric, using the Koenig reaction or modifications [16,47] or biological and microbiological assays [47]. Previous to analysis, nicotinamide is often hydrolyzed to nicotinic acid to obtain a more stable reaction product with higher absorbance (see Sec. 1.2). These methods lack specificity since they are unable to distringuish between the acid and the amide. The procedures are complicated, and strict attention must be paid to the preparation of reagents (because of their instability or caustic properties) to obtain reproducible analytical data.

2.2 Paper and Thin-Layer Chromatography

The separation of nicotinic acid and nicotinamide in pharmaceutical preparations can be effected by paper chromatography and thin-layer chromatography (TLC) [47,50]. In Table 7.1 the separation of vitamin combinations as is customary in pharmaceutical preparations is given. The separation is effected on silica gel, aluminum oxide, or on ion-exchange resins. The detection is usually done by ultraviolet light or spray reagents.

A specific quantitative separation of nicotinic acid and nicotinamide was described by Frei et al. [51]. The substances were applied to cellulose thin-layer and chromatographed with *n*-butanol saturated with 3% ammonia as mobile phase. The spots were detected by ultraviolet reflectance and fluorescence and fluorescence quenching after spraying with Koenig reagent (cyanogen bromide) and *p*-aminobenzoic acid. The coefficient of variation is of the order of 10%, and the detection limit per spot (substance amount that can be qualitatively detected) was 0.001 μg (fluorescence) and 0.2 μg (fluorescence quenching) for nicotinic acid and 0.002 and 0.2 μg for nicotinamide, respectively. Ismaiel [52] separated nicotinamide from other vitamins of the B group by TLC on silica gel. The corresponding areas were scraped from the plate; the quantitation was done by measuring the ultraviolet absorption at 261 nm in 0.1 mol/liter hydrochloric acid after dissoltuion.

2.3 Gas-Liquid Chromatography

No reports are available that indicate any application of gas-liquid chromatography (GLC) to pharmaceutical formulations. Yet, some examples of GLC are given, in which mixtures of isomers are separated. Nicotinic acid cannot be analyzed by direct injection into the gas chromatograph. It is too polar and not volatile enough to achieve a separation in packed columns in microgram amounts. It has therefore been derivatized to the methyl ester, ethyl ester, nicotinamide, *N*-ethylnicotinamide [11,12], or trimethylsilyl derivative [14,53]. Nicotinamide may be chromatographed as such or converted to the ethyl ester [13] or trimethylsilyl derivative [14] or dehydrated with fluoroacetic anhydride to 3-cyanopyridine [54].

Liliedahl [11] described the separation of the esters with the liquid phases 0.75% Marlophen 87 isotherm at 150°C and with Carbowax 400 at

TABLE 7.1 R_f Values of Water-Soluble Vitamins

Thin-layer	Silica gel	Silica gel	Aluminum oxide	Ion-exchange resin Wofatit CP 300
Mobile phase	Water	Glacial acetic acid-acetone-methanol-benzene (5:5:20:70)	Glacial acetic acid-acetone-methanol-benzene (5:5:20:70)	Ethanol-water (10:90)
B_1 (thiamine)	0.5	0	0.54	0
B_2 (riboflavine)	0.4	0.3	0.24	0.42
Nicotinic acid	0.78	0.75	—	1
Nicotinamide	0.49	0.55	0.62	0.7
B_6 (pyridoxine)	0.52	0.13	0.26	0.35
B_{12} (cyanoco-balamin)	0.22	0	0.23	1
C (ascorbic acid)	0.96	0.27	0	—

Source: Modified from Ref. 50.

100°C. Sheppard and Prosser [13] used 2.5% neopentylglycol succinate-10% silicone SE 30 as stationary phase at 180°C for the analysis of the ethyl ester, the amide, and the *N*-ethylnicotinamide. Vessman and Stroemberg used 20% Carbowax 20M-terephthalic acid at 155°C for the separation of nicotinonitril [54].

2.4 Liquid Chromatography

The gas-liquid chromatographic procedures are not established, and therefore the problem of the simultaneous analysis of nicotinic acid and its derivatives in low concentrations must be solved in another way. Another approach is liquid chromatography. Nicotinic acid and nicotinamide appear in pharmaceutical preparations as monosubstances or mixed with several other vitamins and with minerals. Various solutions were used to separate and quantitate all the vitamins in a single chromatographic run. The first publications used ion-exchange materials, such as Zipax SCX columns (1 m × 2.1 mm inner diameter, ID) filled with glass beads (27-37 μm particles) coated with 1% sulfonated fluorocarbon) with 0.1 mol/liter sodium nitrate or 0.1 mol/liter phosphoric acid as mobile phase using ultraviolet absorption at 254 nm for detection. Several monosubstituted pyridine isomers were separated, such as nitriles, amides, and acids [55].

Water-soluble vitamins including nicotinic acid and nicotinamide were separated on Zipax SCX and SAX (strong cation and anion exchangers) by Williams et al. [56], with 0.01-0.1 mol/liter phosphate buffers (pH 4.4) as eluents. The ultraviolet absorbance was monitored at 254 nm. The separation and determination of the vitamins B_1, B_2, B_6, and nicotinamide in commercial vitamin preparations were effected by Callmer and Davies [57] (see Fig. 7.2). They used HS-Pellionex SCX (pellicular strong cation exchanger, 44-53 μm particles, 1 m × 2.1 mm ID column) using 0.025 mol/liter nitric acid as mobile phase with ultraviolet absorption detection at 254 and 280 nm. Recently, Nucleosil 10-SB (spherical microparticulate strong anion exchanger, 10 μm particles) was used for the separation of nicotinamide, L glutaminc, and panthothenate in pharmaceutical preparations by Nishimoto et al. [58]. The application of adsorption liquid chromatography by Caude and LeXuan [59] permitted in the same chromatographic run the separation of several drugs, namely, phenobarbitone, caffeine, and nicotinamide. The separation was effected on a Spherosil 5 μm column (microparticulate spherical silica) with isopropylether-methanol-50% ethylamine (93: 6.86:0.14, v/v) as mobile phase and detected by ultraviolet absorption at 254 nm. A chromatographic run lasted 10 min.

Recently, the analysis was done with microparticulate reversed-phase materials (for a review, see Horvath and Melander [60]). They are formed by reactions of organochloro- or alkoxysilane with the surface hydroxyl (silanol) groups of silica gel. The stationary phase

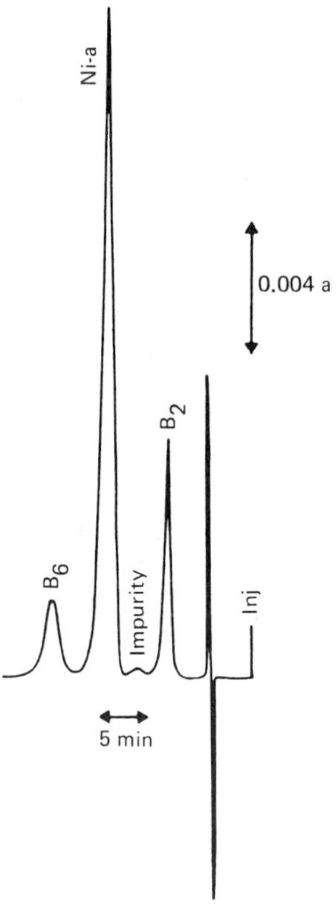

Figure 7.2 Chromatogram of a multivitamin tablet extract. Column: 1 m × 2.1 mm HS Pellionex SCX. Mobile phase: 0.025 mol/liter nitric acid. Flow rate: 20 ml/hr. (Ni-a) nicotinamide, (B2) vitamin B_2, (B6) vitamin B_6. (From Ref. 57.)

becomes hydrophobic, usually possessing an octadecyl or octyl functional group. The mobile phase is polar: water mixed with water-miscible organic solvents, such as methanol or acetonitrile, can handle a great many of the applications.

Very polar compounds and especially ionic or ionizable compounds often elute very quickly or give poorly defined peaks, usually with severe tailing. Only the un-ionized form of acids and bases give sharp peaks. For weak acids and bases this can be achieved by the suppression of ionization by adjusting the pH value. Strong acids and bases cannot be handled by this technique since many are completely ionized in the pH range 2-8. By forming an ion pair with a suitable counterion, however, these strong acids and bases can be made into electrically neutral compounds and therefore retained on a reversed-phase column. The reversed-phase ion-pair technique is more easily mastered than is ion exchange. Unlike conventional ion exchange, the technique can separate nonionic and ionic compounds in the same sample. For extensive reviews, see Gloor and Johnson [61] and Bidlingmeier [62]. The ion-pair reagents are added to the mobile phase regularly to give final concentrations of 0.005-0.02 mol/liter. The technique of gradient elution and the programmed change of wavelength during a run results in an improvement of selectivity, sensitivity, and efficiency of the separation and quantitation of complex mixtures.

By varying the amount of the modifiers, such as sodium heptane sulfonate, sodium camphor sulfonate, and dioctyl sodium sulfosuccinate (each 2×10^{-3} mol/liter) in aqueous methanol the separation of nicotinic acid and its amide was achieved that was previously not well accomplished otherwise. The stationary phase of Sood et al. [63] was μBondapak-C18 (a microparticulate octadecylsilane column, 30×0.39 cm ID, particle size 10 μm). Nicotinic acid and nicotinamide were quantified by monitoring, at 254 nm, the wavelength of the widespread fixed wavelength detectors equipped with mercury lamps; they are well separated from other constituents of the formulation as vitamins, minerals, coloring materials, artificial flavors, preservatives, and excipients (see Fig. 7.3). All nicotinamide preparations showed the presence of nicotinic acid in amounts of 2-15%, which could not be detected by the cyanogen bromide method.

Kirchmeier and Upton [64] used a C18 reversed-phase column (μBondapak-C18 or equivalent) and 5×10^{-3} mol/liter sodium hexane sulfonate as modifier in water-methanol-acetic acid (75:25:1, v/v) for the separation and quantitation of multivitamin blends (nicotinic acid, nicotinamide, and vitamins B1, B_2, and B_6). The flow rate was 0.5 ml/min. The sample pretreatment consisted of dissolution of the multivitamin blends in the aqueous portion of the mobile phase. The agreement with the official method was good, with the advantage of considerably shortening the analysis time (see Fig. 7.4). Walker et al. [65] applied the same mobile phase and column to multivitamin formulations (tablets, fluids, and injectable products). The hexane sulfonic acid concentration was adjusted to $3\text{-}8 \times 10^{-3}$ mol/liter to optimize chromatography. The quantitation was improved by the addition of an internal standard (*p*-hydroxybenzoic acid).

To quantitate the four vitamins nicotinamide and vitamins B_1, B_2, and B_6 in one chromatographic run, a dual-wavelength system was adopted (254 and 280 nm) (see Fig. 7.5). The optimization of the resolution of several compounds may also be accomplished by mixing two modifiers in different proportions (e.g., pentane sulfonate and heptane sulfonate), as different substances modify their capacity factors in different ways according to the nature of the modifier [66] (see Fig. 7.6). Sodium dodecylsulfate has been used as ion-pair reagent in

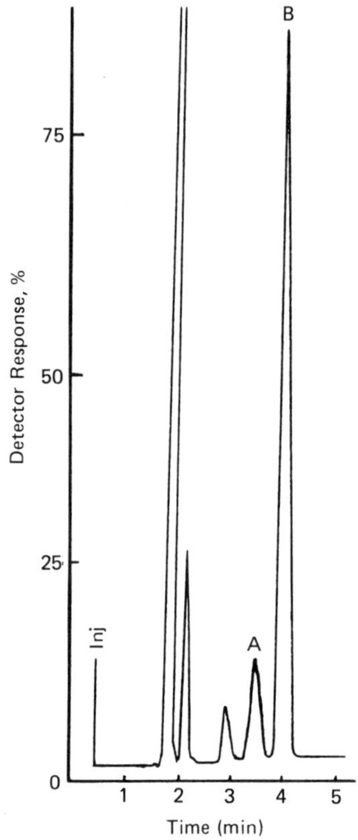

Figure 7.3 Chromatogram of a multivitamin product. (A) nicotinic acid, 0.5 mg per dosage unit, and (B) nicotinamide, 9.3 mg per dosage unit. Column: μBondapak C_{18}, 30 cm × 4 mm; UV detector at 254 nm; mobile phase: 2×10^{-3} mol/liter dioctyl sodium sulfosuccinate in water (300 ml), pH adjusted to 2.5 with formic acid, and methanol (300 ml). Flow rate: 2.5 ml/min. (From Ref. 63; reproduced with permission of the copyright owner.)

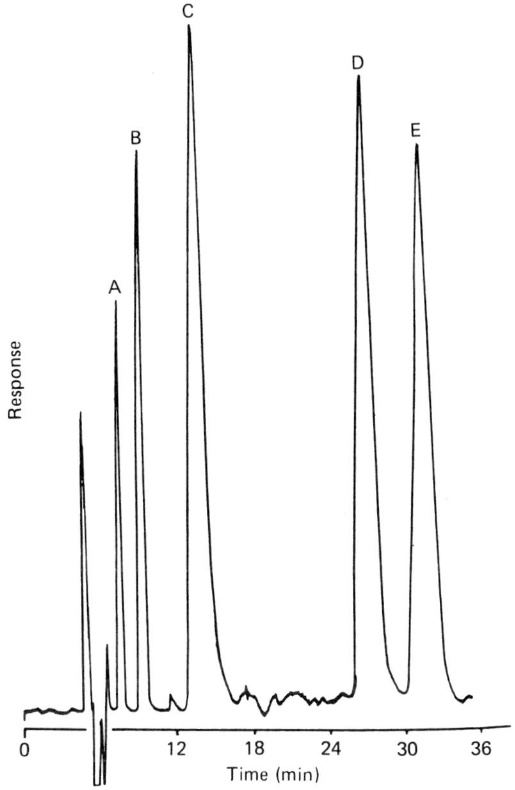

Figure 7.4 Chromatogram of a standard mixture of (A) nicotinic acid, (B) nicotinamide, (C) vitamin B_6, (D) vitamin B_1, and (E) vitamin B_2. Column: μBondapak C17, 10 μm, 30 cm × 4 mm ID. Mobile phase: 25% methanol-75% 0.005 mol/liter sodium hexane sulfonate in 1% acetic acid. Flow rate: 0.5 ml/min. UV detection at 270 nm, and chart speed, 25.4 cm/hr. (From Ref. 64; reproduced with permission of the copyright owner.)

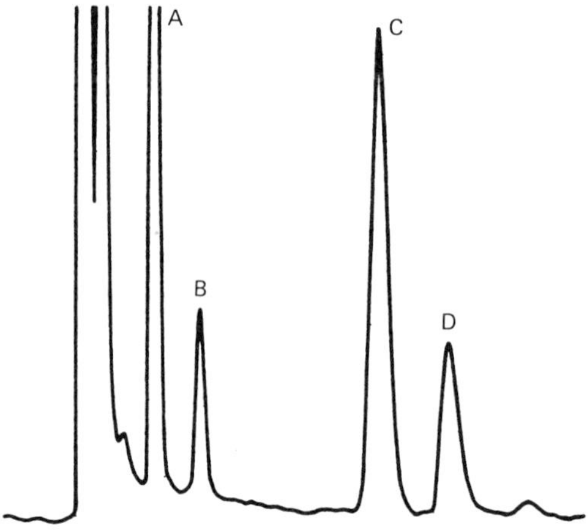

Figure 7.5 Chromatogram of a multivitamin-mineral sample immediately after extraction with the mobile phase containing 4 mg diethylenetriamine penta-acetic acid sodium salt per milliliter: (A) nicotinamide, (B) vitamin B_6, (C) vitamin B_2 and (D) vitamin B_1 were monitored at 280 nm. Mobile phase: 0.14% sodium hexane sulfonate in water-methanol-acetic acid (73:27:1). Flow rate: 1 ml/min. (From Ref. 65; reproduced with permission of the copyright owner.)

the separation of Terweij-Groen et al. [67]. The column contained reversed-phase octylsilane-bonded silica (LiChrosorb RP-8, 10 μm, 15 × 0.3 cm ID). The mobile phase consisted of 6% (w/v) *n*-propanol, 0.3% (w/v) sodium dodecylsulfate, 0.02 mol/liter malic acid, and 0.08 mol/liter total sodium ions, pH 2.5. Wills et al. [68] employed tetrabutylammonium phosphate, pH 7.5, in water-methanol (70:30, v/v) to separate seven water-soluble vitamins (nicotinic acid, nicotinamide, vitamins B_1, B_2, B_{12}, vitamin C, and folic acid) on a C18 reversed-phase system (see Fig. 7.7). The ion-pair modifier was much more effective than various buffers at different pH values. The order of elution from the C18 reversed phase was in the reversed order in comparison with a microparticulate amino phase with 0.125% citrate-methanol (20:80, v/v). The analysis time is halved to 12 min on the amino phase. The vitamins, which were not well separated on the one column, were better separated on the other (Fig. 7.8). As shown by Melander et al. [69], the selectivity can be varied with several phosphate buffers mixed with various amines in changing concentrations for the separation of nicotinamine and vitamins B_1 and B_6.

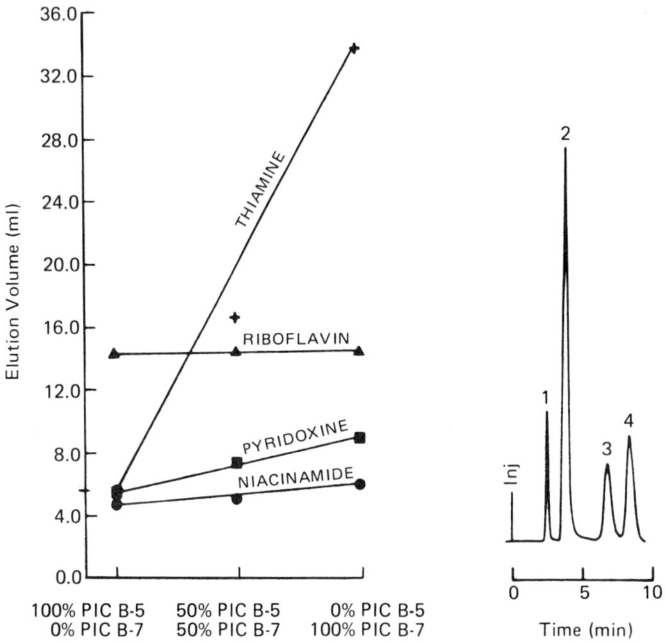

Figure 7.6 Separation of the water-soluble vitamins: (1) nicotinamide, (2) vitamin B_6, (3) vitamin B_2, and (4) vitamin B_1. The diagram on the left shows the retention volume versus counterion composition. PIC B 5 is pentane sulfonate and PIC B 7 is heptane sulfonate. The figure on the right shows a chromatogram of vitamins on μBondapak C18 (30 cm × 4 mm) using 0.0025 mol/liter pentane sulfonate and 0.0025 mol/liter heptane sulfonate with 1% acetic acid in methanol-water (50:50, v/v) as mobile phase. (From Ref. 66.)

For the analysis of a tablet containing eight water-soluble vitamins (vitamins B_1 and B_2, nicotinic acid, nicotinamide, vitamins B_6, B_{12} and C, and pantothenic acid), a two-step isocratic technique was employed by Schuster [70]. In the first step six vitamins were separated using a mixture of acetonitrile-phosphate buffer (80:20, v/v). In the second step the mobile phase ratio was changed to 73:27, which enabled vitamins B_{12} and C to be analyzed. As an alternative, a continuous linear gradient can be used. Figure 7.9 shows the analysis of eight water-soluble vitamins, changing the mobile phase composition from 86.5 to 63% acetonitrile during 8.7 min. Each vitamin was detected at its optimal wavelength, with a programmed variable wavelength ultraviolet detector. The applied amino phase has the character of a weak anion

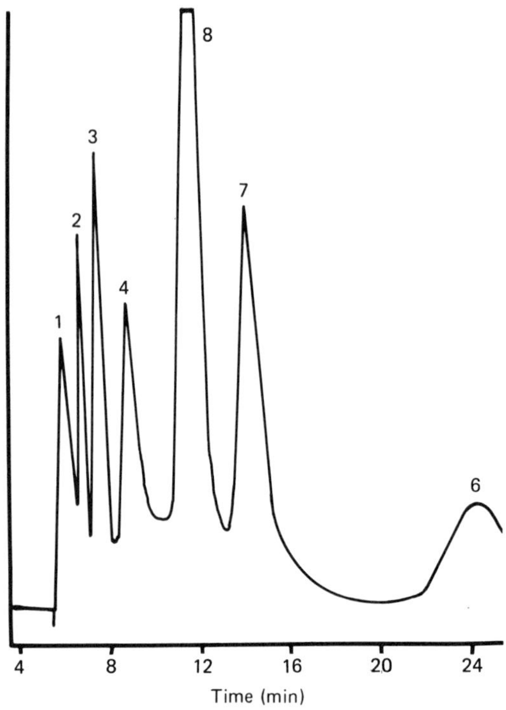

Figure 7.7 Chromatogram of water-soluble vitamins: (1) vitamin C, (2) nicotinic acid, (3) nicotinamide, (4) folic acid, (6) vitamin B_1, (7) vitamin B_2, and (8) vitamin B_{12}. Mobile phase: 0.005 mol/liter tetrabutylammonium phosphate in 70:30 water-methanol on μBondapak C18. Detection was at 254 nm; flow rate 1 ml/min. (Reproduced from Ref 68 by permission of Preston Publications, Inc.)

exchanger very sensitive of pH changes. A careful column conditioning is therefore mandatory to achieve reproducible results. Gradient elution on a C18 reversed-phase column with methanol-0.25 mol/liter sodium dihydrogen phosphate, pH 3.5 (phase A) and methanol (phase B) with a flow rate of 1.6 ml/min (Fig. 7.10) permits the separation of six water-soluble vitamins (nicotinamide and vitamins B_1, B_2, B_6, B_{12}, and C) in one run [71].

Kwok et al. [72] determined nicotinamide in the presence of vitamins B_1, B_2, and B_6 in multivitamin and multivitamin-multimineral tablets. They applied reversed-phase chromatography using a ion-pair reagent. The column contained phenyl-bonded silica (μBondapak phenyl, 10 μm, 30 × 0.39 cm ID). The mobile phases were 0.0025 mol/liter 1-hexane sulfonic acid in water (phase A) and in methanol (phase B). A linear

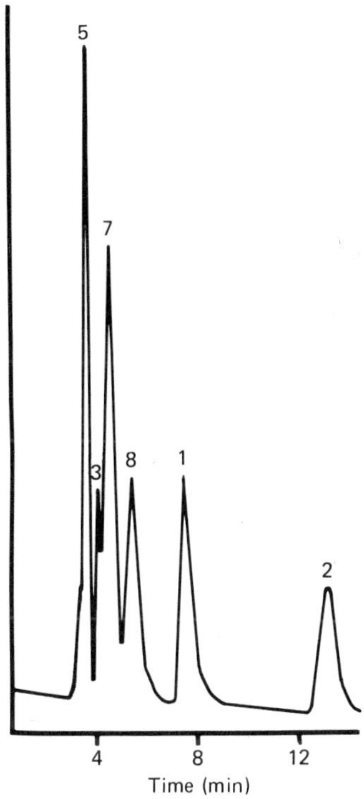

Figure 7.8 Chromatogram of water-soluble vitamins: (1) vitamin C, (2) nicotinic acid, (3) nicotinamide, (5) vitamin B_6, (7) vitamin B_2, and (8) vitamin B_{12}. Mobile phase: methanol-0.125% citrate. Column: μBondapak NH2. Detection and flow rate as in Fig. 7.7. (From Ref. 68; reproduced with permission of the copyright owner.)

gradient was run at a flow rate of 2 ml/min for 18 min. Solvent conditions varied from 0 to 80% methanol. The ultraviolet absorbance detector was set at 280 nm. Hydrochloric acid, 0.1 mol/liter, was used to extract the vitamins from the multivitamin preparations and dimethylsulfoxide containing anhydrous citric acid to disperse the multivitamin-multimineral preparations, since vitamin B_6 is not completely extracted either by 0.1 mol/liter HCl or dimethylsulfoxide due to absorption of the vitamin to the minerals. The extraction of nicotinamide was not impaired by the addition of citric acid to dimethylsulfoxide.

The chromatographic behavior of vitamins in pharmaceutical preparations is varied. An optimal separation of a set of vitamins can be achieved by selecting:

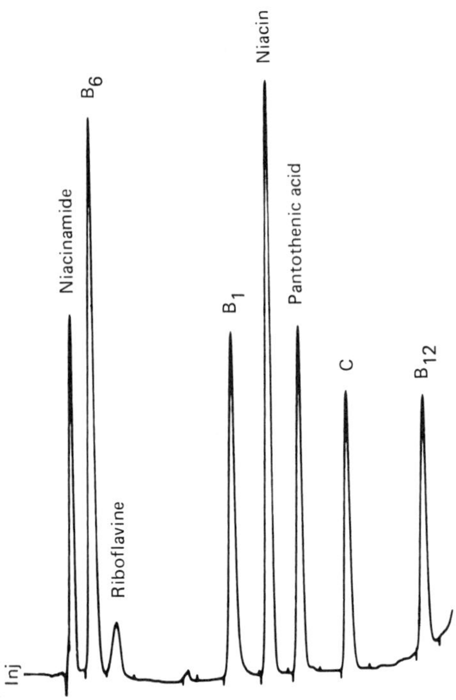

Figure 7.9 Gradient analysis of water-soluble vitamins (standard solution): (Niacinamide) nicotinamide, (B6) vitamin B_6, (Riboflavine) vitamin B_2, (B1) vitamin B_1, (Niacin) nicotinic acid, (C) vitamin C, and (B12) vitamine B_{12}. Column: 25 cm × 4.6 mm ID Lichrosorb NH2, 10 μm. Mobile phase: A, 0.005 mol/liter potassium dihydrogen phosphate; B, acetonitrile. Gradient elution from 86.5% acetonitrile to 63% acetonitrile during 8.7 min. Temperature: 65°C. Each vitamin was detected at its optimal wavelength. Flow rate: 3 ml/min. (From Ref. 70.)

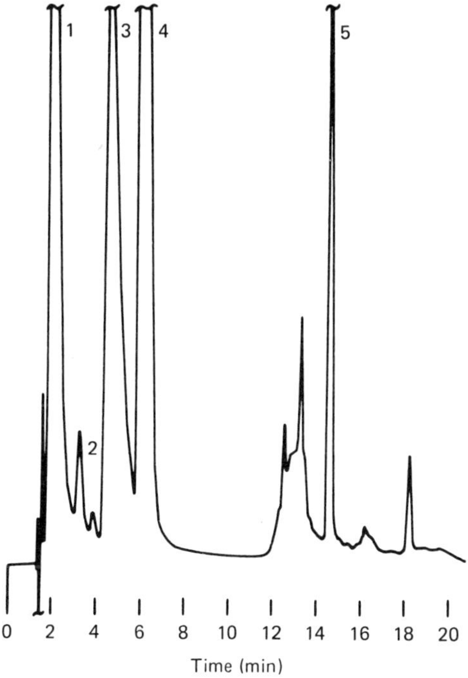

Figure 7.10 Tablet extract of water-soluble vitamins: (1) vitamin C, (2) vitamin B_6, (3) vitamin B_1, (4) nicotinamide, and (5) vitamin B_2. Column: Zorbax ODS, 25 cm × 4.6 mm ID. Mobile phase: A, 0.25 mol/liter sodium dihydrogen phosphate in water, pH 3.5; B, methanol. Gradient elution: 4-8% B (8 min), exponential-4, 8-44% B (3 min), 44% B for 9 min. Flow rate: 1.6 ml/min; temperature 40°C; UV detection at 254 nm. (From Ref. 71.)

1. The column material
2. The composition of the mobile phase
3. Ion-pair reagents
4. Gradient systems
5. Wavelength changing during a run

The best choice depends on the individual problem. A general best column or mobile phase cannot be given. Depending on the local conditions and the formulation to be analyzed, the solution can be different.

3. FOODS

3.1 Introduction

Chemical and microbiological methods are still used for the analysis of nicotinic acid and nicotinamide in food [47,73]. Nicotinic acid and derivatives are found in almost all food products, but in variable concentrations, e.g., yeast (20-60 mg per 100 g), liver (17-21 mg per 100 g), heart (6 mg per 100 g), and muscle meat contain appreciable amounts in general; plant material, with the exception of unpealed rice (6 mg per 100 g), and cereals contain less [2,7]. Kodicek [74] showed that most of the nicotinic acid present in maize is in the form of nicotinyl esters, which are not hydrolyzed on digestion, so that although nicotinic acid is present, it is not available. Chemical hydrolysis prior to assay for nicotinic acid and nicotinamide breaks the ester bonds, releasing the total nicotinic acid content of the food for assay. The same is true for other cereals: in general there is little free nicotinic acid in cereals; most is present as various esters. In wheat bran, much of the nicotinic acid is bound to macromolecules (molecular weight between 1500 and 17,000)—about 60% is bound to polysaccharides and the remaining 40% to peptides and glycopeptides [75]. Roasted coffee contains high amounts (up to 40 mg per 100 g), which originates from the demethylation of trigonelline during the heating process [2]. Nicotinic acid and its amide are heat resistant, but may be appreciably dissolved from food in the water used for cooking.

3.2 Paper and Thin-Layer Chromatography

For the analysis of foods, acid or alkaline pretreatment is generally necessary to liberate the nicotinic acid, which is then analyzed [47]. A method for the determination of nicotinic acid in coffee has been described by Katsui [48]. After extraction with hot 1.1 mol/liter hydrochloric acid and decoloration with basic lead carbonate and filtration, the solution was applied to chromatographic paper and chromatographed with *n*-butanol-ammonia-water (100:2:16, v/v). After drying, spots were visualized using cyanogen bromide and benzidine. The intensity of the spots was compared with standards.

3.3 Gas-Liquid Chromatography

Determinations of nicotinic acid and nicotinamide by gas-liquid chromatography applied to foods are described by Bertling and Tietz [76].

3.4 Liquid Chromatography

Wills et al. [68] recommended application of their method for pharmaceuticals to foods, although the presence of interfering materials should be checked. Simultaneous quantitative analysis of nicotinic acid and vitamins B_1 and B_2 in rice and rice products was described recently by Toma and Tabekhia [77] using a reversed-phase C18 column (μBondapak C18, 30 × 0.39 cm ID). The mobile phase consisted of water-methanol-acetic acid (60:39:1, v/v) with the ion-pair modifiers pentane sulfonate and heptane sulfonate in 0.005 mol/liter concentration. The ultraviolet detector was set at 254 nm. The method was compared with wet chemistry procedures of the AOAC [73]. The liquid chromatographic method proved to be rapid, accurate, and in good agreement with the official method. Cereal extracts were also analyzed by Shuster and Hayward [78] with an octadecylsilane reversed-phase, but with tetrabutylammonium phosphate as modifier in water-methanol (95:5, v/v) with ultraviolet detection at 254 nm.

Hoffmann [79] determined nicotinic acid, L-ascorbic acid, and D-isoascorbic acid in curing adjuncts. The separation was effected on an amino-modified silica gel column (LiChrosorb NH2, 10 μm, 25 × 0.4 cm ID). The mobile phase was 0.01 mol/liter potassium phosphate, pH 3.8, and acetonitrile (15:85, v/v). The flow rate was 1 ml/min; the ultraviolet absorption was monitored at 254 nm. The temperature of the column was kept at 35°C. Skurray [80] determined small amounts of nicotinic acid and vitamins B_1 and B_2 in foods with a C18 reversed-phase column, using heptane sulfonate as ion-pair modifier in acetate buffer. Detection was reached by ultraviolet absorption at 254 nm. Kral [81] determined acid hydrolyzable nicotinic acid derivatives in fruit juices. After acid hydrolysis of nicotinic acid derivatives with sulfuric acid in autoclaves, nicotinic acid was enriched on cation-exchange columns (AG 50W-X8, Bio-Rad Lab.) and separated on an amino-modified silica gel column (Nucleosil 5-NH2, 5 μm, 30 × 0.4 cm ID), preceded by a precolumn of Nucleosil NH2, 10 μm, 2.5 cm. The mobile phase was methanol-water-Na-acetate buffer, 1M, pH 4.6 (300:570:100, v/v); the flow rate 1.2 ml/min. The detection was electrochemical (amperometric) at a stationary mercury dropping electrode at -1.45 V, pulse mode. The method was compared with UV detection. In pure standard solutions the limit of detection is better for UV monitoring; in complex matrices the amperometric detection is better. The linear range was 0.3-123 μm/ml; the limit of detection 40 ng.

4. BIOLOGICAL MATERIALS (BLOOD, PLASMA, URINE, AND TISSUES)

4.1 Introduction

The determination of the concentration of nicotinic acid, nicotinamide, and its metabolites in biological fluids is important to elucidate biochemical pathways (metabolism of tryptophan and biogenesis of nicotinamide adenine dinucleotide and related nucleotides) and the pharmacokinetic behavior (liberation, absorption, distribution, metabolism, and excretion). The determination of the concentration in blood, urine, and tissues was for a long time done only with conventional methods. Microbiological methods, e.g., Goldsmith and Miller [47], colorimetric methods by Carlson [82], Schraepler et al. [83], and Pinder et al. [49], and fluorometric methods for nicotinamide by Clark and coworkers [20,21], have been used. All these methods are often unspecific, not sensitive enough, or subject to interference by related compounds. The isolation of unchanged nicotinic acid and nicotinamide from a bulk material is a major problem. Due to the polarity of those substances, polar solvents must be used for extraction. These solvents are not selective, and interference from metabolites or endogenous compounds is likely, making chromatography necessary.

4.2 Paper and Thin-Layer Chromatography

The results obtained in the identification and quantitation of the metabolic compounds after application of labeled nicotinic acid or nicotinamide to humans, animals, and plants using mainly paper chromatography were reviewed by Fumagalli [25], Weiner [32], and Dawson et al. [24]. The most common mobile phase for the analysis of urine excretion products after application of nicotinic acid and nicotinamide was *n*-butanol saturated with water, ammonia, or acid [84-87]. Several chromatographic systems for the differentiation of nucleotides in tissues derived from nicotinamide or nicotinic acid were devised [49]. Table 7.2 summarizes the R_f values and the solvent systems for intermediates of NAD+ and NADP+ synthesis.

Pinder et al. [49] used both paper chromatography and thin-layer chromatography for tissue extracts (Table 7.2); Frei et al. [51] applied only thin-layer chromatography. The plates were coated with cellulose. Silica gel was used for the identification of the tryptophan metabolites in urine with a bidimensional technique for the resolution of 32 compounds in which nicotinic acid and nicotinamide were also represented [88]. Kala et al. [89] investigated urine for nicotinic acid and its metabolites after application of nicotinyl alcohol. They used TLC with silica gel; the mobile phase consisted of chloroform-ethanol-25% ammonia in various proportions. For the quantitation of nicotinic acid in human plasma Carlson [82] used a modified Koenig reaction (chloramine T and

TABLE 7.2 R_f Values of Nicotinic Acid, Nicotinamide and Other Intermediates of NAD^+ and $NADP^+$ Synthesis

	Thin-layer chromatography[a,b] (R_f)			Paper chromatography[a,c] (R_f)			
Compound	1	2	3	1	2	3	4
Nicotinic acid	0.82	0.77	0.55	0.73	0.72	0.56	0.34
Nicotinamide	0.88	0.87	0.45	0.85	0.80	0.53	0.85
$NaMN^+$	0.47	0.13	0.79	0.31	0.17	0.83	—
$NaAD^+$	0.52	0.15	0.57	0.35	0.10	0.51	—
NAD^+	0.61	0.13	0.58	0.48	0.15	0.53	—
$NADP^+$	0.50	0.03	0.70	0.34	0.02	0.64	—
NMN^+	0.63	0.11	0.73	0.48	0.18	0.84	—

[a]Mobile phases for PC and TLC: (1) isobutyric acid-concentrated ammonia-water (66:1.7:33, v/v), pH 7.4; (2) 95% ethanol-1 mol/liter ammonium acetate (70:30, v/v), pH 5.0; (3) 600 g ammonium sulfate dissolved in 1 liter 0.1 mol/liter sodium phosphate, pH 6.8; (4) *n*-butanol-water-concentrated ammonia (660:114:6, v/v).
[b]Ascending technique on plates coated with MN 300 G cellulose (0.25 mm thick) for 2-3 hr.
[c]Descending technique on Whatman #3 MM paper for ∿ 16 hr.

potassium cyanide) after protein precipitation with acetone, treatment with chloroform, and separation of the aqueous phase. For lower concentrations of nicotinic acid the aqueous extract was submitted to thin-layer chromatography on silica gel, the corresponding spot of nicotinic acid scraped off and eluted, and the eluate determined colorimetrically. This method is still used by several authors for pharmacokinetic studies [26-28,90].

4.3 Gas-Liquid Chromatography

Takikawa et al. [53] investigated the behavior of pentaerythritol tetranicotinate, a nicotinic acid prodrug, in the gastrointestinal tract. They developed a gas chromatographic method for determining pentaerythritol tetranicotinate and its hydrolysates (nicotinic acid, pentaerythritol trinicotinate, dinicotinate, and mononicotinate) in the gastric loop of rats. The gastric content was extracted with acetone-chloroform (similar to Carlson [82]), the organic layer taken to dryness, and the aqueous layer freeze dried. To the organic residue 20 μl of the internal standard (dioctylphthalate) in pyridine was added. To 10 μl of that solution 5 μl trimethylsilyl (TMS) agent (hexamethyldisilazane-trimethylchlorosilane, 5:2) was added and heated to 60°C for 30 min. The aqueous residue was dissolved in 10 μl of the internal standard (diphenylamine) in pyridine and 70 μl of the TMS agent added and heated to 60°C for 30 min. Both reaction mixtures, 2 μl each, were injected into the gas chromatograph. The glass column was packed with 1.5% silicone XE 60 on Chromosorb G HP. The oven, injection port, and detector temperatures were 50-190°C (20°C/min), 230°C, and 230°C for pentaerythritol, nicotinic acid, and pentaerythritol mononicotinate, respectively. For the other substances the gas chromatographic conditions were different.

A gas chromatographic method with a mass fragmentographic detection was reported by Sautebin et al. [91] to estimate the bioavailability of a polymeric bound nicotinic acid preparation in humans. To 0.5 ml plasma 2-methylnicotinic acid was added as internal standard. After acidification to pH 4-5 with 0.1 mol/liter hydrochloric acid followed by saturation with sodium chloride the mixture was extracted with 1.5 ml portions of *n*-butanol. The organic phases were evaporated to dryness and trimethylsilylated with BSTFA (N,O-bis(trimethylsilyl)trifluoroacetamide). The detection was done by multiple selected ion monitoring. The blank value was found to be 30 ng/ml in rats. The method was linear, from 0.06 to 2 μg/ml.

4.4 Liquid Chromatography

Pagella and Spina [92] determined free nicotinic acid in rat plasma. After extraction of nicotinic acid according to Carlson [82], the samples were separated on a strong anion-exchange column (microparticulate packing of Partisil SAX, 25 × 0.46 cm ID). The mobile phase consisted of 0.1% sodium dihydrogen phosphate, pH 4.9; the flow rate was 2 ml/min. The temperature of the column was kept at 40°C; the ultraviolet absorbance was measured at 261 nm. Nicotinamide, nicotinuric acid, and 1-methylnicotinamide did not interfere. The linear range was 1-64 μg/ml. At the limit of quantitation (1 μg/ml) the coefficient of variation was 7.8%. The recovery was 70%. No internal standard was used. Takikawa et al. [93] determined nicotinic acid, nicotinamide, and nicotinuric acid after administration of nicotinic acid to humans. Plasma samples were deproteinized with acetone and extracted with chloroform. The remaining aqueous layer was acidified and evaporated to dryness. After addition of methanol and reevaporation the residue was dissolved in acetone containing quinaldic acid as internal standard. The substances were separated on a strong cation-exchange column (Zipax SCX, 25-37 μm, 50 × 0.21 cm ID, du Pont de Nemours). The mobile phase consisted of 0.02 mol/liter sodium phosphate buffer, pH 2.6, for nicotinic acid and of 0.02 mol/liter sodium phosphate buffer, pH 2.8, for the metabolites. The flow rate was 1 ml/min; the ultraviolet detection was achieved at 260 nm. The column was kept at 45°C in a water bath. The method was linear in the range 0.5-5 μg/ml for nicotinic acid, nicotinamide, and nicotinuric acid. Nicotinamide was not detectable (<0.5 μg/ml) after a single oral dose of 500 mg nicotinic acid.

Hengen et al. [22] applied reversed-phase chromatography to the quantitation of nicotinic acid and the metabolite nicotinuric acid in human plasma and urine. For the determination of plasma, a similar extraction procedure as used by Carlson [82] was applied. The aqueous solution was evaporated and the methanolic extract of the residue injected in a chromatographic system, separating the compounds with a reversed-phase μBondapak C18 column (10 μm, 30 × 0.39 cm ID), using 0.005 mol/liter tetrabutylammonium phosphate as ion-pair modifier in water-methanol (90:10, v/v), pH 7.5. The flow rate was 2 ml/min; the ultraviolet absorption was monitored at 254 nm and the peak area calculated for quantitation. Isonicotinic acid was used as internal standard. The urine was purified through extraction with chloroform; the aqueous phase evaporated and injected in two columns in series, the solvents and detectors being the same as for plasma. Precision and accuracy were good, in the range 0.1-20 μg/ml for plasma and 0.5-1000 μg/ml for urine. The maximal coefficient of variation was 7.5% (for 0.5 μg/ml plasma and 20 μg/ml urine, the number of aliquots in both cases n = 20); an average of 74% was recovered. The limit of quantitative detection was 0.1 μg/ml for plasma and 0.5 μg/ml for urine. A total of 0.5 ml plasma or urine is needed for a determination. Other metabolites and structurally related compounds did not interfere. Figure 7.11 shows representative chromatograms for pure substances, plasma samples, and urine samples. Nicotinamide in plasma and urine cannot be determined by the method of Hengen et al. [22]. The

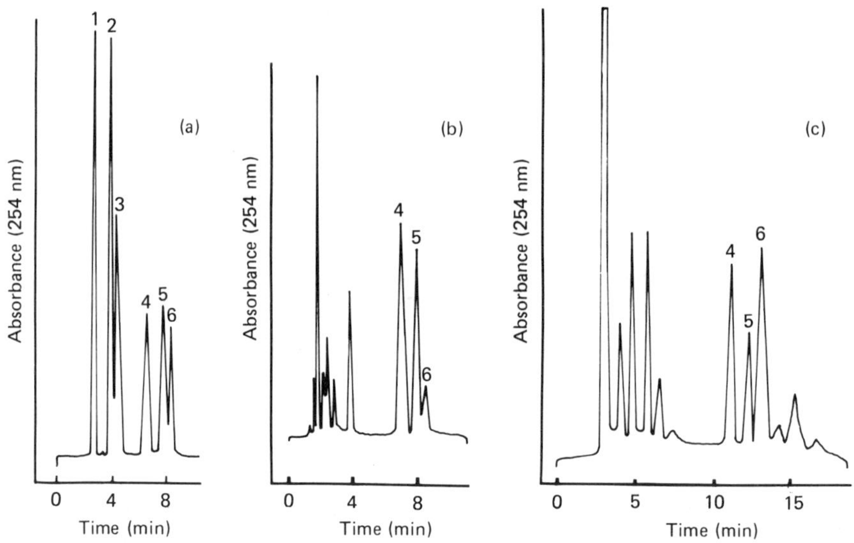

Figure 7.11 Chromatogram of (a) a mixture of (5) nicotinic acid, (6) nicotinuric acid, (1) nicotinamide-*N*-oxide, (2) 2-hydroxypyridine-5-carboxylic acid, (3) nicotinamide, and (4) isonicotinic acid. (b) A plasma extract after ingestion of nicotinic acid. Abbreviations as in (a). (c) A urine extract after ingestion of nicotinic acid. Abbreviations as in (a). Mobile phase: 0.005 mol/liter tetrabutylammonium phosphate in water-methanol (90:10, v/v); UV detection at 254 nm. (From Ref. 22, © 1978, *Clinical Chemistry*.)

extracts contain many unknown endogenous, interfering substances near the retention volume of nicotinamide.

McKee et al. [94] determined nicotinic acid, nicotinamide, nicotinuric acid (Structure 7.17), 1-methylnicotinamide (Structure 7.13), and 1-methyl-6-pyridone-3-carboxamide (Structure 7.22) in human and rat urine. The urine samples were filtered through 0.2 μm Metrical Gelman membrane filters and analyzed on a reversed-phase C18 column (Ultrasphere-ODS, 5 μm, 15 × 0.46 cm ID). The separation was performed under gradient elution conditions. The mobile phase A consisted of sodium 1-pentane sulfonate (PSA), tetramethylammonium chloride (TMA), and potassium dihydrogen phosphate, 10 mmol/liter each. Mobile phase B consisted of PSA and TMA, 1 mol/liter each, in water-acetonitrile (10:90, v/v). The flow rate was 2 ml/min; the temperature was 24.5°C. The gradient was run linearly from 0% B to 6% B in 6 min; the ultraviolet detector was set at 254 nm. An internal standard was not used. The calibration curve was linear in the range 5-50 μg/ml. The pattern of the metabolites is described in Sec. 5.2.3.

De Vries et al. [95] devised a purification procedure, using a small column (1 cm) filled with C18 reversed-phase material (Sep-pak C18), washing with water, and eluting with the chromatographic solvent. An aliquot was directly injected into a C18 reversed-phase column. The mobile phase consisted of water-methanol (58:42, v/v) with 0.004 mol/liter dioctyl sodium sulfoscuccinate as ion-pair modifier and the pH adjusted to 2.3 with formic acid. The flow rate was 2 ml/min; the ultraviolet detector was set at 254 nm. The urine extract had to be separated using two columns. Isonicotinamide was used as internal standard. Precision and accuracy had a coefficient of variation of 6.8%; the recovery was 91% for the range 0.1-10 μg/ml for plasma and 1-50 μg/ml in urine. Metabolites, structurally similar compounds, other vitamins and drugs, and also a large excess of nicotinic acid did not interfere. Figure 7.12 gives representative examples for the pure substances, plasma, and urine.

The use of the ion-pair approach for the analysis of nicotinic acid in plasma was anticipated in 1975 by Wahlund [96], who identified

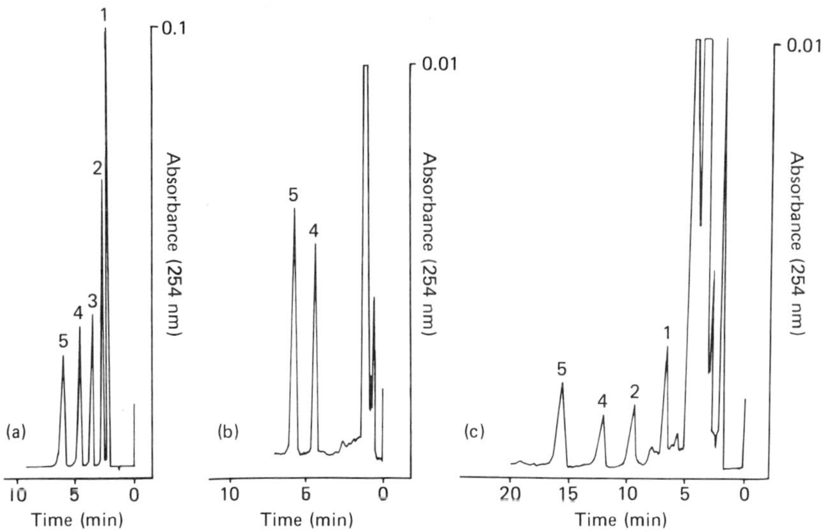

Figure 7.12 Chromatogram of (a) a mixture of (1) isonicotinic acid, (2) nicotinic acid, (3) nicotinuric acid, (4) nicotinamide, and (5) the internal standard isonicotinamide; each 1 μg. (b) A plasma extract after nicotinamide ingestion: (4) nicotinamide (concentration 2.1 μg/ml); (5) isonicotinamide (concentration, 4 μg/ml); 100 μl injection. (c) Urine extract spiked with (1) nicotinic acid, (2) nicotinuric acid, (4) nicotinamide (each 20 μg/ml), and (5) isonicotinamide (50 μg/ml); 10 μl injcction. (From Ref. 95.)

nicotinic acid in spiked serum, employing a silanized silica gel column saturated with 1-pentanol and using tetrabutylammonium phosphate as eluent. Several publications deal with the quantitation of nicotinic acid, nicotinamide, and its metabolites together with other endogenous substances in urine using ion-exchange chromatography and are designed for comprehensive metabolic studies [97-99]. Since they are more time consuming, they are not intended for routine serial determinations of only nicotinic acid or nicotinamide. They are discussed in Sec. 5.

Tarr [100] described the analysis of tryptophan and nine of its metabolites, including nicotinic acid and nicotinamide, in urine. Urine (1-2 ml) was extracted according to the method of Bieleski and Turner [101]. The chromatography was performed on a C18 reversed-phase column (μBondapak C18, 10 μm, 30 × 0.39 cm ID). The mobile phase consisted of 10 mmol/l sodium acetate, pH 4.84. The flow rate was 1 ml/min; after 4.2 min it was increased to 2 ml/min. The compounds were detected by ultraviolet absorption at 254 nm. The analysis of the 10 compounds needed about 60 min. The method was linear in the range 0.05-50 nmol (6 ng to 6.2 μg). The method is intended to study tryptophan metabolism and is therefore not further considered.

5. METABOLITES

5.1 Excretion Metabolism of Nicotinic Acid and Nicotinamide

The biotransformation of nicotinic acid and nicotinamide is shown in Scheme 7.3. In animal tissues nicotinic acid and nicotinamide are converted to several compounds, some of which are readily excreted. The metabolic excretion pattern is different in avians and mammals. Chicken excrete principally β-dinicotinoylornithine (Structure 7.19) and β-nicotinoylglucuronate (Structure 7.18) but no methylated products.

In mammals, the qualitative and quantitative pattern depends on the species and on the dose [25,32,33]. In Scheme 7.3 the main routes of metabolism are given. 1-Methylnicotinamide (1-MN) (Structure 7.13) is recognized as a notable end product of nicotinamide and nicotinamide coenzyme metabolism [102,103]. It results from the methylation of nicotinamide with *S*-adenosylmethionine. Its normal urinary excretion is known to be influenced by dietary intake of its presursors (nicotinamide, nicotinic acid, tryptophan, etc.). The basal concentration of 1-MN in human serum is 3.5 nmol/liter (63 ng/ml) as mean value according to Clark and coworkers [20,21]. In part it is eliminated in the urine as such and the rest oxidized to 1-methyl-4-pyridone-3-carboxamide (4-pyridone, Structure 7.23) and 1-methyl-6-pyridone-3-carboxamide (6-pyridone, Structure 7.22) with aldehyde oxidase in the liver. The metabolites of nicotinamide reflect the nutritional status.

The weight ratio of 6-pyridone to 1-MN is regarded as an important index of nicotinic acid status [104]. De Lange and Joubert [105] defined a 6-pyridone to 1-MN ratio of 1.3:4.0 as normal and and of less than 1.0 as indicative of latent nicotinic acid deficiency. Attempts have been made to correlate the excretion pattern with disease states (schizophrenia states, depression, and mental retardation) [97,106], burns [107], and chemical carcinogens [108].

The administration of nicotinic acid or nicotinamide in therapeutic doses higher than needed as vitamin substitution leads to an enhanced elimination of the above metabolites (1-MN, 4-pyridone, and 6-pyridone) and to measurable amounts of nicotinuric acid (Structure 7.17), nicotinamide-1-oxide (Structure 7.20), and nicotinamide [25] (compare Scheme 7.3).

5.2 Determination of the Excretion Metabolites of Nicotinic Acid and Nicotinamide

5.2.1 Paper and Thin-Layer Chromatography

Many of the earlier metabolic studies were done with radiolabeled nicotinic acid and nicotinamide, separating the metabolites by paper chromatography. van Eys et al. [109] isolated from rat urine nicotinic acid glucuronide (Structure 7.18) as a normal constitutent, using paper chromatography with several mobile phases. Cation-exchange chromatography was used in the prepurification of the urine. The glucuronide was not found in human urine. Chang and Johnson [110] isolated at that time (1959) the new metabolite 1-methyl-4-pyridone-3-carboxamide (4-pyridone, Structure 7.23), along with 1-methylnicotinamide (1-MN, Structure 7.13), 1-methyl-6-pyridone-3-carboxamide (6-pyridone, Structure 7.22), and nicotinamide (Structure 7.2) from rat urine after administration of radioactive [7-^{14}C]nicotinic acid. Electrophoretic mobilities at different pH values were used for the identification.

4-Pyridone (Structure 7.23) was identified by Chang and Johnson [111] in human urine after nicotinic acid application. By paper chromatography with the solvent system 1-butanol-acetone-water (45:5:50, v/v), 1-MN, 4 pyridone, and 6-pyridone, were detected in the urine of monkey. In a study of the metabolism of nicotinamide in the urine of mice, radiolabeled metabolites were separated by paper chromatography in the mobile phase 1-butanol saturated with 3% ammonia and rechromatographed in 1-butanol-acetic acid (60:1) [85]. Abelson et al. [112-115] detected 6-pyridone and 4-pyridone in human plasma under basal conditions. They extracted with an excess of chloroform after alkalization with sodium hydroxide solution and chromatographed on paper with *n*-butanol-ammonia-water (300:4:46, v/v) and *n*-butanol-water (86:14, v/v). Mohran [86] proposed a method for estimating the intake of nicotinamide by determination of 4-pyridone. After oral administration of 2-100 mg to humans, the plasma was extracted for 4-pyridone with chloroform and the solution subjected to paper chromatography with the mobile phase 1-butanol-water (86:14, v/v). The corresponding fluorescent spot formed on the paper after irradiation with ultraviolet light was eluted, and the fluorescence was determined.

Quinn and Greengard [116] separated as enzymatic oxidation products of 1-MN, 4-pyridone, and 6-pyridone, in liver homogenates through paper chromatography with the solvent system 1-butanol-water-ammonia (680:114:6, v/v). They were located on paper by scanning for radioactivity and as spots absorbing ultraviolet light. The appropriate areas were cut out and the amount of radioactivity estimated with a liquid scintillation counter. The ratio of 4-pyridone to 6-pyridone is constant for a given species, after feeding nicotinamide. The urine metabolites were separated by paper chromatography with 1-butanol-3% ammonia as mobile phase [117]. Gluecksohn-Waelsch et al. [118] studied the genetic variation of the oxidase from rat liver which transforms 1-MN to 4-pyridone and 6-pyridone. Paper chromatography was used with 1-butanol-water-concentrated ammonia (680:114:6, v/v) as mobile phase.

Bonavita et al. [84] separated the metabolites found in urine after injection [7-^{14}C]nicotinamide in mice in the presence or absence of azaserine. The separation was achieved on paper with 1-butanol saturated with water as mobile phase. 1-Methylnicotinamide, nicotinic acid, 6-pyridone, nicotinamide, and nicotinamide-*N*-oxide, were found. The latter compound was isolated and identified as a new metabolite of nicotinamide. An isotope dilution analysis for the quantitation of nicotinamide-*N*-oxide (Structure 7.20) in human urine was developed by Sparthan and Chaykin [119,120]. After purifying the radiolabeled urine extract by paper chromatography with 1-butanol saturated with water as mobile phase, the peak was eluted and reduced to nicotinamide with a xanthine oxidase-catalyzed oxidation of xanthine. The entire reaction mixture was chromatographed as before. The nicotinamide was eluted, the radioactivity counted, and nicotinamide determined with the Koenig reaction. In humans only trace amounts were found under basal conditions. After ingestion of 200 mg nicotinamide significant quantities are detected in human urine.

Sastry and Waller [121] identified 6-pyridone in a plant extract from *Trewia nudiflora* by preparative thin-layer chromatography with the mobile phase methanol-chloroform (1:50, v/v, and consecutively 2:5, v/v). Nicotinic acid and nicotinamide were proved with radiolabeling as specific precursors during the biosynthesis. The metabolite *N*-methylnicotinic acid, well known in plants, was also identified. Lee et al. [122] fed rats with nicotinamide or nicotinic acid and identified in the urine two new metabolites, namely 6-hydroxynicotinic acid (Structure 7.16) and 6-hydroxynicotinamide (Structure 7.21) with the following methodology. Thin-layer chromatography was performed on cellulose layers with 1-butanol saturated with ammonia, electrophoresis at pH 9.3, ion exchange on Dowex 50 formate, and paper chromatography. Paper and thin-layer chromatographic systems have been described by Pinder et al. [49] for the separation and characterization of nicotinic acid, nicotinamide, and their nucleotides. Table 7.2 gives a compilation from various solvent systems, modified according to Pinder et al. [49]. These compounds may also be separated by ion-exchange chromatography [123].

5.2.2 Gas Chromatography and Gas Chromatography-Mass Spectrometry

Holmen et al. [124] quantitated *N'*-methylnicotinamide (*N'*-methyl-3-pyridinecarboxamide) in human urine by gas chromatography and also by gas chromatography-mass spectrometry (GC-MS). This compound was recently found and should not be confused with N_1-methylnicotinamide (Structure 7.13) which contains the methyl group at the ring nitrogen, whereas in *N'*-methylnicotinamide the methyl is at the side chain. Urine (2 ml) was extracted with 10 ml chloroform, evaporated, and the residue subjected to gas chromatography on a 1 m column containing 10% Silar 10C on Gas Chrom Q. The column temperature was 250°C. The internal standard was *N',N'*-diethylnicotinamide. The calibration curve was linear in the range 2-300 μg. The results were confirmed by GC-MS. In 4 of 20 volunteers no *N'*-methylnicotinamide was found (less than 2 μg/ml). The mean value for the others was 51 μg/ml.

5.2.3 Liquid Chromatography

Ion exchangers have been frequently used for the prepurification of metabolites before their determination with photometric methods. High-resolution ion-exchange liquid chromatography was reported by Mrochek et al. [97]. These authors used a 150 × 0.6 cm column containing the anion-exchange resin Aminex A 27 (12-15 μm particles). In most cases an Aminex A5 cation-exchange resin (12-15 μm particles) was placed on the top of the main anion-exchange column to produce a coupled column system for better resolution at the front of the chromatogram. A gradient elution was applied with butter of increasing pH. Detection was effected at 254 and 280 nm for the quantitation of the excretion in urine. They applied 1000 mg nicotinic acid or nicotinamide to healthy volunteers and schizophrenic patients. 1-Methylnicotinamide, nicotinamide-1-oxide, 6-pyridone, 4-pyridone, nicotinuric acid, nicotinamide, and nicotinic acid were quantitated. The recovery from these compounds averaged 99.8%, with a coefficient of variation of 5.1%. No significant differences could be found in the excretion pattern of normal volunteers and schizophrenic patients.

Lis et al. [106] used the same anion-exchange resin as Mrochek et al. [97]. An ammonium acetate-acetic acid buffer gradient of 0.015-6.0

mol/liter (pH 4.4) was the eluent. They analyzed 4-hydroxyhippuric acid, 1-MN, and 6-pyridone, in the urine of autistic children. Abnormalities in the excretion of these compounds were found. Bernofsky [98,99] separated several nicotinic acid and nicotinamide metabolites in urine: 1-MN, trigonelline, nicotinamide-1-oxide, 4-pyridone, 6-pyridone, nicotinamide, nicotinic acid, nicotinuric acid, and the 6-hydroxy derivatives of nicotinamide and nicotinic acid (Structure 7.16) (see Fig. 7.13). The chromatographic column consisted of 73.6 × 0.9 cm glass tube filled with Dowex 1 × 4, 200-400 mesh (37-74 μm diameter). Gradient elution was done with ammonium formate-formic acid of various pH and ion strength. Detection was with ultraviolet absorption at 254 nm. Pyridine nucleotides could also be separated by this system. All these ion-exchange systems can separate a large number of substances, but are very time consuming.

Shaikh et al. [102] quantitated 1-MN in rat urine. The method employs a single column chromatographic step with cation-exchange resin AG 50W-X4 (Bio-Rad Lab.). Elution was performed with 1 mol/liter ammonium formate, pH 8.9. The liquid chromatographic separation was effected on a 25 × 0.46 cm Partisil-10 SCX column with the mobile phase 0.25 mol/liter ammonium dihydrogen phosphate, pH 4.3, at a flow rate of 1 ml/min. For monitoring the ultraviolet absorption the wavelength was set at 254 nm. Linearity was found in the range 20 ng to 2 μg/ml urine with a correlation coefficient of 0.995.

Sandhu and Frazer [125] measured 1-MN, 6-pyridone, and trigonelline (*N*-methylnicotinic acid) in rat urine. The urine was obtained from nicotinic acid-deficient rats. The urine, with added isonicotinic acid as internal standard, was extracted with chloroform. The residue after evaporation was extracted with methanol and used for liquid chromatography. This was performed on a cation-exchange column (Partisil

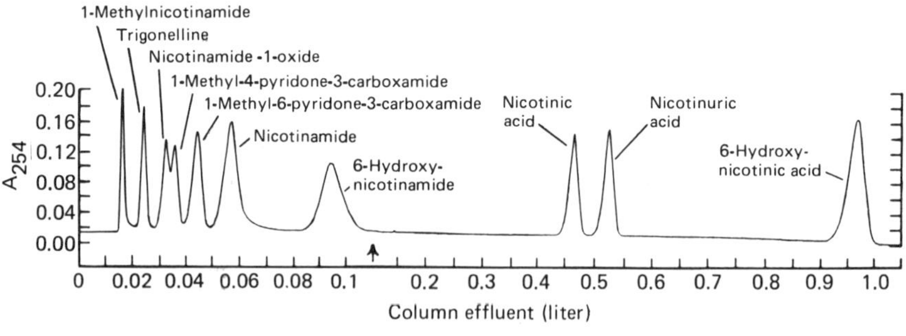

Figure 7.13 Chromatographic separation of known mammalian urinary metabolites of nicotinamide and nicotinic acid on a conventional column of Dowex-1 formate. Elution was conducted at 0.2 ml/min for the first 110 ml (arrow) and at 1 ml/min thereafter. (From Ref. 98.)

10 SCX, 10 μm, 50 × 0.45 cm ID) with a precolumn of Vydac-201, 33-44 μm, 3 × 0.39 cm, with the mobile phase 15 mmol/liter potassium citrate, pH 2, containing 5% methanol. The flow rate was 2.5 ml/min, the detection at 254 nm. The method was linear in the range 5-100 μg/ml, the detection limit 10 ng. The recovery was 81-89%. A decrease was found both in the excretion of 1-MN and 6-pyridone as well as in the ratio of these two nicotinic acid metabolites. The ability to diagnose nicotinic acid deficiency in the rat was established.

The advent of reversed-phase materials permitted the use of these columns to separate and quantitate the metabolites of nicotinic acid and nicotinamide. They have the advantage in comparison to the ion-exchange columns that the equilibrium is achieved in minutes and is readily maintained. Shaikh and Pontzer [126] quantitated 1-MN in urine by direct injection of the urine in a reversed-phase column. The mobile phase was methanol-water (475:522, v/v) modified with 0.005 mol/liter sodium dodecyl sulfate as ion-pair reagent, adjusted to pH 2.5 with sulfuric acid. 1-Methylnicotinamide is well separated from the other peaks. The method was used in excretion studies of nicotinamide in humans [127] (see Fig. 7.14). The oxidation of 1-MN produces 6-pyridone and 4-pyridone, which can also be separated on a reversed-phase C18 column (μBondapak C18, 10 μm, 30 × 0.39 cm or equivalent) with the mobile phase water-methanol-acetic acid (940:50:10, v/v). The peaks were monitored by ultraviolet absorption at 254 nm (see Fig. 7.15).

As mentioned in Sec. 4.4 on liquid chromatography, Hengen et al. [22] determined nicotinuric acid in plasma and urine, a main metabolite of nicotinic acid after high doses of nicotinic acid, with a reversed-phase column. For chromatographic details see Sec. 4.4. McKee et al. [94] determined, as described in Sec. 4.4, besides nicotinic acid and nicotinamide, the metabolites 1-MN, 6-pyridone, and nicotinuric acid in human and rat urine. They found in rats on a diet of 90 mg nicotinic acid per kilogram large quantities of nicotinamide and 1-MN and relatively small amounts of nicotinuric acid and 6-pyridone. In humans without the diet supplement, McKee et al. [94] found in urine large amounts of 1-MN, 6-pyridone, and nicotinuric acid, but little or no nicotinic acid or nicotinamide. The sampling periods for the amounts found are not given, so that a relation to a definite time interval is not possible.

Carter [128] quantified 1-MN and 6-pyridone, the two major metabolites of nicotinic acid. Both metabolites were isolated from human urine by cleanup with an anion-exchange resin (Dowex 1 × 8 resin, 100-200 mesh, Bio-Rad Lab.). The resin (4 ml) was poured into the lower half of the custom-made columns (1 × 10 cm Econo-column, Bio-Rad Lab.). Under each column was placed a flask containing 1 ml pH 7 buffer (0.5 mol/liter potassium phosphate and 0.25 mol/liter octane sulfonate). At neutral pH neither 1-MN nor 6-pyridone was retained on the anion-

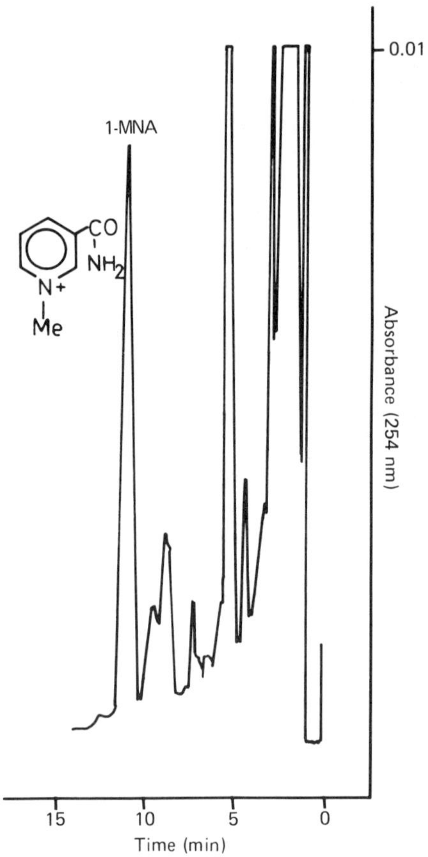

Figure 7.14 Chromatogram of a urine sample 150-210 min after a single dose of 200 mg nicotinamide tablet. (1-MNA) 1-methylnicotinamide, concentration 135 μg/ml. Column: μBondapak C18, 30 × 0.39 cm ID. Mobile phase: methanol-0.005 mol/liter sodium dodecyl sulfate-sulfuric acid (450:550:0.1, v/v), pH 2.5. Flow rate: 2 ml/min. (From Ref. 127.)

exchange resin; in contrast, 70-80% of the compounds in urine absorbing in the 240-270 nm range were anionic in character and bound to the resin. The chromatographic separation was performed on an ion-pair reversed-phase column (Ultrasphere IP, 5 μm, 25 × 0.46 cm ID, Beckmann), fitted with a guard column with reversed-phase material (RP-18, Spheriguard, 5 μm, 3 × 4.6 cm ID). The mobile phase for MN consisted of water-acetonitrile (90:10, v/v) containing 10 mmol/

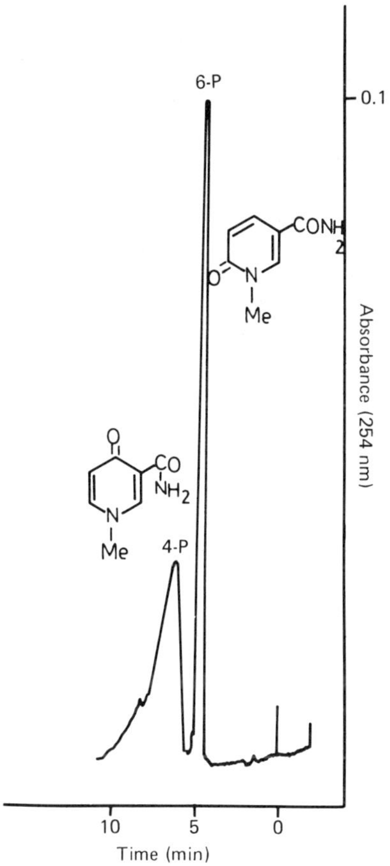

Figure 7.15 Separation of (6P) 1-methyl-6-pyridone-3-carboxamide and (4-P) 1-methyl-4-pyridone-3-carboxamide (1 μg each). Column: μBondapak C18. Mobile phase: water-methanol-acetic acid (940:50:10). Flow rate: 2 ml/min. (From Ref. 127.)

liter dipotassium hydrogen phosphate and 5 mmol/liter sodium 1-octane sulfonate, adjusted to pH 7 with phosphoric acid. The mobile phase for 6-pyridone consisted of water-acetonitrile (98:2, v/v) containing 10 mmol/liter dipotassium hydrogen phosphate, adjusted to pH 7 with phosphoric acid. The flow rate was 1.5 ml/min for both substances; the detection wavelength was 264 nm for 1-MN and 258 nm for 6-pyridone. The linear range for 6-pyridone was 0.5-15 μg/ml; the detection limit 0.05 μg/ml. The overall coefficient of variation of the method was 6.3%. The mean recovery added was 95%. No internal standard was used. The linear range for 1-MN was 0.5-15 μg/ml, the detection limit 0.2 μg/ml, and the coefficient of variation 6.8%. The mean recovery of added substance was 92%. The results of these chromatographic methods were compared with those obtained by colorimetry (6-pyridone) and fluorometry (1-MN). They agree well with each other. The grand mean value of human urine for 1-MN was 5.0 mg per 24 hr (n = 112) and, for 6-pyridone, 9.13 mg per 24 hr.

Terry and Simon [104] developed an assay for 1-MN and 6-pyridone in urine. Both metabolites were isolated from urine using separate ion extractions for both substances and two internal standards. 6-Pyridone was eluted sequentially from a cation-exchange (4 × 0.7 cm ID AG 50W-X80 to an anion-exchange column (4 × 0.7 cm ID AG 1-X8); 1-MN from a weak cation-exchange column (4 × 0.7 cm Bio-Rex 70). They and their two internal standards, 1-methyl-5-carboxyl-[*N*-(2-ethoxy)]amide-2-pyridone and 1-methyl-3-carboxyl-[*N*-(2-ethoxy)]amidepyridine chloride, were quantitated simultaneously by liquid chromatography using a reversed-phase ion-pairing technique. The combined extracts were injected in the chromatograph. The chromatography was performed on a C18 reversed-phase column (Ultrasphere IP, 5 μm, 15 × 0.46 cm ID) preceded by a guard column (Lichrosorb RP 18, 10 μm, 3 × 0.46 cm ID). The temperature of the column was kept at 35°C. The mobile phase was 91% (0.15 mol/liter sodium acetate buffer, pH 5, and 0.001 mol/liter sodium heptane sulfonate) and 9% methanol. The flow rate was 1 ml/min; detection was by ultraviolet absorbance at 254 nm. The detection limit for both compounds was 0.3 μg/ml urine. An exact linear range was not given, but the method seems to be linear, from 5 to 100 μg/ml for 6-pyridone and from 1 to 20 μg/ml for 1-MN, with coefficients of variation from day to day ≤ 5.1%. The normal ranges were 6-51.3 mg per 24 hr (6-pyridone) and 1.6-14.8 mg (1-MN). The 6-pyridone to 1-MN weight ratio was 1.76:5.9.

Nicotinamide-1-oxide has been quantitated in studies on the microsomal N-oxidation of nicotinamide and pyridine compounds by Blaauboer and Paine [129] in rat liver microsomes. The samples were chromatographed on a reversed-phase Partisil-10 ODS column (25 × 0.46 cm ID) preceded by a Whatman Co-Pell-ODS guard column (8 × 2.1 cm ID). An isocratic elution system was used with 10 mmol/liter sodium potassium phosphate buffer at pH 7.4, containing 10% methanol or 20 mmol/liter hydrochloric acid containng 2% methanol. The column pressure was 6900 kPa, and the flow rate 2 ml/min.

Heard and Tritz [130] separated a series of biological pyridines that belong to the pyridine nucleotide cycle on a C18 reversed-phase (μBondapak C18, 10 μm, 30 × 0.39 cm ID) with 5 mmol/liter pentane sulfonic acid or heptane sulfonic acid buffered to pH 2.5 or 3.5, respectively, with potassium dihydrogen sulfate.

ACKNOWLEDGMENT

The authors wish to thank Professor Ellen Weber, Medical Clinic, Section Clinical Pharmacology, University of Heidelberg, for revising the manuscript and for her interest.

REFERENCES

1. D. A. Bender, in *Vitamins in Medicine*, Vol. 1, 4th ed. (B. M. Barker and D. A. Bender, eds.), W. Heinemann Medical Books, London, 1980, p. 315.
2. H. Kating and G. Drephal, in *Fermente, Hormone, Vitamine*, Vol. III/1 (R. Ammon and W. Dirschel, eds.), G. Thieme Verlag, Stuttgart, 1974, pp. 727-765.
3. D. E. Danford and H. N. Munro, in *The Pharmacological Basis of Therapeutics*, 6th ed. (L. S. Goodman and A. Gilman, eds.), MacMillan Publishing, New York, 1980, pp. 1567-1570.
4. J. Goldberger, G. A. Wheeler, R. D. Lillie, and L. M. Rogers, *Public Health Rep.*, *43*, 1385 (1928).
5. T. D. Spies, C. Cooper, and M. A. Blankenhorn, *J. Amer. Med. Ass.*, *110*, 622-627 (1938).
6. C. A. Elvehjem, R. J. Madden, F. M. Strong, and D. W. Wooley, *J. Amer. Chem. Soc.*, *59*, 1767 (1937).
7. W. Martindale, *The Extra Pharmacopeia*, 27th ed. (A. Wade, ed.), Pharmaceutical Press, London, 1980, p. 1691.
8. R. Schoental, *Biochem. Soc. Trans.*, *8*, 147 (1980).
9. M. Windholz, (ed.), 9th ed., *Merck Index*, Merck & Co., Rahway, N.J., 1976.
10. L. Fuller, *Methods Enzymol.*, *66*, 3 (1980).
11. H. Liliedahl, *Acta Chem. Scand.*, *20*, 95 (1966).
12. A. R. Prosser and A. J. Sheppard, *J. Pharm. Sci.*, *57*, 1004 (1968).
13. A. J. Sheppard and A. R. Prosser, *Methods Enzymol.*, *18B*, 17 (1971).
14. J. Roboz, R. Suzuki, and T. F. Yue, *J. Chromatogr.*, *147*, 337 (1978).
15. W. Koenig, *J. Prakt. Chem.*, *69*, 105 (1904).
16. B. Kakac and Z. J. Vejdelek, *Handbuch der photometrischen Analyse organischer Verbindungen*, Vol. 2, Verlag Chemie, Weinheim, p. 771; suppl., 1977, p. 321.
17. P. Karlson, *Kurzes Lehrbuch der Biochemie*, E. Thieme Verlag, Stuttgart, 1980.
18. Committee on Dietary Allowances, Food and Nutritional Board (eds.), *Recommended Dietary Allowances*, 9th ed., National Academy of Sciences, Washington, D.C., 1980, p. 92.

19. P. B. Collins and S. Chaykin, *Biochem. J.*, *125*, 117 P (1971).
20. B. R. Clark, R. M. Halpern, and R. A. Smith, *Anal. Biochem.*, *68*, 54-61 (1975).
21. B. R. Clark, *Methods Enzymol.*, *66*, 5 (1980).
22. N. Hengen, V. Seiberth, and M. Hengen, *Clin. Chem.*, *24*, 1740 (1978).
23. J. Keller, M. Liersch, and H. Grunicke, *Eur. J. Biochem.*, *22*, 263 (1971).
24. R. F. Dawson, D. R. Christman, and R. U. Byerrum, *Methods Enzymol.*, *18B*, 90 (1971).
25. R. Fumagalli, in *Metabolic Effects of Nicotinic Acid and Its Derivatives* (K. F. Gey and L. A. Carlson, eds.), Hans Huber Publishers, Berne, 1971, p. 33.
26. L. A. Carlson, in *Metabolic Effects of Nicotinic Acid and Its Derivatives* (K. F. Gey and L. A. Carlson, eds.), Hans Huber Publishers, Berne, 1971, p. 157-165.
27. N. Svedmyr, L. Harthon, and L. Lundholm, in *Metabolic Effects of Nicotinic Acid and Its Derivatives* (K. F. Gey and L. A. Carlson, eds.), Hans Huber Publishers, Berne, 1971, p. 1085.
28. N. Svedmyr, L. Harthon, and L. Lundholm, *Clin. Pharmacol. Ther.*, *10*, 559 (1969).
29. L. Kaijser, B. Eklund, A. G. Olsson, and L. A. Carlson, *Med. Biol.*, *57*, 114-117 (1979).
30. R. I. Levy, in *The Pharmacological Basis of Therapeutics*, 6th ed., (L. S. Goodman and A. Gilman, eds.), MacMillan Publishing, New York, 1980, p. 834.
31. R. I. Levy, *Clin. Chem.*, *27*, 653 (1981).
32. M. Weiner, *Drug Metabol. Rev.*, *9*, 99 (1979).
33. M. N. Cayen, *Drug Metabol. Rev.*, *11*, 291 (1980).
34. B. Petrack, P. Greengard, and H. Kalinski, *J. Biol. Chem.*, *241*, 2367 (1966).
35. Y. C. Lee, J. L. R. Chandler, and R. K. Gholson, in *Metabolic Effects of Nicotinic Acid and Its Derivatives* (K. F. Gey and L. A. Carlson, eds.), Hans Huber Publishers, Berne, 1971, p. 133.
36. O. N. Miller, J. G. Hamilton, and G. A. Goldsmith, *Am. J. Clin. Nutr.*, *8*, 480 (1960).
37. L. A. Carlson, L. Oro, and J. Ostman, *Acta Med. Scand.*, *183*, 457 (1965).
38. M. Lesne, X. Sturbois, and M. Mercier, *Pharm. Acta Helv.*, *51*, 367 (1976).
39. A. Subissi and W. Murmann, *Arzneim. Forsch.*, *28*, 1143 (1978).
40. R. Testa, M. Gambini, and C. Canestrini, *Farmaco*, *33*, 463 (1978).
41. M. N. Cayen, W. T. Robinson, J. Dubuc, and D. Dvornik, *Biochem. Pharmacol.*, *28*, 1163 (1979).

42. R. Altschul, in *Niacin in Vascular Disorders and Hyperlipemia* (R. Altschul, ed.), Charles C. Thomas, Springfield, Ill., 1964, p. 3.
43. R. Kluthe and B. Schaefer, *Med. Welt, 26*, 2159 (1975).
44. N. Svedmyr and L. Harthon, *Acta Pharmacol. Toxicol., 28*, 66 (1970).
45. G. Wolfram, C. Keller, S. Kilani, and N. Zollner, *Dtsch. Med. Wochenschr., 101*, 76 (1976).
46. E. Marmo, A. P. Caputi, C. Vacca, A. Imperature, and S. Cataldi, *Clin. Ter., 68*, 519 (1974).
47. G. A. Goldsmith and O. N. Miller, in *The Vitamins, Chemistry, Physiology, Pathology, Methods*, Vol. 7 (P. Gyoergy and W. N. Pearson, eds.), Academic Press, New York, 1967, p. 137.
48. G. Katsui, in *Pharmaceutical Applications of Thin-Layer and Paper Chromatography* (K. Macek, ed.), Elsevier, Amsterdam, 1972, p. 470.
49. S. Pinder, J. B. Clark, and A. L. Greenbaum, *Methods Enzymol., 18B*, 20 (1971).
50. H. R. Bolliger and A. Koenig, in *Duennschichtchromatographie*, 2nd ed. (E. Stahl, ed.), Springer Verlag, Heidelberg, 1967, p. 382.
51. R. W. Frei, A. Kunz, G. Pataki, T. Prims, and H. Zuercher, *Anal. Chim. Acta, 49*, 527 (1970).
52. S. A. Ismaiel and D. A. Yassa, *Analyst, 98*, 816 (1973).
53. K. Takikawa, K. Miyazaki, and T. Arita, *J. Pharm. Dyn., 4*, 123 (1981).
54. J. Vessman and S. Stroemberg, *J. Pharm. Sci., 64*, 311 (1975).
55. C. P. Tally, *Anal. Chem., 43*, 1512 (1971).
56. R. C. Williams, D. R. Baker, and J. A. Schmit, *J. Chromatogr. Sci., 11*, 618 (1973).
57. K. Callmer and L. Davies, *Chromatographia, 7*, 644 (1974).
58. N. Nishimoto, Y. Mitani, and S. Hayashi, *J. Chromatogr., 176*, 448 (1979).
59. M. Caude and P. LeXuan, *Chromatographia, 9*, 20 (1976).
60. C. Horvath and W. Melander, *J. Chromatogr. Sci., 15*, 393 (1977).
61. R. Gloor and E. L. Johnson, *J. Chromatogr. Sci., 15*, 413 (1977).
62. B. A. Bidlingmeier, *J. Chromatogr. Sci., 18*, 525 (1980).
63. S. P. Sood, D. P. Wittmer, S. A. Ismaiel, and W. G. Haney, *J. Pharm. Sci., 66*, 40 (1977).
64. R. L. Kirchmeier and R. P. Upton, *J. Pharm. Sci., 67*, 1444 (1978).
65. M. C. Walker, B. E. Carpenter, and E. L. Cooper, *J. Pharm. Sci., 70*, 99 (1981).
66. Waters Associates, Inc., Milford, Mass., Bull. F61, May 1976.
67. C. P. Terweij-Groen, S. Heemstra, and J. C. Kraak, *J. Chromatogr., 161*, 69 (1978).

68. R. B. H. Wills, C. G. Shaw, and W. R. Day, *J. Chromatogr. Sci.*, *15*, 262 (1977).
69. W. R. Melander, J. Stoveken, and C. Horwath, *J. Chromatogr.*, *185*, 111 (1979).
70. R. Schuster, Application Note AN 232-6, September 1978, Hewlett-Packard GmbH, Boeblingen, Germany.
71. du Pont Analytical Instruments, Wilmington, DE, *Multivitamin Analysis Using High Performance Liquid Chromatography. LC Report* E-25475.
72. R. P. Kwok, W. P. Rose, R. Tabor, and T. S. Pattison, *J. Pharm. Sci.*, *70*, 1014 (1981).
73. *AOAC Official Methods of Analysis*, 12th ed., Association of Official Analytical Chemists, Washington, D.C., 1975.
74. E. Kodicek, *Biblio. Nutr. Dieta, 4*, 109 (1962).
75. J. B. Manson, H. Gibson, and E. Kodicek, *Brit. J. Nutr.*, *30*, 297 (1973).
76. L. Bertling and I. Tietz, *Fleischwirtschaft, 58*, 621 (1978).
77. R. B. Toma and M. M. Tabekhia, *J. Food Sci.*, *44*, 263 (1979).
78. N. V. Shuster and S. Hayward, *J. Liquid Chromatogr.*, *3*, 269 (1980).
79. B. Hoffman, *Lebensm. Gericht. Chem.*, *34*, 114 (1980).
80. G. R. Skurray, *Food Chem.*, *7*, 77 (1981).
81. K. Kral, *Fresenius. Z. Anal. Chem.*, *314*, 479 (1983).
82. L. A. Carlson, *Clin. Chim. Acta, 13*, 349 (1966).
83. P. Schraepler, E. Schulz, and H. Scholer, *Med. Welt, 28*, 1900 (1977).
84. V. Bonavita, S. A. Narrod, and N. O. Kaplan, *J. Biol. Chem.*, *236*, 936 (1961).
85. S. Chaykin, M. Dagani, L. Johnson, and M. Samli, *J. Biol. Chem.*, *240*, 932 (1965).
86. Y. Z. Mohran, *Nutr. Dieta, 9,* 177 (1967).
87. K. Murashige, D. McDaniel, and S. Chaykin, *Biochim. Biophys. Acta, 118*, 556 (1966).
88. C. Haworth and T. A. Walmsley, *J. Chromatogr.*, *66*, 311 (1972).
89. H. Kala, R. Danz, and H. Moldenhauer, *Pharmazie, 33*, 803 (1978).
90. W. Kruse, W.Kruse, H. Raetzer, C. C. Heuck, P. Oster, B. Schellenberg, and G. Schlierf, *Eur. J. Clin. Pharmacol.*, *16*, 11 (1979).
91. L. Sautebin, L. Galli, M. G. Puglisi, M. G. Ciapponi, and R. Paoletti, *Pharmacol. Res. Commun.*, *13*, 141 (1981).
92. P. G. Pagella and M. Spina, *Farmaco, Ed. Prat.*, *36*, 336 (1981).
93. K. Takikawa, K. Miyazaki, and T. Arita, *J. Chromatogr. Biomed. Appl.*, *233*, 343 (1982).
94. R. W. McKee, Y. A. Kang-Lee, M. Panaqua, and M. E. Swendseid, *J. Chromatogr. Biomed. Appl.*, *230*, 309 (1982).
95. J. X. De Vries, W. Guenthert, and R. Ding, *J. Chromatogr. Biomed. Appl.*, *221*, 161 (1980).

96. K. G. Wahlund, *J. Chromatogr.*, *115*, 411 (1975).
97. J. E. Mrochek, R. L. Jolley, D. S. Young, and W. J. Turner, *Clin. Chem.*, *22*, 1821 (1976).
98. C. Bernofsky, *Anal. Biochem.*, *96*, 189 (1979).
99. C. Bernofsky, *Methods Enzymol.*, *66*, 23 (1980).
100. J. B. Tarr, *Biochem. Med.*, *26*, 330 (1981).
101. R. L. Bieleski and T. A. Turner, *Anal. Biochem.*, *17*, 278 (1966).
102. B. Shaikh, N. J. Pontzer, S. S. Huang, and W. L. Zielinski, *J. Chromatogr. Sci.*, *15*, 215 (1977).
103. L. S. Dietrich, *Amer. J. Clin. Nutr.*, *24*, 800 (1971).
104. R. C. Terry and M. Simon, *J. Chromatogr. Biomed. Appl.*, *232*, 261 (1982).
105. D. J. DeLange and C. P. Joubert, *Amer. J. Clin. Nutr.*, *15*, 169 (1964).
106. A. W. Lis, D. I. McLaughlin, R. K. McLaughlin, E. W. Lis, and E. G. Stubbs, *Clin. Chem.*, *22* , 1528 (1976).
107. B. Barlow, J. L. Sutton, and A. W. Wilkinson, *Clin. Chim. Acta*, *75*, 337 (1977).
108. B. R. Clark, P. A. Murai, P. A. Pomeranz, R. A. Halpern, and R. A. Smith, *Cancer Res.*, *35*, 1727 (1975).
109. J. van Eys, O. Touster, and W. J. Darby, *J. Biol. Chem.*, *217*, 287 (1955).
110. M. L. W. Chang and B. C. Johnson, *J. Biol. Chem.*, *234*, 1817 (1959).
111. M. L. W. Chang and B. C. Johnson, *J. Biol. Chem.*, *236*, 2096 (1961).
112. D. Abclson, *Fed. Proc.*, *21*, 466 (1962).
113. D. Abelson, A. Boyle, C. Depatie, and H. Seligson, *Clin. Chim. Acta*, *8*, 603 (1963).
114. D. Abelson and A. Boyle, *Fed. Proc.*, *22*, 232 (1963).
115. D. Abelson, A. Boyle, and H. Seligson, *J. Biol. Chem.*, *238*, 717 (1963).
116. G. P. Quinn and P. Greengard, *Arch. Biochem. Biophys.*, *115*, 146 (1966).
117. R. L. Felsted and S. Chaykin, *J. Biol. Chem.*, *242*, 1274 (1967).
118. S. Gluecksohn-Waelsch, P. Greengard, G. P. Quinn, and L. S. Teicher, *J. Biol. Chem.*, *242*, 1271 (1967).
119. M. Sparthan and S. Chaykin, *Anal. Biochem.*, *31*, 281 (1969).
120. M. Sparthan and S. Chaykin, *Methods Enzymol.*, *18B*, 46 (1971).
121. S. D. Sastry and G. R. Waller, *Phytochemistry*, *11*, 2241 (1972).
122. Y. C. Lee, R. K. Gholson, and N. Raica, *J. Biol. Chem.*, *244*, 2277 (1969).
123. J. Preiss and P. Handler, *J. Biol. Chem.*, *233*, 488 (1958).
124. H. Holmen, H. Egsgaard, J. Funck, and E. Larsen, *Biomed. Mass Spectrom.*, *8*, 122 (1981).

125. J. S. Sandhu and D. R. Frazer, *Int. J. Vitaminol. Nutr. Res.*, *51*, 139 (1981).
126. B. Shaikh and N. J. Pontzer, *J. Chromatogr.*, *162*, 596 (1979).
127. J. X. De Vries, W. Guenthert, W. Sturn, and R. Ding, in *Koenigsteiner Chromatographie Tage 1980* (U. J. Kaiser and K. H. Franzen, eds.), Waters GmbH, Koenigstein, Germany, 1980, p. 210.
128. E. G. A. Carter, *Am. J. Clin. Nutr.*, *36*, 926 (1982).
129. B. J. Blaauboer and A. J. Paine, *Xenobiotica*, *10*, 655 (1980).
130. J. T. Heard and G. J. Tritz, *J. Liquid Chromatogr.*, *4*, 325 (1981).

8

VITAMIN B_1: THIAMINES

TAKASHI KAWASAKI and HIROSHI SANEMORI

Department of Biochemistry, Hiroshima University School of Medicine, Hiroshima, Japan

1. INTRODUCTION

1.1 Chemical Properties

The chemical structure of thiamine is shown in Fig. 8.1. A pyrimidine moiety (2-methyl-4-amino-5-hydroxymethylpyrimidine) and a thiazole moiety (4-methyl-5-hydroxyethylthiazole) are connected by a methylene group. The double-salt form of thiamine with hydrochloric acid

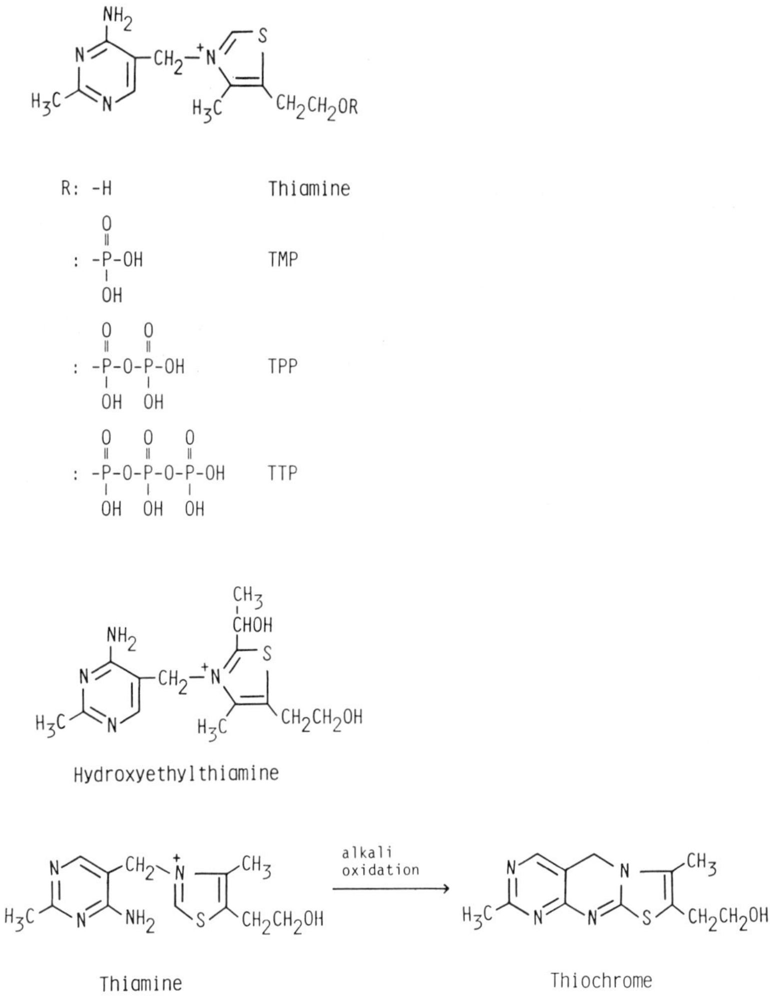

Figure 8.1 Structures of thiamine and its related compounds.

($C_{12}H_{17}N_4OSCl \cdot HCl$; molecular weight, 337.28) is readily soluble in water, less soluble in methanol and glycerol, nearly insoluble in ethanol, and insoluble in ether and benzene [1].

Thiamine in water is most stable between pH 2 and 4 and unstable at alkaline pH and is heat labile with its decomposition dependent on pH and exposure time to heat [2].

Phosphate esters of thiamine are also shown in Fig. 8.1. Thiamine monophosphate (TMP), thiamine pyrophosphate (TPP), and thiamine triphosphate (TTP) are commonly found in organisms. From 80 to 90% of the total thiamine content in cells is TPP, the coenzyme form of thiamine. Thiamine pyrophosphate in dried state is stable for several months when stored at low temperature in the dark. In solution, TPP is unstable and partially decomposes to TMP and/or thiamine when stored for several months at pH 5 and 38°C. However, TPP in solution at pH 2-6 and at 0°C is stable for 6 months [3].

For chemical determination of thiamine, thiochrome (Thc) is the most important compound, and is shown in Fig. 8.1. Thiochrome is quantitatively derived from thiamine by alkaline oxidation with cyanogen bromide [4] or potassium ferricyanide [5-8]. It is a highly fluorescent compound. Thiamine phosphate esters are also quantitatively converted to thiochrome phosphate esters without degradation of phosphate bond: thiochrome mono-, pyro-, and triphosphate are referred to as ThcMP, ThcPP, and ThcTP, respectively.

Thiochrome and its phosphates are fluorescent at pH greater than 8, and all of these have identical excitation maxima at 375 nm and nearly identical fluorescence maxima at 432 to 435 nm [9,10] as shown in Fig. 8.2.

Alkaline solutions of thiochrome and its phosphates above pH 9 are stable at least for 3 days at room temperature. ThcPP and ThcTP in 0.1 N HCl solution are converted quantitatively to ThcMP on heating above 95°C for 10 min (Sanemori and Kawasaki, unpublished observation).

Hydroxyethylthiamine, a derivative of thiamine, is found in cells in the form of hydroxyethyl-TPP [11]. It is converted to the corresponding thiochrome derivative by alkaline potassium ferricyanide oxidation but not by alkaline BrCN oxidation [12]. This difference in the oxidation property of hydroxyethylthiamine is useful in its determination.

1.2 Biochemistry

1.2.1 Absorption

Intestinal thiamine transport has been studied mainly by the everted sac method or tissue accumulation method [13,14]. The results obtained suggest that thaimine is absorbed actively against a concentration gradient at low concentrations, less than 2 μM, and that thiamine

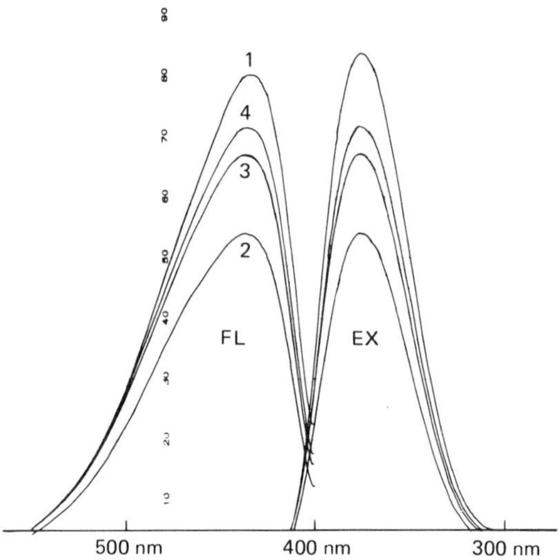

Figure 8.2 Excitation and fluorescence spectra of thiochrome compounds. Thiochrome (Thc) derivatives were prepared from the corresponding thiamine derivatives by alkaline BrCN oxidation: (1) Thc; (2) ThcMP; (3) ThcPP; and (4) ThcTP. (From Ref. 10.)

at concentrations above 2 μM is absorbed by diffusion [15-17]. The relationship of thiamine transport across the brush-border membrane and its phosphorylation to TPP has not been clarified [14]. Although several studies support the idea that there is a close relationship between transport and phosphorylation [18-21], other studies question the role of phosphorylation in transport [16,22). Resolving this relationship using everted intestinal sacs or rings is difficult technically since coupling of thiamine metabolism to the transport process should always be taken into consideration.

To study the mechanism of thiamine entry through the membrane, brush-border membrane vesicles were prepared from guinea pig jejunum. The vesicles contained no thiamine pyrophosphokinase [E.C. 2.7.6.2] but has a Na^+ gradient-dependent amino acid transport system [23]. With these preparations, Hayashi et al. [24] obtained evidence that thiamine is transported by simple diffusion and that a specific carrier system is not involved in thiamine transport in the small intestinal brush-border membrane of the guinea pig. After passing through the brush-border membrane by simple diffusion, thiamine is accumulated as TPP by thiamine pyrophosphokinase in the cytoplasm [16,18,25].

1.2.2 Metabolism

Thiamine is metabolized to TPP by thiamine pyrophosphokinase in animal cells. This enzyme has been purified from liver [26], brain [27-29], and heart [30,31]. It is also present in parsley leaves [32], yeast [33], and *Paracoccus denitrificans* [34,80]. In *Escherichia coli*, thiamine is metabolized to TPP by a two-step reaction catalyzed by thiamine kinase (E.C. 2.7.1.89] [35,36] and TMP kinase (E.C. 2.7.4.10] [36, 37]. Thiamine pyrophosphate is further metabolized to TTP in yeast [38] and rat brain [39,40], but its enzymatic conversion by the brain preparation is disputed by Schrijver et al. [41]. A net increase in the TTP level in rat liver after thiamine administration is shown in Fig. 8.3.

1.2.3 Physiological Functions

Thiamine is required to be converted to TPP before exerting its physiological functions as a coenzyme in cells. Thiamine pyrophosphate functions as a coenzyme for several important enzymes in carbohydrate and amino acid metabolism [42], including pyruvate dehydrogenase [E.C. 1.2.4.1], oxoglutarate dehydrogenase [E.C. 1.2.4.2], transketolase [E.C. 2.2.1.1], branched-chain α-keto acid dehydrogenase [E.C. 1.2.4.4], pyruvate decarboxylase [E.C. 4.1.1.1], and carboxylate carboligase [E.C. 4.1.1.47].

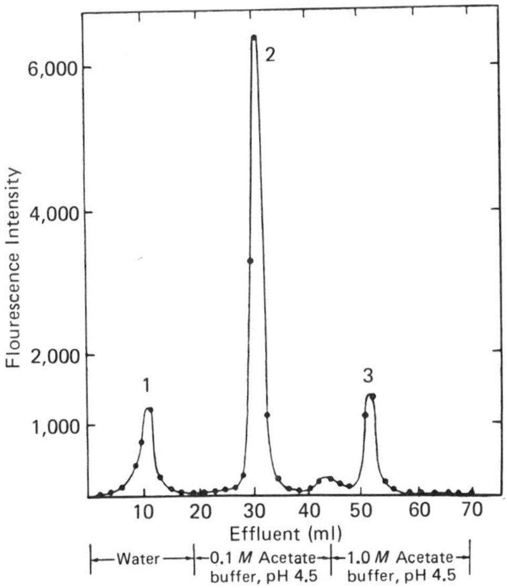

Figure 8.3 Chromatogram of thiamine and its phosphate esters in the rat liver: (1) thiamine and TMP; (2) TPP; and (3) TTP. (From Ref. 54.)

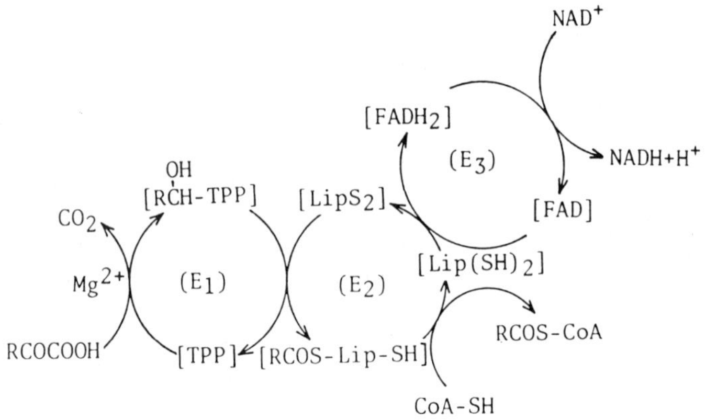

Figure 8.4 The mechanism of oxidative decarboxylation reaction of α-keto acids. E_1, α-keto acid dehydrogenase; E_2, lipoate acyltransferase; E_3, lipoamide dehydrogenase. R: CH_3-, HOOC-CH_2CH_2-; R-CH(OH)-TPP: "active aldehyde"; $LipS_2$: lipoate; $Lip(SH)_2$: dihydrolipoate; RCOS-Lip SH: 6-acyldihydrolipoate; [ ]: coenzymes bound to the enzyme proteins.

In oxidative decarboxylation of pyruvate and oxoglutarate, as shown in Fig. 8.4, TPP functions as a carrier of "active aldehyde" to form the intermediates, hydroxyethyl-TPP, and α-hydroxy-β-carboxypropyl-TPP, and are finally transferred to CoA to form acetyl-CoA and succinyl-CoA, respectively.

Thiamine is known to affect nerve functions as an antipolyneuritis factor. There are many reports on the effects of thiamine on nerve conductance or neurotransmission [39,43-46]; however, the exact mechanism of neurological function is not known.

2. LIQUID CHROMATOGRAPHY

2.1 Introduction

Paper chromatography [47,48] and thin-layer chromatography [49-52] have been described and used for the separation and determination of thiamine compounds. These are not reviewed, however, because of their rather qualitative characteristics.

The separation and assay of thiamine phosphate esters by column chromatography, which was carried out before the use of high-pressure liquid chromatography (HPLC) is briefly reviewed.

2.2 Ion-Exchange Chromatography

A systematic study to separate thiamine phosphate esters by ion-exchange chromatography was carrier out by Siliprandi and Siliprandi [48]. They described conditions for the separation of chemically synthesized TMP, TPP, and TTP in detail, and proposed the use of paper chromatography for qualitative separation and ion-exchange chromatography for quantitative separation of these phosphate esters. However, both fractions of TPP and TTP separated by the ion-exchange chromatography were shown to be highly active on yeast apodecarboxylase [48], which indicated TPP contamination of the TTP fraction. Furthermore, in this method, high concentrations of the solution (millimolar levels) were used. This procedure is therefore not directly applicable to the separation of thiamine phosphate esters in animal and plant tissues that normally have micromolar amounts of the phosphate esters.

Quantitative assay conditions for thiamine and its phosphates in animal tissues were first described by Rindi and de Giuseppe [53]. Thiamine and TMP passed through the column together and were then separately determined after incubation in the presence and absence of phosphatase. Their method consisted of two procedures: removal of impurities in the extracted materials by chromatography on a charcoal column treated with cholesteryl stearate, and subsequent separation of thiamine compounds by anion-exchange column chromatography on Dowex-1 X-8. The recovery of thiamine obtained after treatment on the charcoal column averaged 95%. The contents of thiamine and its phosphate esters in rat tissues are given in Table 8.1.

The analytical method of Rindi and de Giuseppe [53] was modified by Koike et al. [54]. Because they found the conditions described by Rindi and de Giuseppe inconsistent in separating TTP from TPP, they successfully used Dowex-1 X-4, as shown in Fig. 8.3. Thiamine and TMP eluted together in peak 1 of Fig. 8.3. They were then separately determined by the thiochrome method [4] after incubation of the eluate with or without Taka-diastase. The contents of thiamine, TMP, TPP, and TTP in rat liver were 0.75, 0.77, 7.37, and 1.09 μg/g wet weight, respectively, which results are in good agreement with those of Rindi and de Giuseppe [53]. The recovery of thiamine and of its phosphates added separately to tissue extracts was 91.0% and 81.0-83.3%, respectively [54].

The analytical method described by Koike et al. [54], however, is not adequate for the analysis of many biological samples. The charcoal column pretreatment and ion-exchange column chromatography are technically difficult and require a fairly long completion time. The charcoal treatment requires special techniques that can cause problems of reproducibility of results.

A similar separation method, especially for TPP and TTP in liver, was reported by Parkhomenko et al. [55] using a Sephadex cation-exchange column without charcoal pretreatment. Thiamine monophosphate and

TABLE 8.1 Contents of Thiamine and Its Phosphate Esters in Rat Tissues[a]

	Brain (5)		Heart (5)		Kidney (7)		Liver (6)	
Compound	μg/g	%	μg/g	%	μg/g	%	μg/g	%
Thiamine	0.11 ± 0.08	4.4 ± 0.10	0.16 ± 0.01	2.3 ± 0.35	0.22 ± 0.02	6.2 ± 0.45	0.24 ± 0.01	3.5 ± 0.15
TMP	0.30 ± 0.09	11.5 ± 0.25	0.41 ± 0.05	5.9 ± 0.91	0.35 ± 0.06	9.8 ± 1.42	0.66 ± 0.01	9.4 ± 0.43
TDP	2.61 ± 0.15	78.9 ± 1.18	7.44 ± 0.44	86.0 ± 1.06	3.47 ± 0.13	78.6 ± 1.62	6.81 ± 0.44	77.9 ± 1.28
TTP	0.19 ± 0.02	5.0 ± 0.19	0.57 ± 0.04	5.6 ± 0.48	0.27 ± 0.0±	5.2 ± 0.24	0.92 ± 0.13	9.0 ± 1.48

[a]Values are expressed as micrograms of each compound per gram weight of tissue. The values given are means ± SE. The numbers in parentheses are the number of determinations. The percentages show the thiamine content of each fraction as a percentage of the total thiamine of the tissue.

Source: From Ref. 53.

thiamine eluted into the same fraction, although the authors stated that the separation of TMP and thiamine by this method was not their aim. This method, using a Sephadex cation exchanger, also requires 24 hr for completion of the elution of TMP with the recovery of 90-100% of TPP and TTP applied to the column. The minimum amount of thiamine phosphate esters detected by this method was calculated to be about 1 μg (approximately 2 nmol for each ester).

3. HIGH-PRESSURE LIQUID CHROMATOGRAPHY

3.1 Introduction

High-pressure liquid chromatography has developed in recent years for the analysis of drugs, biological materials, and metabolites. The analysis of thiamine with this method has been reported for pharmaceutical preparations [56-58], foods [59-62], urine [63,64], and serum [64-66]. These methods, however, involved analýsis of free thiamine and not thiamine phosphate esters. Since thiamine exists mostly as TPP in foods and serum, samples were first treated with Taka-diastase to hydrolyze TPP to thiamine [59,61,62,64-66].

The analysis of thiamine and its phosphates by LC was reported by Gubler and Hemming [67], Ishii et al. [10,68], Sanemori et al. [69], and Kimura et al. [70].

3.2 Analytical Systems

3.2.1 Stationary Phase

There are many kinds of stationary phase used for analysis by LC. Porous silica particles chemically bounded to a monomolecular layer of octadecyltrichlorosilane, such as μBondapak C_{18}, are most frequently used in reversed-phase LC. Strong cation exchange resin was also used in the early 1970s for high-speed chromatographic separation of thiamine in pharmaceutical preparations [71,72].

3.2.2 Mobile Phase

The mobile phase used in HPLC is dependent on the hydrophobicity of the solid phase. In reversed-phase LC, appropriate mixtures of methanol and water (or buffer solution), with or without ion-pair chromatography (IPC) reagents, are used. For both mobile and solid phases, appropriate chemical compounds must be individually determined.

3.2.3 Detection

The detection of thiamine compounds in the eluate is carried out either spectrophotometrically, usually at 254 nm, or fluorometrically,

In fluorometric detection, two different procedures are used. First, samples containing thiamine compounds are converted to thiochrome compounds and then chromatographed [10,63-65,68,69]. In the second procedure, samples are first directly chromatographed and then thiamine compounds in the eluate are converted to thiochrome compounds using reagents for alkaline oxidation [62,67,70]. Although different postcolumn fluorogenic oxidation systems were described by Gubler and Hemming [67,73] and by Kimura et al. [66,70], the principle is the same. For example, the effluent from the column was mixed with 0.01% potassium ferricyanide in 15% NaOH, which was sent to a mixing coil by a proportioning pump at the rate of 0.3 ml/min. Thiochrome fluorescence in the mixture was then measured fluorometrically [66,70].

The excitation wavelength for the fluorometric detection is usually 375 nm, but the emission wavelength is slightly different: 425 nm [65], 430 nm [10,68,69], 435 nm [64], and 450 nm [66], respectively. In exceptional cases, excitation of 362 nm and emission of 464 nm were used by Van de Weerdhof et al. [62]. A fluorometer equipped with one primary filter (peaks at 360 nm) and two secondary filters (peaks at 436 nm) was used by Gubler and Hemming [67].

The detection limit depends on the method used, and the fluorometric assay is much more sensitive than the spectrophotometric assay. The latter method is suitable for analysis of large quantities of thiamine in pharmaceutical preparations and foods: the detection limit is approximately 30 ng or 90 pmol as thiamine hydrochloride [60]. On the other hand, the minimum detection by the fluorometric method is 0.1 pmol or 33.7 pg as thiamine hydrochloride [69]. Fluorometric detection is more suitable for the analysis of thiamine in biological materials such as cells, serum, and urine. For example, total serum thiamine of 3 ng/ml is the lowest limit of detection by fluorometry [64,65]. Kimura et al. [66], using only 100 μl blood, can analyze the total thiamine content fluorometrically.

The time required for completion of chromatography is usually less than 10 min, and in some cases less than 25 min [56-68].

3.2.4 Precision

In all analytical systems listed above, the recovery of thiamine added to the sample is 100 ± 10%, and the coefficient of variation is 1-5%.

3.2.5 Internal Standard

As internal standard, either sodium salicylate [65] or anthracene [64] has been used. These compounds, which are not related to thiamine in structure but have fluorescence characteristics identical to those of thiochrome, have been used only to calibrate the volume injected into the column. In other studies, thiochrome compounds derived from authentic thiamine and its phosphates are used as calibration standards for the analysis of thiamine phosphates [10,68,69], since the fluorescence intensity of thiochrome phosphates varies considerably (see Fig. 8.2) [10].

3.3 Thiamine

3.3.1 Pharmaceutical Preparations

Reversed-phase HPLC systems were recently reported for the simultaneous assay of thiamine with other water-soluble vitamins in multivitamin preparations. No special treatment of the sample was required before chromatography.

In the system described by Kirchmeier and Upton [56], a μBondapak C_{18} column was used. The mobile phase was composed of 250 ml methanol with 750 ml 5 mM sodium hexane sulfonate-1% acetic acid, and detection was carried out spectrophotometrically at 270 nm. Niacin, niacinamide, pyridoxine, thiamine, and riboflavine were separated within 35 min, and the detection limit of thiamine was 90 ng. A chromatogram for this separation is shown in Fig. 8.5. The precision of this HPLC method for the assay of thiamine as well as the other vitamins in five multivitamin blends, was good. The percentage relative standard deviation was less than 3.0 for thiamine, and the percentage recovery of thiamine added in an amount equal to that already present in the sample was 99%. Comparative analytical data indicated that the average deviation of the HPLC method from the current official analytical procedures was 6.4% for thiamine, and that the HPLC assay tended to give lower analytical values by 0.5% for thiamine.

The other HPLC system described by Walker et al. [57] for the analysis of thiamine in multivitamin preparation was similar to that shown above: niacinamide, thiamine, riboflavine, and pyridoxine were separated by a reversed-phase C_{18} (μBondapak) column with the mobile phase containing 25% methanol-0.1% hexane sulfonate-1% acetic acid. The vitamins in eluates were monitored at 254 nm.

3.3.2 Foods

Thiamine in foods was also analyzed by HPLC. These include rice [61], cereals [60], meat [59], and other by-products. In these systems, thiamine could be determined simultaneously with riboflavine and niacin.

Samples, in either 0.1 N HCl or 0.1 N H_2SO_4, were autoclaved for 30 min and then subjected to enzymatic hydrolysis with Taka-diastase and papain [59,61]. The filtrate was then chromatographed. The system used for rice and rice products [61] was as follows: column, μBondapak C_{18}; detection, ultraviolet (UV) 254 nm; mobile phase, methanol-glacial acetic acid-H_2O-PIC 5-PIC 7 (390:10:600:25:25, v/v). A chromatogram obtained by this HPLC system is shown in Fig. 8.6.

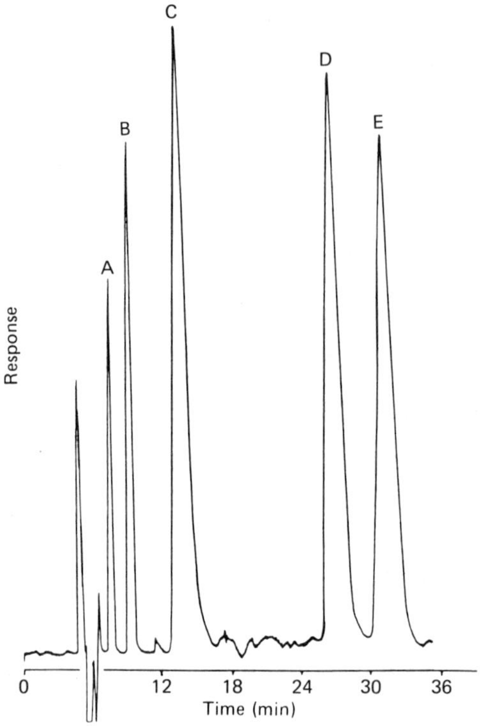

Figure 8.5 Chromatogram of thiamine and other B vitamins in multivitamin preparations: (A) niacin; (B) niacinamide; (C) pyridoxine; (D) thiamine; and (E) riboflavine. (From Ref. 56.)

Recovery data for samples of rice measured by both the HPLC and the American Association of Cereal Chemists (AACC) [74] methods showed no significant variation between the HPLC and AACC methods at levels of both 1 and 5% probability. The detection limit for thiamine was 30 ng.

The HPLC system for cereals [60] is quite similar to that for rice [61], with minor exceptions. The mobile phase contained 12.5% acetonitrile-87.5% 10 mM phosphate buffer (pH 7.0) containing 5 mM heptane sulfonate.

The HPLC system was modified to analyze thiamine in meat and meat products [59]. The sample filtrate was oxidized to thiochrome by the alkaline ferricyanide method and then thiochrome was extracted with isobutyl alcohol. Aliquots of the extract were chromatographed using Spherisorb silica in the column and 90% chloroform-10% methanol for the mobile phase and determined fluorometrically. The average recovery values were 84.4-94.2%, and the lowest detection limit for thiamine was 0.05 ng (0.15 pmol).

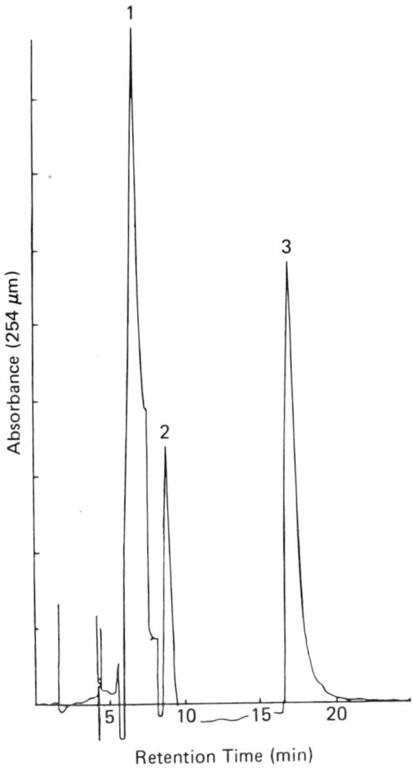

Figure 8.6 Elution profile of B vitamins in rice: (1) niacin; (2) riboflavine; and (3) thiamine. (From Ref. 61.)

3.3.3 Clinical Specimen

The HPLC system is useful for the analysis of thiamine in clinical specimens of urine and blood because it is quick and highly sensitive to small amounts of thiamine. The urinary thiamine assay by HPLC was first reported by Roser et al. [63]. In this assay, urinary thiamine was passed through a Decalso cation-exchange column to remove interfering compounds, eluted with 3.4 M KCl, and then converted to thiochrome by alkaline potassium ferricyanide. The reaction mixture was extracted with isobutanol. The extract was then chromatographed by the system: a silica (LiChrosorb) column; mobile phase, methanol-diethyl ether (22:88, v/v); a fluorometer.

This assay method includes the complicated sample treatments used prior to the development of HPLC [75], in which thiamine in the isobutanol extract is determined directly by a fluorometer. This means that the HPLC step in this assay is not necessary except to save time or to be used in detecting low concentrations of thiamine.

Hisaoka et al. [65] and Yasuda et al. [64] described an HPLC assay method of total thiamine in blood or plasma. They treated the samples in the same manner as Roser et al. [63] with Taka-diastase. The free thiamine thus formed was converted to thiochrome by alkaline BrCN oxidation reaction. The thiochrome extracted into *n*-butyl alcohol was subjected to reversed-phase (μBondapak C_{18}) HPLC with fluorometric detection. The mobile phase was acetonitrile-H_2O (70:30, v/v) [65] and methanol-H_2O (95:5, v/v) [64], respectively. Both methods resulted in good recovery values, 98 ± 3% [65] and 98.6-104.8% [64] with coefficients of variation of 3 and 5.6%, respectively, and gave the identical detection limit of 3 ng thiamine per milliliter blood or plasma.

Kimura et al. [66] described the HPLC method for blood thiamine. Samples were hydrolyzed with Taka-diastase, chromatographed through an adsorbent (Shodex-OH) column with a mobile phase of 0.2 M NaH_2PO_4, and then detected fluorometrically after the automated oxidation reaction of the eluate with alkaline potassium ferricyanide. This method requires only 0.1 ml blood and showed 90-100% recovery of added thiamine and good correlation ($r = 0.89$) of the data to those obtained by the ordinary BrCN assay method [4].

The three methods required Taka-diastase pretreatment of the samples before chromatography, which may be useful clinically. However, it is best to avoid enzymatic hydrolysis of samples and use the analytical system for thiamine phosphates described by Ishii et al. [10,68] or as described in this chapter as the improved HPLC method.

Indeed, a direct measurement of TPP in erythrocyte hemolysates was recently reported by Warnock [76], although a minimum of 2 ml of the packed cells is required for analysis. Thiamine pyrophosphate in the extract of human erythrocytes was converted to ThcPP by alkaline oxidation with BrCN and then chromatographed by a Bondapak-NH_2 column with a mobile phase of methanol-100 mM potassium phosphate buffer (pH 7.5) (50:50, v/v). The effluent was determined fluorometrically (excitation, 365 nm; emission, 460 nm). A mean TPP value of 123 ± 27 ng/ml packed normal human erythrocytes was obtained, which is consistent with those reported previously using the enzymatic method [77].

3.4 Thiamine Phosphate Esters

3.4.1 *Straight-Phase Liquid Chromatography*

The first complete and very sensitive separation of thiamine and thiamine phosphate esters in straight-phase LC was reported by Ishii et al. [10]. The analytical system used a LiChrosorb-NH_2 (250 × 4.6 mm inner diameter, ID; 10 μm) for the stationary phase and acetonitrile (CH_3CN)-90 mM potassium phosphate buffer, pH 8.4 (60:40, v/v) for the mobile phase. Detection was carried out at room temperature fluorometrically (excitation, 375 nm; emission, 430 nm).

Applied samples were thiochrome, ThcMP, ThcPP, and ThcTP, which were converted from thiamine and the corresponding phosphate esters by alkaline BrCN oxidation. This oxidation was carried out by mixing 1.6 ml sample with 0.2 ml 0.3 M BrCN and then 0.2 ml 1 M NaOH. It is also important to prepare a blank by reversing the order of BrCN and NaOH according to the method of Fujiwara and Matsui [4]. Since the intensity of thiochrome fluorescence depends on pH and reaches a steady level at pH above 8 [9,10], the mobile phase should contain a buffer of pH above 8. The detection limit of this system was 1 pmol for each of the thiamine phosphates [10]. However, this sensitivity was insufficient to determine quantities of TTP in rat brain unless 40 μl of the sample were injected [68].

Improved instrumentation to reduce the detection limit is now used in our laboratory. The main change has been the replacement of the fluorometric detector, which improved the detection limit to 0.05 pmol for thiamine and its phosphates. The stationary and mobile phases are identical to those described above, and a spectrofluorometer equipped with a mercury lamp as a light source is used at 363 nm for excitation and 430 nm for emission.

The capacity factor (k') is calculated when concentrations of potassium phosphate buffer (pH 8.4) are varied at a constant ratio (60:40, v/v) of acetonitrile to the buffer. As shown in Fig. 8.7, the k' values obtained between 75 mM and 100 mM buffer concentrations are satisfactory for the separation of thiamine phosphates, and 90 mM buffer is therefore preferential in the standard analytical system. When the ratio of acetonitrile to 75 mM potassium phosphate buffer (pH 8.4) was varied from 10:90 to 60:40, the k' values for each thiamine and its phosphates remained less than 5.0, but the fluorescence intensity of thiochrome compounds increased 2.5-fold. Based on these results, the mobile phase for the straight-phase LC system is composed of acetonitrile-90 mM potassium phosphate buffer, pH 8.4 (60:40, v/v).

A chromatogram for Thc, ThcMP, ThcPP, and ThcTP (1 pmol each) obtained by the improved straight-phase LC, is illustrated in Fig. 8.8. The resolution factors (R_S) calculated from the widths and retention times are 2.4 for Thc-ThcMP, 1.6 for ThcMP-ThcPP, and 2.1 for ThcPP-ThcTP, respectively. These values of R_S show a complete separation of each peak, as seen in Fig. 8.8. The detection limit of thiamine and its phosphates is 0.05 pmol, which is consistent with that in the reversed-phase HPLC [69] equipped with the same fluorometer.

3.4.2 *Reversed-Phase Liquid Chromatography*

A reversed-phase LC has been reported by Sanemori et al. [69], in which ThcTP elutes first, followed by ThcPP and then ThcMP, and finally Thc. Thiamine compounds are converted to thiochrome compounds

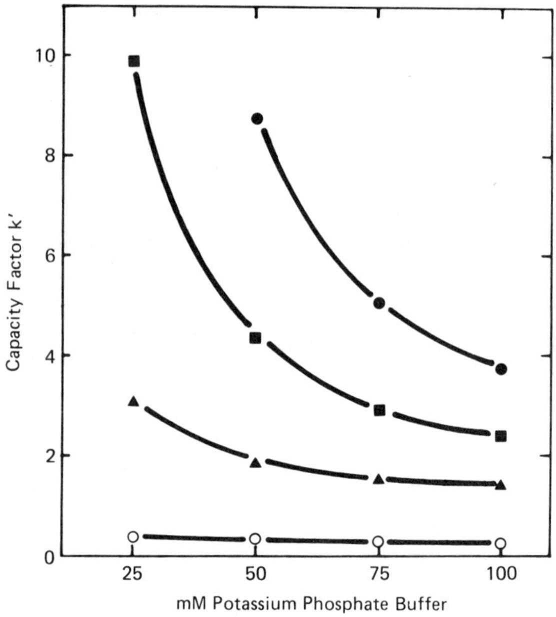

Figure 8.7 Capacity factor (k') for thiochrome and its phosphates separated by the improved straight-phase LC. Concentrations of potassium phosphate buffer (pH 8.4) are varied at a fixed ratio of acetonitrile to the buffer (60:40, v/v): o, thiochrome (Thc); ▲, ThcMP; ■, ThcPP; ●, ThcTP.

by alkaline oxidation prior to separation. In the reversed-phase system, TSK-GEL LS410 (ODS; 300 × 4 mm ID; 10 μm) is used for the stationary phase and 2.5% *N,N*-dimethylformamide (DMF)-25 mM potassium phosphate buffer (pH 8.4) is used in the mobile phase. The detector is identical to that used in the improved straight-phase LC. After completion of ThcMP elution, the mobile phase is changed to 25% DMF-25 mM potassium phosphate buffer (pH 8.4) for elution of Thc.

The capacity factors (k') were calculated for ThcMP, ThcPP, and ThcTP by varying the DMF concentrations in a fixed concentration (25 mM) of potassium phosphate buffer (pH 8.4) or by varying the buffer concentration in a fixed concentration of DMF (2.5%). The k' values ranged from 1.8 to 6.0 in 2.5% DMF-25 mM potassium phosphate buffer (pH 8.4) [69] showing a suitable property of the mobile phase for separation of thiamine phosphates. The resolution factors (R_s) for ThcTP-ThcPP and ThcPP-ThcMP by this system were 1.6 and 2.2, respectively, when calculated from the chromatogram shown in Fig. 8.9. The detection limit of thiamin and its phosphates was 0.05 pmol [69].

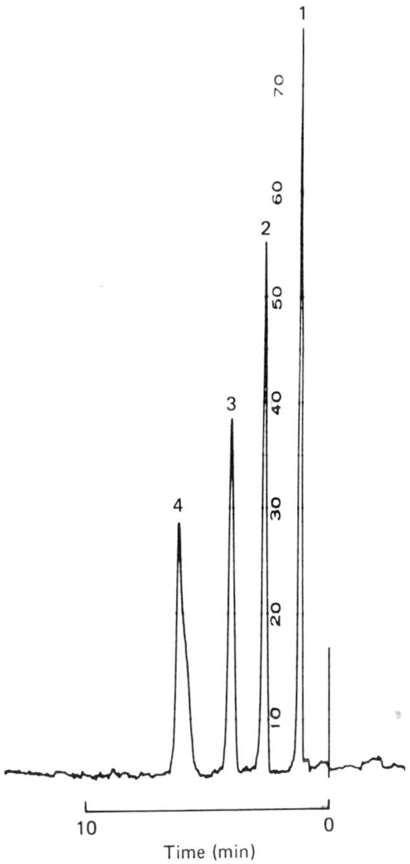

Figure 8.8 Chromatogram of thiochrome and its phosphate esters analyzed by the improved straight-phase LC. Thiochrome (Thc) and its phosphates were derived from thiamine and its phosphates by alkaline BrCN oxidation: (1) Thc; (2) ThcMP; (3) ThcPP; and (4) ThcTP (1 pmol each).

A reversed-phase LC of thiamine phosphates followed by postcolumn fluorogenic oxidation was recently described by Kimura et al. [78], though the application data for biological materials were not included. A μBondapak C_{18} column was used with a mobile phase of a 0.2 M sodium phosphate-phosphoric acid buffer (pH 4.3). The effluent was oxidized by the system described [66,70].

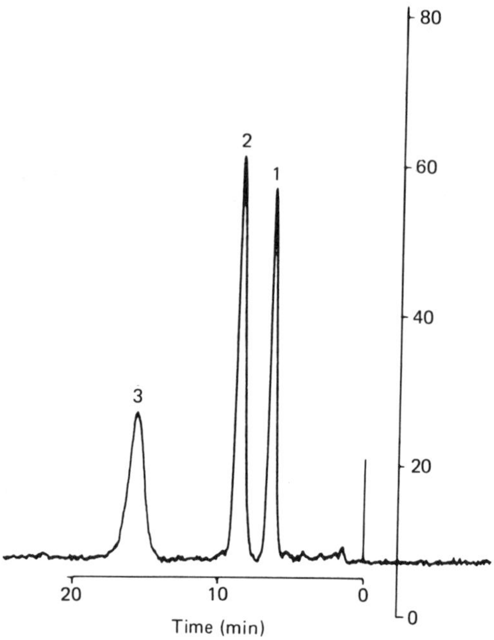

Figure 8.9 Chromatogram of thiochrome and its phosphate esters separated by the reversed-phase LC. Thiochrome (Thc) phosphates were prepared from thiamine phosphates by alkaline BrCN oxidation: (1) ThcTP; (2) ThcPP; and (3) ThcMP. (From Ref. 69.)

3.5 Applications

3.5.1 Tissue Contents of Thiamine and Its Phosphates

To determine thiamine and its phosphates at low concentrations in cells and tissues, pretreatment with charcoal is necessary to concentrate the compounds and remove impurities that interfere with the analysis [53,54]. The method to detect thiamine in food and serum involves enzymatic hydrolysis of thiamine phosphates to thiamine before separation using LC. The straight-phase and reversed-phase LC systems are preferred over charcoal or enzymatic treatments because they can analyze tissue thiamine compounds simply by conversion of thiamine to thiochrome before chromatography. An example of the procedures to determine the tissue content of thiamine and its phosphates is described.

Male Wistar rats weighing 250-300 g were sacrificed by decapitation, and the liver and brain were immediately removed. Approximately 200 mg of the tissue was weighed and homogenized in 5 volumes of 10% trichloroacetic acid (TCA) by use of Polytron 20 ST at 4°C. After centrifugation for 15 min at 16,000 × *g*, the supernatant obtained was extracted at least three times with the same volume of ether to remove TCA. The water layer obtained, 0.4 ml, was mixed with 50 μl 0.3 M BrCN and then with 50 μl 1 M NaOH. In blank experiments, 50 μl 1 M NaOH were first added to 0.4 ml of the water layer and mixed, followed by the addition of 50 μl 0.3 M BrCN. An aliquot of both the oxidized sample and the blank was injected onto the column and chromatographed by the improved straight-phase LC.

Chromatograms of thiamine and its phosphates analyzed by the improved straight-phase LC are shown in Fig. 8.10 (liver) and 8.11 (brain). Good separation of peaks 1, 2, 3, and 4, representing thiamine, TMP, TPP, and TTP, respectively, is evident. Although unidentified peaks can be seen in the sample, no peaks appear in the region of TTP. Blank sample peaks are subtracted from test sample peaks as

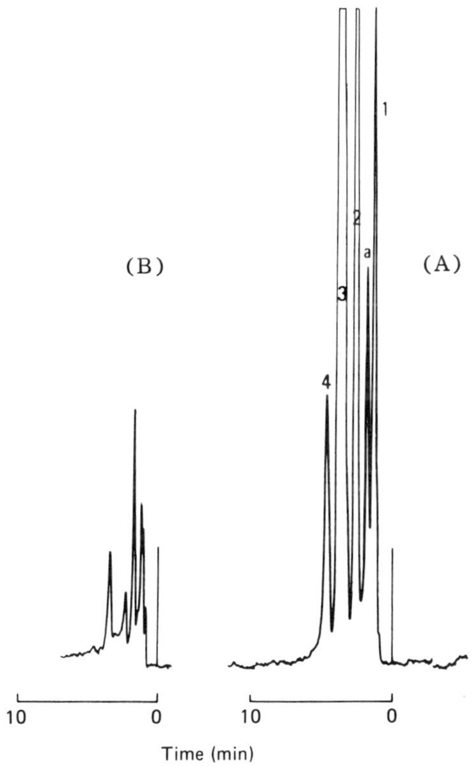

Figure 8.10 Elution pattern of thiamine and its phosphates in the rat liver analyzed by the improved straight-phase LC: (A) sample; (B) blank; (1) thiamine; (2) TMP; (3) TPP; (4) TTP; and (a) unidentified peak.

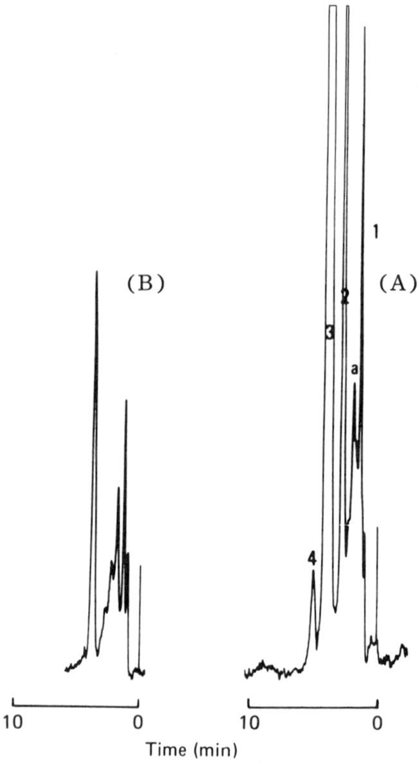

Figure 8.11 Elution profile of thiamine and its phosphates in the rat brain analyzed by the improved straight-phase LC: (A) sample; and (B) blank. Symbols are same as in Fig. 8.10.

described by Fujiwara and Matsui [4]. Cellular concentrations of thiamine and its phosphates can then be calculated (Table 8.2).

Chromatograms of TTP, TPP, and TMP in rat liver and brain analyzed by the reversed-phase LC are shown in Fig. 8.12 and 8.13, respectively. Peaks 1, 2, and 3 correspond to TTP, TPP, and TMP, respectively. Although not shown, thiamine was eluted by changing the mobile phase to 25% DMF-25 mM potassium phosphate buffer (pH 8.4). This system allows detection of a sharp peak of TTP without disturbing the peak in the blank. When 0.2 pmol ThcTP was prepared from TTP and added to the oxidized sample and then chromatographed, the ThcTP peak became higher, with the recovery of 103% (Fig. 8.12). Cellular concentrations of TTP, TPP, and TMP found by reversed-phase LC are shown in Table 8.3.

TABLE 8.2 Contents of Thiamine and Its Phosphate Esters in Rat Tissues[a]

Compound	Liver (nmol/g wet weight)	Brain (nmol/g wet weight)
Thiamine	1.45 ± 0.72 (4.8)	0.34 ± 0.13 (5.3)
TMP	3.98 ± 1.87 (13.2)	1.27 ± 0.60 (19.6)
TPP	24.02 ± 4.49 (79.7)	4.83 ± 0.49 (74.5)
TTP	0.68 ± 0.1± (2.3)	0.04 ± 0.02 (0.6)
Total thiamine	30.13	6.48

[a]Values represent the means ± SD for five rats. The numbers in parentheses refer to percentages of total thiamine.

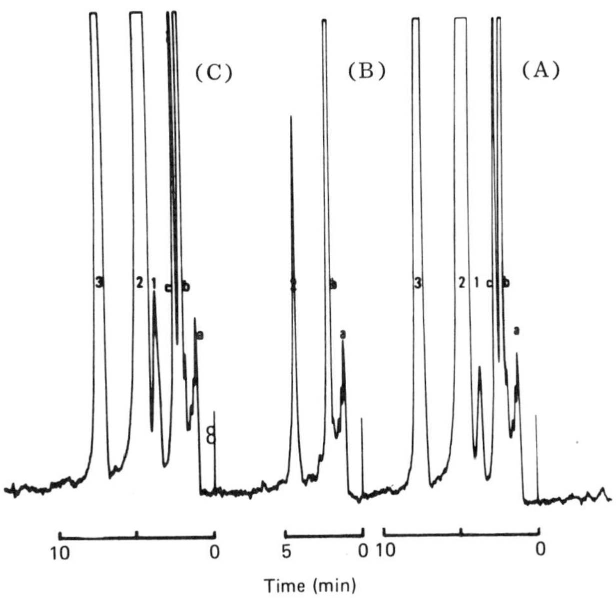

Figure 8.12 Chromatogram of thiamine phosphate esters in the rat liver: (A) sample; (B) blank; (C) sample plus 0.2 pmol ThcTP; (2) TPP; (3) TMP; (a and b) peaks in the blank; and (c) unidentified peak. (From Ref. 10.)

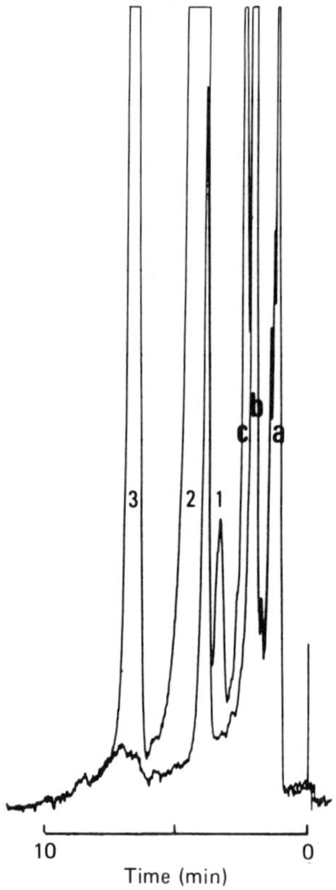

Figure 8.13 Chromatogram of thiamine phosphates in the rat brain analyzed by the reversed-phase HPLC. Chromatogram of rat brain sample and the blank were overlapped on the chart: (1) TTP; (2) TPP; (3) TMP; (a and b) peaks found in the blank; and (c) unidentified peak in the sample. A blank peak is also found in sample peak 2 region.

TABLE 8.3 Contents of Thiamine Phosphate Esters in Rat Tissues[a]

Compound	Liver (nmol/g wet weight)	Brain (nmol/g wet weight)
TMP	3.53 ± 0.12	0.52 ± 0.10
TPP	22.7 ± 0.85	6.45 ± 0.70
TTP	0.38 ± 0.15	0.06 ± 0.02

[a]Values represent the mean ± SD for three rats.

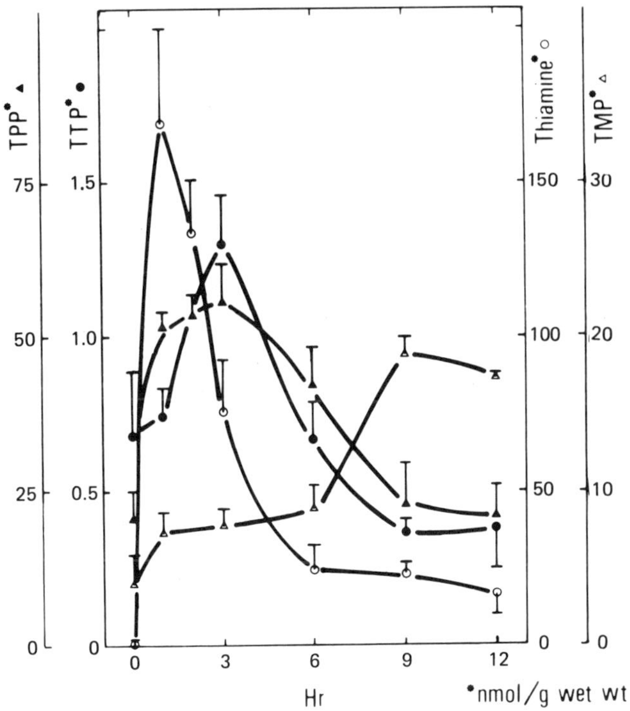

Figure 8.14 Profile of changes in content of thiamine and its phosphates in the liver of rats injected intraperitoneally with thiamine tetrahydrofurfuryldisulfide: ○, thiamine; △, TMP; ▲, TPP; ●, TTP. (From Ref. 79).

The recovery of thiamine and its phosphates added before homogenization was 98 ± 5% in both the analytical methods used, showing accurate determination of these compounds in animal tissues. Because of the variability of fluorescence intensity of thiochrome and its phosphates, as seen in Fig. 8.2, standards consisted of injecting known amounts of thiamine and thiamine phosphates, which were converted to thiochromes, onto the column.

The results of the quantities of thiamine phosphates in rat tissues by LC using either the straight-phase or the reversed-phase LC (Tables 8.2 and 8.3) method show good agreement. Thiamine triphosphate in total thiamine was detected at 0.6% in the brain and 2.3% in the liver, substantially lower than the results reported by Rindi and de Giuseppe [53], as shown in Table 8.1.

3.5.2 Thiamine Metabolism in Animal Tissues

To study the cell metabolism of thiamine and its phosphates, a disulfide derivative of thiamine, thiamine tetrahydrofurfuryldisulfide (TTFD) hydrochloride, was injected intraperitoneally into rats at 25 mg/kg body weight. The animals were sacrificed at various times after injection, and their livers were assayed for changes in thiamine and thiamine phosphate levels. Concentration levels were determined by straight-phase LC. Patterns of changes in the contents of thiamine compounds are shown in Fig. 8.14.

The concentration of thiamine rapidly rose to a peak level 1 hr after injection and then decreased fairly rapidly. This was followed by a rise in TPP levels one-third lower than the maximal thiamine level. Thiamine pyrophosphate remained at the level for 2 hr before gradually decreasing. The concentration of TTP, after an initial lag time, rose to a maximum peak concentration 3 hr after injection. These results indicate a net increase of TTP in vivo, followed by a rapid reduction in concentration [79]. It is clear that the HPLC technique used in this experiment was quite adequate to study thiamine metabolism.

4. CONCLUSIONS

High-pressure liquid chromatography is a useful technique in the quantitative analysis of thiamine and its phosphate esters in pharmaceutical preparations and biological materials. Although it is still in the developmental stage, effort should be made to simplify the LC system. For example, serum thiamine, primarily in the form of TPP, was assayed by HPLC after enzymatic hydrolysis of the sample [64-66]. This time-consuming treatment can be avoided by directly measuring thiamine phosphates with the improved straight LC. Additionally, the LC system should be capable of simultaneously separating thiamine from other vitamins, if possible, in biological materials.

REFERENCES

1. *Merck Index*, 9 th ed. (M. Windholz, ed.), Merck & Co., Rahway, New Jersey, 1976, p. 1199.
2. B. C. P. Jansen, in *The Vitamins*, 2nd ed., Vol. 5 (W. H. Sebrell, Jr. and R. S. Harris, eds.), New York, Academic Press, 1972, p. 99.
3. S. Yurugi, in *Experimental Procedures in Biochemistry*, Vol. 13, *Vitamins & Coenzymes*, Japan Biochemical Society, Kagakudojin, Tokyo, 1975, p. 60.
4. M. Fujiwara and K. Matsui, *Anal. Chem.*, *25*, 810 (1953).
5. F. Bergel and A. R. Todd, *Chem. Ber.*, *68*, 2257 (1935).
6. F. Bergel and A. R. Todd, *J. Chem. Soc.*, 1601 (1936).

7. F. Bergel and A. R. Todd, *J. Chem. Soc.*, 365 (1937).
8. D. J. Hennesy and L. R. Cerecedo, *J. Am. Chem. Soc.*, *61*, 179 (1939).
9. W. Hamanaka, *Vitamins*, *32*, 576 (1965).
10. K. Ishii, K. Sarai, H. Sanemori, and T. Kawasaki, *Anal. Biochem.*, *97*, 191 (1979).
11. M. Morita, Y. Nishibe, and T. Mineshita, *J. Vitaminol. (Kyoto)*, *14*, 67 (1968).
12. Y. Shibata, *J. Biochem.*, *59*, 76 (1966).
13. G. Rindi and U. Ventura, *Physiol. Rev.*, *52*, 821 (1972).
14. R. C. Rose, *Annu. Rev. Physiol.*, *42*, 157 (1980).
15. A. M. Hoyumpa, H. M. Hiddleton, F. A. Wilson, and S. Schenker, *Gastroenterology*, *68*, 1218 (1975).
16. T. Komai, K. Kawai, and H. Shindo, *J. Nutr. Sci. Vitaminol.*, *20*, 163 (1974).
17. U. Ventura and G. Rindi, *Experientia*, *21*, 645 (1965).
18. V. Basilico, G. Ferrari, G. Rindi, and G. D'Andrea, *Arch. Int. Physiol. Biochim.*, *87*, 981 (1979).
19. G. Rindi and G. Ferrari, *Experientia*, *33*, 211 (1977).
20. G. Rindi and U. Ventura, *Experientia*, *23*, 175 (1967).
21. U. Ventura, G. Ferrari, R. Tagliabue, and G. Rindi, *Life Sci.*, *8*, 699 (1969).
22. T. Komai and H. Shindo, *J. Vitaminol. (Kyoto)*, *18*, 55 (1972).
23. K. Hayashi, S. Yamamoto, K. Ohe, A. Miyoshi, and T. Kawasaki, *Biochim. Biophys. Acta*, *601*, 654 (1980).
24. K. Hayashi, S. Yoshida, and T. Kawasaki, *Biochim. Biophys. Acta*, *641*, 106 (1981).
25. G. Cusaro, G. Rindi, and G. Sciorelli, *Int. J. Vitaminol. Nutr. Res.*, *47*, 99 (1977).
26. Y. Mano, *J. Biochem.*, *47*, 283 (1960).
27. L. R. Johnson and C. J. Gubler, *Biochim. Biophys. Acta*, *156*, 85 (1968).
28. J. W. Peterson, C. J. Gubler, and S. A. Kuby, *Biochim. Biophys. Acta*, *397*, 377 (1975).
29. T. Wakabayashi, *Vitamins*, *52*, 223 (1978).
30. M. Hamada, *J. Jpn. Biochem. Soc.*, *41*, 837 (1969).
31. Z. Suzuoki, K. Furuno, K. Murakami, T. Fujita, T. Matsuoka, and K. Takeda, *J. Nutr.*, *94*, 427 (1968).
32. H. Mitsuda, Y. Takii, K. Iwami, and K. Yasumoto, *J. Nutr. Sic. Vitaminol.*, *21*, 189 (1975).
33. Y. Kaziro, *J. Biochem.*, *46*, 1523 (1959).
34. H. Sanemori, Y. Egi, and T. Kawasaki, *J. Bacteriol.*, *126*, 1030 (1976).
35. A. Iwashima, H. Nishino, and Y. Nose, *Biochim. Biophys. Acta*, *258*, 333 (1972).
36. H. Nakayama and R. Hayashi, *J. Bacteriol.*, *112*, 1118 (1972).

37. H. Nishino, A. Iwashima, and Y. Nose, *Biochem. Biophys. Res. Commun.*, *45*, 363 (1971).
38. T. Yusa, *Plant Cell. Physiol.*, *2*, 471 (1961).
39. Y. Itokawa and J. R. Cooper, *Biochim. Biophys. Acta*, *196*, 274 (1970).
40. J. H. Pincus, Y. Itokawa, and J. R. Cooper, *Neurology*, *19*, 841 (1969).
41. J. Schrijver, T. Dias, and F. A. Hommes, *Neurochem. Res.*, *3*, 699 (1977).
42. L. O. Krampitz, *Annu. Rev. Biochem.*, *38*, 213 (1968).
43. R. L. Barchi, in *Thiamine* (C. J. Gubler, M. Fujiwara, and P. M. Dreyfus, eds.), New York, John Wiley and Sons, 1976, p. 283.
44. L. Eder and Y. Dunant, *J. Neurochem.*, *35*, 1278 (1980).
45. H. P. Gurtner, *Helv. Physiol. Pharmacol. Acta (Suppl.)*, *11*, 1 (1961).
46. A. Von Muralt, *Vitamins Horm.*, *5*, 93 (1947).
47. A. Rossi-Fanelli, N. Siliprandi, and P. Fasella, *Science*, *116*, 711 (1952).
48. D. Siliprandi and N. Siliprandi, *Biochim. Biophys. Acta*, *14*, 52 (1954).
49. H. Ahrens and W. Korytnyk, *Anal. Biochem.*, *30*, 412 (1969).
50. S. A. Ismaiel and D. A. Yassa, *Analyst*, *98*, 816 (1973).
51. C. Levorato and L. Cima, *J. Chromatogr.*, *32*, 771 (1968).
52. Z. Z. Ziporin and P. P. Waring, *Methods Enzymol.*, *18A*, 86 (1970).
53. G. Rindi and L. de Giuseppe, *Biochem. J.*, *78*, 602 (1962).
54. H. Koike, T. Wada, and H. Minakami, *J. Biochem.*, *62*, 492 (1967).
55. J. M. Parkhomenko, A. A. Rybina, and A. G. Khalmuradov, *Methods Enzymol.*, *62*, 59 (1979).
56. R. L. Kirchmeier and R. P. Upton, *J. Pharm. Sci.*, *67*, 1444 (1977).
57. M. C. Walker, B. E. Carpenter, and E. L. Cooper, *J. Pharm. Sci.*, *70*, 99 (1981).
58. R. B. H. Wills, C. G. Shaw, and W. R. Day, *J. Chromatogr. Sci.*, *15*, 262 (1977).
59. C. Y. W. Ang and F. A. Moseley, *J. Agr. Food Chem.*, *28*, 483 (1980).
60. J. F. Kamman, T. P. Labuza, and J. J. Wartesen, *J. Food Sci.*, *45*, 1497 (1980).
61. R. B. Toma and M. M. Tabekhia, *J. Food Sci.*, *44*, 263 (1979).
62. T. Van de Weerdhof, M. L. Wiersum, and H. Reissenweber, *J. Chromatogr.*, *83*, 455 (1973).
63. R. L. Roser, A. H. Andrist, Y. H. Harrington, H. K. Naito, and D. Lonsdale, *J. Chromatogr.*, *146*, 43 (1978).
64. K. Yasuda, R. Ikeda, and A. Kawada, *Clin. Pathol.*, *29*, 564 (1981).
65. M. Hisaoka, T. Terao, T. Kojima, and T. Morioka, *Ann. Rep. Sankyo Res. Lab.*, *32*, 98 (1980).

66. E. Kimura, T. Fujita, and Y. Itokawa, *Clin. Chem.*, *28*, 29 (1982).
67. C. J. Gubler and B. C. Hemming, *Methods Enzymol.*, *62*, 63 (1979).
68. K. Ishii, K. Sarai, H. Sanemori, and T. Kawasaki, *J. Nutr. Sci. Vitaminol.*, *25*, 517 (1979).
69. H. Sanemori, H. Ueki, and T. Kawasaki, *Anal. Biochem.*, *107*, 541 (1980).
70. E. Kimura, T. Fujita, S. Nishida, and Y. Itokawa, *J. Chromatogr.*, *188*, 417 (1980).
71. K. Callmer and L. Davies, *Chromatographia*, *7*, 644 (1974).
72. R. C. Williams, D. R. Baker, and J. A. Schmit, *J. Chromatogr. Sci.*, *11*, 618 (1973).
73. B. C. Hemming and C. J. Gubler, *J. Liquid Chromatogr.*, *3*, 1697 (1980).
74. AACC, *Approved Methods of the AACC*, 7th ed., American Association of Cereal Chemists, St. Paul, Minnesota, 1962.
75. A. Fujita, *Adv. Enzymol.*, *15*, 389 (1954).
76. L. G. Warnock, *Anal. Biochem.*, *126*, 394 (1982).
77. L. G. Warnock, C. R. Prudhomme, and C. Wagner, *J. Nutr.*, *108*, 421 (1978).
78. E. Kimura, B. Panijpan, and Y. Itokawa, *J. Chromatogr.*, *245*, 141 (1982).
79. H. Sanemori and T. Kawasaki, *Experientia*, *38*, 1044 (1982).
80. H. Sanemori and T. Kawasaki, *J. Biochem.*, *88*, 223 (1980).

9

FLAVINS

MICHAEL A. MARLETTA

Department of Nutrition and Food Science, Massachusetts Institute of Technology, Cambridge, Massachusetts

DAVID R. LIGHT[*]

Department of Chemistry, Massachusetts Institute of Technology, Cambridge, Massachusetts

[*] *Present affiliation:* Biocatalysis Department, Genentech Inc., San Francisco, California

1. INTRODUCTION

Methods to separate and analyze flavins appeal to a reasonably diverse audience, and therefore what we have attempted to do in this chapter is to present a comprehensive literature survey summarizing methods to the present. We shall concentrate on the more recent methods, in particular the advances in liquid chromatography (LC) technology that have been applied to this problem and have met with considerable success. This review will divide the various methods into three general areas: paper and thin-layer chromatography (TLC), liquid chromatography, and affinity chromatography. The chapter will, in our opinion, be useful to biochemists, nutritionists, clinical chemists, and enzymologists.

2. NATURAL FLAVINS

Riboflavin, vitamin B_2 [7,8-dimethyl-10-(1'-D-ribityl)isoalloxazine] structurally is composed of an isoalloxazine ring with a ribityl side chain at the nitrogen at position 10 (Fig. 9.1). The major biological forms of the flavins are riboflavin and derivatives of it, namely, flavin mononucleotide (FMN) and flavin-adenine dinucleotide (FAD). As will be discussed below, there are a number of other natural favvins, some of which appear to represent very specialized functions. The primary function of the flavin moiety in biochemistry is its position as a coenzyme in redox reactions. The flavins occupy a crucial position in redox biochemistry because they have the ability to carry out 1- or 2-electron processes. This, along with their reduction potential, allows them to act as electron shuttles between those coenzymes that are limited to 2-electron chemistry, such as NADH, and those limited to 1-electron chemistry, such as the various cytochromes. Reduced flavins are extremely reactive toward molecular oxygen yielding the oxidized flavin and hydrogen peroxide.

As mentioned above, the most common derivatives of riboflavin are FMN and FAD (Fig. 9.1). However, structural variants of riboflavin exist, including roseoflavin, which does not act as a coenzyme but as

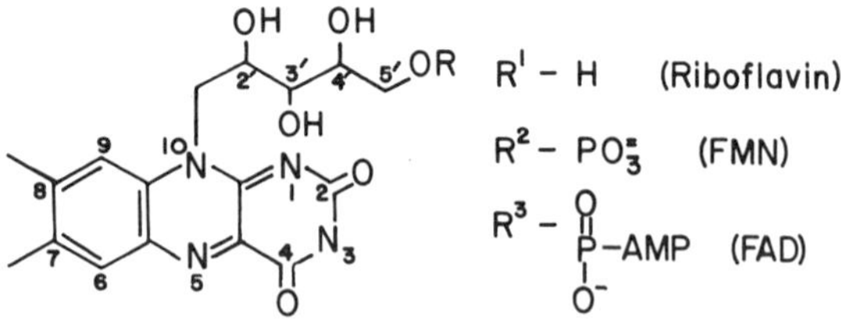

Figure 9.1 Common derivatives of riboflavin.

an antibiotic produced by *Streptomyces davawensis*. In this molecule the 8-methyl group is replaced by a dimethylamino group. Recent investigations with a bacterium that carries out methanogenesis implicates a novel flavin in the reduction of CO_2 to CH_4 [1]. The same group determined the structure, with the major novelty that the molecule is a 5-deazaflavin. The structural alterations provide the proper reduction potential to enable the coenzyme to reduce CO_2.

Flavins are bound to proteins principally by either tight noncovalent association or through a covalent bond. Examples of covalent attachment are attachment via the 8-methyl group with a histidyl residue of the protein backbone, as in succinate dehydrogenase [8α-(N_3-histidyl)-FAD], or with a cystinyl residue again at the 8-methyl group in liver monoamine oxidase (8α-[(*S*)-cysteinyl]FMN).

3. CHEMICAL PROPERTIES

Flavins are easily detected during chromatographic procedures owing to their strong absorption of ultraviolet (UV) light and their intense fluorescense. Neutral solutions of riboflavin are greenish yellow. The absorption spectrum shows maxima at 223, 268, 359-375, 446, and 475 nm, and the absorption in the visible spectrum has been used to quantitatively measure riboflavin [2]. The absorption spectra of FMN and FAD are very similar to that of riboflavin. All three lose their absorbance in the visible region when they are reduced by two electrons. Riboflavin, FMN, and FAD all show fluorescence emission at around 530 nm when excited at 440-500 nm.

The most important chemical property of flavins, which also adds to the problem of analysis, is their sensitivity to light. In fact, riboflavin is relatively heat stable if protected from light. Riboflavin, therefore, is resistant to destruction during cooking [3]; however, milk left in sunlight for 2 hr loses more than 50% of the riboflavin [4, 5]. Flavins are also labile in alkali. Riboflavin in an alkaline solution with exposure to UV light decomposes to the relatively more fluorescent compound lumiflavin, and this phenomenon has been used to enhance the detection in chromatographic procedures [2]. Other chemical properties of flavins are discussed in the sample preparation section where flavin stability is of the utmost importance.

This chapter deals primarily with the naturally occurring flavins and concentrates on summarizing reports concerning TLC and LC methods. A significant problem associated with analytical and preparative chromatography of flavins is sample preparation, and we have devoted a major portion of this chapter to that problem. We have divided the sample preparation section into specific areas of the analysis of vitamins, food, tissue, and the flavin cofactor of purified flavoproteins. This we think covers most of the problems facing researchers in any area of flavin chemistry. Our emphasis on TLC is because it is an easy, effective, and inexpensive method and therefore has found wide applicability. The newest developments, as in most analytical problems, come in the area of LC, where this rapidly developing technology is being quickly applied to many complex mixtures. Significant advances have been made with LC and flavins, and we have provided references to a number of general procedures.

4. SAMPLE PREPARATION FOR FLAVIN DETERMINATION

The success of an analytical procedure, whether it be the unambiguous identification of a labile flavin derivative in a complex biological matrix or the quantitative determination of riboflavin in a relatively well defined vitamin mixture is very much dependent on the method of sample preparation used before analytical chromatographic methods are attempted. No single method of sample preparation will be useful in all cases, and each application will require special procedures depending on the sample characteristics and the analytical information desired. However, certain generalities apply to all cases (for example, precaution must be taken to protect photolabile flavins from exposure to light) and most cases fall into one of several general categories. The remainder of this section will summarize methods of preparing samples from the following sources: multivitamin preparations, urine, food, tissue, and flavin coenzymes attached to purified enzymes (noncovalent and covalent). Of these, the first three are of obvious importance from a commercial, nutritional, and medical standpoint. However, with the increasing interest in the development of the area of biotechnology and enzyme engineering [6], the latter method will play an increasingly important role from the standpoint of research and development and of quality assurance of the final product.

4.1 Analysis of Multivitamin Preparations

Multivitamin preparations contain a variety of components in addition to riboflavin, including iron, salts, sweeteners, artificial colorings, flavorings, and more than 10 other vitamins. In order to assure accurate quantitation the number of steps and transfers in an extraction procedure should be kept to a minimum. In order to avoid photolysis of riboflavin, all procedures must be performed in amber or light-resistant containers. The most widely used and simplest procedure is to dissolve a single dosage or a weighed portion of finely ground tablets in a volumetric container. To dissolve the sample, water or the aqueous portion of the chromatographic system to be used for analysis is commonly used. Opinions vary as to the best method of dissolving the preparations and include ultrasonication [7], shaking 1 hr at room temperature [8], swirling for 15 min at 6°C [9], and heating in a boiling water bath for 1 min [10] or longer [11]. Coated vitamins must be heated for at least 20 min at 70°C or to the melting point of the coating to extract the vitamins from the gelatin matrix [12]. Fat-soluble vitamins may be extracted from the aqueous solution with ether [9]. Internal standards may be added to the aqueous solution [10,12], and filtration or centrifugation may be required before analysis [7,9,12].

4.2 Analysis of Food

Procedures for the extraction of meat and other food products have been described in detail [13]. All procedures should be performed in the absence of direct light. In general, several grams of meat or vegetable are placed in 10-100 ml of acid (up to 0.25 N HCl or H_2SO_4) and autoclaved for 30 min at 120°C. The mixture is cooled and the pH adjusted to 4-4.5 with sodium acetate. After incubating the sample with amylases (Taka-diastase) and papain for several hours the proteins are precipitated with trichloroacetic acid (TCA) and heating the sample is filtered or subjected to centrifugation before analysis [14,15]. The combination of acidification and treatment with crude hydrolytic enzymes guarantees that the FAD and FMN levels are hydrolyzed to riboflavin which then can be analyzed [14-16]. An extraction procedure that utilizes less harsh conditions has been reported and involves the extraction of chromophores including riboflavin from a wide variety of foods with combinations of 2% ammonia, 50-90% ethanol, and chloroform [16]. Ang and Moseley [14] demonstrate that much greater sensitivity can be achieved by photolysis of the riboflavin in the sample to lumiflavin before analysis by illuminating the sample with UV light at high pH. This procedure takes advantage of the enhanced detectibility of lumiflavin in the fluorimeter and allows the detection of as little as 20 pg flavin.

4.3 Analysis of Flavin in Urine

The method used to prepare samples of urine for analysis will depend on the components to be analyzed and on the technique to be used for analysis. As always, care should be taken to protect samples from direct light. If the determination of urinary riboflavin is the only objective of an analysis, then raw urine may be directly injected into the high-pressure liquid chromatograph after a brief low speed (1400 × *g*) centrifugation [17,18]. It is advisable to use an expendable precolumn when this technique is employed [18]. A more complicated procedure is required if analysis is to be performed by TLC, since salts and other components in urine will cause the samples to streak if spotted directly. Treatment with zinc acetate and formalin followed by partial purification on a talc column [19] or fractionation on an R-15 resorcinol resin column [20] have been used successfully to pretreat urine for subsequent TLC analysis. In addition, Tillotson and Karcz [20] demonstrated the presence of conjugated riboflavin in rat urine by comparing fractionated samples with samples that were autoclaved with HCl before fractionation.

4.4 Analysis of Tissue Samples

Depending on the question one wishes to answer, it may be necessary to analyze a tissue sample for flavins that exist in a variety of chemical environments. The extraction procedure required for the analysis of free riboflavin, typically 0.3-3% of total flavin [21,22], will be different from the procedure needed to determine total tissue flavin, including riboflavin, FAD, FMN, a number of riboflavinyl glucosides, and flavin coenzymes bound to many different proteins, both noncovalently and covalently, through one of several different linkages. Fortunately, excellent references exist that may be consulted for stepwise accounts of procedures to be used for even the most intractable tissue. The following general points should be kept in mind.

As always, exposure to direct light during the entire procedure must be avoided. Measurement of the leakage of free flavin by bacterial cells can be accomplished by analysis of the filtrate of cells suspected in water for different time intervals [23]; however, when tissue analysis is desired, methods to break up tissue must be employed. Homogenization is best for both plant [24] and animal [25,26] tissue; the latter may be defatted by extraction with organic solvent [25]. Since FMN and FAD are hydrolyzed to riboflavin in acid, the sum total of all three can be determined by extraction with boiling 0.1 N H_2SO_4 [25]. Enzymatic hydrolysis with phosphatase can be used for the same purpose [25]. It is most often the case that one wishes to determine the levels of the intact coenzymes, and this represents a greater challenge. The only acid treatment that can be considered is extraction with 5 or 10% trichloracetic acid, which must be performed for short times (15-20 min) with ice-cold (0°C) samples and reagents [24,25]. Even with these precautions, Massey and Swoboda [24] find that a 10% TCA extraction of submitochondrial particles from beef heart destroys 4-5% of the FAD. Repeated extractions with hot (80°C) H_2O (10 min each), as described by Yagi [21,22], is effective and avoids hydrolysis of FAD and FMN. Extraction with aqueous phenol may also be used [26]. Flavins may be occluded in precipitated protein when samples with high protein content are studied, or it may be desired to determine flavin covalently bound to peptides (see next section) and trypsin digestion can be used [25]. Finally, the crude extract may be partially purified before analysis of the individual flavin components by chromatography on cellulose [26], ion-exchange [26], gel permeation [26], or phenolic resin [25,27] columns.

4.5 Determination of Flavin in Purified Flavoproteins and Enzymes

The accurate determination of the exact nature of the flavin in a purified flavoenzyme is the analytical procedure that probably requires the most care in order to accurately identify the species present. This is because at least six possible coenzymes may be encountered. Extraction procedures for the noncovalently bound coenzymes, FAD and FMN, will be different from the procedures required for the various types of covalently bound coenzymes. Table 9.1 lists the types of covalently bound coenzymes that have been encountered and their sources. The following two sections will briefly outline procedures that have proven to be successful in the extraction of flavin coenzymes. The goal of this type of analysis is twofold: (1) the determination of the exact chemical nature of the flavin coenzyme present, and (2) the accurate quantitation of the stoichiometry of coenzyme to enzyme in the sample. The destructive nature of light must, therefore, be mentioned once again. Extraction and measurement procedures must be performed in subdued or diffuse light, in light-resistant or foil-wrapped tubes, or even in the presence of a red photographic light source.

4.5.1 Extraction of Noncovalently Bound Flavin Coenzymes

Four methods are used to remove noncovalently bound FAD and FMN from flavoproteins. The general strategy is to denature and precipitate the apoenzyme and to retain the coenzyme in the extraction medium for subsequent analysis. The amount of flavoprotein used for analysis will depend on the amount of purified enzyme that can be spared.

TABLE 9.1 Types of Covalently Bound Flavin Coenzymes

Modified flavin	
8α-(N^3-histidyl)FAD	Succinate dehydrogenase [28]; Choline oxidase [29]; D-6-hydroxynicotine oxidase [30]; sarcosine dehydrogenase [30]
8α-(N^1-histidyl)FAD	Thiamine dehydrogenase [30]; gluonolactone oxidase [31]; cyclopiazonate oxidocyclase [30]
8α-(*S*)-cysteinyl-FAD	Monoamine oxidase [32]
6-(*S*)-cysteinyl-FMN	Trimethylamine dehydrogenase [33,34]
8α-(*O*)-tyrosyl-FAD	*p*-Cresol methylhydroxylase [35]

For a typical 50,000 molecular weight enzyme the coenzyme will represent on the order of 1-2% of the mass, so that 0.01-1 mg protein will yield 0.1-20 μg flavin coenzyme.

4.5.1.1 Extraction by Boiling. The simplest method to remove FAD or FMN is to boil the protein in a milliliter or less of water or buffer for 3-10 min in a foil-wrapped centrifuge tube. Precipitated protein is removed by centrifugation. By way of example, the reader is referred to the following successful applications of the method: electron transfer flavoprotein [36]; glycerol-3-phosphate dehydrogenase [37]; dihydrothymine dehydrogenase [38]; glutamate synthase [39]; adrenodoxin reductase [40]; and DNA photolyase [33].

4.5.1.2 Extraction with Ice-Cold Trichloroacetic Acid. The sample is extracted with 5-10% TCA for 5-20 min and maintained at 0°C in the absence of light. The sample is then centrifuged briefly to pellet the precipitated protein, and the supernatant is immediately neutralized with NaOH or NH_4OH. The use of the shortest extraction time possible to accomplish the removal of flavin is recommended, since a 4-5% loss of FAD has been observed when this procedure is used [24]. Examples of the method are the following: glycerol-3-phosphate dehydrogenase [37]; hydrogenase [41]; glutamate synthase [39]; and submitochondrial particles [24].

4.5.1.3 Extraction with Boiling Methanol. The use of up to 80% ethanol in a boiling water bath for several minutes may be more effective than boiling water alone in some cases. Centrifugation and reextraction of the resultant pellet may be required. This method has been used for D-amino acid oxidase [42,43] and D-amino acid dehydrogenase [44] and has the advantage that the supernatant may be injected directly on an analytical reversed-phase LC column for analysis [45].

4.5.1.4 Phenol Extraction. Osumi et al. [46] report the extraction of FAD from purified rat liver acyl-CoA oxidase with phenol. The coenzyme must be subsequently transferred to water before analysis by TLC.

4.5.2 Extraction of Covalently Bound Flavin Coenzymes

Covalently bound FAD was first identified in succinate dehydrogenase by Green et al. [47] as flavin that is not extractable by acidification or boiling. If the sample of flavoprotein to be analyzed contains free FAD or FMN it can be removed by precipitating the protein with ice-cold 5% TCA. The precipitate is pelleted by centrifugation and the acid extractable supernatant discarded. The pellet is quickly neutralized with NaOH or NH_4OH to preserve the integrity of the flavin [28]. Treatment of flavoproteins that have covalently bound coenzymes with proteases such as trypsin and chymotrypsin often generates a peptidyl flavin. Bacterial trimethylamine dehydrogenase yields a dodecapeptidyl FMN [34], choline oxidase treated with pronase E is converted to a heptapeptidyl FAD [29], and succinate dehydrogenase has been converted to a hexapeptide [28]. The dodecapeptidyl FMN fragment of trimethylamine dehydrogenase can be converted to 6-(*S*)-cysteinyl-FMN by further treatment with aminopeptidase M. Since acid will not hydrolyze the nitrogen carbon bond attaching histidine to the 8α-methyl of FAD, acid hydrolysis can be used to hydrolyze the peptide bonds in the protein to generate flavins at the riboflavin level (FAD is also hydrolyzed) [28]. High-voltage electrophoresis can be used to separate 8α-(N^3-histidyl)riboflavin from 8α-(N^1-histidyl)riboflavin [30].

5. THIN-LAYER AND PAPER CHROMATOGRAPHY OF FLAVINS

A wide array of solvent systems have been used to separate flavins. This reflects the diverse chemical characteristics of flavin derivatives found in nature and synthesized in the laboratory. Table 9.2 represents a partial listing of the solvent systems and adsorbents that have been used and a number of excellent reviews exist that suggest alternate systems [26,27,48]. It is very useful to have a number of solvent systems with different separation characteristics in hand, since comigration of a standard and unknown flavin in several different systems is much stronger evidence of identity than comigration in a single system alone. Furthermore, such systems are essential for the establishment of two-dimensional TLC separations [27,53]. The following general comments are offered to improve the utility of the technique.

1. Butanol:acetic acid:water (BAW) (4:1:5) is the most generally useful solvent and is a good starting point for all applications. It must be freshly prepared before use since butylacetate forms within 24 hr and the R_f values obtained with older solutions will vary.
2. The fluorescent background on silica gel plates increases upon storage. Since this will interfere with the detection of flavins, particularly nonpolar flavins, which run near the solvent front, it is a good procedure to prerun the plates in the same solvent system before using them to separate samples.
3. Often, samples from biological sources contain high concentrations of buffers or salts. Cospotting these salts with the flavins in the sample will impede the migration of the polar flavins, especially FAD. It may be necessary to remove the salts before the analysis is performed, and this can be accomplished by chromatography on a Bio Gel P2 polyacrylamide gel column [56]. It is also possible to spot the flavin standard with an equivalent amount of salt-containing

TABLE 9.2 Representative Systems for TLC and Paper Chromatography of Flavins

Abbreviation	Composition	Ratio (v/v)	Support	Comment
BAW	*n*-Butanol-acetic acid-water	4:1:5	Silica gel; cellulose; paper	Most widely used solvent system for paper chromatography and silica and cellulose TLC [22,24,26,30,33, 42,48-50]
		4:1:5	Silica gel	1. Separation of mono, di, tri, and tetra butryates of riboflavin [51] 2. Separation of riboflavin, 10-hydroxymethylflavinacetate, and lumichrome [52] 3. Separation of numerous photolysis products of riboflavin [53,54]
		4:1:5	Paper	1. Separation of 2',5'-anhydroriboflavine, 8α-(N^3-histidyl)riboflavin, and riboflavin [55] 2. Purification of peptidylriboflavin [24]
		4:1:5	Cellulose	1. Separation of riboflavin, 10-hydroxymethylflavinacetate, and lumichrome [52] 2. Identification of FAD [38]
		4:2:2	Silica and paper	Preparations of peptidyl and aminoacyl riboflavins [24,29]
		4:2:4	Cellulose	Separation of peptidylriboflavins and photolysis products [29,31]
		4:3:3	Cellulose or silica	Useful variant [33]
		12:3:5	Silica	Useful variant [26,56]
		5:2:3	Cellulose	Separation of riboflavin phosphates and acetylated derivatives [57]
Phosphate	5% NaH_2PO_4 in water		Paper	1. Separation of FAD and FMN [47,48] 2. Separation of mono and di methylesters of FMN [58]

			Cellulose	Separation of FAD and FMN [33]
	0.7 M potassium phosphate (pH 6.8)		silica	Separation of FAD and FMN and riboflavin [26,40]
BPAW	*n*-Butanol-pyridine-acetic acid-water	3:3:1:3	Cellulose	Separation of riboflavin, FAD, and FMN [30]
			Paper	Same [48]
		15:10:3:12	Cellulose	Preparation of 8α-(N^3-histidyl)riboflavin [29]
AB	Acetone-benzene	1:1	Silica gel	1. Separation of mono, di, tri, and tetra butyrates of riboflavin [51] 2. Acetone containing solvents give variable R_f values [27]
PAW	Pyridine-acetic acid-water	19:2:79	Silica gel	1. Riboflavin in urine [59] 2. Separation of riboflavine from riboflavinyl glycosides and photolysis products [27] 3. Tailing of FAD and poor separation of riboflavin; FMN well separated [60]
			Talc	Separation of FAD and FMN (streaks) [36]
WMBC	Water-methanol-butanol-chloroform	2:2:8:1.5	Silica gel	[27]
AM	Ammonia-methanol	3:7	Silica gel	Separation of various coenzymes from FMN [23]
BW	Water-saturated *n*-butanol	3:7	Silica gel	Same [23]

TABLE 9.2 (*Continued*)

Abbreviation	Composition	Ratio v/v	Support	Comment
AKMB	Acetic acid-2-butanone-methanol-benzene	5:5:20:70	Silica gel	Separation of photolysis products of riboflavin [27,54]
AAW	*n*-Amyl alcohol-acetic acid-water	3:1:3	Silica gel	Separation of numerous photolysis products of riboflavin [53]
FAW	Formic acid-*t*-amyl alcohol-water	1:3:1	Cellulose	1. Identification of FAD [38] 2. Solvents containing formic acid may give varying R_f values [27]
BMP	*n*_Butanol-methanol-5% $Na_2HPO_4 12H_2O$ in water	65:15:30	Cellulose	Same [38]
	3% Ammonium acetate		Silica gel	1. Identification of FAD [37,39] 2. Ammonia forms hydrate with lumichrome [27]
CMW	Chloroform-methanol-water	18:1:1	Silica gel	Separates riboflavin from 2',5'-anhydroriboflavin [34]
BPW	*n*-Butanol-pyridine-water	1:1:1	Cellulose	Same [34]
CEA	Chloroform-ethylacetate-acetic acid	1:2:1	Silica gel	Recommended for lumichrome [27]

ZAW	Benzyl alcohol-acetic acid-water	3:1:3	Silica gel	Separation of lumichrome and riboflavin [27]
AABM	Acetone-acitic acid-benzene-methanol	1:1:14:4	Cellulose	Separation of water-soluble vitamins from riboflavin [9]
CM	Chloroform-methanol	0.5:9	Silica gel	Separation of flavin phosphates and derivatives [57]
CEW	Chloroform-ethanol-water	50:25:1	Silica gel	Separation of riboflavin from vitamins; FMN and FAD not separated [11]
CME	Chloroform-methanol-ethylacetate	5:5:2	Silica gel	Separation of flavin phosphates and derivatives [57]
TAM	Toluene-acetic acid-methanol	50:4:45	Silica gel	Separation of flavins in urine; FAD and FMN remain at origin; riboflavin separated from 10-hydroxy-ethyflavin, lumiflavin, 10-formylmethylflavin, and lumichrome [19]
BMBA	Benzene-methanol-butanol-acetic acid	70:20:5:5	Silica gel	Same [19]
PB	*n*-Propanol-1% sodium borate	1:1	Paper	Separation of riboflavin from sepiapterin [61]
CoW	Collidine-water	3:1	Cellulose	Separation of riboflavin phosphates and acetylated derivatives [57]

buffer as a control; however, care should be taken to watch the amount of flavin in the unknown and standard since the degree of retardation and streaking is dependent on the amount of FAD applied.

4. Development of TLC plates and paper chromatograms should be performed in the dark or subdued light. A simple precaution is to wrap the tank in aluminum foil.
5. The most sensitive method for visualizing flavins on a plate or filter paper is by their fluorescence upon exposure to UV light. Often a plate with no visible yellow spots is found to have many fluorescent spots under a hand-held mineral lamp. However, not all flavin analogs are fluorescent, and 1-deazaflavin in particular cannot be visualized by this method. Furthermore, photolysis of riboflavin, FAD, or FMN to lumiflavin by UV light may lead to an overestimation of the latter compound if plates are exposed to UV light for prolonged periods.
6. Yagi [22,49,51] reports a sensitive method for detecting flavins on TLC plates or paper chromatograms, which involves extracting the spots with water and converting the riboflavin, FAD, and FMN to lumiflavin in NaOH. The lumiflavin is quantitated by fluorescence. A similar method utilizes the fluorescence of flavins directly [52]. Analogs of flavin can be quantitated by extraction and measurement of their absorbance if their extinction coefficients are known [62]. Chromatograms can be quantitated directly by using an automatic recording and integrating reflectance desitometer [19]. Finally, peptidyl flavins are readily identified since they yield a positive ninhydrin test [30,34].

6. LIQUID CHROMATOGRAPHY OF FLAVINS

A number of LC systems for the separation and analysis of flavins are presented in Table 9.3. Reverse-phase LC has a number of advantages that recommend its use.

1. The method provides rapid separation of vitamins [7,12,63] and a wide variety of flavins and flavin analogs [45]. When a gradient is used, reequilibration of the column is also very rapid (5 min) when compared with ion-exchange chromatography [64].
2. Components in complex vitamin samples may be retained on a reversed-phase column and alter its chromatographic abilities [7]. However, such contaminates are readily removed by washing with 100% methanol. In contrast, analogous contamination of normal-phase columns by aqueous extracts of food resulted in a total column lifetime of a few days to a few weeks [15].
3. For preparative purposes the use of a volatile buffer (ammonium acetate) in aqueous methanol allows rapid removal of the solvent

TABLE 9.3 Comparison of HPLC Systems Used to Separate Flavins

Classification	Column	Mobile phase	Comment
Reverse phase	C18 (silica bonded to octadecyltri-chlorosilane)	1. Methanol-aqueous 0.005 M ammonium acetate (pH 6.0); 20:80, 30:70, 40:60, 50:50	Analytical separation of up to 40 different flavin analogs, including riboflavin, FAD, and FMN [45]
		2. Methanol-aqueous 0.01 M KH_2PO_4 (pH 5.0); 35:65	Direct injection of raw urine with pre-column allows separation of riboflavin from FMN [18]
		3. Linear gradient of 10% methanol up to 70% methanol in aqueous 0.005 M ammonium acetate (pH 6.0)	Preparative separation of many flavins and flavin analogs from ATP, ADP, and AMP; use of volatile salt in buffer simplifies sample recovery [45]
		4. Isocratic, methanol-water (34:66)	Determination of urinary riboflavin; detection by fluorescence [17]
Ion-pair reverse phase	C18 (silica bonded to octadecyltri-chlorosilane)	1. Methanol-aqueous 1% acetc acid with up to 0.008 M hexane sulfonic acid (25:75)	Separation of riboflavin from other water-soluble vitamins (nicotinamide, pyridoxol, and thiamine) [7,12,63]
		2. Methanol-aqueous 0.2% camphorsulfonic acid (pH 7.0), 48:52	Separation of thiamine and riboflavin [63]

TABLE 9.3 (*Continued*)

Classification	Column	Mobile phase	Comment
Normal phase	Silica	1. Chloroform-methanol-5% sodium acetate (pH 4.0), 60:28:4.5	Separation of riboflavin from other multivitamin mixture [10]
		2. 0.1 M Sodium acetate (pH 4.8)	Detection of riboflavin and thiamine in food
		3. Chloroform-methanol (90:10)	Total flavin content of foods converted to lumiflavin; system used to quantitate and separate lumiflavin
Ion exchange	Pellicular anion exchange	Linear gradient between 0.01 M and 0.55 M salt (KH_2PO_4 and KCl), pH 3.45, 70°C	Separation of FAD, FMN, riboflavin, and AMP [69]
	Strong cation exchange	0.01 M NaH_2PO_4 with H_3PO_4	At pH 4.5 nicotinic acid elutes before riboflavin; at pH 2.6 the order of elution reversed

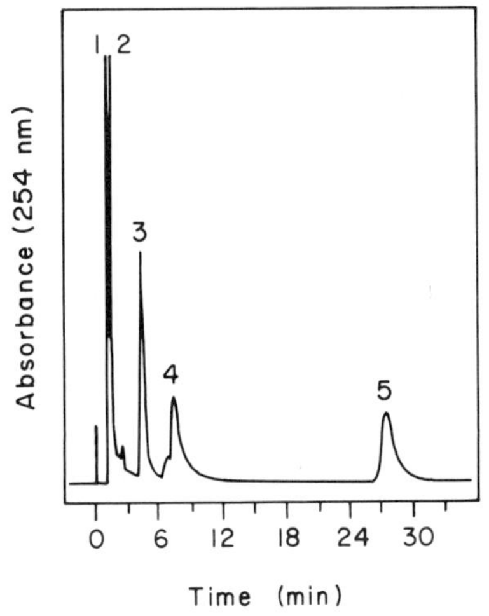

Figure 9.2 Analytical LC of flavins. Sample consisted of (1) ATP and ADP; (2) AMP; (3) FAD; (4) FMN; and (5) riboflavin. ATP, ADP, and AMP were included because ATP is used in the enzymatic synthesis of flavine analogs and all three are found at the end of the reaction. Conditions: μBondapak C18 column (Waters Assoc., Inc.) run isocratically at 2 ml/min with 20% methanol-80% water (5 mM ammonium acetate, pH 6.0).

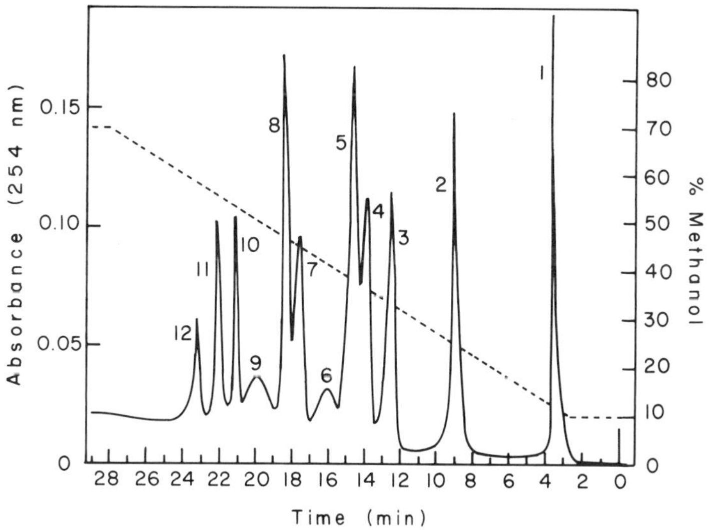

Figure 9.3 Preparative LC of flavins. Sample consisted of (1) ATP; (2) 8-hydroxy-FAD; (3) FAD; (4) 1-deaza-FAD; (5) 5-deaza-FAD; (6) FMN; (7) 1-deaze-FMN; (8) riboflavin; (9) 5-deaza-FAD; (10) 1-deazariboflavin; (11) 6-methylriboflavin; and (12) 5-deazariboflavin. Conditions: LiChrosorb RP-18 preparative column (Altex, 1 × 25 cm), with gradient elution of 10-70% methanol and water (r mM ammonium acetate, pH 6.0) at 4 ml/min.

and buffer salts. As opposed to TLC systems, buffer salts from biological samples do not interfere with reverse-phase LC and are washed out in the void volume.

For riboflavin determination the agreement between the LC method and the USP fluorometric method is excellent. The average deviation of the LC method (percentage relative to the USP method) was 4.5%. The LC assay (percentage relative to the USP analytical result) gave analytical values that were 1% high for riboflavin. However, the deviations per se are not that important when consideration of the interferences associated with fluorometric measurements are taken into account. In fact, the LC assav should be even more valuable [7,12].

Flavins can be detected using a UV wavelength detector (210, 254, and 280 nm) or a visible wavelength detector (450 nm for FAD, FMN, and riboflavin). The latter wavelength eliminates interference by trace UV-absorbing contaminants. Fluorescence detection (excite, 440-450 nm; detect, 520-530 nm) also removes interference from contaminants and is very sensitive (3-30 ng riboflavin easily detected in urine [65]. By first converting riboflavin to lumiflavin (UV irradiation for 30 min at pH 10-12) and analyzing the lumiflavin by normal-phase chromatography, Ang and Moseley [14] can detect as little as 20 pg riboflavin in meat, although this method requires more laborious sample treatment and extractions. A typical separation of some natural flavins and flavin analogs is illustrated in Figs. 9.2 and 9.3.

7. AFFINITY CHROMATOGRAPHY OF FLAVINS

The speed and ease of separation of riboflavin from FMN and FAD and of analogs of riboflavin from their respective analogs of FMN and FAD by reverse-phase LC [45] is expected to replace the preparative DEAE chromatography method of Massey and Swoboda [24] and the use of Bio Gel P2 columns for desalting flavins [56]. However, one chromatographic method deserves mention because of its great utility. Commercial preparations of FMN are a mixture of riboflavine 4'-phosphate (4'-FMN) [20-26] and the natural coenzyme riboflavine 5'-phosphate (5'-FMN, 74-80%) [66]. Pure 5'-FMN and 5'-FMN flavin analogs can be obtained by the enzymatic synthesis of FMN and ATP and riboflavin [56, 67]. However, Meyhew and Strating [66] reported a biospecific chromatographic method of purifying 5'-FMN from contaminating 4'-FMN that utilizes covalently immobilized apoflavodoxin from *Peptostraeptococcus elsdenii*, The FMN mixtures are applied in 0.1 M potassium phosphate, pH 7.4, and 5'-FMN is retained. Elution of 5'-FMN is achieved by using 5% TCA, which denatures the bound flavodoxin and releases the coenzyme. This method has allowed the purification of 5'-FMN and 5'-FAD flavin analogs, which are 97-99.4% pure, from mixtures containing 15-30% 4'-FMN. A similar method for purifying FAD has been developed;

however, since it does not distinguish between different analogs of FAD, it cannot be recommended over reverse-phase LC [68].

ACKNOWLEDGMENTS

We would like to thank Prof. C. Walsh of MIT in whose laboratory we carried out our work on flavin chromatography, and our thanks to Ms. Kimberly Bond for her careful preparation of the manuscript.

REFERENCES

1. D. Eirich, G. Vogels, and R. Wolfe, *Biochemistry, 17*, 4583 (1978).
2. T. Wagner-Jauregg, in *The Vitamins*, 2nd ed., Vol. V (W. H. Sebrell and R. S. Harris, eds.), Academic Press, New York, 1972, pp. 3-42.
3. R. R. Williams and V. H. Cheldelin, *Science, 96*, 22 (1942).
4. W. J. Peterson, F. M. Haig, and A. O. Shaw, *J. Amer. Chem. Soc., 66*, 662 (1944).
5. J. A. Ziegler, *J. Amer. Chem. Soc., 66*, 1039 (1944).
6. A. M. Klibanov, *Science, 219*, 722 (1983).
7. R. L. Kirchmeier and R. P. Upton, *J. Pharm. Sci., 67*, 1444 (1978)
8. S. Ghisla, W. C. Kenney, W. R. Knappe, W. McIntire, and T. P. Singer, *Biochemistry, 19*, 2537 (1980).
9. R. T. Nuttall and B. Bush, *Analyst, 96*, 875 (1971).
10. D. Wittmer and W. G. Haney, Jr., *J. Pharm. Sci., 63*, 588 (1974).
11. S. A. Ismaiel and D. A. Yassa, *Analyst, 98*, 1 (1973).
12. M. C. Walker, B. E. Carpenter, and E. L. Cooper, *J. Pharm. Sci., 70*, 99 (1981).
13. W. Horwitz, in *Official Methods of Analysis of the Association of Official Analytical Chemists*, 11th ed., Association of Official Analytical Chemists, Washington, D.C., 1970, p. 774.
14. C. V. Ang and F. A. Moseley, *J. Agr. Food Chem., 28*, 482 (1980).
15. T. van der Werdhof, M. L. Wiersum, and H. J. Reissenweber, *Chromatography, 83*, 455 (1973).
16. A. Perdih and I. Ekar, *Z. Lebensm. Unters. Forsch., 163*, 100 (1977).
17. V. J. Gatautis and H. K. Naito, *Clin. Chem., 27*, 1672 (1981).
18. M. D. Smith, *J. Chromatogr., 182*, 285 (1980).
19. C. Haworth, R. Oliver, and R. Swaite, *Analyst, 96*, 432 (1971).
20. J. A. Tillotson and M. S. Karcz, Jr., *J. Nutr., 107*, 1269 (1977).
21. K. Yagi, *J. Biochem., 38*, 161 (1951).
22. K. Yagi, *Methods Enzymol., 18B*, 290-296 (1971).

23. R. S. Coles, *J. Gen. Microbiol.*, *86*, 147 (1975).
24. V. Massey and B. E. P. Swoboda, *Biochem. Z.*, *338*, 474 (1963).
25. J. Koziol, *Methods Enzymol.*, *18B*, 253-284 (1971).
26. A. G. Fazekas and K. Kokai, *Methods Enzymol.*, *18B*, 385-398 (1971).
27. G. E. Treadwell, Jr., and D. E. Metzler, *Anal. Biochem.*, *46*, 261 (1972).
28. T. P. Singer, J. Salach, P. Hemmerich, and A. Ehrenberg, *Methods Enzymol.*, *18B*, 416-427 (1971).
29. M. Ohta-Fukuyama, Y. Miyake, S. Emi, and T. Yamano, *J. Biochem.*, *88*, 197 (1980).
30. D. E. Edmondson, W. C. Kenney, and T. P. Singer, *Biochemistry*, *15*, 2937 (1976).
31. H. Nakagawa, A. Asano, and R. Sato, *J. Biochem.*, *77*, 221 (1975).
32. T. P. Singer and D. E. Edmondson, *FEBS Lett.*, *42*, 1 (1974).
33. N. Iwatsuki, C. O. Joe, and H. Werbin, *Biochemistry*, *19*, 1172 (1980).
34. D. J. Steenkamp, W. C. Kenney, and T. P. Singer, *J. Biol. Chem.*, *253*, 2812 (1978).
35. W. McIntire, D. E. Edmondson, T. P. Singer, and D. J. Hopper, *J. Biol. Chem.*, *255*, 5663 (1980).
36. C. L. Hall and H. Kamin, *J. Biol. Chem.*, *250*, 3476 (1975).
37. E. S. Cole, C. A. Lepp, P. D. Holohan, and T. P. Fondy, *J. Biol. Chem.*, *253*, 7952 (1978).
38. T. Shiotani and G. Weber, *J. Biol. Chem.*, *256*, 219 (1981).
39. R. E. Miller and E. R. Stadtman, *J. Biol. Chem.*, *247*, 7407 (1972).
40. R. P. Foster and L. D. Wilson, *Biochemistry*, *14*, 1477 (1975).
41. F. Jacobson, L. Daniels, J. A. Fox, C. Walsh, and W. H. Orme-Johnson, *J. Biol. Chem.*, *257*, 3385 (1982).
42. K. Horiike, Y. Nishina, Y. Miyake, and T. Yamano, *J. Biochem.*, *78*, 57 (1975).
43. D. J. T. Porter, J. G. Voet, and H. J. Bright, *J. Biol. Chem.*, *248*, 4400 (1973).
44. P. J. Olsiewski, G. J. Kaczorowski, and C. Walsh, *J. Biol. Chem.*, *255*, 4487 (1980).
45. D. R. Light, C. Walsh, and M. A. Marletta, *Anal. Biochem.*, *109*. 87 (1980).
46. T. Osumi, T. Hashimoto, and N. Ui, *J. Biochem.*, *87*, 1735 (1980).
47. D. E. Green, S. Mii, and P. M. Kohout, *J. Biol. Chem.*, *217*, 551 (1955).
48. K. Maslowski, *J. Chromatogr.*, *18*, 609 (1965).
49. K. Yagi, *J. Biochem.*, *38*, 161 (1951).
50. T. Yubisui and M. Takeshita, *J. Biol. Chem.*, *255*, 2454 (1980).
51. K. Yagi and N. Oishi, *J. Vitaminol. (Kyoto)*, *17*, 49 (1971).
52. I. M. Hais, J. Cerman, V. Lankašov̌, and M. Skranová, *J. Chromatogr.*, *78*, 105 (1973).

53. G. E. Treadwell, W. L. Cairns, and D. E. Metzler, *J. Chromatogr.*, *35*, 376 (1968).
54. J. Cerman and I. M. Hais, *J. Chromatogr.*, *36*, 393 (1968).
55. D. E. Edmondson, *Biochemistry*, *16*, 4308 (1977).
56. R. Spencer, J. Fisher, and C. Walsh, *Biochemistry*, *15*, 1043 (1976).
57. G. Scola-Nagelschneider and P. Hemmerich, *Eur. J. Biochem.*, *66*, 567 (1976).
58. C. Arsenis and D. B. McCormick, *J. Biol. Chem.*, *241*, 330 (1966).
59. U. Kramer, R. Bitsch, and D. Hotzel, *Klin. Wochenschr.*, *55*, 243 (1977).
60. E. J. Faeder and L. M. Siegel, *Anal. Biochem.*, *53*, 332 (1973).
61. J. D. Taylor and G. J. Proksch, *J. Chromatogr.*, *27*, 509 (1967).
62. C. Walsh, J. Fisher, R. Spencer, D. W. Graham, W. T. Ashton, J. E. Brown, R. D. Brown, and E. F. Rogers, *Biochemistry*, *17*, 1942 (1978).
63. D. P. Wittmer and W. G. Haney, Jr., in *GLC and HPLC Determination of Therapeutic Agents*, Pt. 3 (K. Tsuji, ed.), Marcel Dekker, New York, 1979, p. 1235.
64. H. W. Shmukler, *J. Chromatogr. Sci.*, *10*, 137 (1972).
65. W. Slavin, A. T. Williams, and R. F. Adams, *J. Chromatogr.*, *134*, 121 (1977).
66. S. G. Mayhew and M. J. Strating, *Eur. J. Biochem.*, *59*, 539 (1975).
67. A. H. Merrill, Jr., and D. B. McCormick, *J. Biol. Chem.*, *255*, 1335 (1980).
68. V. Massey and L. D. Mendelsohn, *Anal. Biochem.*, *95*, 156 (1979).

10

VITAMIN B_6 ANALYSIS: SAMPLE PREPARATION, EXTRACTION PROCEDURES, AND CHROMATOGRAPHIC SEPARATIONS

JOSEPH T. VANDERSLICE, STELLA G. BROWNLEE, MARIANN E. CORTISSOZ, and CATHERINE E. MAIRE

Nutrient Composition Laboratory, Beltsville Human Nutrition Research Center, U.S. Department of Agriculture, Beltsville, Maryland

---

Mention of a trademark or proprietary products does not constitute a guarantee or warranty of the product by the U.S. Department of Agriculture and does not imply its approval to the exclusion of other products that may also be suitable.

1. INTRODUCTION

1.1 Principal Metabolic Functions

It is well known that vitamin B_6 participates as a coenzyme in amino acid metabolism through decarboxylation, transamination, deamination, desulfhydration, and racemization [1,2]. In addition, all compounds implicated as neurotransmitters are either synthesized or metabolized with the aid of vitamin B_6-dependent enzymatic reactions. Thus, this vitamin plays a critical role in the functional integrity of the brain and the peripheral and central nervous systems [3]. Vitamin B_6 also acts along with vitamin E to control the metabolism of unsaturated fats and is involved in carbohydrate metabolism as an essential part of the enzyme, glycogen phosphorylase [4]. In fact, it has been estimated that 50% of the body stores (20-30 mg) of the vitamin are found in this enzyme [5].

1.2 Nutritional Considerations

A deficiency of vitamin B_6 leads to lower concentrations of coenzyme A in the blood, reduced absorption of vitamin B_{12}, an increased excretion of vitamin C, and impaired antibody production and thus a decreased immune response to antigens [4-6]. A vitamin B_6-deficient diet first manifests itself by a reduction in body pool size, followed by the lowering of the vitamin and its metabolite concentrations in the blood, urine, and tissues [7]. Lowered activity of vitamin B_6-dependent enzyme systems soon follows, along with early signs of metabolic disturbances and clinical central nervous system changes, together with abnormal electroencephalograms and irritability. Finally, severe metabolic disturbances, anemia, seborrheic dermatitis, convulsive seizures, and hyperacusis are observed [7].

1.3 Need for Additional Data

Studies to determine the average daily intake of this nutrient are hampered by the lack of reliable data on foods as consumed. In addition, information on the bioavailability of the different forms of vitamin B_6, particularly when bound to other molecules, is limited. The best estimates that can be made, however, do indicate that large segments of the average population in the United States are not receiving adequate daily intakes of vitamin B_6 [8]. The situation is further complicated by the fact that the intake of drugs, including alcohol, significantly increases the body's demand for the vitamin. Drugs either couple with the active metabolic form, thus inactivating the coenzyme, or block the metabolic interconversion of the different forms of the vitamin [9,10]. Hydrazide-type drugs are particularly potent inhibitors as they form strong complexes with the aldehyde forms of the vitamin. Hydrazide structures are also present in many foods, and it is possible that these can be released during food preparation and may interfere with vitamin B_6 activity [11]. However, there is only one case in which hydrazides in food have been shown to be a significant problem [12,13].

Environmental hazards are an unknown problem. Hydrazine, a rocket fuel, may cause an added demand for vitamin B_6 in workers constantly exposed to it. A question has been raised as to whether EDTA, a food additive and chelating agent, can interfere with the metabolism of this vitamin through complex formation [10]. Clearly, these questions will only be answered when satisfactory extraction and quantitation procedures are developed for analyses of natural products.

1.4 Necessity of Identifying Various Forms

There are six free forms of vitamin B_6 but only one metabolic product, pyridoxic acid. The biochemistry of this vitamin is quite complex, so for metabolic studies, it would be quite helpful to know the exact forms and amounts of each vitamer present in different tissues and in blood or plasma. In fact, a perusal of the published proceedings of three recent symposia on vitamin B_6 analysis, metabolism, and nutritional requirements indicates a real need for a quick, reliable, and accepted analytical methodology [14-16]. Nutrient composition analyses of foods present their own problems. Although there is some eivdence that the different free forms of the vitamin are equally available to the body [17, 18], there is no certainty that bound forms are equally extracted from food as eaten and transported equally across the intestinal wall [19]. Such bioavailability studies can only be done with a method capable of analyzing all six forms of the vitamin.

1.5 Focus of Chapter

Chemical methods, such as modern gas or liquid chromatography, do have the capability of quantifying different forms of the vitamin. They must, however, meet stringent conditions of extraction. Frequently, they are not as specific as microbiological assays [20], and thus all interfering compounds must be eliminated from real samples during the extraction procedures. Both gas and liquid chromatography have been successfully applied to the partial or complete separation of the different vitamins from a mixture of pure standards. It is rare, however, to see clean chromatographic traces from biological samples. More effort has to be made to develop extraction procedures that are compatible with the analytical method, release bound forms of the vitamers without destruction, and yield 100% recovery of all vitamin forms from biological samples. Methods described in this chapter show that these goals are realistic.

This chapter will focus on sample collection and handling, extraction and sample preparation procedures, separation and quantitation of the vitamin forms, and automation of chromatographic systems, as these are important factors in the quantitation of any component in samples as complex as food or biological materials.

2. CHEMISTRY AND BIOCHEMISTRY FOR THE ANALYST

2.1 Derivatization, Internal Standards, and Molecular Structures

The structures of six free forms that exhibit vitamin B_6 activity (PMP, PM, PNP, PN, PLP, and PL)* are shown in Fig. 10.1 together with their metabolic product, pyridoxic acid (4-PA). Also shown are the structures of two derivatives of these pyridoxine compounds, pyridoxal semicarbazone (PL-SC) and pyridoxic acid lactone, that have been used to increase sensitivity when using a fluorescence detector in liquid chromatography [21-23]. Semicarbazide reacts with the aldehyde forms, PL and PLP, to yield the more fluorescent semicarbazone. Formation of the 4-PA lactone from 4-PA requires tedious chemistry, which makes automation of the system difficult. Finally, structures for two frequently used internal standards, 3-hydroxypyridine (HOP) and 4-deoxypyridoxine (4-DP) are given. Note that these have the pyridine

---

*Vitamin B_6 compounds have been abbreviated, according to published recommendations (IUPAC-IUB Commission on Biochemical Nomenclature, 1970) as follows: PMP, pyridoxamine phosphate; PM, pyridoxamine; PNP, pyridoxine phosphate; PN, pyridoxine; PLP, pyridoxal phosphate; and PL, pyridoxal.

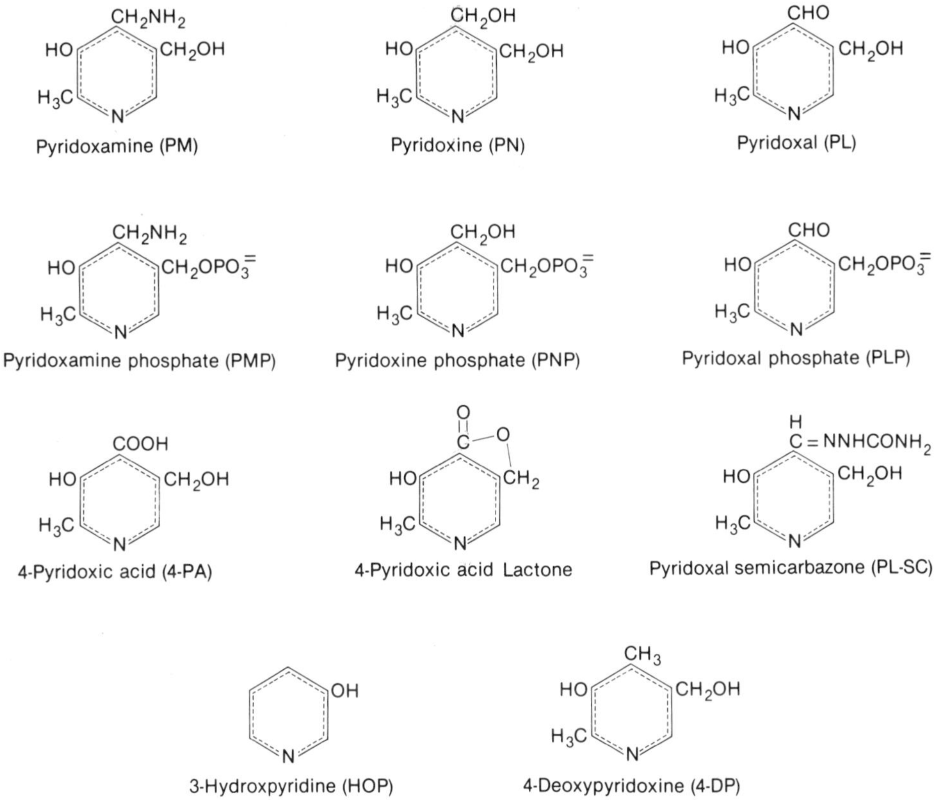

Figure 10.1 Vitamer B_6 and related structures.

ring structures and should behave through extraction, cleanup, and analysis procedures in a manner similar to the free forms of the vitamin. The structures of 4-DP and PL-SC are of interest in another connection. As mentioned earlier, the main antagonists of vitamin B_6 are either analogs, such as 4-DP, which compete with the coenzyme for binding sites on the apoenzymes, or hydrazides, which form hydrazones with PLP similar to the PL-SC structure and inactivate the coenzyme.

2.2 Stability

The vitamers are quite unstable to near-ultraviolet and UV radiation, so either red or yellow light must be used for illumination when handling samples. All the vitamers are relatively stable in acid solution; PL and PLP are unstable in basic solution [24]. In addition, it has been reported that, in solutions containing less than 2 ng/ml of PLP, a 50%

loss in PLP will occur within an hour unless the solution has been previously degassed [25]. It has also been noted that neutral solutions must be kept at 4°C to prevent any loss of PLP over a 24 hr period [25].

Although interconversions between the different forms of vitamin B_6 can be controlled in a solution of pure standards, such is not necessarily the case when dealing with tissues or body fluids. Snell [26] has detailed the problems that the analyst faces when confronting real samples. He notes, for example, that he observed traces of PM and PL when heating PN in bacteriological media. Conversion of PL and PLP to PM and PMP can easily occur through the nonenzymatic transamination reaction:

$$\alpha\text{-Amino acid} + \text{PL} \rightleftarrows \alpha\text{-keto acids} + \text{PM}$$

Transamination is known to occur, for example, in the cooking of meats [27]. Raw meat contains mostly PLP, but after cooking, the principal vitamin form is PMP [28]. Enzymes, such as the kinases, phosphatases, and oxidases, can readily interconvert the various forms if they are not immediately deactivated [29-33]. Experiments with rat liver have shown that these conversions take place within minutes after the liver is excised [34], so rapid laboratory extraction procedures are required to ensure integrity of the sample.

2.3 Binding and Bioavailability

In biological systems, the different forms of the vitamin are often bound to other molecules that cause additional problems for the analyst. The most common bond for PLP is the formation of a Schiff's base with the ε-amino group of a lysine residue [35,36]. PN, on the other hand, forms the glucoside, 5'-*O*'-(β-D-glucopyranosyl), in rice bran (Fig. 10.2) [37]. The latter is available as vitamin B_6 to the rat [38], perhaps because it can be cleaved by β-D-glucosidase. Evidence persists that both the Schiff's bases and the glycosides are easily hydrolyzed by acid extraction with autoclaving. The Schiff's bases appear to be readily hydrolyzed by other deproteinating agents, such as perchloric acid or sulfosalicylic acid, but it has not been ascertained that the same is true for the glucosides.

Snell has observed that PL can form with amino acids the reaction products shown in Fig. 10.3 in biological systems, all of which exhibit reduced or no vitamin activity in animals or humans [26]. It would be interesting to know whether the various extraction procedures currently used release the vitamin from these structures. If so, the content as determined by analysis would be greater than the amount nutritionally available. Food processing can cause a loss of vitamin activity through the formation of reaction products that are unavailable to the

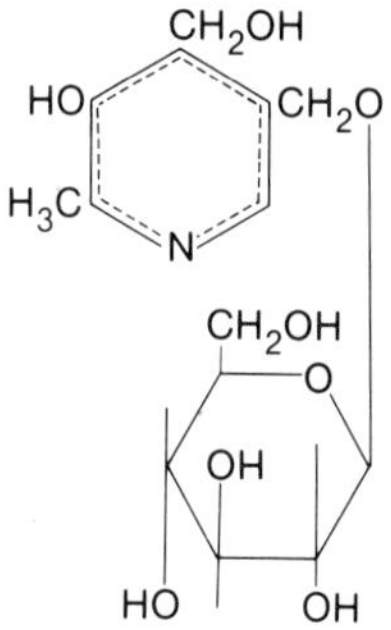

Figure 10.2 The structure of 5'-*O*'-(β-D-glucopyranosyl), a bound form of vitamin B_6 in rice bran. (From Ref. 37.)

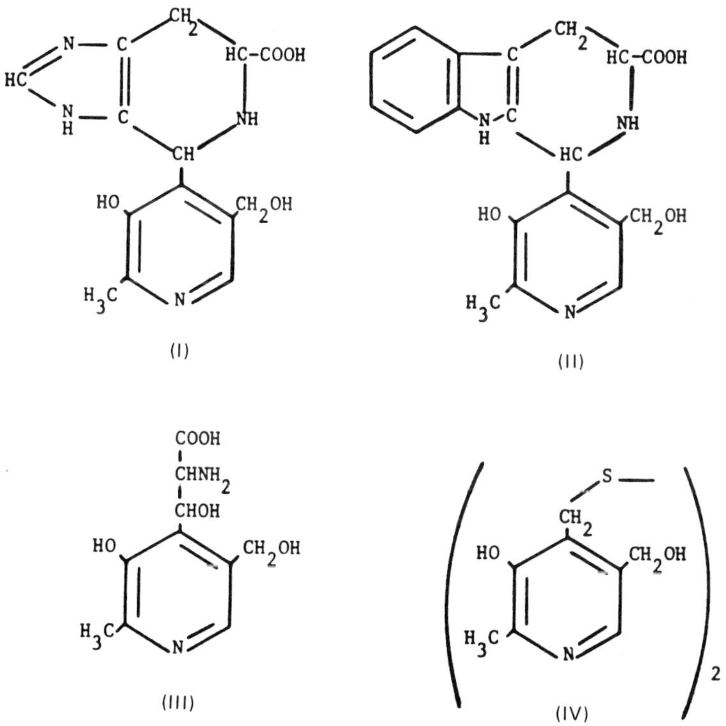

Figure 10.3 Structures of reaction products between PL and (I) histidine, (II) tryptophan, (III) glycine, and (IV) cysteine. (From Ref. 26.)

organism. For example, the cysteine-PL product shown in Fig. 10.3 can be formed in stored, heat-sterilized samples of milk [39-41]. Gregory and Kirk, using a model food system, have shown that ε-pyridoxyllysine is formed from vitamin B_6 during storage [42]. It exhibits reduced vitamin activity and accounts for approximately 50% binding of PLP during storage. This same reaction product is also found in such processed foods as evaporated milk, sardines, canned tuna, and dried beef [43]. Compounds of a similar chemical nature have been synthesized and their properties studied [26,44]. They exhibit partial or complete vitamin activity in rats, in which the amine oxidases permit regeneration of PM or PL. Linseed meal added to a diet for chickens inhibited growth seriously [12]. The effect could be counteracted by the addition of PN to the diet. The responsible antagonist was found to be 1-amino-D-proline, a secondary hydrazine that forms stable hydrazones with PL and PLP [13]. These hydrazones are, however, easily broken up with water treatment and drying. Thus, extraction followed by subsequent analysis would have indicated a diet adequate in vitamin B_6 when, in fact, it was mostly unavailable to the animals. Nelson et al. [45] have shown that the vitamers PN and PL exist in bound states (90%) in frozen orange juice. The responsible compound or compounds have not as yet been identified, but the reported vitamin activity is only 40% of the free forms [46]. Finally, the vitamer forms in tissue, PMP and PLP, are bound to proteins as Schiff's bases. Although strong acid extraction will free these forms, it is not at all certain that, under physiological conditions, all the bound substances will be released [26]. The determination of bioavailability as opposed to vitamin content does present real problems for analysts in that the extraction procedures must mirror the actual extraction by the animal or human digestive system.

2.4 Forms in Biological Samples

In animal and fish tissue, the principal forms of vitamin B_6 found are PLP and PMP, with the former predominating the raw samples. The ratio of PLP to PMP varies from organ to organ. If samples are cooked or processed, transamination occurs, with PMP becoming the dominant form of the vitamin. PN appears to be the principal form in most cereals, dried beans, fresh and dried fruits, and nuts. Eggs and milk contain mostly PLP and PMP; cheese contains mostly PMP. Good food sources of vitamin B_6 usually contain from 1 to 5 μg/g. Some typical values taken from Polansky et al. [47-49] are shown in Table 10.1. The values reported here are taken from composite samples and thus do not show variations from sample to sample. We have found that variations from identical cuts of pork from different carcasses, for example, or from different cans of tuna fish, can be significant. In pork the variation can be as much as a factor of 2 [28].

TABLE 10.1 B_6 Contents (μg/g) of Selected Foods[a]

Food	PN	PL	PM	Total
Wheat, dark hard winter, 1961	2.78	0.57	0.43	3.78
Rice, white, long grain	1.17	0.34	0.27	1.78
Corn, whole kernel yellow, Iowa 1962	0.43	3.14	1.34	4.91
Barley, malting, Western 1961	2.18	0.49	0.68	3.87
Flour, all purpose	0.32	0.14	0.12	0.58
Egg noodles, uncooked	0.38	0.37	0.13	0.88
Bananas, Jamaica	3.40	0.46	1.43	5.29
Oranges, naval (Florida)	0.30	0.16	0.05	0.51
Melon, honeydew (California)	0.17	0.10	0.29	0.56
Raisins, dried, seedless	2.22	0.23	0.09	2.54
Peanuts, roasted (Virginia)	3.11	0.31	1.67	4.49
Walnuts, English (California)	1.97	4.17	0.28	6.42
Lima beans, dry	5.05	0.97	0.70	6.72
Eggs, whole, fresh	0.02	0.44	0.55	1.01
Milk, homogenized, pasteurized	0.01	0.26	0.07	0.34
Cheese, natural Swiss	0.03	0.10	0.41	0.54
Cheese, processed pasteurized American	0.03	0.06	0.58	0.67
Beef, chunk, raw	0.30	1.46	1.46	3.22
Chicken, breast, raw skinned	0.44	5.08	1.31	6.83
Pork, ham, cooked	0.33	0.52	2.34	3.19
Pork, loin, raw	0.58	2.68	0.92	4.18
Tuna fish, canned, water packed	0.05	0.25	3.48	3.78

[a]Note: Values given are for nonphosphorylated and hydrolyzed phosphorylated forms of the vitamin.
Source: From Refs. 47, 48, and 49.

Excised tissues from newly sacrificed animals contain principally PLP and PMP but in different proportions depending on the tissue. Table 10.2 gives typical values.

Plasma presents an interesting problem. It is generally accepted that the principal vitamer present is PLP bound to serum albumin [50]. A recent chromatographic analysis, however, indicated that plasma contained a substantial amount of PN [51]. Values from two different investigations are shown in Table 10.3. This discrepancy will be addressed later in this chapter.

2.5 Summary

With regard to vitamin B_6 analyses of biological materials, the analyst must be cognizant of the stability of the different vitamers during sample handling and extraction. There are indications that the different forms can be interconverted or degraded by enzymatic activity or by nonenzymatic transamination reactions. One must be aware of the types or vitamer binding that can exist in samples. For total vitamer content, it must be ascertained whether the extraction procedures permit quantitation in their naturally occurring forms in the tissue or physiological fluid. In bioavailability studies, bound forms, which exhibit different vitamin activity to an organism than do the free forms, must be either extracted in proportion to their activity or appropriate corrections must be included.

TABLE 10.2 PMP and PLP Values (nmol/g) in Mature Rat Tissue

Tissue	PMP	PLP	PLP/PMP
Liver	20.3	31.5	1.55
Kidney	28.8	11.9	0.41
Brain	9.0	8.0	0.89

Source: From Ref. 34.

TABLE 10.3 Vitamin B_6 Content (pmol/ml) of Human Plasma

Source	PMP	PM	PN	PLP	PL	Total
Vanderslice et al. [51]	25	12	216	77	[a]	329
Lumeng and Li [50]	5	6.2	18.6	76.9	17.4	124.1

[a]Concentration below detection limit.

3. GENERAL CONSIDERATIONS

When collecting and preparing samples for analyses of vitamin B_6 content, a number of factors may affect the final results. These are sample collection, sample uniformity, time and conditions of storage, extraction and cleanup procedures, derivatization, percentage recoveries, and the use of an internal standard. They are discussed in general terms simply to highlight their importance.

3.1 Sample Handling

3.1.1 Sample Collection

In some situations, procurement of a representative sample is clearly defined. Such is the case for plasma samples from individuals. In other situations, obtaining a representative sample is a major research problem in itself. Suppose, for example, one is interested in the average daily intake of vitamin B_6 by individuals in the United States from the consumption of pork loins. The selection of samples for analysis involves a study of hog-feeding diets, major production sources, packing plant procedures, selection of grades of meat, marketing analysis, and consumption patterns. Without a careful statistical analysis of these different factors, an enormous number of samples would have to be analyzed to obtain reliable representative numbers. Clearly, this is a complex undertaking and is beyond the discussion in this chapter. But it should be considered when choosing samples for analysis.

3.1.2 Sample Uniformity

Liquid samples normally do not present a problem; solid samples often do. A cut of pork may weigh 250 g or more from which the analyst might need 1 g. It has been shown from our experience that a Robot-Coupe food processor, or a similar system, gives the desired uniformity for vitamin B_6 analysis. It is fast and convenient and yields homogenous samples.

3.1.3 Sample Storage

After homogenization, some samples can be frozen for several months without appreciable loss of vitamin content. Muscle tissue falls into this category, as does fortified breakfast cereal [9,52]. Human breast milk can be kept overnight without deproteination and still retain its vitamin activity [53]. However, there is a small conversion of PL to PLP. If deproteinated, however, a loss of 50% in PLP is observed in the rethawed sample. Newly excised animal tissue other than muscle must be immediately deproteinated and analyzed as soon as possible (see Ref. 34 for more detail). All new classes of samples should be checked for "best" storage conditions. In some cases, immediate analysis is necessary.

3.2 Extraction and Cleanup Procedures

In favorable situations, vitamin B_6 is found in samples at concentrations of about 1 part per million. An ideal extraction procedure should remove this amount from an organic matrix without loss or destruction and provide extracts for final analysis that are free of interfering compounds.

3.2.1 Autoclaving Versus Nonautoclaving Procedures: Sample Cleanup

The early extraction procedure of autoclaving samples in various concentrations of hydrochloric acid was developed by Rabinowitz and Snell [54] and others [20] and is still being used today. However, it is not normally favored by chromatographers as it does hydrolyze the phosphorylated forms of the vitamin and is a relatively slow procedure. Extraction with trichloroacetic (TCA), perchloric, tungstic, metaphosphoric, or sulfosalicylic (SSA) acids in the 5% (w/v) range is usually preferred [55,56]. Centrifugation and filtration normally follow.

3.2.2 Compatibility with Analytical System

Sometimes, as in anion-exchange chromatography with fluorescent detection and SSA extraction, it is necessary to use a precolumn procedure to remove the acid from the extract, as it fluoresces strongly and eventually elutes from the analytical column if not previously removed [56]. It is important that the method of extraction be compatible with the final analytical determination. We have found that extraction with TCA, for example, leaves undesirable residues on anion-exchange analytical columns, whereas the use of SSA leaves the column free of these residues.

3.2.3 Derivatization Versus Nonderivatization

Because PLP is so unstable, it is desirable in many cases to complete the extraction procedure rapidly and inject onto the final column as quickly as possible. This emphasizes the need for sensitive detectors so that sample derivatization is not required. Derivatization is, of course, essential in gas chromatography, but it has also been used in liquid chromatography to increase sensitivity and specificity [21,22]. In such cases, it is imperative that the operator check for 100% recovery from spiked samples.

3.2.4 Recoveries

Reliable methods consistently give recoveries of 100 ± 10% for all vitamers from spiked samples; otherwise, the extraction and analytical procedures should be viewed with suspicion. There have been reports of recoveries that vary from 50 to 130%, depending on the vitamer and the sample analyzed. The necessary large corrections to obtain "recognized" values for the concentrations of the vitamers are unacceptable in our opinion; the analytical procedures should be further developed and refined to give consistent, accurate recoveries.

3.2.5 *Internal Standards*

The use of an internal standard has certain obvious advantages. If added to the original homogenized sample before any treatment, it automatically corrects for any volume changes during processing and analysis. It corrects for day-to-day variations of system components and operator error. It facilitates the final calculation of vitamer concentrations and is practically essential for a fully automated system.

An internal standard serves another useful purpose. When the protein in a sample is denatured during extraction, as much as 5-10% of the vitamers can be occluded in the precipitate. If the internal standard is added before denaturation and is equally occluded, the loss is automatically accounted for. Otherwise, washing of the precipitate becomes a necessity.

An ideal internal standard has the following characteristics. It must be stable; it must behave identically during the extraction and cleanup procedures to the vitamers; it must elute as a noninterfering peak during the analysis step; and finally, in the case when postcolumn reactions occur, it must react with the reagents in the same manner as the vitamers.

3.3 Analytical Systems

It is desirable that any chromatographic system be rapid, reproducible, sensitive, and capable of automation. Rapidity is not an overriding concern here. The slow step in dealing with real samples is sample collection and preparation along with extraction and cleanup. The critical time is from extraction of sample to injection onto the analytical column. Elution times should not vary from sample to sample, as this makes automation virtually impossible. Reproducibility of concentrations for the entire procedure should be better than 10% and preferably nearer to 5%. It is obvious that detection be as sensitive as required. We prefer to achieve this through instrumentation rather than postcolumn chemistry.

Biological samples are so numerous that automation has become a necessity. The present development of methodology normally requires components from different manufacturers, each of which has its own internal microprocessor. Thus, it is advantageous to have an external microcomputer control that can bypass the different internal logics of the various components and be under operator control. Equipment on the market today makes this external control possible.

4. GAS-LIQUID CHROMATOGRAPHY

Because it normally provides good resolutions and short analysis times and is available in many laboratories, a suitable gas-liquid chromatographic procedure for the analysis of vitamin B_6 in real samples would be valuable. Although the initial investigations appear promising, difficulties in forming volatile derivatives of the phosphorylated forms of the vitamin still persist. Early work was hampered by a lack of sensitivity, but this situation has improved through the use of electron capture and conductivity detectors. Still to be developed are suitable extraction procedures that would provide clean samples for final analysis. There is only one recent report of a gas-liquid chromatographic analysis of a biological sample (blood serum) [57].

4.1 Derivatizations

Volatile derivatives of the nonphosphorylated forms, PM, PN, and PL, have been formed with acetyl [57-61], isopropylidene [58], 3-*O*-benzyl [58], trimethylsilyl [58,62-65], trifluoroacetyl [66], and heptafluorobutyryl reagents [67]. In one case [65], an internal standard, deoxypyridoxine, which should behave properly during extraction, was also derivatized. In general, derivatization occurs rapidly. Care must be taken to keep all derivatives free of water.

There has been one report on a trimethylsilyl derivatization of PLP and subsequent gas chromatographic analysis [68]. However, there are indications that the extraction procedures are not sufficiently developed to permit analysis of natural samples.

4.2 Stationary Phases

The stationary phases employed were either methyl, phenyl, or 2-cyanoethyl silicone, but in one case polyethylene glycol was used [64]. Sheppard and Proser combined neopentylglycol succinate along with methylsilicone to form the stationary phase. A typical gas chromatogram is shown in Fig. 10.4.

4.3 Recoveries

No difficulties are experienced with recovery experiments on PN and PM from standard solutions. However, there were problems with the derivatives of PL unless PL was converted to either a hemiacetal or to a methyloxime before derivatization [58,66]. A few investigators [57, 58,66] reported the derivatization of the metabolic product, 4-PA, but for two of the investigators [58,66], this acid had to be converted to its lactone form before derivatization.

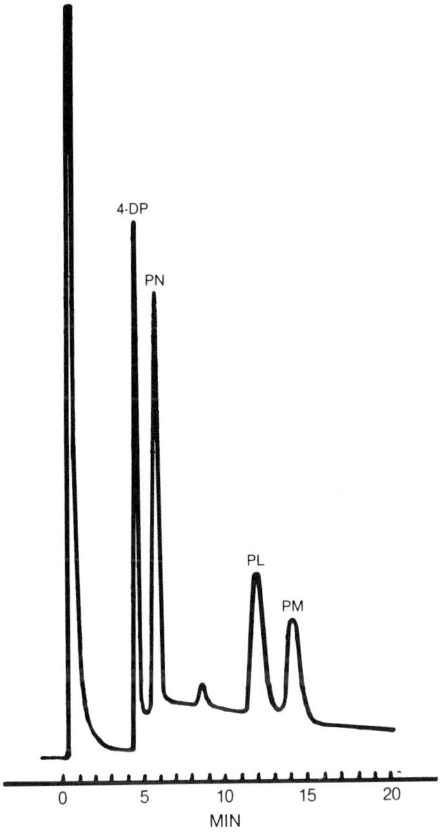

Figure 10.4 Gas chromatogram of vitamin B_6 standard solution. Stationary phase 5% silicone oil DC-550 on Chromosorb P AW DMCS, 80-100 mesh; Injection port, 220°C; column temperature, 150°C; carrier gas, argon at 40 ml/min. (From Ref. 65.)

4.4 Detection

Early workers used flame ionization or β-argon ionization detectors, and their lack of sensitivity restricted gas chromatography to the analysis of vitamin pills or pharmaceutical products. Recently, the use of electron capture or conductivity detectors has so increased the sensitivity that concentrations of the order of 1 ng/ml or less can be detected [66,67].

4.5 A Biological Sample

Hilker and Sotelo [57] have recently described a gas chromatographic analysis of vitamin B_6 in blood serum. The extraction procedure involves deproteination with TCA, removal of TCA by ether washing, a precolumn (Dowex 50W-X8) cleanup of the samples, and hydrolysis of PLP to PL followed by drying. The dried sample is derivatized with acetic anhydride and chromatographed on a Tracor Mt 200 instrument equipped with a Hall conductivity detector. The reported sensitivity is 1-5 ng for PL, PN, PM, and 4-PA. There is no indication that 4-PA had to be lactonized before analysis. No report on recoveries was given, and a chromatographic trace was not presented in the published abstract so that the efficiency of the separations could not be evaluated.

4.6 Summary

Gas-liquid chromatography may soon be able to separate and quantify all forms of the vitamin plus pyridoxic acid. Successful volatile derivatizations of the phosphorylated forms are essential, as well as the introduction of a suitable internal standard. The development of extraction and cleanup procedures is perhaps even more difficult than in liquid chromatography. The recent introduction of SSA as an extraction agent together with the associated cleanup procedure may yield clean samples for gas chromatographic analysis.

5. LIQUID CHROMATOGRAPHY

Recent advances in instrumentation and liquid chromatographic techniques have made it possible to detect all six forms of vitamin B_6 plus 4-PA at plasma level concentrations. Extraction and cleanup procedures have been so refined that relatively clean samples can be obtained from such diverse sources as food, animal tissues, and plasma. The use of highly sensitive fluorescence detectors enables the detection of all forms without recourse to postcolumn chemistry.

The first liquid chromatographic separations of all the vitamers were accomplished with cation-exchange columns [69,70], but attempts to modernize this chromatography have failed to separate all six forms. (The authors are aware of one unpublished cation-exchange separation of all forms of the vitamin but do not know the specific details [71].) Although reversed-phase chromatography with and without ion pairing attempted to separate these vitamers, the only successful separation of all forms has been with ion-exchange chromatography. Indeed, Thompson [72] has attempted normal-phase, reversed-phase, ion-pairing, and ion-exchange separations and concluded that the last is the most durable and practical, as it can withstand bulky samples and will handle hydrolyzed food samples. The authors are inclined to agree with this view, as their anion-exchange columns have routinely survived more than 2 years of continuous analysis of different complex samples.

At this time, it is fair to say that modern liquid chromatographic techniques are on the verge of offering unprecedented opportunities for research in the areas of food composition and metabolic studies.

5.1 Reversed-Phase Chromatography

5.1.1 Ion Pairing

Ion pairing sometimes permits the separation of samples containing ionic species on non-ion-exchange columns. This approach consists of adding to the mobile phase aggressive reagents that complex with the ionic species to form ion pairs that behave as nonpolar molecules and thus can be separated. The first ion-pair reversed-phase separation of the nonphosphorylated forms of the vitamin was reported by Wittmer and Haney [73]. They used a μBondapak C18 column with sodium dodecyl sulfate as the ion-pairing agent and a water-methanol mobile phase. A similar separation was reported by Williams [74], who used sodium-1-heptane sulfonate as the pairing agent and added acetic acid to the water-isopropanol mobile phase. μBondapak C18 was again used for the column packing material. Neither of the above procedures included an internal standard.

Morita and Mizuno [75] reported the separation of PNP, PLP, 4-PA, and an internal standard, 4'-propoxypyridoxine-5'-phosphate (4'-PPNP), by ion pairing with cetyltrimethylammonium bromide as the reagent. A Lichrosorb RP-8 column was used, and a phosphate buffer was incorporated in the methanol-water mobile phase.

All these investigators used UV absorption near 280 nm and were limited by sensitivity constraints. The reported sensitivities were approximately 1 nmol. Wittmer and Haney and Williams reported results only for standard solutions, but Morita and Mizano indicated that 4-PA in urine could be detected with their procedure, although no chromatographic trace was presented. The ion-pair separation of Wittmer and Haney is shown in Fig. 10.5.

5.1.2 Solvophobic Separations with Reversed Phase

In 1976 and 1977, Horvath and his associates [76,77] showed that weak acids and bases could be separated on a nonpolar packing material without the presence of ion-pairing reagents or organic compounds in the mobile phase. They were able to interpret their results on the basis of hydrophobic effects and used the theory of Sinanoglu [78], with Abdulnur [79], to develop theoretical predictions that agreed with the experimental results—at least with octadecylsilica columns. Shortly after this, Gregory and Kirk [42] achieved a separation of the non-

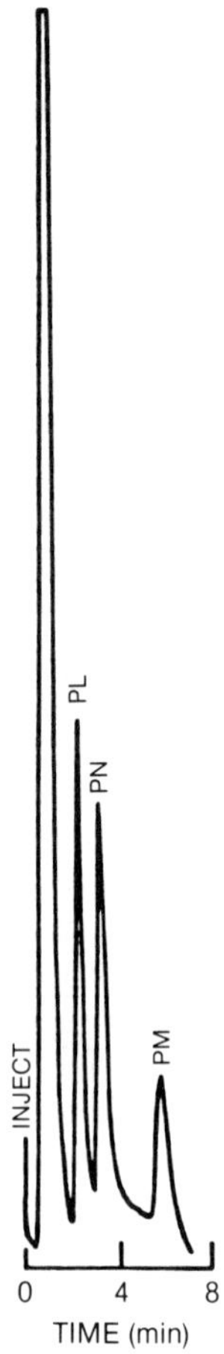

Figure 10.5 Ion-pair reversed-phase separation of vitamin B_6 from a pharmaceutical preparation. See text for conditions. (From Ref. 73.)

phosphorylated forms of the vitamin with a μBondapak C18 column. A phosphate buffer at pH 2.2 was the mobile phase. Fluorescence detection was used so that the sensitivity was adequate for the detection of spiked food samples. Trichloroacetic acid was used to deproteinate the solution, and no interferences were reported from other compounds even though no cleanup procedure was used. No internal standard was used. Gregory [80] used basically the same procedure for the determination of PN in fortified breakfast cereals and compared the results with those from a microbiological assay. The lower limits of sensitivity for this method were not reported, and recoveries ranged from 75% for PM to 95% for PN.

In the same year, Yoshida et al. [81] reported on the separation of PN, PM, PL, 4-PA, and 4-DP, an internal standard, by the use of a Toya Soda Gel LS-160 column and a mobile phase of 0.0045 M $NaClO_4$-0.04 M KH_2PO_4-2 M NaCl (pH 5.5). This packing material is a polyether bonded to a silica gel and can be used in either a size exclusion mode or a reversed-phase mode. It is not clear whether there is some ionic character to the packing material. In this case, the column was used in its reversed-phase mode. The lower limits for fluorescence detection were 0.2 nmol/ml, which is not adequate for normal levels in plasma. Yoshida, et al., however, were able to analyze the blood plasma from rabbits and dogs that were previously fed oral doses of vitamin B_6. The chromatographic traces were remarkably free from background noise and as good as those obtained from standard solutions. It is to be hoped that similar clean traces will be observed as the sensitivity is increased. Trichloroacetic acid was used to deproteinize the plasma after it had been treated with acid phosphatase to hydrolyze the phosphates. Recoveries of spiked samples ranged from 90 to 100% for all vitamers except PLP. Here the recoveries varied from 85 to 90%. It does take 90 min to hydrolyze the phosphates, and it may be that some interconversions or degradation occurred during this time as previously noted in the case of rat liver. Nevertheless, this work is a valuable contribution to this field. Traces are shown for standard solutions and rabbit plasma in Fig. 10.6.

In 1980, Lim et al. [82] used a system very similar to Gregory's to separate PM, PN, and PL using 4-DP as the internal standard. Even though UV detection was used, the lower limits of detection were reported to be about 30 pmol. The method was applied to milk extracts that had been run through a Sep-pak C-18 cleanup column. There was no mention of the use of a deproteinating agent, and no percentage recoveries were reported. The elution times for the milk extract varied from those obtained with pure standards. The internal standard was not used in the milk extract analysis. Since the baseline was not sufficiently clean, more improvement in extraction and cleanup procedures is necessary.

O'Reilly et al. [83] also used a system similar to Gregory's and reported relative retention times for PM, PMP, PL, PLP, PN, and 4-PA, but no trace of the standards was shown. From the traces shown for plasma and urine and the reported retention times, it appears that the phosphorylated forms cannot be separated from the nonphosphorylated. Recoveries were checked only for PN and were reported to be 100% for both urine and plasma. Perchloric acid was used as the deproteinating agent, and the observed vitamer peaks were reported to be PL, PN, and 4-PA. Since UV detection was used, the lowest observable levels of the vitamers were of the order of 2 nmol/ml, and the method was

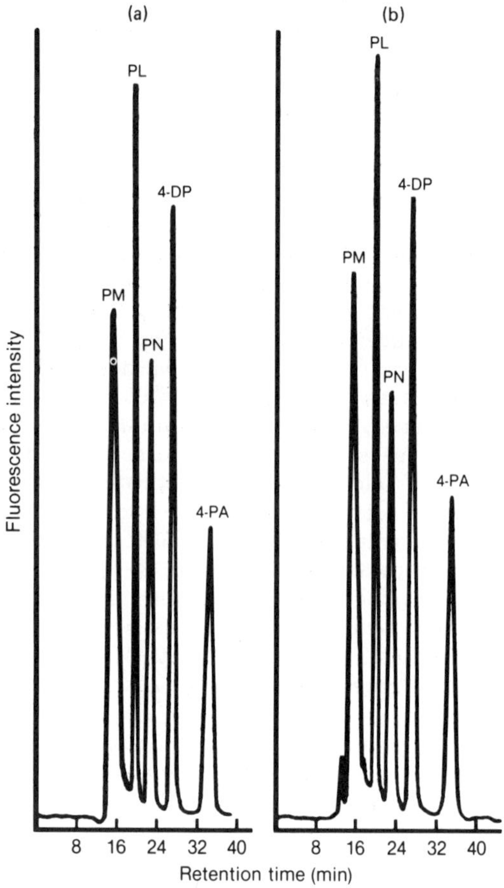

Figure 10.6 Reversed-phase separation of vitamin B_6 from standard solution (a) and from rabbit blood (b). Recovery tests. (From Ref. 81, © 1978 Pharmaceutical Society of Japan.)

used only on patients with high-dose vitamin intake. No internal standard was used, and the separations shown could be improved upon for better accuracy.

5.1.3 Current Problems with Reversed-Phase Analyses

At a recent conference, Gregory and Kirk [84] showed typical separations for the phosphorylated and nonphosphorylated vitamer forms by reversed-phase chromatography. It is clear from their diagram that the six forms have not been separated with reversed phase, although separation of the aldehyde forms was achieved by first forming the semicarbazone derivatives. All these reported methods do not have the required sensitivity for handling normal levels of the vitamer in plasma, but it may be possible to improve sensitivity through the use of fluorometric detectors, such as the Perkin-Elmer 650-40. With reversed-phase separations, there is sometimes a problem of changing elution times from one complex sample to another, and these times differ from those for a mixture of pure standards. Difficulties can arise in automated systems in which the vitamers are identified by elution times and concentrations calculated with the use of an internal standard. We suspect that the problem stems from the ionic strength of extracts from real samples, which may be considerably higher than those of pure standards. Note that Yoshida et al. [81], who used a high ionic strength mobile phase, did not observe any shifts in elution times from standard to complex sample.

Although chromatographic traces from extracts are adequate for the concentrations of vitamers so far investigated, there are indications in many cases that interferences will develop as sensitivity is increased to handle lower levels of vitamer concentrations. It is likely that further cleanup procedures will be required. There has been limited success in using reversed-phase to separate all forms of the vitamer.

5.2 Ion-Exchange Chromatography

5.2.1 Cation Exchange

The first modern chromatographic separation of the nonphosphorylated forms of the vitamin was accomplished with a pellicular cation-exchange column [85]. Also, the first successful semiautomatic analysis of food samples was reported by Yasumoto et al. [86] using a macroporous cation-exchange column. All published cation-exchange separations involve only the nonphosphorylated forms. The authors recently have seen a cation-exchange separation of all forms of the vitamin plus 4-PA from a plasma extract but are not cognizant of any details [71]. Clearly, cation-exchange chromatography has analytical possibilities for biological samples.

The first separation of PN, PM, and PL was accomplished by Williams et al. [85] using a Zipax SCX pellicular cation-exchange column with a pH 4, 0.1 M KH_2PO_4 mobile phase. Ultraviolet detection was used, so the sensitivities are not sufficient for most practical applications other than vitamin tablets. No internal standard was incorporated into the procedure, and no natural products were studied.

Williams and Cole [87] later used an Aminex A-5 cation-exchange column with a formate buffer to separate the three forms by both isocratic and gradient elution. Again, no internal standard was used and the sensitivity was insufficient for biological samples. The investigators did analyze vitamin tablet preparations but had difficulty with interfering peaks for such samples as chicken, dried beans, and whole wheat.

In the same year, Yasumoto et al. [86] made a major advance by partially automating a cation-exchange separation (Aminex A-5) with phosphate buffers as the mobile phase. A three-step buffer mobile phase was used, as was postcolumn chemistry to enhance the sensitivity. 4-Deoxypyridoxine was used as the internal standard and reacted with the color-producing reagent, diazide 5-chloraniline 2,4-disulfonyl chloride, as did PN, PM, and PL. The postcolumn chemistry took 2 min and detection was by UV. For biological materials such as yeast, wheat flour, rat liver and rice bran, the vitamers were extracted by autoclaving the sample at 130°C in 0.4 N HCl for 1 hr. These biological materials yielded a number of interfering peaks, but the concentrations of PL, PN, and PM in the sample could still be quantified. However, further sample cleanup is desirable (Fig. 10.7). No recoveries were reported for spiked biological samples. The lower limit of detection was approximately 100 nmol. Yasumoto et al. emphasized the value of an internal standard because it enabled the investigators to cope with mechanical losses during extraction and correct for possible differences in analytical sensitivity among individual runs. Such errors are otherwise difficult to eliminate in actual practice.

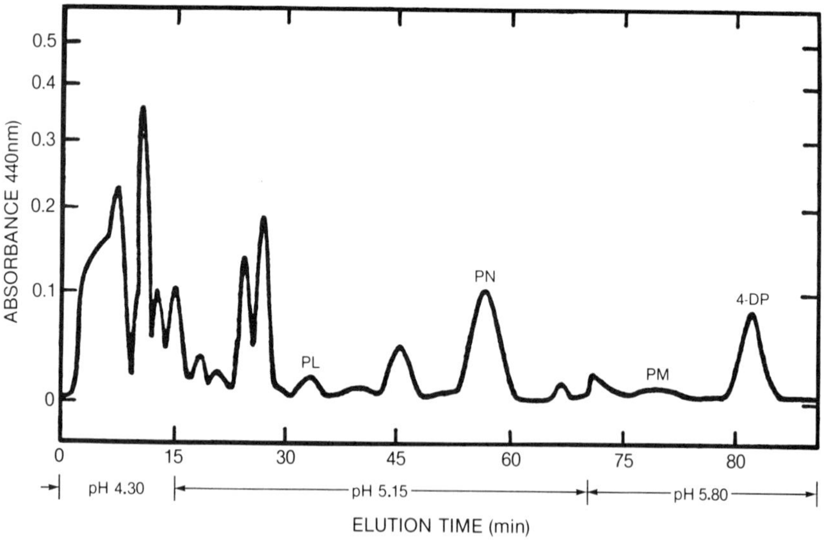

Figure 10.7 Cation-exchange chromatogram of vitamin B_6 from rice bran extract. (From Ref. 86, © 1975 Vitamin Society of Japan.)

In 1978, Wong [88] used a slightly modified version of the Williams, Baker, and Schmit separation in an attempt to determine the vitamin B_6 content of fruit and vegetable extracts. The method was too insensitive for the task, and heavily spiked samples yielded recoveries of 60% for PL, 85% for PM, and 100% for PN. A Dowex AG 50-X4 precolumn cleanup was used prior to final analysis.

Thompson [72], after experimenting with different chromatographic methods, finally settled on cation exchange for the separation of the nonphosphorylated forms. Aminex A-5 and A-9 resins were used along with ammonium formate buffers at pH 5.5 as the mobile phase. For milk samples, a gradient of molarity (0.01 M-1 M) was necessary to achieve separation and quantitation. Extraction and hydrolysis of the phosphorylated forms was accomplished with dilute acid. No recoveries were reported, and no internal standard was used.

5.2.2 Anion Exchange

The first separation of all forms of the vitamin plus 4-PA by modern chromatographic procedures was accomplished with anion-exchange chromatography [89]. The vitamers were eluted isocratically from an Aminex A-25 column at 55°C with a 0.4 M NaCl-0.01 M glycine, pH 10, mobile phase. A stepwise gradient to pH 2.5 eluted 4-PA. The original separation used only one column adn required two fluorometric detectors with different filters to achieve sensitivity. An improved version [22] of this procedure makes use of two columns at different temperatures with a Perkin-Elmer 650-40 fluorescence detector. The sensitivity is adequate for normal plasma level concentrations. This system is illustrated in Fig. 10.8. The method uses 3-hydroxypyridine as an internal standard, and recoveries are close or equal to 100%.

By the introduction of sulfosalicylic acid (SSA) as a deproteinating agent [56], the method proved applicable to a wide variety of biological materials, including cereals, human breast milk, meats, fish, rat tissue, plasma, and fruits [28,34,51,53,56,90,91]. In all cases, recoveries were close to 100%. Typical traces are shown in Figs. 10.9 through 10.18. Note the relatively clean baselines for most samples.

The use of SSA as an extraction agent does require a precolumn (DOWEX AG 2 X8) to remove this reagent from the sample extract before final analysis, since it is highly fluorescent and does bind to the anion-exchange resin. The choice of 3-hydroxypyridine as an internal standard was made after considerable testing of its stability and its ability to behave identically to the vitamers throughout all sample preparation, extraction, and cleanup procedures. The incorporation of this internal standard immediately after weighing of the sample automatically corrects for any sample dilutions and daily variations in instrument sensitivities. It also makes full automation of the system easy, since the integrator uses the known weight of the internal standard to automatically calculate the concentrations of the vitamers after each run.

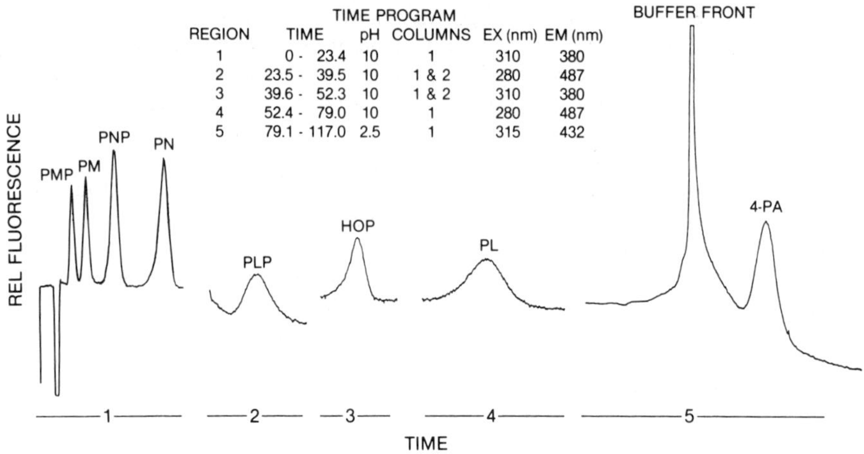

Figure 10.8 Anion-exchange separation of the B_6 vitamers. (From Ref. 22; separation first reported in Ref. 89.)

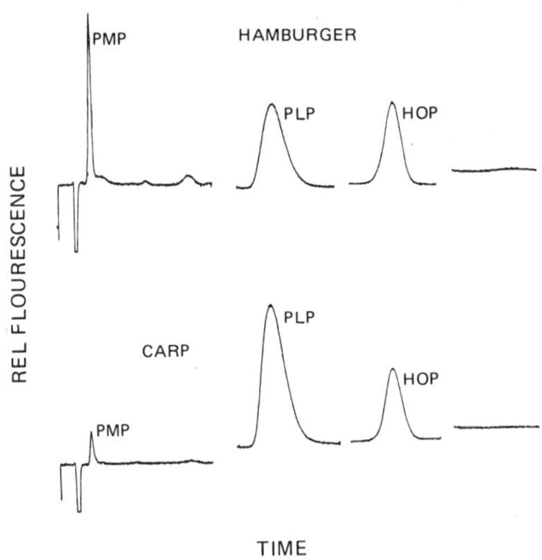

Figure 10.9 Anion-exchange separation of vitamin B_6 extract from hamburger and carp. (From Ref. 56.)

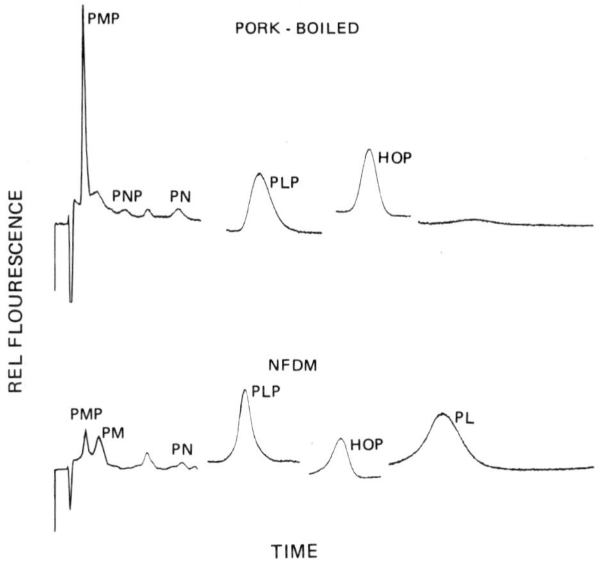

Figure 10.10 Anion-exchange separation of vitamin B_6 extract from broiled pork and nonfat dry milk (NFDM). (From Ref. 56.)

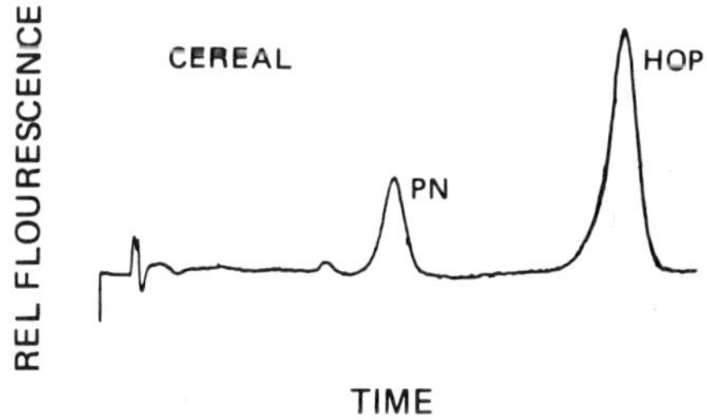

Figure 10.11 Anion-exchange separation of vitamin B_6 extract from fortified breakfast cereal. (From Ref. 90.)

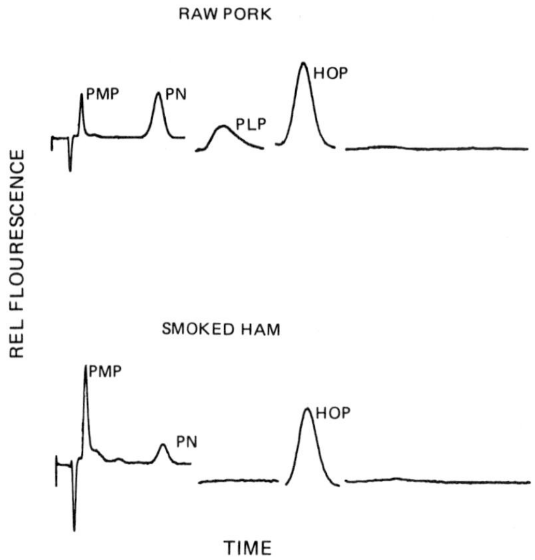

Figure 10.12 Anion-exchange separation of vitamin B_6 extract from raw pork and smoked ham. (From Ref. 91.)

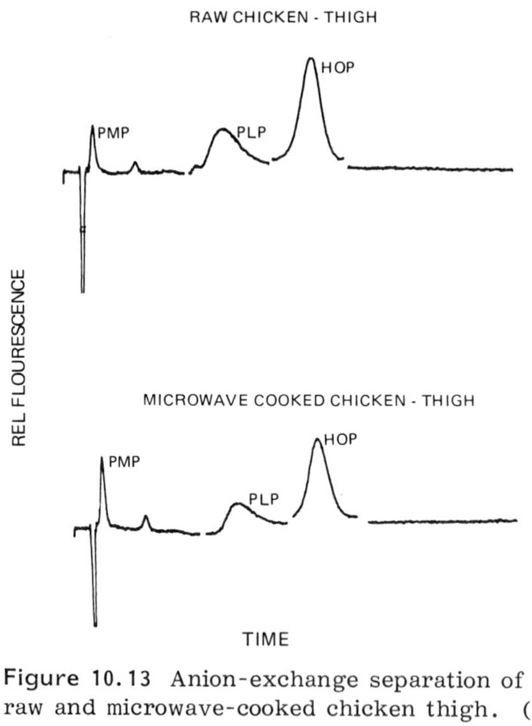

Figure 10.13 Anion-exchange separation of vitamin B_6 extract from raw and microwave-cooked chicken thigh. (From Ref. 91.)

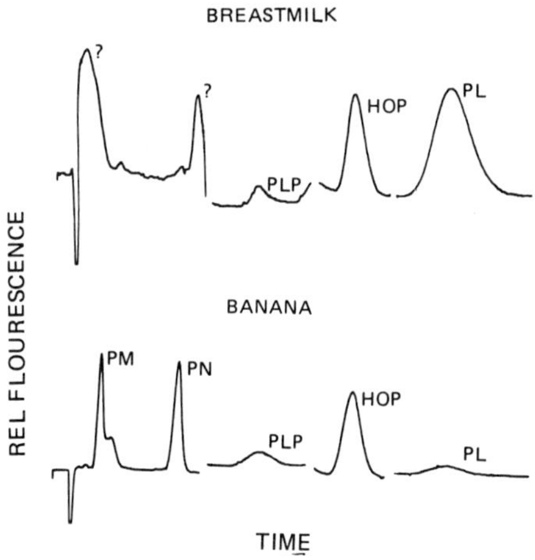

Figure 10.14 Anion-exchange separation of vitamin B_6 extract from human breastmilk from subject on unsupplemental diet [53]. Of the vitamer content, 90-95% is in the form of PL. Also shown is a chromatogram of a vitamin B_6 extract from banana. The principal form found here is PM (55-60%), in contrast to Polansky et al. [48], who reported PN to be the principal form. The total B_6 contents by both methods were comparable.

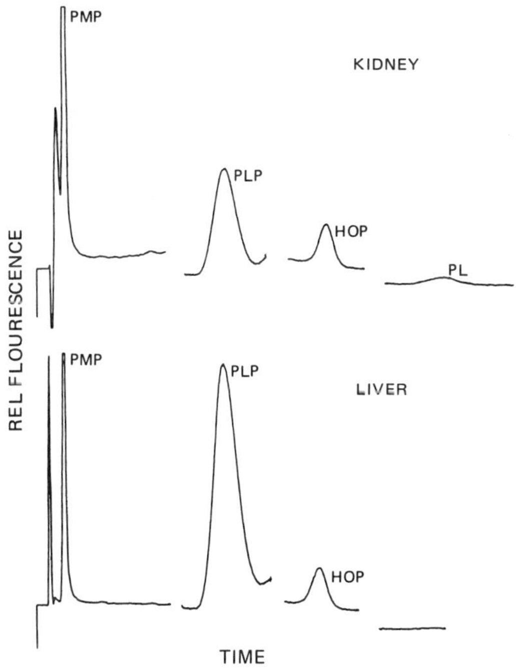

Figure 10.15 Anion-exchange separation of vitamin B_6 extract from rat kidney and rat liver. (From Ref. 34.)

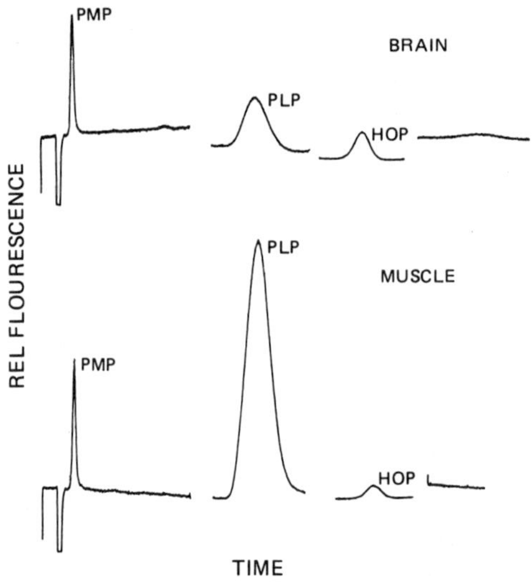

Figure 10.16 Anion-exchange separation of vitamin B_6 extract from rat brain and rat muscle. (From Ref. 34.)

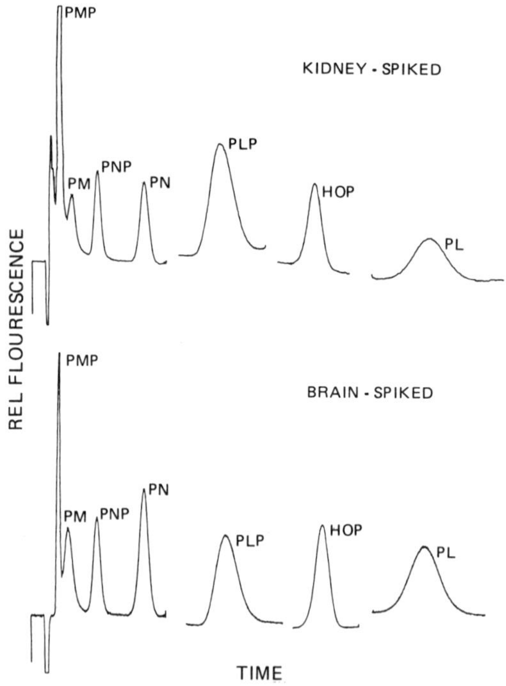

Figure 10.17 Anion-exchange separation of vitamin B_6 extract from spiked rat kidney and rat brain. Recoveries are close to 100%. (From Ref. 34.)

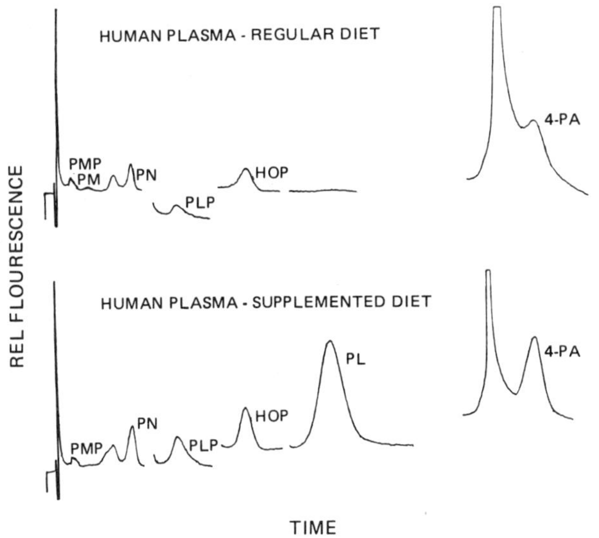

Figure 10.18 Anion-exchange separation of vitamin B_6 extract from human plasma. (From Ref. 34.)

The method has the advantages of quick extraction, easy cleanup, and only 30 min maximum elapsed time between sample weighing to final injection onto the analytical column. The use of 0.4 M NaCl in the mobile phase adjusts the ionic strength to approximately that found in sample extracts as they elute from the precolumn, and thus the elution times do not vary from one form of sample to another. Column lifetimes are another advantage, as they routinely last for more than 2 years even with the many diverse samples being analyzed. The columns are cleaned regularly with a 0.4 M NaCl-0.01 M glycine, pH 2.5, solution and occasionally with a 0.4 M NaCl, pH 11.5, solution to remove compounds left by some samples.

It has been our experience that not enough attention has been paid to sample preparation and handling. These vitamers can be rapidly destroyed either by enzyme activity or simple degradation. The procedure for each new sample type must be checked for recoveries before finally settling on the best possible one. For example, it was found that, in freshly excised rat livers, the vitamers are rapidly interconverted unless the sample is immediately deproteinated [34]. Thus, the deproteinating agent had to be added during the initial homogenization. For more details, see Ref. 34.

In brief, this anion-exchange procedure offers an excellent opportunity to observe all forms of the vitamin in a sample at any one time. The general applicability of the method, however, is dependent on the quality control exerted by the resin manufacturer. Indeed, the growing sophistication of liquid chromatographic separations will require better resin quality control by all manufacturers. For example, for over 5 years we have had extraordinary reproducibility from different batches of A-25 resin, but the latest batch had entirely different characteristics and did not separate the phosphorylated from the nonphosphorylated forms under our previously used conditions. As a consequence, we are being forced to investigate alternative resins that may or may not have better reproducibility from batch to batch.

5.2.3 *Critique of Ion-Exchange Chromatography*

Ion-exchange chromatography has several major advantages. The polymeric packing materials normally used can withstand bulky, complex samples and high ionic strengths without deterioration. Both cation [71] and anion-exchange [22] systems are capable of separating the six vitamers and their metabolite, 4-PA. Compatible extraction procedures have been developed for biological samples, such as animal tissue and plasma [34]; however, both our laboratory and Yasumoto et al. [86] have found additional refinements of extraction procedures are necessary for samples from grains and vegetables.

One of the disadvantages of ion-exchange chromatography is the relatively long retention times for the vitamers. Automation allows a large number of samples to be analyzed; however, degradation may occur while extracts await injection and analysis. Because of this vitamer instability, extreme care must be exercised in handling samples awaiting analysis. In the future, this problem may be minimized by reducing resin particle size and miniaturizing the systems.

5.2.4 *Remarks on Detection*

In the liquid chromatography systems for vitamin B_6, detection has been either by UV absorption or by fluorescence. The latter is the preferred choice as it provides the needed sensitivity and selectivity for samples of biological interest. Unfortunately, it is not possible to list all the fluorescent characteristics of the different B_6 vitamers or pyridoxic acid and its lactone, as these depend strongly on the solvent, ionic strength, and pH. Bridges et al. [92] have made extensive studies of these variations for many of the compounds of interest, and the reader is best referred to their article to obtain approximate characteristics for their own particular conditions.

5.3 Selected Vitamer Analysis

Arguments can be made that it is sometimes necessary to analyze for specific vitamers or 4-PA alone. For example, 4-PA in urine is sometimes used to determine the vitamin status of an individual. On the other hand, if it has previously been determined that the principal forms of the vitamin—as in meats or tissue—are PLP and PMP, it may be worthwhile to develop a simpler assay for only these two. PLP in plasma is considered by some to be a good indication of the vitamin B_6 status of an individual. A simple assay for this would be useful.

5.3.1 Determination of 4-PA in Urine

Gregory and Kirk [93] have developed a reversed-phase procedure for the determination of 4-PA in urine. TCA is used to precipitate the protein in the sample. The extract, after centrifugation, is then injected into the chromatographic system, which is composed of a μBondapak C18 column and a mobile phase of 0.033 M KH_2PO_4-5% methanol at pH 2.2. Detection is by fluorescence, and the analysis time is approximately 5 min. At normal levels of 4-PA concentration in urine (0.5-8 μg 4-PA per milliliter) the signal-to-noise ratio was better than 5:1. Recovery spikes always yielded values greater than 100%. Whether this is due to the formation of 4-PA lactone during sample preparation and analysis is open to question. Arend and Brown [94] have compared this method with the older, more tedious method of Reddy et al. [95] and concluded that both methods yield comparable results. To compensate for the lack of an internal standard, a spiked sample is analyzed along with the unspiked sample.

5.3.2 Determination of Amine and Aldehyde Forms in Biological Materials

By forming carbazone derivatives of PLP and PL, Gregory [21] has been able to separate these forms on a reversed-phase column, Lichrosorb RP-8, with a mobile phase of 0.033 M KH_2PO_4-2.5% acetonitrile at pH 2.2. A typical separation is shown in Fig. 10.19. The method has been applied to the analysis of PLP and PL in rat tissue using perchloric acid as an extraction reagent. The method has recently been extended to include the determination of PMP and PM concentrations in biological samples [96]. PMP and PM are converted to PLP and PL with sodium glyoxalate, and the concentrations of the four vitamers are determined by differential analysis. To analyze one sample, four determinations must be made, one unspiked and one spiked without glyoxalate, and one spiked and one unspiked with glyoxalate added. The absence of an internal standard requires these four analyses and the resuspension of the initial protein precipitate in perchloric acid to guarantee recoveries. In some other procedures proper selection of an internal standard eliminates these problems. Recoveries range from 60 to 130%, depending on the individual vitamer, and differ from sample to sample. Part of the problem may be that the samples were homogenized before the deproteinating reagent was added. We have found that it was necessary to homogenize the sample in a solution containing

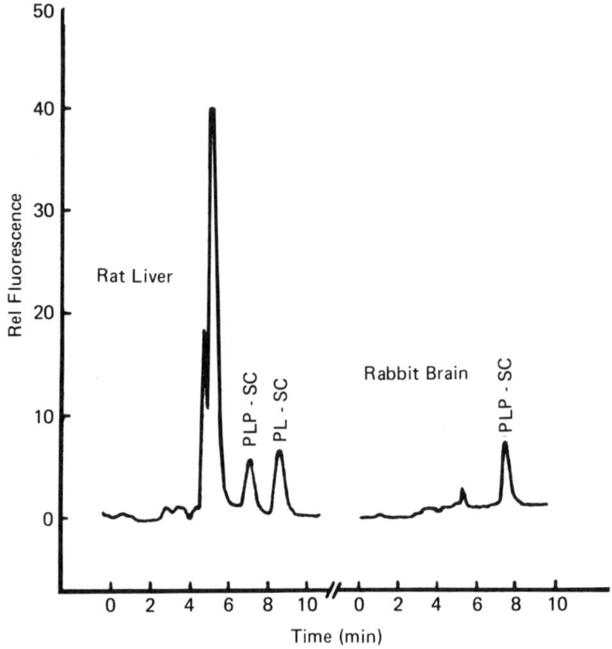

Figure 10.19 Reversed-phase separation of PLP-SC and PL-SC extract from rat liver and rabbit brain. (From Ref. 21.)

the deproteinating reagent because interconversions were found to take place, particularly with rat liver, presumably due to enzyme activity. This overall procedure offers no advantages in time or simplicity over the anion-exchange separation and will not detect the PN or PNP which sometimes appear in samples.

5.3.3 Determination of PLP in Plasma

Allemmark et al. [97] have developed a sensitive assay for the determination of PLP in plasma that appears to be superior to the radioenzymatic procedure previously used. It uses the tyrosine apodecarboxylase activity toward L-3,4-dihydroxyphenylalanine (L-DOPA), which is PLP dependent. The amount of dopamine product formed is proportional to the original concentration of PLP. The plasma was deproteinated with TCA and diluted. Tyrosine apodecarboxylase was added and the solution preincubated in a sodium acetate buffer for 1 hr. L-DOPA was then added and the reaction allowed to run for a given amount of time (∿45 min). The reaction was quenched with perchloric acid and injected into a 10 μ Nucleosil SA pellicular cation-exchange column. The mobile phase was a citrate-acetate buffer at pH 5.2, and an electrochemical detector was used to determine the amount of dopamine formed. The method is sensitive to 1 ng PLP or less.

5.3.4 *Determination of PL and PLP in Blood*

Schrijver et al. [98] have used a version of the Lim et al. chromatographic procedure [82] to analyze for PL and PLP in whole blood. Separation of these forms was achieved, although the other forms of vitamin B_6 could not be separated from these two. However, under the experimental conditions used, the fluorescence detection sensitivity of the other forms was low enough to cause minimal interference. Sample preparation and TCA extraction took about 1 hr, and relatively clean samples were obtained. Postcolumn derivatization with semicarbazide was used, and the lower limits of detection were 5 nmol/liter. Recoveries from spiked samples were 100%. Automation of the system allowed 125 samples to be run in 48 hr. No internal standard was used.

5.3.5 *Miscellaneous Applications*

Korytnyk et al. [99] have studied the inhibition by analogs of PLP and PL on pyridoxal kinase and pyridoxine phosphate oxidase activity. The substrate was ATP, and the production of ADP was monitored by anion-exchange chromatography. A Bondapak AX Corasil resin was used with a mobile phase of 0.125 M $NH_4H_2PO_4$ buffered at pH 3. Ultraviolet detection was used. Gregory [100] used a similar system to study the activity of pyridoxal kinase activity by ε-pyridoxyllysine, a vitamin B_6 degradation product in food. Both fluorometric and ultraviolet detectors were used in series to determine the concentration of reactants and products.

5.4 Automation

With recent improvements in extraction procedures, it is now possible to analyze a variety of biological samples with modern chromatographic techniques. The number of samples to be considered can be enormous, and some form of automation is required to release the operator for daily sample collection and preparation. It is our opinion that liquid chromatographic techniques for vitamin B_6 analysis are still in the developmental stages. The most successful systems are composed of modular components, each of which has its own internal logic and can be obtained from different manufacturers. To successfully automate such a complete system, the controlling processor either has to deal with the individual idiosyncrasies of each component or be able to bypass them. We believe the latter approach is simpler, cheaper, and more versatile.

A recent automation of the anion-exchange chromatographic system described earlier is shown in Fig. 10.20 [25,101]. An inexpensive

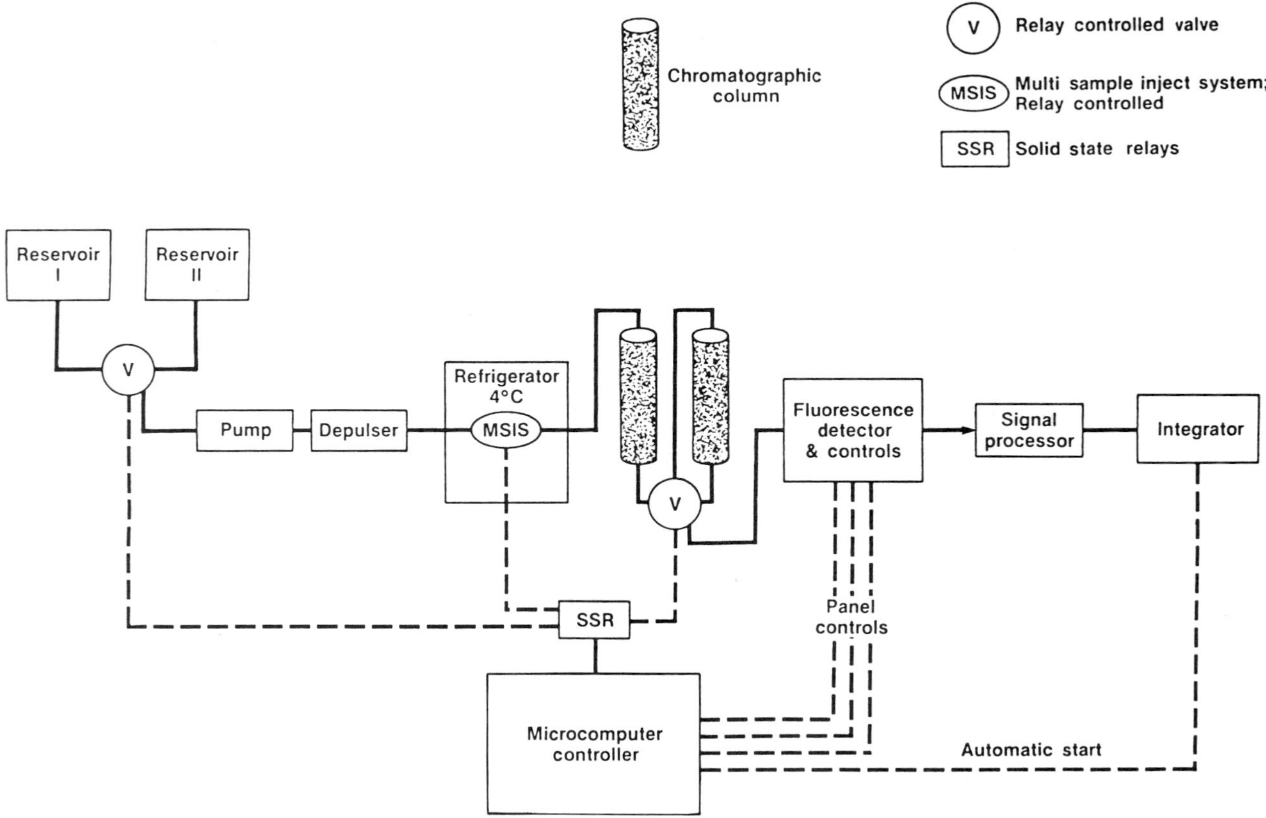

Figure 10.20 Block diagram of automated anion-exchange chromatographic system. Dashed lines correspond to electrical leads from the microcomputer to various components. (From Ref. 25.)

microcomputer (Rockwell AIM-65) with 4000 bytes of memory is used to control all valves, integrator starts and stops, and keys on the fluorescence spectrophotometer. The valves in the system are solenoidal and require high power. Each is activated by an optically isolated solid-state relay, which in turn is controlled by a microcomputer input-output buffer line. Input-output buffer lines are hard-wired to the detector panel board and are programmed to drain a small current to ground at a given time. This simulates pressing of the keys on the panel, and thus they are remotely controlled by the microcomputer. The integrator start-stop command is done in the same manner. Thus, an inexpensive microcomputer can be made to control an array of highly technical instruments. The multiple sample inject system will accommodate up to 19 samples that can be run automatically over a 24 hr period.

With vitamin B_6, one has to be concerned with the stability of the vitamers over this time period. Solutions of pure standards have been found to be stable for 1 day if kept at 4°C. However, if the concentration of PLP falls below 2 ng/ml, it is necessary to perform all procedures using degassed solutions; otherwise, serious losses of this vitamer are encountered [25]. It will be necessary to investigate the stability of the vitamers in extracts from each type of biological sample analyzed.

5.5 Comparison of Liquid Chromatographic and Microbiological Results

Concentration values of vitamin B_6 in biological materials obtained by liquid chromatographic techniques are normally in agreement with those obtained by other assay procedures. Gregory and his coworkers [42, 80,96] have made numerous comparisons between microbiological and chromatographic assays and have concluded that the results confirm the general validity of the chromatographic procedures. In fact, the latter is often superior to the former in both precision and ease.

In our anion-exchange work we have made fewer comparisons than Gregory and his coworkers. The greatest discrepancies have occurred for fortified breakfast cereals. From the wide range of cereals we studied several were randomly selected for microbiological assay by a commercial laboratory (AOAC method) and by Guilarte et al. [102], who have developed a microbiological-radioassay. The results were inconsistent, and the reasons for the discrepancies are not clear [90, 102].

For animal tissue, the agreement between different methods is quite good. We obtain excellent agreement for the PLP content of rat liver with the enzymatic assay of Reynolds [34,103]. The analysis of pork gave interesting results. Polansky's microbiological assay results agree within experimental error to those obtained by us on raw pork loin [52]. However, when these same samples were analyzed by a commercial laboratory using the AOAC microbiological assay, presumably the same as Polansky's, the results were consistently lower than ours [28]. There was a definite correlation between the discrepancy and the amount of PLP in the sample.

The values obtained by the anion-exchange method on rat tissues gave values consistent with those reported by others [34]. However, different analytical methods were not tried on the same sample.

There is a definite discrepancy in the case of blood plasma [51]. The anion-exchange analysis, using SSA as the extraction agent, yields much higher values of PN than have been found by other methods using different extraction procedures [71,104]. The spectral scan of the eluting peak indicated only the presence of PN. Whether this observation is real or just an artifact must be checked by further comparative studies on identical samples.

In brief, much of the effort up to now has been put into developing good extraction procedures so that relatively clean extracts would be available for final analysis. Since this has been accomplished, a closer scrutiny of the results obtained by different procedures is in order.

6. SUMMARY AND RECOMMENDATIONS

Liquid chromatography has advantages over gas chromatography for the analysis of vitamin B_6. There are liquid chromatographic systems that are capable of analyzing for all forms of the vitamin in metabolically important samples. Gas chromatography could be as successful in the analysis of this vitamin in natural products, if the phosphorylated forms of the vitamin can be derivatized and interfering compounds removed.

Both reversed-phase and ion-exchange chromatography have been developed to the stage where they can routinely handle samples of biological interest. However, studies still have to be undertaken to refine the various extraction procedures to obtain "clean" samples from all types of biological systems. Further, extensive efforts are still required in the whole area of bioavailability. Not enough checks have been done on sample handling procedures and their effect on the analytical results, and more comparative studies still have to be made on identical samples between the different successful procedures. With these provisions in mind, one can only quote Esmond Snell [26], who stated "Provided the extraction problem can be satisfactorily solved, these procedures should give us analytical tools of unprecedented precision and sensitivity for future investigations of vitamin B_6 distribution and metabolism."

ACKNOWLEDGMENTS

The authors are indebted to Gary Beecher and Betty Li for their helpful editorial comments. This work was supported in part by an Interagency Reimburseable Agreement No. 2Y01-HB 60041-05 from the National Heart, Lung, and Blood Institute, NIH.

POSTSCRIPT

Literature review for this manuscript was completed in mid-1981 so recent important contributions are not included.

REFERENCES

1. E. E. Snell, *Physiol. Rev.*, *33*, 509 (1953).
2. E. E. Snell, P. M. Fasella, A. E. Braunstein, and A. R. Fanelli (eds.), *Chemical and Biological Aspects of Pyridoxal Catalysis* (International Union of Biochemistry Symposium 30), Macmillan, New York, 1963, pp. 1-594.
3. M. Ebadi, in *Human Vitamin B_6 Requirements: Proceedings of a workshop, Letterman Army Institute of Research, Presidio of San Francisco, California, June 1976*, National Research Council-National Academy of Sciences, Washington, D.C., 1978, p. 129.
4. R. J. Kutsky, *Handbook of Vitamins and Hormones*, Van Nostrand Reinhold, New York, 1973, pp. 60-61.
5. L. C. Robson, M. R. Schwartz, and W. D. Perkins, in *Human Vitamin B_6 Requirements: Proceedings of a Workshop, Letterman Army Institute of Research, Presidio of San Francisco, California, June 1976*, National Research Council-National Academy of Sciences, Washington, D.C., 1978, p. 162.
6. L. C. Robson and M. R. Schwartz, in *Vitamin B_6 Metabolism and Role in Growth* (G. P. Tryfiates, ed.), Food and Nutrition Press, Westport, Conn., 1980, p. 205.
7. H. E. Sauberlich, in *Methods in Vitamin B_6 Nutrition: Analysis and Status Assessment* (J. E. Leklem and R. D. Reynolds, eds.), Plenum Press, New York, 1981, p. 232.
8. E. M. Pao and S. J. Mickle, *Food Technol.*, *35*, 58 (1981).
9. J. C. Bauernfeind and O. N. Miller, in *Human Vitamin B_6 Requirements: Proceedings of a Workshop, Letterman Army Institute of Research, Presidio of San Francisco, California, June 1976*, National Research Council-National Academy of Sciences, Washington, D.C., 1978, p. 78.
10. F. Rosen, E. Mihich, and C. Nichol, *Vitamins Horm.*, *22*, 609 (1964).
11. H. J. Klosterman, *J. Agr. Food Chem.*, *22*, 13 (1974).
12. F. H. Kratzer, D. E. Williams, B. Marshall, and P. N. Davis, *J. Nutr.*, *52*, 555 (1954).
13. H. J. Klosterman, G. L. Lamoureaux, and J. L. Parsons, *Biochemistry*, *6*, 170 (1967).

14. *Human Vitamin B_6 Requirements: Proceedings of a Workshop, letterman Army Institute of Research, Presidio of San Francisco, California, June 1976*, National Research Council-National Academy of Sciences, Washington, D.C., 1978
15. G. P. Tryfiates (ed.), *Vitamin B_6 Metabolism and Role in Growth*, Food and Nutrition Press, Westport, Conn., 1980.
16. J. E. Leklem and R. D. Reynolds (eds.), *Methods in Vitamin B_6 Nutrition: Analysis and Status Assessment*, Plenum Press, New York, 1981.
17. E. E. Snell, *J. Biol. Chem.*, *154*, 313 (1944).
18. R.-H. Yamada and T. Tsuji, in *Vitamin B_6 Metabolism and Role in Growth* (G. P. Tryfiates, ed.), Food and Nutrition Press, Westport, Conn., 1980, p. 335.
19. J. F. Gregory and J. R. Kirk, in *Human Vitamin B_6 Requirements: Proceedings of a Workshop, Letterman Army Institute of Research, Presidio of San Francisco, California, June 1976*, National Research Council-National Academy of Sciences, Washington, D.C., 1978, p. 72.
20. M. M. Polansky, in *Methods in Vitamin B_6 Nutrition: Analysis and Status Assessment* (J. E. Leklem and R. D. Reynolds, eds.), Plenum Press, New York, 1981, p. 21.
21. J. F. Gregory, *Anal. Biochem.*, *102*, 374 (1980).
22. J. T. Vanderslice and C. E. Maire, *J. Chromatogr.*, *196*, 176 (1980).
23. S. F. Contractor and B. Shane, *Clin. Chim. Acta*, *21*, 71 (1968).
24. E. Cunningham and E. E. Snell, *J. Biol. Chem.*, *158*, 491 (1945).
25. J. T. Vanderslice, J. F. Brown, G. R. Beecher, C. E. Maire, and S. G. Brownlee, *J. Chromatogr.*, *216*, 338 (1981).
26. E. E. Snell, in *Methods in Vitamin B_6 Nutrition: Analysis and Status Assessment* (J. E. Leklem and R. D. Reynolds, eds.), Plenum Press, New York, 1981, p. 1.
27. M. M. Polansky and E. W. Toepfer, *J. Agr. Food Chem.*, *17*, 1394 (1969).
28. J. T. Vanderslice, S. G. Brownlee, and M. E. Cortissoz, *J. Ass. Offic. Anal. Chem.* Sept. (1984).
29. E. E. Snell, *Vitamins Horm.*, *16*, 77 (1958).
30. D. B. Coursin, *Am. J. Clin. Nutr.*, *9*, 306 (1961).
31. H. Wada and E. E. Snell, *J. Biol. Chem.*, *236*, 2089 (1961).
32. G. M. Brown and J. J. Reynolds, *Annu. Rev. Biochem.*, *32*, 447 (1963).
33. H. E. Sauberlich, in *The Vitamins*, Vol. II (W. H. Sebrell and R. S. Harris, eds.), Academic Press, New York, 1968, p. 32.
34. J. T. Vanderslice, C. E. Maire, and G. R. Beecher, in *Methods in Vitamin B_6 Nutrition: Analysis and Status Assessment* (J. E. Leklem and R. D. Reynolds, eds.), Plenum Press, New York, 1981, p. 123.
35. E. E. Snell, *J. Amer. Chem. Soc.*, *67*, 194 (1945).

36. E. H. Fisher, A. B. Kent, E. R. Snyder, and E. G. Krebs, *J. Amer. Chem. Soc.*, *80*, 2906 (1958).
37. K. Yasumoto, H. Tsuji, K. Iwami, and H. Mitsuda, *Vitamins (Kyoto)*, *50*, 327 (1976).
38. H. Tsuji, J. Okada, K. Iwami, K. Yasumoto, and H. Mitsuda, *Vitamins (Kyoto)*, *51*, 153 (1977).
39. F. W. Bernhart, E. D'Amato, and R. M. Tomarelli, *Arch. Biochem. Biophys.*, *88*, 267 (1960).
40. G. Wendt and F. W. Bernhart, *Arch. Biochem. Biophys.*, *88*, 270 (1960).
41. V. Srncová and J. Davidek, *J. Food Sci.*, *37*, 310 (1972).
42. J. F. Gregory and J. R. Kirk, *J. Food Sci.*, *43*, 1801 (1978).
43. J. F. Gregory and J. R. Kirk, *Nutr. Rev.*, *39*, 1 (1981).
44. D. Heyl, S. A. Harris, and K. Folkers, *J. Amer. Chem. Soc.*, *74*, 414 (1952).
45. E. W. Nelson, C. W. Burgin, and J. J. Cerda, *J. Nutr.*, *107*, 2128 (1977).
46. E. W. Nelson, H. Lane, and J. J. Cerda, *J. Nutr.*, *106*, 1433 (1976).
47. M. M. Polansky, E. W. Murphy, and E. W. Toepfer, *J. Ass. Offic. Agr. Chem.*, *47*, 750 (1964).
48. M. M. Polansky and E. W. Murphy, *J. Amer. Diet. Ass.*, *48*, 109 (1966).
49. M. M. Polansky and E. W. Toepfer, *J. Agr. Food Chem.*, *17*, 1394 (1969).
50. L. Lumeng and T.-K. Li, *Fed. Proc.*, *37*(3), 588 (1978).
51. J. T. Vanderslice, C. E. Maire, and G. R. Beecher, *Amer. J. Clin. Nutr.*, *34*, 947 (1981).
52. J. T. Vanderslice, S. G. Brownlee, and M. E. Cortissoz *J. Ass. Offic. Anal. Chem.* Sept. (1984).
53. J. R. Vanderslice, S. G. Brownlee, C. E. Maire, R. D. Reynolds, and M. Polansky, *Amer. J. Clin. Nutr.* *37*, 867 (1983).
54. J. C. Rabinowitz and E. E. Snell, *Ind. Eng. Chem. Anal. Ed.*, *19*, 277 (1947).
55. Y. H. Loo and L. Badger, *J. Neurochem.*, *16*, 801 (1969).
56. J. T. Vanderslice, C. E. Maire, R. F. Doherty, and G. R. Beecher, *J. Agr. Food Chem.*, *28*, 1145 (1980).
57. D. M. Hilker and F. O. Sotelo, *Fed. Proc.*, *40*(3 part 2), 863 (1981).
58. W. Korytnyk, G. Fricke, and B. Paul, *Anal. Biochem.*, *17*, 66 (1966).
59. W. Korytnyk, *Methods Enzymol.*, *18A*, 500 (1970).
60. A. R. Proser, A. J. Sheppard, and D. A. Libby, *J. Ass. Offic. Anal. Chem.*, *50*, 1348 (1967).
61. L. T. Senello and C. J. Argoudelis, *Anal. Chem.*, *41*, 171 (1969).
62. A. J. Sheppard and A. R. Proser, *Methods Enzymol.*, *18A*, 494 (1970).

63. Y. Ohnishi, Z. Horii, and M. Makita, *Yakugaku Zasshi, 87*, 747 (1967).
64. K. Richter, M. Becchi, W. Vetter, and W. Walther, *Helv. Chim. Acta, 50*, 364 (1967).
65. E. M. Patzer and D. M. Hilker, *J. Chromatogr., 135*, 489 (1977).
66. T. Imanari and Z. Tamura, *Chem. Pharm. Bull., 15*, 896 (1967).
67. A. K. Williams, *J. Agr. Food Chem., 22*, 107 (1974).
68. B. E. Haskell, *Fed. Proc., 27*(2), 554 (1968).
69. E. A. Peterson and H. A. Sober, *J. Amer. Chem. Soc., 76*, 169 (1954).
70. H. G. Tiselius, *Clin. Chim. Acta, 40*, 319 (1972).
71. S. P. Coburn, Personal communication.
72. J. N. Thompson, in Symposium Proceedings: *Application of High Pressure Liquid Chromatographic Methods for Determination of Fat Soluble Vitamins A, D, E, and K in Foods and Pharmaceuticals*, presented by Association of Vitamin Chemists in conjunction with Waters Associates, 1978, p. 62.
73. D. P. Wittmer and W. G. Haney, Jr., *GLC and HPLC Determination of Therapeutic Agents* (K. Tsuju, ed.), Marcel Dekker, New York, 1976, p. 1235.
74. A. K. Williams, *Methods Enzymol., 62D*, 415 (1979).
75. E. Morita and N. Mizuno, *J. Chromatogr., 202*, 134 (1980).
76. C. Horvath, W. Melander, and I. Molnár, *J. Chromatogr., 125*, 129 (1976).
77. C. Horvath, W. Melander, and I. Molnár, *Anal. Chem., 49*, 142 (1977).
78. O. Sinanoglu, in *Molecular Associations in Biology* (B. Pullman, ed.), Academic Press, New York, 1968, pp. 427-445.
79. O. Sinanoglu and S. Abdulnur, *Fed. Proc., 24*(2, Suppl. 15), S-12 (1965).
80. J. F. Gregory, *J. Agr. Food Chem., 28*, 486 (1980).
81. T. Yoshida, N. Yunoki, Y. Nakazima, T. Kaito, and T. Anmo, *Yakugaku Zasshi, 98*, 1319 (1978).
82. K. L. Lim, R. W. Young, and J. A. Driskell, *J. Chromatogr., 188*, 285 (1980).
83. W. J. O'Reilly, P. J. M. Guelen, M. J. A. Hoes, and E. van der Kleyn, *J. Chromatogr., 183*, 492 (1980).
84. J. F. Gregory and J. R. Kirk, in *Methods in Vitamin B_6 Nutrition: Analysis and Status Assessment* (J. E. Leklem and R. D. Reynolds, eds.), Plenum Press, New York, 1981, p. 149.
85. R. C. Williams, D. R. Baker, and J. A. Schmit, *J. Chromatogr. Sci., 11*, 618 (1973).
86. K. Yasumoto, K. Tadera, H. Tsuji, and H. Mitsuda, *J. Nutr. Sci. Vitaminol., 21*, 117 (1975).
87. A. K. Williams and P. D. Cole, *J. Agr. Food Chem., 23*, 915 (1975).

88. F. F. Wong, *J. Agr. Food Chem.*, *26*, 1444 (1978).
89. J. T. Vanderslice, K. K. Stewart, and M. M. Yarmas, *J. Chromatogr.*, *176*, 280 (1979).
90. J. T. Vanderslice, C. E. Maire, and J. E. Yakupkovic, *J. Food Sci.*, *46*, 943 (1981).
91. J. T. Vanderslice, S. G. Brownlee, and M. E. Cortissoz, *J. Ass. Offic. Anal. Chem.*, Sept. (1984).
92. J. W. Bridges, D. S. Davies, and R. T. Williams, *Biochem. J.*, *98*, 451 (1966).
93. J. F. Gregory and J. R. Kirk, *Amer. J. Clin. Nutr.*, *32*, 879 (1979).
94. R. A. Arend and R. D. Brown, *Amer. J. Clin. Nutr.*, *34*, 1984 (1981).
95. S. K. Reddy, M. S. Reynolds, and J. M. Price, *J. Biol. Chem.*, *233*, 691 (1958).
96. J. F. Gregory, D. B. Manley, and J. R. Kirk, *J. Agr. Food Chem.*, *29*, 921 (1981).
97. S. Allenmark, E. Hjelm, and U. Larsson-Cohn, *J. Chromatogr.*, *146*, 485 (1978).
98. J. Schrijver, A. J. Speek, and W. H. P. Schreurs, *Int. J. Vitaminol. Nutr. Res.*, *51*, 216 (1981).
99. W. Korytnyk, M. T. Hakala, P. G. G. Potti, N. Angelino, and S. C. Chang, *Biochemistry*, *15*, 5458 (1976).
100. J. F. Gregory, *J. Biol. Chem.*, *255*, 2355 (1980).
101. J. F. Brown, J. T. Vanderslice, C. Maire, S. G. Brownlee, and K. K. Stewart, *J. Automat. Chem.*, *3*, 187 (1981).
102. T. Guilarte and P. McIntyre, *J. Nutr.*, *111*, 1861 (1981).
103. M. S. Sloger, L. G. Scholfield, and R. D. Reynolds, *J. Nutr.*, *108*, 1355 (1978).
104. L. Lumeng and T.-K. Li, in *Vitamin B_6 Metabolism and Role in Growth* (G. P. Tryfiates, ed.), Food and Nutrition Press, Westport, Conn., 1980, p. 27.

11

BIOTIN

FRANÇOIS FRAPPIER and MICHEL GAUDRY

Laboratorie de Chemie Organique Biologique, Pierre and Marie Curie University, Paris, France

1. INTRODUCTION

Biotin (*cis*-tetrahydro-2-oxothieno[3,4-*d*]imidazoline-4-valeric acid) (Fig. 1.11) has been isolated from egg yolk in crystalline form by Kögl [1], with Tönnis [2]. The structure of that vitamin, named also bios II, factor X, coenzyme R, anti-egg white injury, or vitamin H, was established in 1942 by Melville et al. [3] and its synthesis was achieved by Harris et al. a year later [4,5]. The absolute configuration of biotin was elucidated by x-ray determination on biotin by Traub [6,7] and confirmed by Bonnemere et al. [8] and Trotter and Hamilton [9], who worked on a N_1'-carboxy derivative of biotin. These determinations have been confirmed and extended more recently by De Titta et al. [10,11].

2. CHEMICAL PROPERTIES

2.1 Stability

In contrast to the reputation of lability of vitamins, biotin is very stable and can be autoclaved without being affected. This molecule even resists autoclaving in concentrated sulfuric acid (4 mol/liter, 120°C, 2 hr), conditions commonly used to extract biotin from biological samples [12,13]. However, biotin can be easily oxidized into biotin sulfoxides. This process, which is negligible in concentrated solutions, can become predominant in very dilute solutions, for instance, during anion-exchange column chromatographic purification of radiolabeled biotin [14]. This can be prevented by use of carefully degassed solvents [14]. In addition to this oxidation process, microbial growth can drastically lower the concentration of nonsterile dilute solutions of the vitamin.

2.2 Chemical Reactivity

Biotin exhibits normal reactivity at the potentially reactive centers of the molecule. The carboxyl group has a normal pK and can be esterified normally, whereas the sulfur atom can be oxidized with various

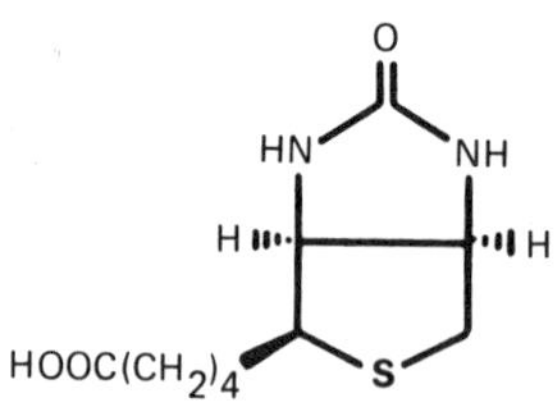

Figure 11.1 Structure of biotin.

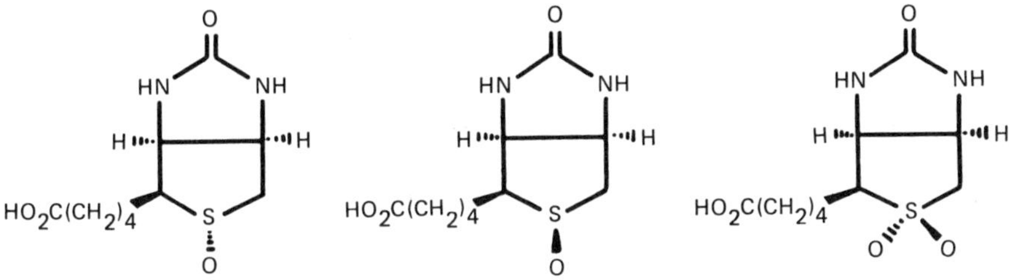

Figure 11.2 Structures of *d*-biotin sulfoxide, *l*-biotin sulfoxide, and biotin sulfone.

reagents yielding a mixture of isomeric *d*- and *l*-biotin sulfoxides [15-17], which can further react, leading to biotin sulfone (Fig. 11.2) [18]. Reactions of the ureido ring occur almost exclusively at the N_1'-atom upon nitrosation [19] or methoxycarboxylation [20], whereas both N_1' and N_3' atoms are methylated by a mixture of formaldehyde and formic acid [19].

3. BIOCHEMICAL PROPERTIES

3.1 Attachment of Biotin to the Apoenzymes

After entering the cell, biotin becomes covalently bound to the apoenzymes of biotin-dependent enzymes during the last step of their biosynthesis [21]. In all the enzymes examined so far, the biotin has been shown to be linked through an amide bond of the ε-amino group of a lysin (Fig. 11.3). This reaction occurs in two steps, an activation step yielding biotinyl AMP (Eq. 1), which reacts with the lysin ε-amino group, yielding the active holoenzyme (Eq. 2):

$$\text{ATP} + \text{biotin} \rightarrow \text{biotinyl-AMP} + \text{PPi} \quad (1)$$

$$\text{Biotinyl-AMP} + \text{apoenzyme} \rightarrow \text{holoenzyme} + \text{AMP} \quad (2)$$

3.2 Biotin-Dependent Enzymes

The biotin-dependent enzyme group includes three classes of enzymes: carboxylases, transcarboxylase, and decarboxylases [22,23]. In all the reactions catalyzed by these enzymes, biotin acts as a CO_2 carrier either to activate CO_2 and transfer it to a substrate, as in carboxylases, or to transfer a CO_2 moiety between two substrates, as in transcarboxylase, or to abstract the CO_2 from a substrate as in decarboxylases. In the three cases, *N*-carboxybiotin (Fig. 11.4) is an intermediate.

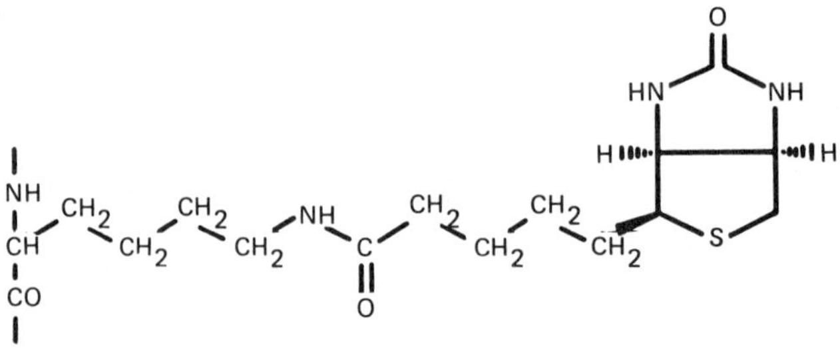

Figure 11.3 Structure of biotin covalently linked through the ε-amino group of lysin to its apoprotein.

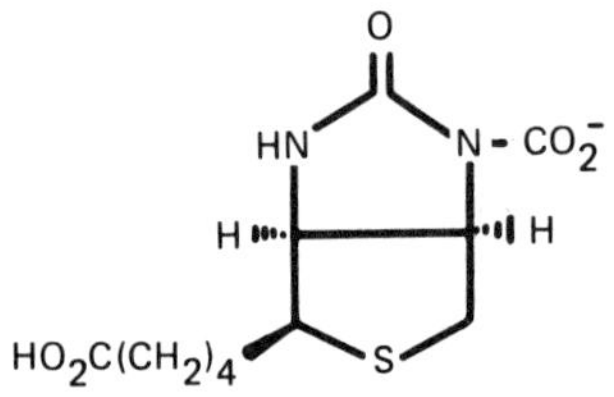

Figure 11.4 Structure of *N*-carboxybiotin.

3.2.1 Carboxylases

The carbon dioxide ligase group comprises six enzymes. Acetyl-CoA carboxylase [E.C. 6.4.1.2] catalyzes the formation of malonyl-CoA from acetyl-CoA and is involved in fatty acid synthesis (Eq. 3) [24].

$$CH_3COSCoA + ATP + HCO_3^- \rightarrow {}^-O_2CCH_2COSCoA + ADP + Pi \qquad (3)$$

Propionyl-CoA carboxylase [E.C. 6.4.1.3] converts propionyl-CoA into methylmalonyl-CoA [25]. This enzyme is essential in isoleucine degradations (Eq. 4):

$$CH_3CH_2COSCoA + ATP + HCO_3^- \rightarrow CH_3(CO_2^-)CHCOSCoA + ADP + Pi \qquad (4)$$

β-Methylcrotonyl-CoA carboxylase [E.C. 6.4.1.4], which transforms β-methylcrotonyl-CoA into β-methylglutaconyl-CoA (Eq. 5), is a key enzyme in leucine degradation [26]:

$$(CH_3)_2C{=}CHCOSCoA + ATP + HCO_3^- \rightarrow {}^-O_2CCH_2(CH_3)C{=}CHCOSCoA + ADP + Pi \quad (5)$$

Pyruvate carboxylase [E.C. 6.4.1.1] is an important regulatory enzyme since it converts pyruvate into oxaloacetate (Eq. 6) [27]:

$$CH_3COCOOH + ATP + HCO_3^- \rightarrow {}^-O_2CCH_2COCOOH + ADP + Pi \quad (6)$$

These four carboxylases are found in animals, and a deficiency in one or all of these activities leads to serious metabolic troubles.

Urea carboxylase [E.C. 6.3.4.6] is induced in microorganisms growing on urea as nitrogen source [28]. It catalyzes the transformation of urea in allophanate, which is further hydrolyzed to carbon dioxide and ammonia by an allophanate hydrolase.

Geranyl-CoA carboxylase [E.C. 6.4.1.5] is found in microorganisms and is involved in the degradation of isoprenoid compounds [29].

3.2.3 *Transcarboxylase*

The carboxyltransferase group contains only one enzyme, the methylmalonyl-CoA transcarboxylase [E.C. 2.1.3.1], which is induced in *Propionibacterium shermanii* and catalyzes the transcarboxylation between methylmalonyl-CoA and pyruvate [30,31] (Eq. 7). This enzyme is important in the metabolism of propionate:

$$CH_3(CO_2^-)CHCOSCoA + CH_3COCOOH \rightarrow CH_3CH_2COSCoA + {}^-O_2CCH_2COCOOH \quad (7)$$

3.2.3 *Decarboxylases*

Two enzymes, the methylmalonyl-CoA decarboxylase [E.C. 4.1.1.41] [32] and the oxaloacetate decarboxylase [E.C. 4.1.1.3] [33] are encountered in microorganisms.

3.3 Biotin Deficiency

Daily requirements are very low, and biotin deficiency can be induced in animals by feeding them with raw egg whites, which contain avidin, a glycoprotein that has a very high affinity for biotin. Cases of spontaneous biotin deficiency in humans were not detected before 1976, but since that time several cases of this deficiency have been diagnosed [34-36]. The clinical symptoms (alopecia, skin rashes, and presence of metabolites in urine), linked to low or even absent carboxylase activities, can be eliminated by biotin absorption. Patients have been classified into two categories, the neonatal and the late-onset forms [37]. The former is characterized by an unusually low holocarboxylase synthetase activity [38-40], whereas the latter is characterized by a low level of plasmatic biotin resulting probably from an absence of biotinidase activity [41,42].

3.4 Complexation with Avidin

Avidin, a glycoprotein from egg white, is composed of four identical subunits and exhibits a very high affinity for biotin ($K_D = 10^{-15}$ M) [43]. This property is also shared by streptavidin, a glycoprotein isolated from *Streptomycetes* [44,45]. This high affinity for biotin has been used extensively to induce biotin deficiencies, in affinity chromatography [46], as a tool in molecular biology [46], or in biotin determinations.

4. BIOTIN DETERMINATIONS

The classic methods for biotin determination are not chromatographic techniques but take advantage of the biochemical properties of the vitamin. Depending on whether avidin is used, they can be classified into two categories.

4.1 Methods Without Avidin

4.1.1 Reaction with Paradimethylaminocinnamaldehyde

Upon reaction with this compound [47,48] in acid medium, a pink color develops, which is characteristic of the ureido ring and can be used to determine the biotin concentration spectrophotometrically at 533 nm. This method is easy to use but its sensitivity is low, in the microgram range.

4.1.2 Enzymatic Determination

This assay is based on the formation of holopyruvate carboxylase from apopyruvate carboxylase and biotin followed by the determination of the pyruvate carboxylase activity, the activity being proportional to the amount of biotin available to form the active holoenzyme [49]. This method is not very easy to use, but its sensitivity is very good (as low as 0.5 ng).

4.1.3 Microbiological Determinations

These determinations have been used from the beginning and are among the most sensitive assays for biotin (0.1 ng). Several microorganisms have been tested [50], but the more commonly used are *Saccharomyces cerevisiae* (Fleischmann strain 139 ATCC 9896), whose growth allows the determination of "total biotin," i.e., biotin, biotin-

d-sulfoxide, biocytine, dethiobiotin, and pelargonic acid [51], and *Lactobacillus plantarum* 17.5 (ATCC 8014), which responds to "true biotin," i.e., biotin and biotin-*d*-sulfoxide [52]. These determinations are easy to use, although time consuming, and are well adapted to repetitive routine tests.

4.2 Methods Involving Avidin

4.2.1 2-(4'-Hydroxyazobenzene)benzoic Acid Method (HABA)

This method [43,53,54] takes advantage of the formation of a complex between avidin and HABA. Biotin, which has a higher affinity for the avidin than HABA, quantitatively displaces the dye from the complex. The titration is generally monitored spectrophotometrically at 500 nm. This method is easy to use, but its sensitivity is low (around 1 μg).

4.2.2 Spectral Shift Method

Upon complexation with biotin, the avidin absorbance at 233 nm is modified [43,53,54]. The sensitivity of this method is similar to that of the HABA method.

4.2.3 Fluorescence Determinations

Two methods have been described so far that monitor the titration of biotin with avidin using fluorescence. The first one measures the quenching of the avidin tryptophan fluorescence by biotin [55,56], whereas the second involves the increase of fluorescence of a fluorescently labeled avidin upon addition of biotin [57]. The sensitivities are, respectively, 20 and 0.5 ng.

4.2.4 Chemiluminescence

This method [58] is based on the competition between biotin and a biotin luminol derivative for avidin. The titration is monitored using the chemiluminescence of the luminol derivative.

4.2.5 Isotope Dilution Methods

These methods, which depend on the availability of labeled biotin, have been developed recently and are based on the dilution of the biotin to be estimated by a known amount of labeled biotin, followed by the determination of the specific activity of the avidin complexed and/or of the free biotin after thorough separation of both species. [^{14}C]biotin [59-62], [^{3}H]biotin [63], and even a ^{125}I-labeled derivative of biotin [64] have been used. The separation of free and avidin-complexed biotin has been achieved with bentonite [63], zinc hydroxide [59-61], or dextran-coated charcoal [62]. The sensitivity depends on the specific activity of the labeled biotin and can range from 20 ng with [^{14}C]biotin, to 1 ng with [^{3}H]biotin, down to less than 0.1 ng with the ^{125}I-labeled derivative.

5. CHROMATOGRAPHIC STUDIES

5.1 Ion-Exchange Chromatography

Since anion-exchange resins are available with high chemical purity, ion-exchange chromatography can be used for the purification and concentration of traces of weakly acidic materials such as biotin. This technique has been applied to isolate and purify biotin and biotin vitamers from the culture filtrate of many microorganisms, including molds, yeasts, and bacteria.

Ogata [13,65] used a Dowex 1X2 formate column. After washing with deionized water, the biotin vitamers [biotin, biotin sulfoxide (Fig. 11.1), and dethiobiotin (Fig. 11.5)] were eluted with formic acid (0.012 mol/liter). This has also been achieved on Dowex 1-acetate column [66]. After washing with water, the column was developed with increasing concentrations of acetic acid (0.5-2 mol/liter). The biotin, biotin sulfoxide, and dethiobiotin were eluted together with acetic acid (1.5 mol/liter).

In order to purify biotin and its sulfoxides extracted from the culture medium of *Aspergillus niger*, Tepper et al. [67] and Im et al. [68] converted them to biotin sulfone by treatment with hydrogen peroxide in excess and chromatographed them on a Dowex 1X8 formate column. Biotin sulfone was eluted with a linear gradient of ammonium sulfate (0-1 mol/liter). The ammonium ion has been removed on Dowex 50, a sulfonic acidic cation-exchange resin.

The same Dowex 1X2 chromatography technique proved to be a very good technique to monitor and titrate biotin and vitamers in biological medium during an investigation of the biosynthesis of biotin [69-71]. The elution pattern in Fig. 11.6 illustrates, for instance, the chromatographic separation of a mixture of [^{3}H]dethiobiotin and [^{14}C]biotin.

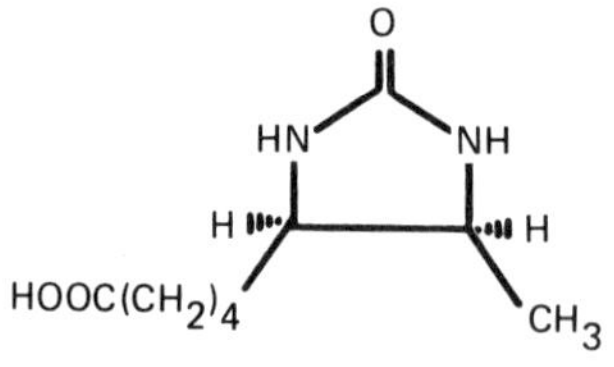

Figure 11.5 Structure of dethiobiotin.

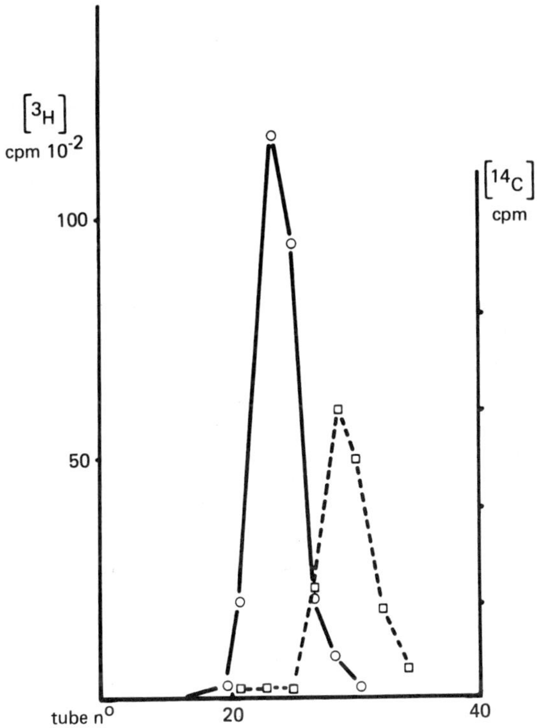

Figure 11.6 Elution pattern from column chromatography of [^{3}H]-dethiobiotin (50 μg SA, 200 mCi/mmol) and [^{14}C]biotin (5 μg SA, 31 mCi/mmol). The mixture was poured over a column 1.5 × 16.5 cm of Dowex 1X2 (formate) and eluted with a linear gradient of formic acid (0-1 mol/liter, 600 ml). Fractions (12 ml) were collected.

5.2 Paper Chromatography

So far, only few examples of identification of biotin by paper chromatography have been described in the literature. Ascending paper chromatography has been carried out with *n*-butanol-HCl, 1 mol/liter (6:1) [72], *n*-butanol-acetic acid-water (60:15:25) [66], *n*-butanol-acetic acid-water (2:1:1) [73], or *n*-butanol-formic acid-water (4:1:1) as developing solvents. More recently, Shrift [74] compared biotin and selenobiotin by descending paper chromatography using Whatman 1 filter paper with the solvent system ethanol-*t*-butanol-formic acid-water (24:8:2:6).

5.3 Thin-Layer Chromatography

5.3.1 TLC and Visualization Conditions

Thin-layer chromatography (TLC) has been applied to analyze biotin in biotin vitamers or multivitamin mixtures. This technique is more convenient and gives better results than paper chromatography. Silica gel plates have been generally used as support, but cellulose powder has been also proposed. The developing solvents are listed in Table 11.1.

Several detection reagents have been proposed to detect biotin [76]: 1% potassium permanganate, 1% dimethylaminobenzaldehyde in HCl (1 mol/liter) or iodine vapor. The detection limits obtained with such reagents are relatively high, about 5 μg. Better results have been obtained by spraying with mixtures of *o*-toluidine-potassium iodide [75] or *o*-toludine-chlorine [77]; biotin is revealed as a deep blue spot. The detection limit was found to be 0.3 μg [75].

However, the best detection agent is *p*-dimethylaminocinnamaldehyde (*p*-DACA). Shimada et al. observed that upon reaction with *p*-DACA in sulfuric acid-methanol mixtures, biotin yielded an intense reddish orange color with maximal absorbance at 533 nm [79]. This reagent is quite specific for biotin, since the ureido ring is a prerequisite for suitable color response. The reaction proceeds by formation of the conjugate imine (Schiff's base) with cyclic compounds [45]. A spray reagent composed of equal volumes of 2% sulfuric acid in ethanol and 0.2% *p*-DACA has been used by McCormick to develop chromatograms [46].

TABLE 11.1 TLC-Developing Solvents

Solvents	Proportions (v/v)	References
Acetone-acetic acid-benzene-methanol	1:1:14:4 1:1:11:11	75-77 77
Butanol-water	1:1	76
Butanol-water-acetic acid	4:5:1	76
Chloroform-methanol-4% formic acid	10:1:1	76
Chloroform-methanol-formic acid	70:40:2	78
Butanol-water-benzene-methanol	2:1:1:1+ε	70,71

5.3.1 *Analysis of Multivitamin Mixtures*

Nuttall and Bush [75] described a chromatographic method for the analysis of multivitamin preparations. After extraction of fat-soluble vitamins, water-soluble materials were separated in three TLC systems. Biotin was resolved with acetone-acetic acid-benzene-methanol (1:1:14:4) as solvent and visualized by spraying *o*-toluidine-potassium iodide. Addition of standard samples could be included if quantitative results are required. However, the reproducibility of the technique has not been tested.

Groningsson and Jansson [78] worked out a method for the determination of biotin in the presence of other water-soluble vitamins. After dissolution of the lyophilized preparation and addition of the internal standard (2-imidazolidone), the sample was applied on a silica gel plate and eluted with chloroform-methanol-formic acid (70:40:2). Biotin was visualized by spraying with *p*-DACA and determined in situ by reflectance measurements. The sensitivity of the method could be increased by spraying with paraffin after the coloring procedure. Under these conditions, the detection limit was 10 ng.

5.4 Gas-Liquid Chromatography Techniques

5.4.1 *GLC*

Biotin is not volatile enough for direct GLC analysis but it is possible to use this technique after substitution of the active hydrogen of the carboxyl group. While developing a general method for the analysis of urinary acids, Horning et al. [80] observed that trimethylsilyl ester-trimethylsilyl ether were the more attractive derivatives for GLC separation of hydroxy acids and water-soluble vitamin acids. They used bis-trimethylsilyl acetamide (BSA) as silylating agent.

Using this procedure, Viswanathan et al. [81] described the quantitative determination of biotin after its conversion into silyl ester. The column was a 2% OV 17 on diatomite (chromosorb G, AW/HMDS treated). *n*-Octosane was added as internal standard, and detection was achieved with a flame ionization detector. The detection limit for biotin trimethylsilyl (TMS) ester was approximatively 3 μg. Viswanathan et al. observed that BSA was suitable for complete silylation, whereas the hexamethyldisilazane-trimethylchlorosilane (2:1) mixture was not. Furthermore, they assumed that the reaction occurred exclusively at the carboxyl group and no reaction took place on the two amido NH groups (Fig. 11.7).

More recently, Johnson and Boyden [82] proved that the proposed structure (Fig. 11.7) was incorrect. In fact, the silyl derivative prepared according to Ref. 81 was not the silyl ester but the trisilyl derivative (Fig. 11.8).

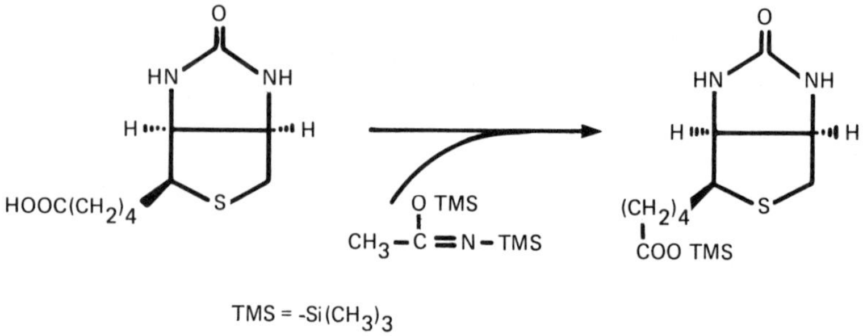

Figure 11.7 TMS esterification of biotin.

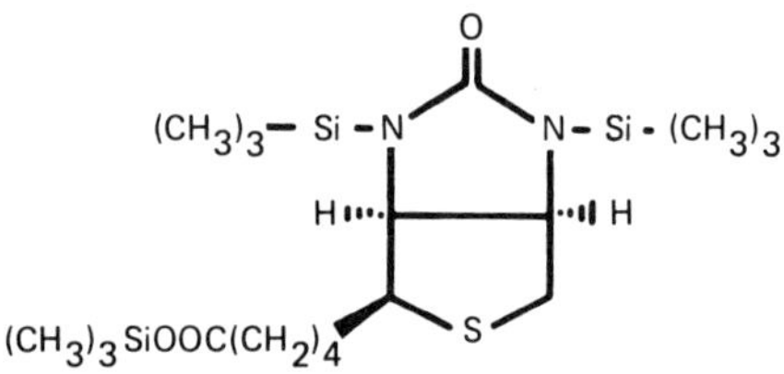

Figure 11.8 Trisilyl derivative of biotin.

This trisilyl derivative can be formed upon heating biotin with a solution of BSA and pyridine (1:1). The same silylating procedure can be applied to the biotin methyl ester in order to prepare the N-silylated methyl ester.

In an earlier study, Janecke and Voege [83] attempted to determine biotin and its oxidation products in commercial multivitamin preparations using the silylation technique. However, the separation of biotin, biotin sulfone, and sulfoxides from a multivitamin mixture was not thorough and the two isomeric sulfoxides could not be separated.

5.4.2 GC-MS

Johnson and Boyden [82] analyzed biotin with gas chromatography-mass spectrometry (GC-MS) using a 3% OV 17 column. They characterized the trisilylated biotin derivative (Fig. 11.8) and the corresponding methyl ester. The mass spectra consisted of the molecular ion and a base peak at m/z = 73, corresponding probably to the loss of one trimethylsilyl group.

In conclusion, the silylation derivatization seems to give the best results for analysis of biotin by GLC. However, this reaction is hardly reproducible and no direct application to the identification or estimation of biotin in biological mixture has been described in literature so far.

5.5 Liquid Chromatography Techniques

5.5.1 LC

In the case of thermostable and involatile compounds such as biotin, LC is a very valuable technique of analysis. However, the limitation lies in the availability of a suitable detection system.

Biotin does not absorb in the UV region, and only refractometer detection can be applied to direct analysis. Biotin methyl ester has been easily identified and separated from its biogenetic precursor dethiobiotin methyl ester using reversed-phase chromatography (C18) with a methanol-water mixture in isocratic mode (4:6) [84]. Satisfactory results, although with the usual low sensitivity due to the detector, have been obtained (Fig. 11.9).

In order to facilitate the analysis of small quantities of biotin and its analogs, a convenient derivatization for UV or fluorimetric detection has been described [85]. Biotin has been converted to UV-absorbing

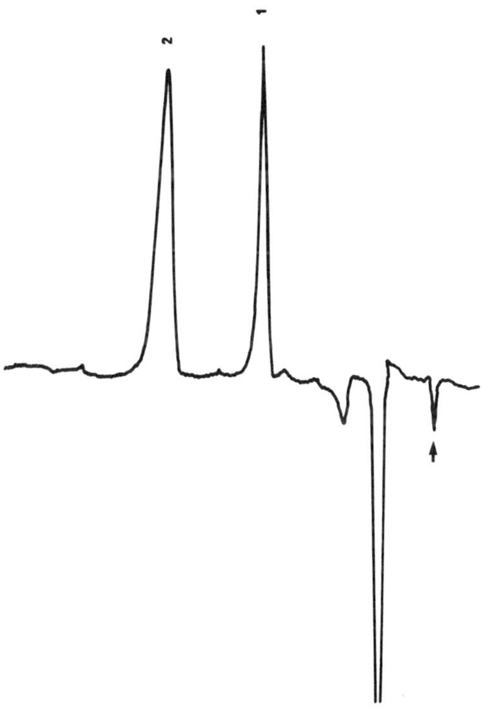

Figure 11.9 Liquid chromatographic separation of biotin methyl ester (peak 1) and dethiobiotin methyl ester (peak 2). Injected quantity: 50 μg; refractometer detection; column C18 bonded; solvent: methanol-water (4:6). (From Ref. 84.)

BAP esters by reaction with ω-4-dibromoacetophenone (DBAP) and to fluorescent Mmc esters by reaction with 4-bromomethoxycoumarin (Br-Mmc). Biotin and dethiobiotin are easily separated using an reversed-phase C18 bonded system with methanol-water or tetrahydrofuran (THF)-water as mobile phase. A continuous THF-water gradient is the best system to achieve a good chromatographic separation of the BAP or Mmc esters of dethiobiotin, biotin, and its oxidative derivatives, sulfone and sulfoxides (Fig. 11.10). With the UV technique (λ = 254 nm), the detection limit was approximately 50 ng, whereas it fell to 5 ng with fluorescence detection.

These LC methods are helpful in qualitative and quantitative analysis. In each case the regressions between the detection response and the injected samples are linear for the concentrations range usually seen in biological samples (from 0.15 to 4 nmol for BAP esters and 0.015 to 1 nmol for Mmc esters).

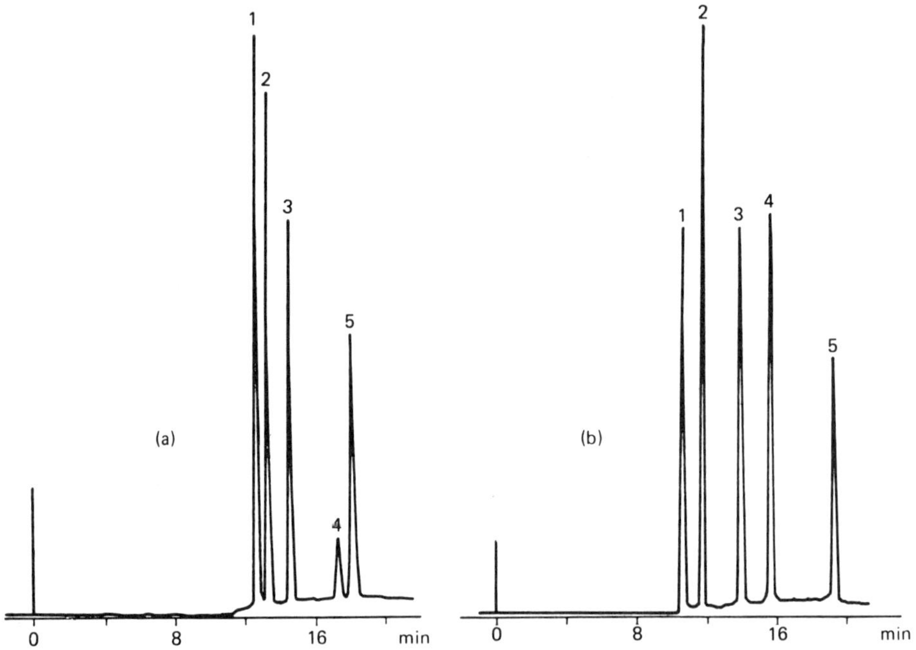

Figure 11.10 Liquid chromatographic separation of biotin and its analogs (a) BAP esters and (b) Mmc esters. Peak 1, biotin-*l*-sulfoxide; peak 2, biotin-*d*-sulfoxide; peak 3, biotin sulfone; peak 4, biotin; peak 5, dethiobiotin. Conditions: 2 × 33 cm C18 bonded columns; gradient: THF-H_2O from 30:70 to 60:40; flow rate, 1 ml/min. (From Ref. 85.)

5.5.2 *LC-MS*

Recently, interesting results were obtained by using the combined LC-MS technique [86]. Mass spectrometry is very specific and appeared as sensitive as UV detection (Table 11.2). After treatment with diazomethane [70,71], biotin has been identified in biological medium as methyl ester [86] using a reversed-phase C8 bonded column and an isocratic system methanol-water (4:6) as mobile phase. The mass spectrometer was operated in positive ion chemical ionization mode, and the limit of detection was lower than 10 ng (Fig. 11.11). The regression between the detector response and the injected sample was found to be linear up to 300 ng.

TABLE 11.2 Comparison of the Detection Limits of Biotin Determination Methods

Method	Detection limit (ng)	References
p-DACA	10^3	47,48
Enzymatic	0.5	49
Microbiological	0.1	50-52
Avidin, HABA	10^3	43,53,54
Avidin spectral shift	10^3	43,53,54
Fluorescence	20	55,56
	0.5	57
Chemiluminescence	1-2	58
Isotope dilution		
^{14}C	20	59-62
^{3}H	1	63
^{125}I	0.1	64
TLC	3×10^2	75
	10	78
GLC	3×10^3	81
LC		
UV	50	85
Fluorescence	5	85
LC-MS	10	86

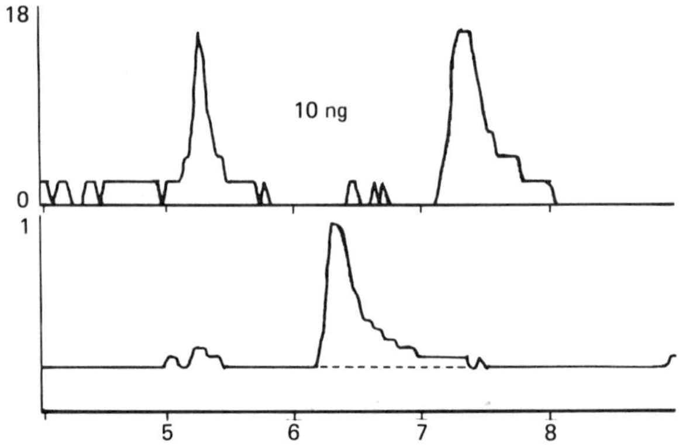

Figure 11.11 Separation of 10 ng biotin methyl ester (M^{+} = 258) and 10 ng of dethiobiotin methyl ester (M^{+} = 228) by LC-MS. Apparatus: Hewlett Packard LC 1084 B coupled with MS 5985 B. Column, C8, 10 cm; flow rate, 1 ml/min; solvent, methanol-water (4:6). Detection mode: positive ion chemical ionization. (From Ref. 86.)

6. CONCLUSION

Survey of the literature reveals that biotin concentrations have never been determined by chromatographic methods. In most cases, titrations have been achieved either by microbiological methods or by use of avidin complexation (Table 11.2).

The most convenient method depends on the range of biotin concentration. The avidin HABA method is very useful and rapid when concentrated solutions of biotin are concerned, whereas microbiological methods are unrivaled for routine tests on samples containing very low amounts of the vitamin. In the intermediate range, isotopic dilution methods are preferable because of their accuracy and ease in practice.

In biological materials, such as blood, bacteria, or cells, biotin exists in two forms, either free or bound, that is, covalently linked to carboxylases, and the determination of total biotin implies the hydrolysis of the peptide bond between biotin and the apoenzyme. This is currently achieved by autoclaving in 4 N sulfuric acid at 120°C for 2 hr. The pH is then brought to 2-3 and biotin is adsorbed on active charcoal. Elution can be performed with an ethanol-ammonium hydroxyde mixture (50% ethanol-28% ammonium hydroxyde, 9:1) [87]. Following concentration under reduced pressure, the syrup is extracted with 75% ethanol and purified on ion-exchange column AG 1x2.

The active charcoal adsorption step can be replaced by successive ethanol precipitations followed by concentration of the supernatant and ion-exchange chromatography [88].

In both cases the crude extract is suitable for the microbiological or avidin titrations.

Further derivatization, which is necessary for the chromatographic analysis as UV-absorbing BAP esters or as fluorescent Mmc esters [85], can be used to improve the purification of biotin or derivatives by chromatography on silica gel columns. The methyl esters, which are easily obtained after treatment with diazomethane for 15 min, can be eluted on silica gel columns with ethyl acetate-methanol (4:1) mixtures.

Among the recent techniques, the LC and especially the LC-MS coupling combine high sensitivity with good reproducibility, and it is likely that these techniques will become leading methods in quantitative biotin determinations in the near future.

REFERENCES

1. F. Kögl, *Chem. Ber.*, *68*, A16 (1935).
2. F. Kögl and B. Tönnis, *Z. Physiol. Chem.*, *242*, 43 (1936).
3. D. B. Melville, A. W. Meyer, K. Hofmann, and V. Du Vigneaud, *J. Biol. Chem.*, *146*, 487 (1942).
4. S. A. Harris, D. E. Wolf, R. Mozingo, and K. Folkers, *Science*, *97*, 447 (1943).
5. S. A. Harris, D. E. Wolf, R. Mozingo, G. E. Arth, R. C. Anderson, N. R. Easton, and K. Folkers, *J. Amer. Chem. Soc.*, *67*, 2096 (1945).
6. W. Traub, *Nature*, *178*, 649 (1956).
7. W. Traub, *Science*, *129*, 210 (1959).
8. C. Bonnemere, J. A. Hamilton, L. K. Steinrauf, and J. Knappe, *Biochemistry*, *4*, 240 (1965).
9. J. Trotter and J. A. Hamilton, *Biochemistry*, *5*, 713 (1966).
10. G. T. De Titta, J. W. Edmonds, W. Stallings, and J. Donohue, *J. Amer. Chem. Soc.*, *98*, 1920 (1976).
11. G. T. De Titta, R. Parthasarathy, R. H. Blesing, and W. Stallings, *Proc. Nat. Acad. Sci. U.S.*, *77*, 333 (1980).
12. C. H. Pai and H. C. Lichstein, *Biochim. Biophys. Acta*, *100*, 28 (1965).
13. K. Ogata, *Methods Enzymol.*, *18A*, 390 (1970).
14. F. Frappier, M. Gaudry, and S. Lavielle, unpublished results (1983).
15. D. B. Melville, *J. Biol. Chem.*, *208*, 495 (1954).
16. R. Lett and A. Marquet, *Tetrahedron*, *30*, 3365 (1974).
17. R. Lett and A. Marquet, *Tetrahedron*, *30*, 3379 (1974).
18. K. Hofmann, D. B. Melville, and V. Du Vigneaud, *J. Biol. Chem.*, *141*, 207 (1941).
19. A. B. A. Jansen and P. J. Stokes, *J. Chem. Soc.*, 4909 (1962).
20. J. Knappe, E. Ringelmann, and F. Lynen, *Biochem. Z.*, *335*, 168 (1961).

21. P. N. A. Murthy and S. P. Mistry, *Biochem. Rev.*, *43*, 1 (1972).
22. J. Moss and M. D. Lane, *Adv. Enzymol.*, *35*, 321 (1971).
23. H. G. Wood, *TIBS*, 4 (1976).
24. A. W. Alberts and P. R. Vagelos, *Enzymes*, *6*, 37 (1972).
25. F. Kalousek, M. D. Darigo, and L. E. Rosenberg, *J. Biol. Chem.*, *255*, 60 (1980).
26. F. Lynen, *Methods Enzymol.*, *71*, 781 (1981).
27. M. C. Scrutton and M. R. Young, *Enzymes*, *6*, 1 (1972).
28. P. A. Whitney and T. G. Cooper, *J. Biol. Chem.*, *247*, 1349 (1972).
29. W. Seubert, E. Fass, and U. Remberger, *Biochem. Z.*, *338*, 265 (1963).
30. H. G. Wood, *Enzymes*, *6*, 83 (1972).
31. H. G. Wood, *CRC Crit. Rev. Biochem.*, *7*, 143 (1979).
32. J. H. Galivan and S. G. H. Allen, *Arch. Biochem. Biophys.*, *126*, 838 (1969).
33. J. R. Stern, *Biochemistry*, *6*, 3545 (1967).
34. A. Munnich, J. M. Saudubray, A. Cotisson, F. X. Coudé, H. Ogier, C. Charpentier, C. Marsac, G. Carré, M. Bourgeay-Causse, and J. Frezal, *Eur. J. Pediatr.*, *137*, 203 (1981).
35. B. M. Charles, G. Hosking, A. Green, R. Pollitt, K. Bartlett, and L. S. Taitz, *Lancet*, *2*, 118 (1979).
36. M. J. Cowan, D. W. Wara, S. Packman, A. J. Ammann, H. Yishino, L. Sweetman, and W. Nyhan, *Lancet*, *2*, 115 (1979).
37. L. Sweetman, *J. Inher. Metab. Dis.*, *4*, 53 (1981).
38. B. J. Burri, L. Sweetman, and W. L. Nyhan, *J. Clin. Invest.*, *68*, 1491 (1981).
39. H. K. Ghneim and K. Bartlett, *Lancet*, *2*, 1187 (1982).
40. M. Zamboni, M. Gaudry, A. Marquet, A. Munnich, J. M. Saudubray, and C. Marsac, *Clin. Chim. Acta.*, *122*, 241 (1982).
41. B. Wolf, R. E. Grier, R. J. Allen, S. I. Goodman, and C. L. Kien, *Clin. Chim. Acta*, *131*, 273 (1983).
42. M. Gaudry, A. Munnich, J. M. Saudubray, H. Ogier, G. Mitchell, C. Marsac, M. Causse, A. Marquet, and J. Frezal, *Lancet*, *2*, 397 (1983).
43. N. M. Green, *Adv. Prot. Chem.*, *29*, 85 (1975).
44. F. Tausig and F. Wolf, *Biochem. Biophys. Res. Commun.*, *14*, 205 (1964).
45. H. C. Lichstein and J. Birnbaum, *Biochem. Biophys. Res. Commun.*, *20*, 41 (1965).
46. E. A. Bayer and M. Wilchek, *Methods Biochem. Anal.*, *26*, 1 (1980).
47. D. B. McCormick and J. A. Roth, *Anal. Biochem.*, *34*, 226 (1970).
48. D. B. McCormick and J. A. Roth, *Methods Enzymol.*, *18A*, 383 (1970).
49. S. Haarasilta, *Anal. Biochem.*, *87*, 306 (1978).

50. R. B. Ferguson and H. C. Lichstein, *J. Bacteriol.*, *75*, 366 (1958).
51. E. E. Snell, R. E. Eakin, and R. J. Williams, *J. Amer. Chem. Soc.*, *62*, 175 (1940).
52. L. D. Wright and H. R. Skeggs, *Proc. Soc. Exp. Biol. Med.*, *56*, 95 (1944).
53. N. M. Green, *Methods Enzymol.*, *18A*, 418 (1970).
54. N. M. Green, *Biochem. J.*, *94*, 23c (1965).
55. H. J. Lin and J. C. Kirsch, *Anal. Biochem.*, *81*, 442 (1977).
56. H. J. Lin and J. C. Kirsch, *Methods Enzymol.*, *62D*, 287 (1979).
57. M. H. H. Al-Hakeim, J. Landen, D. S. Smith, and R. D. Nargessi, *Anal. Biochem.*, *116*, 264 (1981).
58. H. R. Schroeder, P. O. Vogelhut, R. J. Carris, R. C. Boguslaski, and R. T. Buckler, *Anal. Chem.*, *48*, 1933 (1976).
59. R. L. Hood, *J. Sci. Food Agr.*, *26*, 1847 (1975).
60. R. L. Hood, *Methods Enzymol.*, *62D*, 279 (1979).
61. R. L. Hood, *Anal. Biochem.*, *79*, 635 (1977).
62. R. Rettenmaier, *Anal. Chim. Acta*, *113*, 107 (1980).
63. K. Dakshinamurthy and R. Allan, *Methods Enzymol.*, *62D*, 284 (1979).
64. T. Horsburg and D. Gompertz, *Clin. Chim. Acta*, *82*, 215 (1978).
65. K. Ogata, T. Tochikura, S. Inaura, K. Irushima, S. Takasawa, M. Kikuchi, and A. Nishimura, *Agr. Biol. Chem. (Tokyo)*, *29*, 895 (1965).
66. M. A. Eisenberg, *Biochem. Biophys. Res. Commun.*, *8*, 437 (1962).
67. J. P. Tepper, D. B. McCormick, and L. D. Wright, *J. Biol. Chem.*, *241*, 5734 (1966).
68. W. B. Im, D. B. McCormick, and L. D. Wright, *Methods Enzymol.*, *62D*, 385 (1979).
69. G. Guillerm, F. Frappier, M. Gaudry, and A. Marquet, *Biochimie*, *59*, 119 (1977).
70. A. G. Salib, F. Frappier, G. Guillerm, and A. Marquet, *Biochem. Biophys. Res. Commun.*, *88*, 312 (1979).
71. F. Frappier and A. Marquet, *Biochem. Biophys. Res. Commun.*, *103*, 1288 (1981).
72. K. Ogata, Y. Izumi, K. Aoike, and Y. Tani, *Agr. Biol. Chem. (Tokyo)*, *37*, 1093 (1973).
73. S. Iwahara, D. B. McCormick, and L. D. Wright, *Methods Enzymol.*, *18A*, 404 (1970).
74. C. Lindblow Kull, F. J. Kull, and A. Schrift, *Biochem. Biophys. Res. Commun.*, *93*, 572 (1980).
75. R. T. Nuttall and B. Bush, *Analyst*, *96*, 875 (1971).
76. V. M. Svetlaeva, D. S. Danilova, and M. T. Yanotovskii, *Khim. Farm. Zh.*, *12*, 140 (1978); *CA*, *90*, 12365 u (1979).
77. H. Thieleman, *Sci. Pharm.*, *42*, 221 (1974).
78. K. Groningsson and L. Jansson, *J. Pharm. Sci.*, *68*, 364 (1979).

79. K. Shimada, Y. Nagase, and U. Matsumoto, *Yakugaku Zasshi*, *89*, 436 (1969).
80. M. G. Horning, E. A. Boucher, and A. M. Moss, *J. Gas Chromatogr.*, *5*, 297 (1967).
81. V. Viswanathan, F. P. Mahn, V. S. Venturella, and B. Z. Senkowski, *J. Pharm. Sci.*, *59*, 400 (1970).
82. R. N. Johnson and G. R. Boyden, *J. Pharm. Sci.*, *66*, 1212 (1977).
83. H. Janecke and H. Voege, *Z. Anal. Chem.*, *254*, 355 (1971).
84. M. Jouany, A. Olesker, P. L. Desbene, and F. Frappier, unpublished results (1982).
85. P. L. Desbene, S. Coustal, and F. Frappier, *Anal. Biochem.*, *128*, 359 (1983).
86. M. Azoulay, P. L. Desbene, F. Frappier, and Y. Georges, (submitted for publication).
87. K. Ogata, *Methods Enzymol.*, *18A*, 397 (1970).
88. S. Iwahara and Y. Morishita, *Agr. Biol. Chem. (Tokyo)*, *41*, 2087 (1977).

12

VITAMIN B_{12} AND RELATED CORRINOIDS

JAN LINDEMANS and JOHANNES ABELS

Institute of Hematology, Erasmus University, Rotterdam, The Netherlands

1. INTRODUCTION

After the successful treatment of an experimental anemia in dogs [1] and of patients with pernicious anemia [2] with large amounts of raw liver, more than 20 years elapsed before the isolation and identification of the active principle, vitamin B_{12} or cyanocobalamin, ensued in 1948. This red material was isolated in crystalline form from liver almost simultaneously by Folkers and coworkers [3] and Smith [4]. Its corrinoid structure was defined by both chemical analysis [5] and x-ray crystallography [6] in 1956. Once it was recognized that the biologically active forms of vitamin B_{12} were highly photosensitive, Barker et al. [7-9] were able to isolate coenzyme B_{12} (adenosylcobalamin) from *Clostridium tetanomorphum* and from liver. A second biologically active form of vitamin B_{12} was detected in extracts from human plasma and proved to be identical with synthetically produced methylcobalamin [10, 11].

In all these studies the techniques for separation and detection of cobalamins were highly important. Even today these techniques are being continuously expanded and improved; the latest development is the application of high-pressure liquid chromatography (HPLC). It is the aim of this chapter to summarize established and modern techniques for chromatographic separation and detection of vitamin B_{12}, and to weigh their value in present clinical, biochemical, and pharmaceutical investigations.

1.1 Chemistry

1.1.1 Structure

Vitamin B_{12} or cyanocobalamin belongs to the corrinoids. This is a group of compounds having in common a corrin nucleus, i.e., a partially hydrogenated tetrapyrrole of which two of the pyrroles are joined directly rather than through methene bridges, with a central cobalt atom bound by coordinate linkages to the nitrogen atoms of the four pyrroles. Vitamin B_{12} is further characterized by a specific number of methyl, propionamide, and acetamide side chains attached to the pyrroles, and one side chain on ring D, in which the propionic acid is amidated with 1-amino-2-propanol, which, in its turn, is esterified with α-D-ribofuranosyl-(5,6-dimethylbenzimidazole)-3'-phosphate. The second N atom of the 5,6-dimethylbenzimidazole forms the fifth coordinate linkage with the cobalt atom in the α position (Fig. 12.1). The β position in the naturally occurring cobalamins is occupied by a CN^- (cyanocobalamin), OH^- (hydroxocobalamin), H_2O (aquocobalamin), CH_3 (methyl-

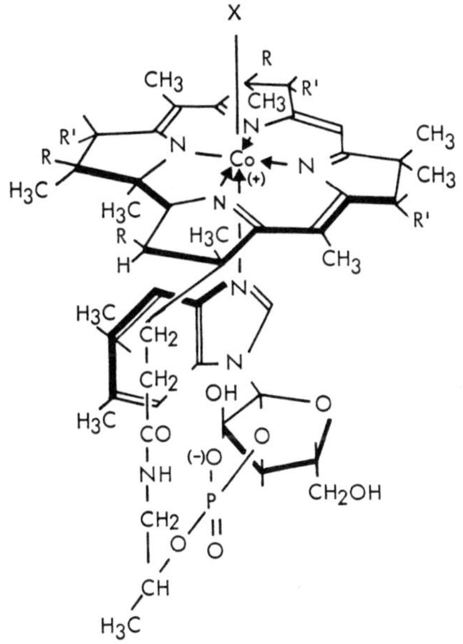

- X	name	
- CN	cyanocobalamin	(CN-Cbl)
- OH	hydroxocobalamin	(OH-Cbl)
- H_2O	aquocobalamin	(HOH-Cbl)
- deoxy-adenosyl	coenzyme B_{12}	(AdoCbl)
- CH_3	methylcobalamin	(MeCbl)

R = CH_2CONH_2
R' = $CH_2CH_2CONH_2$

Figure 12.1 Structural formula of cobalamin.

cobalamin, or the desoxyadenosyl group (adenosylcobalamin or coenzyme B_{12}). Other members of the corrinoid group, described in the section on biosynthesis, are depicted in Fig. 12.2. For an extensive treatise on corrinoid nomenclature the reader is referred to the IUPAC-IUB reports on this subject [12,13].

1.1.2 Stability

Cyanocobalamin is at room temperature fairly soluble in water (12 g/liter) and also in lower alcohols and phenol, but almost insoluble in acetone, ether, and chloroform. The molecular weight is 1355.4 ($C_{63}H_{88}O_{14}N_{14}PCo$) and the molecule is neutral in water. Between pH 4 and 7 cyanocobalamin is stable in aqueous solution and can be heated at 120°C without significant loss.

Alkaline hydrolysis (0.1 M) at 100°C induces the formation of "dehydrovitamin B_{12}," a biologically inactive form with a lactam ring fused to ring B of the cobamide nucleus [14]. Cyanide can be split off on exposure to light with the production of hydroxocobalamin. This process is stimulated by the addition of acid [15]. Hydroxocobalamin occurs in neutral or acid solution as aquocobalamin. Hydroxocobalamin can be converted to cyano-, sulfito-, chloro-, cyanato-, nitrito-, bromo-, thiocyanato-, and azidocobalamin by the addition of appropriate reactants under proper conditions [16]. At alkaline pH and excess of

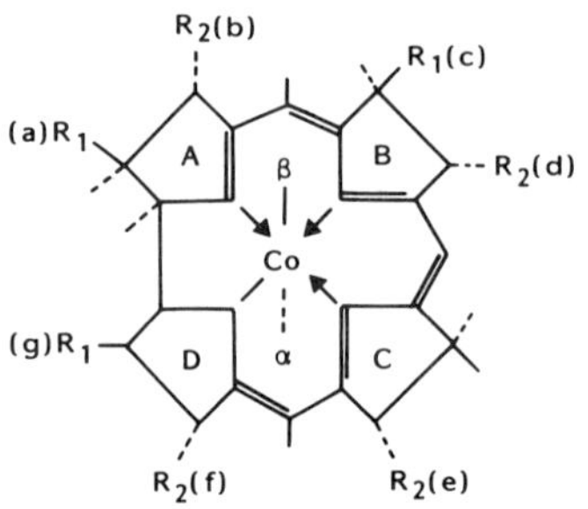

Cobyrinic acid:	R_1	= $-CH_2COOH$
	R_2	= $-CH_2CH_2COOH$
Cobyric acid:	R_1	= $-CH_2CONH_2$
	R_2(bde)	= $-CH_2CH_2CONH_2$
	R_2(f)	= $-CH_2CH_2COOH$
Cobynic acid:	R_1	= $-CH_2COOH$
	R_2(bde)	= $-CH_2CH_2COOH$
	R_2(f)	= $-CH_2CH_2CO\text{-}NH\text{-}CH_2CHOH\text{-}CH_3$
Cobinamide:	R_1	= $-CH_2CONH_2$
	R_2(bde)	= $-CH_2CH_2CONH_2$
	R_2(f)	= $-CH_2CH_2CO\text{-}NH\text{-}CH_2CHOH\text{-}CH_3$

Figure 12.2 Structural formulas of cobalamin precursors.

cyanide stable dicyanocobalamin is formed. Mild acid hydrolysis of corrinoids causes the loss of the peripheral amides from the propionamide side chains and the formation of mono-, di-, and tricarboxylic acids [17]. The most vulnerable is the propionamide in the e position. Deamidation of the more stable acetamide side chains requires more drastic conditions, in which the nucleotide moiety is also released together with the isopropylamine group at position f. The susceptibility to acid hydrolysis depends also on the ligands in α and β position.

Cyanocobinamide, 1-α-D-ribofuranosyl-5,6-dimethylbenzimidazole, and phosphate are formed during alkaline hydrolysis of cyanocobalamin in a suspension of cerous hydroxyde, pH 8-9, at 100°C for 2 hr [18].

In cyano- and hydroxocobalamin the cobalt atom is in the 3^+ state. Reduction to the Co^{2+}, and even the Co^{1+} state, can be brought about by electrochemical and purely chemical means [19]. The color then changes from red (Co^{3+}) to brown (Co^{2+}) and gray-green (Co^{1+}). Co^{2+}-cobalamin is also formed when aquocobalamin is treated with monothiols [20,21].

The naturally occurring organocorrinoids, AdoCbl and MeCbl, share extreme photosensitivity with most other artificially synthesized organocorrinoids. This makes it necessary to perform all analytical work with these compounds in the dark or under dim-red light.

In the absence of oxygen the photodecomposition of adenosylcobalamin leads to the formation of Co^{3+}-cobalamin [22] and a 5'-deoxyadenosyl that cyclizes to 8,5'-cyclic-adenosine [23]. In the presence of oxygen, aquocobalamin and adenosine-5'-carboxaldehyde are formed [24].

Photolysis of methylcobalamin occurs very rapidly in aqueous solution, with formation of formaldehyde and aquocobalamin as the major products. In the absence of oxygen the reaction is rather slow and gives rise to the formation of Co^{2+}-cobalamin and methane [25,26]. Remarkably, photolysis of methylcobalamin in the presence of homocysteine yields methionine, a methylation reaction that under aerobic, intracellular conditions occurs only in the presence of an enzyme and a reducing system [27].

The carbon-cobalt bond in methyl- and adenosylcobalamin is stable in neutral aqueous solution in the dark and withstands even heating for 20 min at 100°C [28]. By heating in 0.1 M HCl the carbon-cobalt bond of adenosylcobalamin is cleaved; the alkylcobalamins are relatively stable in dilute acid or alkali. Cyanide forms another threat to the adenosyl-cobalt bond, in particular at alkaline pH and at elevated temperature, forming dicyanocobalamin. The alkylcorrinoids are more resistant to the effect of cyanide, but are spliced through electrophilic attacks on the methyl group, for instance by ionic mercury or thallium, into aquocobalamin and the methylmercury cation or methylthallic dication [29].

Treatment of Co^{3+}-corrinoids with sodium bisulfite renders sulfitocorrinoids, which behave very similarly to the alkylcorrinoids with respect to photodecomposition and reactions with various compounds [30,31].

1.1.3 Spectroscopy

All corrinoids have distinct colors, varying from yellow, red, and purple to brown-green and blue [28], which can be utilized advantageously to identify the various compounds. The most reliable standard for the determination of the extinction coefficient of a corrinoid is the γ band of the compounds dicyanocobalamin or dicyanocobinamide, which are readily formed by cyanide, alkaline pH, and light from all other forms of cobalamin and cobinamide. The molar extinction coefficient of the sharp γ band at 367-368 nm amounts to $30.8 \times 10^3\ M^{-1}\ cm^{-1}$ [28,32].

The adsorption spectra of some representative corrinoids are presented in Figure 12.3.

1.2 Biosynthesis of Cobalamin

The biosynthesis of cobalamin is restricted to specific microorganisms, and most of our knowledge on biosynthetic pathways, extensively reviewed by Friedman [33,34] and Battersby and McDonald [35] and summarized in Fig. 12.4, stems from studies with cultures of *Streptomyces griseus*, *Propionibacterium shermanii*, and *Clostridium tetanomorphum*. The first steps in cobalamin biosynthesis follow those of porphyrin biosynthesis. At the stage of uroporphyrinogen III, seven methyl groups, derived from *S* adenosylmethionine, are attached and a hydrolytic ring contraction takes place by which the methene bridge between rings A and D is expelled. Subsequently, the cobalt atom is inserted and several side-chain modifications take place. The acetate side chain of ring C is decarboxylated and becomes a methyl group. At this stage the intermediate is called cobirynic acid. By amidation of the remaining carboxylic groups, except the propionic acid side chain of ring D, cobyric acid is formed, which is converted to cobinamide by the attachment of 1-amino-2-propanol, stemming from threonine, to the last propionic acid side chain. Cobinamide phosphate is formed at the expense of ATP through a cobinamide kinase activity. The next step in the biosynthesis

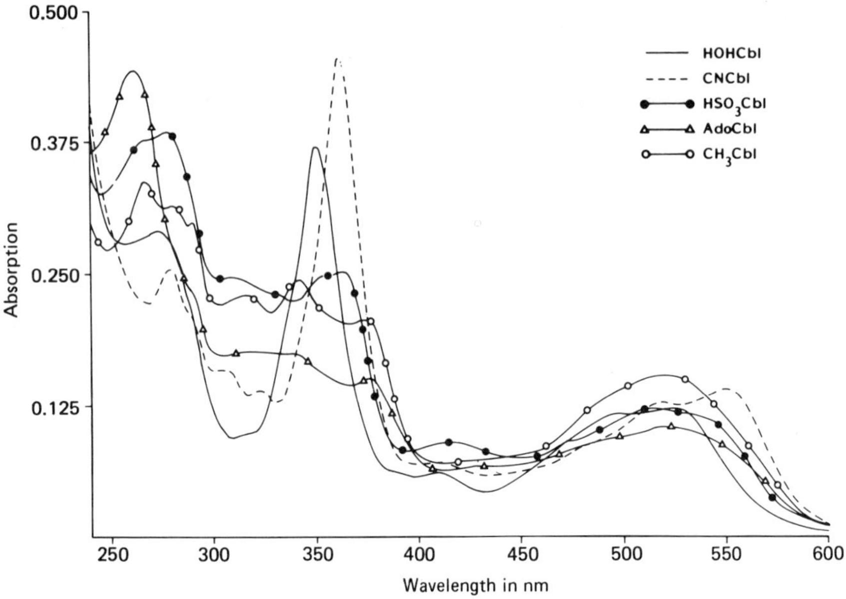

Figure 12.3 Absorption spectra of the naturally occurring cobalamins

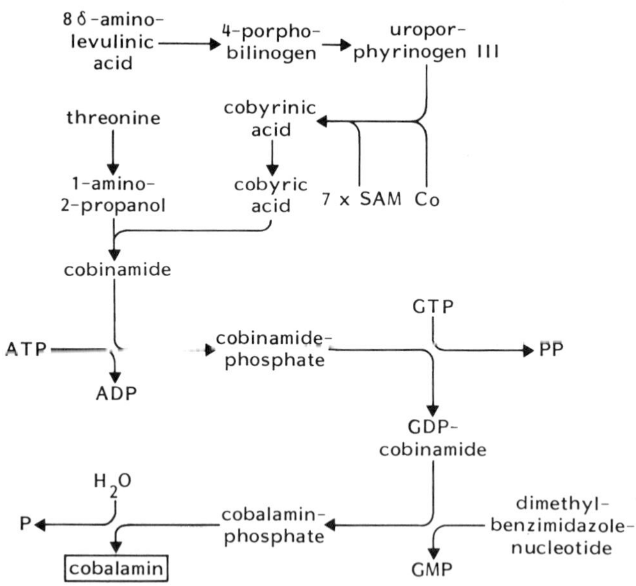

Figure 12.4 Biosynthesis of cobalamin.

is the formation of an anhydride linkage between the *N*-β-glycosidic ribonucleoside-5'-phosphate of GTP and the phosphate of cobinamide phosphate to form GDP-cobinamide. The GMP-moiety is subsequently exchanged for 5,6-dimethylbenzimidazole-*N*-α-ribose-5'-phosphate. The latter compound, with its rather unique *N*-α-glycosidic bond, is synthesized by the action of *trans*-*N*-glycosidase or nicotinate nucleotide phosphoribosyltransferase [E.C. 2.4.2.21]. This enzyme has very little base specificity, and it is possible to induce the formation of different corrinoids by feeding the growing bacteria all kinds of different bases. The final product of the base-exchange reaction, cobalamin phosphate, is ultimately dephosphorylated to cobalamin, preferentially in the "base-off" condition.

1.3 Cobalamin-Dependent Metabolism

The natural, metabolically active, forms of vitamin B_{12} are adenosylcobalamin and methylcobalamin, both of which can be formed from hydroxocobalamin. Most of our knowledge about the conversion of hydroxo- and cyanocobalamin into the active coenzyme forms stems from bacterial systems [36,37], but there is evidence that it is representative for the mammalian system as well [15,38,39]. The first step in the conversion of hydroxocobalamin is the two-step reduction of the Co^{3+} atom to the Co^{1+} state at the expense of NADH, by two different but cooperative reductases, both flavoproteins. After this reduction an adenosylating enzyme catalyzes the transfer of the adenosyl moiety from ATP to the Co^{1+}-cobalamin, yielding deoxyadenosylcobalamin and tripolyphosphate. The formation of methylcobalamin probably occurs in the process of the complex methyltransferase reaction catalyzed by N^5-methyltetrahydrofolate homocysteine methyltransferase [E.C. 2.1.1.13] [40], which is described later.

Adenosylcobalamin-dependent reactions always involve the exchange of a hydrogen atom of one carbon atom with another group (R) on an adjacent carbon atom. The only exception is the AdoCbl-dependent ribonucleotide reductase reaction in bacteria and the eukaryotic flagellate *Euglena gracilis*. An overview of AdoCbl-dependent reactions is presented in Table 12.1 [41].

Of these reactions only the methylmalonyl-CoA mutase [E.C. 5.4.99.2] reaction also takes place in mammalian metabolism and forms the link between the metabolism of odd-chain fatty acids, cholesterol, isoleucine, valine, threonine, methionine, and propionic acid, and the tricarboxylic acid cycle through succinyl-CoA (Fig. 12.5). The reaction is of special importance for ruminants, in which propionate, arising from cellulose degradation, forms a major source of energy. The enzyme involved, localized in the mitochondria, is composed of two nonidentical subunits, one of which binds adenosylcobalamin. A deficiency of adenosylcobalamin or the apoenzyme results in the accumulation of

TABLE 12.1 Adenosylcobalamin-Dependent Reactions

Enzyme	Substrate	Transferred group (R)	Product
Glutamate mutase	L-glutamate	Glycine	L-threo-β-methylaspartate
Methylmalonyl-CoA-mutase	L-methylmalonyl-CoA	-COSCoA	Succinyl-CoA
α-Methyleneglutarate mutase	α-Methyleneglutarate	-C(=CH_2)COOH	Methylitaconate
Dioldehydrase	Ethyleneglycol; 1,2-propanediol	OH	Acetaldehyde Propionaldehyde
Glycerol dehydrase	Glycerol	OH	β-Hydroxypropionaldehyde
Ethanolamine ammonia lyase	Ethanolamine	NH_2	Acetaldehyde + NH_3
β-Lysine mutase	L-β-lysine	NH_2	L-erythro-3,5-diaminohexanoic acid
α-Lysine mutase	D-lysine	NH_2	D-2,5-diaminohexanoic acid
α-Leucine-2,3-aminomutase	α-Leucine	NH_2	β-Leucine
Ornithine mutase	D-Ornithine	NH_2	D-threo-2,4-diaminovaleric acid
Ribonucleotide reductase	Ribonucleotide	—	2'-Deoxyribonucleotide

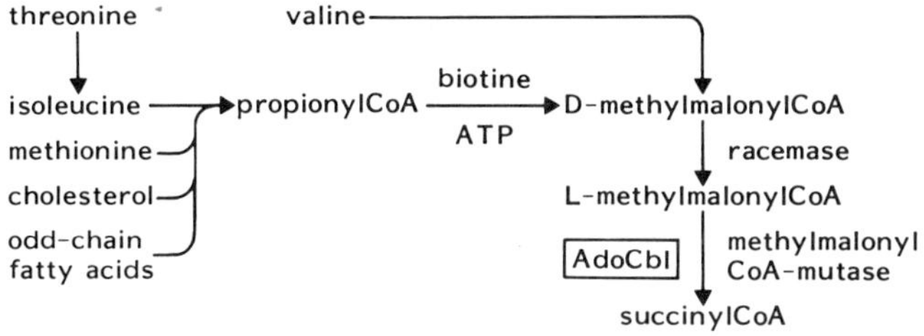

Figure 12.5 Propionic acid pathway.

methylmalonic acid, which is secreted by the kidneys and gives rise to methylmalonic aciduria [42].

Methylcobalamin-dependent reactions, involved in the synthesis of methionine in animals and microorganisms, and in the formation of acetate and methane in bacteria, have been reviewed by Poston and Stadtman [43] and recently by Taylor [40]. As stated before, methylcobalamin is formed from reduced Co^{2+}-cobalamin in the course of the methyltetrahydrofolate homocysteine methyltransferase reaction (Fig. 12.6). Presumably, the Co^{2+}-cobalamin is bound by the apoenzyme, further reduced to Co^{1+}-cobalamin in a $FADH_2$-requiring process, and methylated by 5-methyltetrahydrofolate in the presence of catalytic amounts of *S*-adenosylmethionine. Further transfer of methyl groups from methyltetrahydrofolate to homocysteine occurs independently of *S*-adenosylmethionine. Through this reaction, cobalamin is involved in the regeneration of tetrahydrofolate, an indispensible folate intermediate in the de novo biosynthesis of thymidilate, hence DNA, and in the regeneration of methionine.

Various bacteria from sewage sludge use methanol and acetate as major source of energy. The methyl group is first transferred to cobalamin and second released as methane, as a result of reductive cleavage [43].

In *Clostridium thermoaceticum* acetate is synthesized from CO_2 in a folate- and MeCbl-dependent pathway. CO_2 is first reduced to formate and 10-formyltetrahydrofolate is formed. After further reduction to 5-methyltetrahydrofolate, the methyl group is transferred to cobalamin, thus forming methylcobalamin. Subsequently, two molecules of acetate arise from a reaction between methylcobalamin and pyruvate, with carboxylmethyl-Cbl as a likely intermediate [44].

2. EXTRACTION OF COBALAMIN FROM BIOLOGICAL MATERIALS

In living tissues and organisms the cobalamins are always bound intracellularly, with high affinity, to the enzymes for which they function as cofactors, and in the extracellular fluids of higher animals, to specific transport proteins, such as transcobalamin II and R-type cobalamin-binding protein in human plasma.

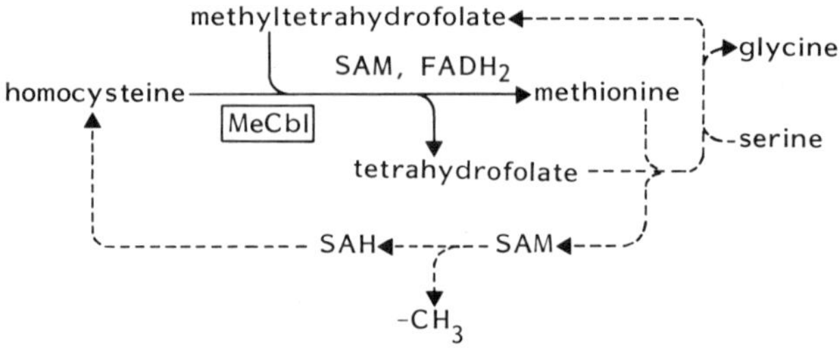

Figure 12.6 MeCbl-dependent regeneration of methionine and tetrahydrofolate.

2.1 Microorganisms

An overview of established extraction methods was given by Pawelkiewicz in 1962 [45] and by Friedrich in 1975 [46]. In summary, cell material is heated either in dilute buffer, 1% acetic acid, or in 75, 80, or 90% ethanol for 20-30 min at 80-100°C. The extraction efficiency improves by the addition of 0.01% cyanide, but then only the cyano forms of the different corrinoids are recovered. For extraction of organocorrinoids, e.g., AdoCbl and MeCbl, the procedure must be carried out under dim-red illumination, to avoid photodecomposition, in the absence of cyanide. After the heating particulate matter is removed by filtration or centrifugation.

Then the corrinoids are adsorbed onto activated carbon, Montmorillonite, or the cation-exchange resin Amberlite IRC-50-H. Elution from activated carbon is performed with alcohol-water mixtures; elution from Montmorillonite requires slightly alkaline buffers or phenol solutions. Cobalamin is released from the Amberlite by elution with water after neutralization of the ion-exchange resin with ammonia or, after washing with dilute acid, by elution with 0.1 M HCl in 75% acetone [47] or with 0.1 M H_2SO_4 in 50% acetone [48]. Further purification ofthe cobalamins in the eluate has generally been performed with phenol extraction [49].

Penley et al. [50] used phenol extraction for the isolation of fluoroalkylcobalamins from a cell-free system derived from *Methanobacterium* (strain MoH), containing aquocobalamin in a Freon-saturated aqueous solution. Koppenhagen et al. [51] applied extraction with hot 90% ethanol containing 0.2% glacial acetic acid, followed by phenol extraction, for the isolation of rhodium analogs of methylcobalamin and of coenzyme B_{12}. The aqueous corrinoid solution was extracted with portions of phenol-chloroform (3:2, w/w) and, after washing with water, the corrinoids were reextracted from the phenol by aqueous *n*-butanol. Hot ethanol extraction, followed by ion-exchange chromatography on DEAE- and CM-cellulose columns, was also used by Quadros et al. [52] for the isolation of ^{57}Co-labeled cobalamin coenzymes with high specific activity from *S. griseus*.

In their study on the biosynthesis of vitamin B_{12}, Ford and Friedman [53] used aqueous 75% acetone at room temperature in the dark to extract various forms of cobyric acid, cobinamide, cobinamide phosphate, and GDP-cobinamide from *P. shermanii*. Removal of salts and other contaminants was obtained by pehnol extraction.

2.2 Tissues and Cells

The methods for the extraction of cobalamins from microorganisms are generally also applicable to the extraction from tissues or tissue culture cells. For instance, Toohey and Barker [9] used hot 80% ethanol for the extraction of coenzyme B_{12} from liver acetone powder. Linnell et al. [54], Mahoney and Rosenberg [15], Fenton and Rosenberg [55], Frenkel et al. [56], Gimsing et al. [57], and Jacobson et al. [58] all applied hot ethanol extraction, four volumes of absolute ethanol on one volume of tissue or cell homogenate, as a primary step in the isolation of cobalamin. Farquharson and Adams [59] also used this procedure to study the forms of vitamin B_{12} in foods, except for the pH 9.6 of the buffer in which the food material was homogenized.

The extraction of cobalamin analogs from animal tissues by Kondo et al. [60] was carried out in one volume of 0.5 M sodium acetate-HCl pH 4.5, and two volumes of saline, in a Waring blender and heating for 45 min at 100°C.

The use of phenol in the subsequent purification step in the isolation of cobalamins has been largely replaced by the nonpolar adsorbent Amberlite XAD-2. As an extension of the work of Vogelmann and Wagner [61], Fenton and Rosenberg [55] developed a method for the desalting and partial purification of cobalamins using Amberlite XAD-2 columns or silanized silica gel 60 H. Columns are prepared in Pasteur pipettes plugged with glass wool or in 10 ml syringes filled with about 5 ml of Amberlite XAD-2 in methanol. The mesh size of the Amberlite must be reduced by crushing in a mortar in order to obtain a suitable flow rate. The silica gel 60 H is ready for use. The column bed is washed subsequently with 10 ml methanol, 10 ml methanol-water-acetic acid (50:50:5, v/v), and 10 ml water-acetic acid (100:5). The water in the solvents must be degassed under vacuum prior to use. The residue obtained by evaporation of an ethanol extraction from tissue homogenate is dissolved in a small volume of 1% acetic acid and applied to the adsorbent column. Salts and other compounds are removed by washing with 1% acetic acid, and cobalamins are eluted with 6-7 ml degassed water-methanol-acetic acid (50:50:5). Recoveries of OH-Cbl, SO_3-Cbl, and AdoCbl have been claimed to range from 90 to 98%, whereas phenol extraction generally yielded 82-93% of the cobalamins in an extract of rat liver mitochondria [55]. Cobalamins can also be eluted from the column with alkaline solvents, for instance, methanolic 1% ammonium hydroxide. However, under these conditions, OH-Cbl is strongly retained by silanized silica gel 60 H [55].

Amberlite XAD-2 adsorption has been used by Koppenhagen et al. [51] in the isolation of rhodium analogs of cobalamin cofactors and by Gimsing et al. [57] in the desalting of chemically prepared radioactive cobalamin forms.

A variant of the Amberlite XAD-2 is the Sep-Pak C18 reversed-phase cartridge (Waters) as applied by Jacobson et al. [58] for the desalting of an extract from L1210 leukemic lymphoblasts. The residue of the ethanol extract is dissolved in water and passed through the cartridge. After washing with water, elution of cobalamins is brought about by 50% acetonitrile.

In all these studies little attention has been paid to the efficiency of the recovery of the individual cobalamin forms in extracts. Only Frenkel et al. have noticed that the recovery of OH-Cbl was decreased and variable in comparison with the almost equal recoveries of CN-, Ado-, and MeCbl. A tentative correction was made for the apparent loss of OH-Cbl on the basis of the difference between total tissue cobalamin content and the sum of collected cobalamin fractions after separation [56].

2.3 Plasma

In most of the work on plasma cobalamin forms, the method initially described by Lindstrand and Ståhlberg in 1963 [10] is still being used. One volume of plasma is poured into four volumes of absolute alcohol and heated for 20 min at 80°C. After removal of precipitate by filtration the alcohol in the filtrate is evaporated under vacuum at 30°C. The residual water phase is extracted three to six times with ether to remove lipid material. In the next step cobalamins are extracted from the water phase into phenol by shaking in four volumes of phenol with 15% water. The phenol phase is shaken with one part of acetone, three parts of ether and a small amount of water. Phenol in the resulting water phase is further removed with ether. The water phase containing the cobalamin is evaporated to remove residual ether. Linnell et al. [62] used essentially the same procedure with smaller volumes of blood plasma.

Unsatisfactory extraction recoveries of cobalamin from plasma were first mentioned by Gimsing et al. [63]. By adding cyano[^{57}Co]cobalamin as an internal standard they showed that a loss of 30-50% of the original amount of cobalamin occurred during the ethanol and phenol extractions. At first they reported equal extraction recoveries for all cobalamin forms present in plasma, but this statement was withdrawn later when an underestimation, due to inadequate extraction of OH-Cbl, was reported [64]. Similar observations had been published by Mahoney and Rosenberg [65] and by Frenkel et al. [56].

Inadequate extraction of one component obviously invalidates the values of the substance concentrations of all other individual forms when calculated on the basis of their relative amounts in a chromatographed serum extract and the total amount of cobalamin in the original serum.

Gimsing et al. [66], who quantified the loss of hydroxocobalamin and also of sulfitocobalamin during ethanol extraction in serum, attributed the loss of these cobalamin forms to binding onto histidine residues of plasma proteins, a phenomenon already described by Bauriedel et al. [67], Taylor and Hanna [68], and Lien and Wood [69]. This binding process is optimal at pH 7.3, has specificity for aquocobalamin but not sulfitocobalamin, increases in the presence of 5 M urea, and is competitively inhibited by Cd^{2+} ions. In the histidine-bound form, Co^{3+}-cobalamine should be more resistant to reductive agents than in the free form.

By utilizing the inhibiting effect of cadmium, Gimsing et al. [66] were able to improve the extraction recovery of OH-Cbl as well as SO_3-Cbl to levels comparable to the extraction of CN-, Ado- and, MeCbl, as calculated from recoveries of added radioactive cobalamin forms.

Studies in our laboratory by Van Kapel et al. [70] confirmed the selective loss of OH-Cbl during plasma extraction with hot ethanol and in 1% acetic acid, showing a recovery of 83-93% in the extract of a plasma sample for added CN-, Ado-, and MeCbl, but only 30-40% for added OH-Cbl and SO_3-Cbl. In this study silanized silica gel 60 H, according to the method of Fenton and Rosenberg [55], was used to adsorb cobalamins from the extraction medium. The extraction efficiency could be improved by selective conversion of OH-Cbl into SO_3-, NO_2-, or N_3-cobalamin as a result of the addition of $Na_2S_2O_5$, $NaNo_2$, or NaN_3, but this made separation of the individual cobalamin forms thereafter impossible. In the course of these studies our attention was drawn by a color change from red to brown in ethanol as well as in dilute acid during the heating step, when this plasma was spiked with about 0.1 mmol/liter OH-Cbl. Since thiols reduce Co^{3+}-corrinoids to brown Co^{2+}-corrinoid-thiol complexes [20,21], it was assumed that this reduction might originate from thiol groups of proteins that had become more accessible by denaturation during heating [71]. Protein-thiol in plasma occurs in a concentration of 0.4-0.6 mmol/liter [72], which is sufficient to tackle a large amount of OH-Cbl. Various thiol-blocking agents were therefore investigated with respect to their ability to prevent the loss of OH- and SO_3-Cbl. In this respect chloroacetophenone, β-hydroxomercuribenzoate, ethylacrylate, *N*-ethylmaleimide, and iodoacetamide indeed improved the OH-Cbl recovery to the level of the other cobalamins, but only ethylacrylate and *N*-ethylmaleimide did not alter the distribution of the individual cobalamins.

We therefore applied the use of four volumes of 12.5 mmol/liter *N*-ethylmaleimide in 1% acetic acid for the extraction of plasma cobalamins. The extraction efficiencies of added cobalamins under the influence of *N*-ethylmaleimide, using silanized silica 60 H for desalting, are presented in Table 12.2.

Concerning the difference in opinion on the mechanism of OH-Cbl binding by plasma proteins between Gimsing et al. [66] and Van Kapel

TABLE 12.2 Extraction Efficiency of Exogenously Added Cobalamins Under the Influence of *N*-Ethylmaleimide

Cobalamin[a]	Control[b]	NEM[b]
OH-Cbl	19.5	77.2
SO_3-Cbl	23.1	73.6
CN-Cbl	78.4	86.8
AdoCbl	78.6	86.8
MeCbl	79.6	82.6

[a]Plasma enriched with about 800 pmol/liter of each individual cobalamin.
[b]Extraction with 4 volumes of 1% acetic acid (+12.5 mmol/liter *N*-ethylmaleimide) followed by desalting on silanized silica gel 60H. Expressed as a percentage of the amount added, as determined by radioisotope dilution assay.

et al. [70], various arguments favor the latter concept of SH binding. Incomplete OH-Cbl extraction has been observed both at pH 4.3 in acetate buffer and at neutral pH during hot ethanol extraction, whereas histidine binding apparently only occurs above pH 5.0 [68]. Moreover, OH-Cbl binding to plasma proteins can be prevented by *N*-ethylmaleimide, *p*-hydroxomercuribenzoate, or iodoacetamide, which are all SH-directed agents. In addition, Cd^{2+} ions not only bind to histidine in proteins, but can also block SH groups [73]. This observation has been confirmed in our laboratory in 1% acetic acid and hot ethanol extracts of human plasma using protein-SH-group detection with 5,5'-dithio-bis-(2-nitrobenzoic acid, DTNB) according to Sedlak and Lindsay [74]. On the contrary, Brewer and Riehm [75] have provided evidence that nonspecific interactions occur between *N*-ethylmaleimide and imidazole in proteins which might be an alternative explanation for the improvement of OH-Cbl extraction with *N*-ethylmaleimide in hot ethanol. Nevertheless, extraction of cobalamins in 1% acetic acid and 10 mmol/liter *N*-ethylmaleimide, directly followed by nonpolar adsorption to silanized silica gel, is much easier than hot ethanol extraction, which requires evaporation of the ethanol before the nonpolar adsorption step. Cadmium acetate is also less applicable in 1% acetic acid extraction because the denatured plasma solution becomes rather gelatinous and difficult to elute over the silica gel columns.

3. DETECTION OF CORRINOIDS IN CHROMATOGRAPHY

3.1 Light Absorption

In preparative chromatography with sufficient amounts of cobalamins, direct visual detection of red, orange and yellow spots or fractions or spectrophotometric monitoring of column eluates or TLC plates is possible. In general, high specificity and reasonable sensitivity are obtained at the γ band of the adsorption spectrum at 350 nm for corrinoids with Co-C bonds and at 365 nm for the cyano forms. With a broad slit width all forms can be monitored together; however, differences in molar extinction coefficients will result in different response factors. The less specific wavelengths at 254 and 280 nm have also been used succesfully, especially in high-pressure liquid chromatography. Frenkel et al. [56] obtained a linear response at 254 nm between 100 ng and 100 μg injected cobalamin.

3.2 Bioautography

Bioautography of chromatograms for analytical purposes was introduced by Lindstrand and Ståhlberg [10]. Paper chromatograms were placed on an agar dish (30 × 30 cm), inoculated with *Escherichia coli* 113-3 (ATCC 10586), and preincubated for 4 hr at 37°C. After 16 hr at 37°C the chromatograms were removed and the visual growth of the bacteria on cobalamin spots registered. Linnell et al. [62] improved the bioautography method by incorporating a growth indicator 2,3,5-triphenyltetrazoliumchloride, in the medium. Sensitivity was improved such that cobalamins extracted from only 2 ml serum could be identified in a thin-layer chromatogram. The colored spots were first evaluated by visually comparing the spots of unknown samples with the localization and color intensity of known standards; later, Linnell et al. [76] used a Johce-Loebl "chromoscan" model J-313 filter 5040 (500-600 nm) for reflected light scanning. Standard curves were preapred with different concentrations of the various cobalamin forms to establish the growth responses of the individual cobalamins. These were found to be equal for AdoCbl, CN-Cbl, and OH-Cbl, but a somewhat lower growth response was found for MeCbl. Gimsing et al. [63], who also used a growth indicator for quantifying individual cobalamin forms in the TLC plates, extracted the growth indicator with absolute ethanol at 37°C for 2 hr and measured the optical density at 485 nm. They reported equal growth rates for all cobalamins in the range 10-80 fmol and used CN-Cbl as a standard. Recently the growth response of *E. coli* in agar plates was evaluated more thoroughly by Gimsing et al. [56] (Fig. 12.7). In their final procedure the chromatograms were exposed to light after application of CN-Cbl and MeCbl standards and the filter paper, put between the TLC plate and the agar, was saturated with 2 mmol/liter sodium bisulfite.

3.3 Radioisotope Dilution Assay

With the introduction of high-pressure liquid chromatography for the separation of cobalamins from tissue extracts, the detection of cobalamin in the effluent fractions became amenable for assaying by competitive

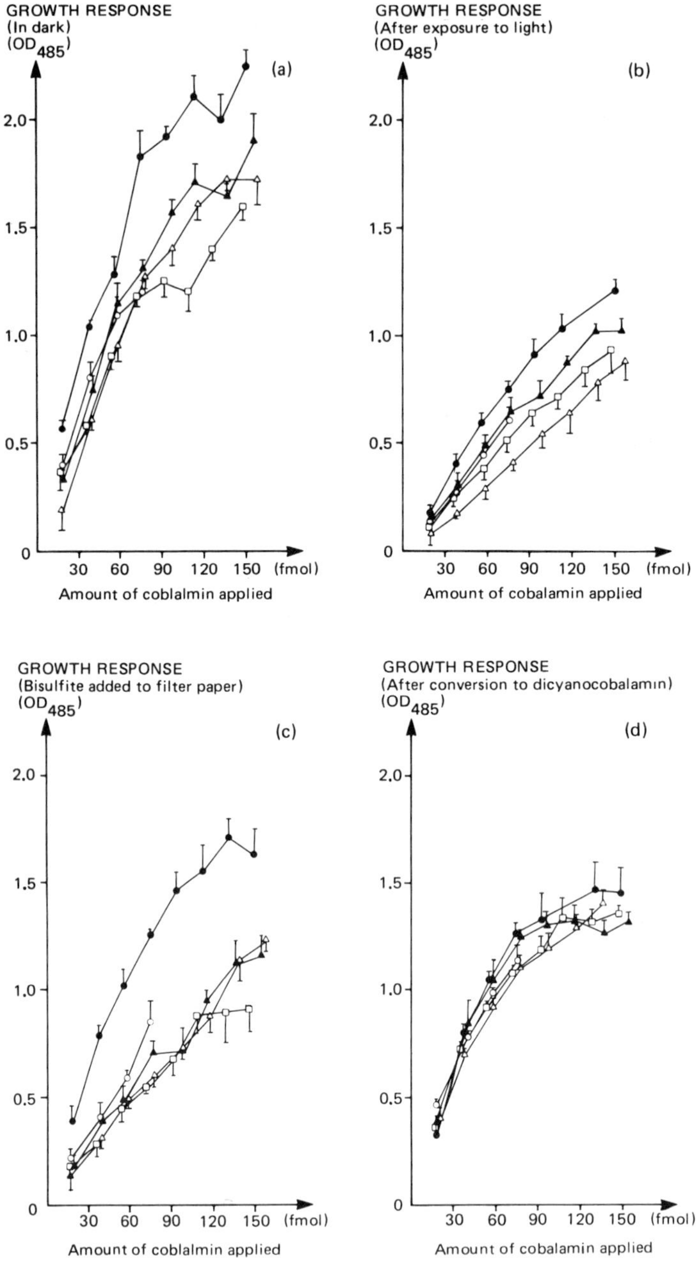

Figure 12.7 Growth responses of *E. coli* strain 113-3 in bioautography on silica gel plates to five cobalamin forms under the indicated conditions, means and standard deviations of fourfold analysis indicated. MeCbl (□), AdoCbl (▲), CN-Cbl (●), OH-/aquo-Cbl (○), SO_3-Cbl (△). (From Ref. 57.)

protein binding [56]. This method is subject to various matrix effects; this is why Van Kapel et al. [70] developed a special modification of the radioisotope dilution assay. Buffer conditions in cobalamin standards and column fractions are equalized by preparing the cobalamin standard solutions in 17% (v/v) acetonitrile and 1.5% (w/w) polyvinylpyrrolidone (PVP; molecular weight 25,000-30,000) in HPLC buffer, i.e., 85 mmol/liter phosphoric acid-triethanolamine, pH 3.3, and by adding PVP to the column eluate fractions in a final concentration of 1.5% (w/v).

Samples of the column fractions are diluted with 750 μl 45 mmol/liter sodium glutamate, 1 mmol/liter KCN, buffer pH 4.10, containing 40 pg CN[^{57}Co]Cbl. Cobalamins in the column eluate are in the free form. Therefore, without previous heating, 100 μl hog intrinsic factor, in which R binder is blocked with 30 nmol/liter $(CN)_2$Cbi, or human salivary R binder protein, both with a binding capacity of about 25 pg CN[^{57}Co]Cbl, is added. After 45 min incubation at room temperature, free cobalamin is removed by the addition of 300 μl dextran-coated charcoal (0.5% and 5% w/v, respectively, in water) and centrifugation for 10 min at 2500 × *g* at 10°C. The protein-bound radioactivity in 0.9 ml supernatant is counted.

The application of this procedure is not limited to the above conditions. However, care should be taken to keep the buffer condition of the standard solutions the same as that of the column solvent. In gradient elution the changes in solvent conditions, in this example the percentage acetonitrile, should not affect the cobalamin-binding of intrinsic factor of R-binder protein. The minimal detection limit of this method per cobalamin form is about 100 fmol. The advantage of this type of cobalamin assay over the microbiological detection is the possibility to discriminate the biologically active forms bound by both intrinsic factor and R-binder protein and biologically inactive cobalamin-analogs bound only by R-binder protein [77].

4. CHROMATOGRAPHY OF CORRINOIDS

4.1 Ion-Exchange Chromatography

Ion-exchange chromatography has been used mainly in preparative separations of cobalamin forms and their precursors. The use of Amberlite resins and cellulose derivatives has been reviewed by Pawelkiewicz [45]. With these materials it was impossible to separate methyl-, cyanoado, and hydroxocobalamin in one step. Tortolani et al. [78] first tried a combination of CM-cellulose and Dowex 50W-X2 columns to reach this goal, but later succeeded in a one-step separation on SP-Sephadex [79]. Subsequently, Gams et al. [80,81] and Gimsing and Hippe [82] used this procedure to separate the various coenzyme forms in L1210 cells, mitochondria, and plasma. Begley and Hall [83] successfully used this technique for measuring the conversion of CN[^{57}Co]Cbl into coenzyme forms in cells in tissue culture. They found that a fifth form of cobalamin, later identified as sulfitocobalamin, almost coeluted with the CN-Cbl peak, but by a small modification of the fraction collection procedure the SO_3-Cbl peak was eluted separately from the CN-Cbl peak (Fig. 12.8).

Separation of adenosyl forms of cobyric acid and various cobinamide and corrinoid intermediates in the biosynthesis of cobalamin in *P. shermanii* on Dowex 50W-X8 (200-400 mesh), by stepwise increase in pH of the eluting 0.05 M acetate buffer, was carried out by Ford and Friedman [53].

Ion-exchange paper chromatography on Whatman CM-82 paper was applied by Lee [84] for the separation of biosynthetic intermediates of cobyrinic acid, isolated from *P. shermanii*. Ion-exchange chromatography seems in particular suitable for the separation of the different carboxylic acid derivatives obtained by mild acid hydrolysis of cyanocobalamin. Each additional hydrolysis of an amide group on the corrin side chain results in longer retention on a QAE-Sephadex-column eluted with an acetic acid gradient in 0.4 M pyridine [85].

Recently, Katada et al. [86] described a novel isomeric form of vitamin B_{12} and its derivatives that could be separated from native cobalamin by DEAE-Sephadex chromatography, on which the isomeric form is retained, but the native cobalamin is not, or by SP-Sephadex chromatography in which the isomeric form always moves just a little faster than its normal counterpart.

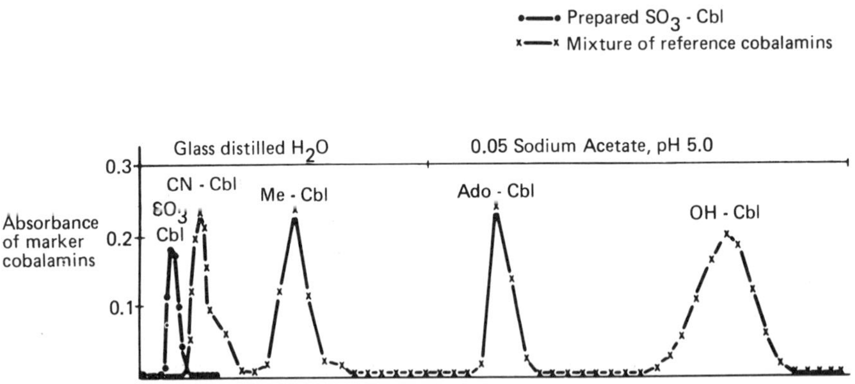

Figure 12.8 SP-Spehadex C-25 column chromatography of marker cobalamins at approximately 100 μg each. (From Ref. 83.)

4.2 Nonpolar Adsorption Chromatography

As already mentioned in the section on corrinoid extraction, the nonpolar adsorbent Amberlite XAD-2 has proven to be very valuable for batch separation of cobalamins from polar materials [55]. This technique is based on the work of Vogelmann and Wagner [61], who collected a vast amount of data on the chromatographic behavior of 50 different corrinoids, including rhodibamides and hydrogenobamides, and on the influence of C8, C10, and C13 substitutions, the axial ligands, and the nucleotide on retention times. Elution was generally performed with 3-20 vol% *tert*-butanol, and the effect of different conditions, such as variations in pH and cyanide concentration, were explored.

The method was used successfully in the isolation of descobaltocobalamin from a crude extract of *Chromatium* by Koppenhagen et al. [87]. An aqueous ethanol-acetic acid extract from 500 g wet cells was applied on an Amberlite XAD-2 column (4.0 × 7.0 cm) and sequentially eluted with: 0, 2, 10, 20, and 50% *tert*-butanol (v/v) in water. Hydrogenobyric acid eluted at 10%, descobaltocobalamin at 20%, and uroporphyrins at 50% butanol. After insertion of the rhodium atom in the descobaltocobalamin, the reaction mixture was subjected to desalting on a XAD-2 column, followed by further purification on CM- and DEAE-cellulose and once more on a XAD-2 column to obtain the pure mono- and dicyanorhodibalamins, eluting at 8 and 10% *tert*-butanol, respectively.

4.3 Paper Chromatography

The earliest analytical separations of plasma cobalamin extracts were carried out on Whatman 2 paper with *sec*-butanol-glacial acetic acid-water (100:3:50) [10] or *n*-butanol-isopropanol-water (10:7:10) [11] as ascending solvent. Afterward, paper chromatography was largely replaced by silica- and cellulose-TLC, but Allen and Majerus are still using ascending paper chromatography for the analysis of the acid hydrolysis products of cyanocobalamin following the procedure of Armitage et al. [17]. Chromatograms were developed on Whatman 3 MM paper at room temperature using 2-butanol-glacial acetic acid-water (440:4.1:150) with about 0.07% KCN as the ascending solvent. A similar system, 2-butanol-glacial acetic acid (440:4.1), saturated with water (about 210 ml) and containing 6.2 nmol KCN, was used by Kolhouse and Allen [88] for the preparation of various cobalamin analogs and by Kondo et al. [60] for the analytical chromatography of cobalamin analogs from animal tissues.

4.4 Thin-Layer Chromatography

In thin-layer chromatography (TLC) of corrinoids two adsorbents have been employed: cellulose and silica gel. The separation of a large number of organocobalamins on cellulose TLC with four different solvent systems has been described by Firth et al. [89] (Table 12.3). These systems have been used, among others, by Penley et al. [50] for the analysis of fluoroalkylcobalamins. Formylmethylcobalamin and related cobalamins were separated on cellulose TLC with 1-butanol-ethanol-H_2O (10:3:7), 0.5% NH_4OH, by Vickrey et al. [90].

The analysis of cobalamins from human plasma became much more sensitive and convenient by the introduction of silica gel-cellulose TLC and the use of a growth indicator for the bioautography [62]. Linnell et al. [91] developed a two-dimensional chromatobioautographic method in which complete separation of all individual plasma cobalamins was obtained. As little as 70 fmol of cobalamin can be spotted on the chromatogram, which is further developed by an ascending solvent, butanol-ammonia-water (75:2:25) in the first direction and, after drying, in a direction perpendicular to the first separation by the ascending solvent water-saturated benzyl alcohol. A representative separation is shown in Fig. 12.9.

Farquharson and Adams [59] used silica gel TLC with *sec*-butanol-propanol-water-ammonia (7:4:3:1) as the ascending developing solvent to analyze the forms of vitamin B_{12} in food. Silica gel TLC was also used by Nexø and Andersen [92] for the distribution analysis of plasma cobalamins bound either to transcobalamin I or to transcobalamin II. The solvent used for separation of the cobalamins was *sec*-butanol-isopropanol-water-ammonia (30:45:25:2). This chromatographic system was further evaluated with regard to R_f values of the individual cobalamin forms by Gimsing et al. [63], with special emphasis on the separate detection of sulfitocobalamin, a cobalamin form of which the presence in plasma is controversial. With the use of this system to analyze plasma and erythrocyte cobalamins after extraction in the presence of cadmium acetate and XAD-2 desalting, Gimsing et al. [57] obtained the following R_f values in duplicate analyses from six separate assays: MeCbl, 0.40; AdoCbl, 0.23; CN-Cbl, 0.35; OH-Cbl, 0.00; SO_3-Cbl, 0.58; (Ado)CN-Cba, 0.22; and $(CN)_2$Cbi, 0.003; the variation coefficients ranged from 0.03 to 0.08. The lower detection limit after bioautography was found to be 1-2 fmol, and the precision 0.15 for a sample of 80 fmol and 0.10 for a sample of 280 fmol MeCbl. The precision figures for relative amounts of cobalamins in a plasma pool, which was analyzed five times in 4 consecutive weeks, are presented in Table 12.4. Unfortunately, the authors did not present the results of recovery studies of individual cobalamins added to plasma.

Fenton and Rosenberg [55] also applied the principle of nonpolar absorption chromatography to the field of TLC separation of cobalamins. They presented R_f values for the five naturally occurring cobalamins in plasma on silanized silica gel 60 H plates developed with various combinations of acetic acid, methanol, and water (Table 12.5). In general, complete separation of all five cobalamin forms requires two-dimensional chromatography to obtain complete resolution of CN- and AdoCbl. Since

TABLE 12.3 Thin-Layer Chromatography of Vitamin B_{12} Derivatives on Cellulose[a]

Axial ligands	Solvent I (R_f 0.25)	Solvent II (R_f 0.23)	Solvent III (R_f 0.31)	Solvent IV (R_f 0.37)
		Cobalamins		
H_2O	0.15-0.75[b]	0.25-0.75[b]	0.30	0.55-0.75[b]
N≡C-	1.00	1.00	1.00	1.00
CH≡C-	1.50	1.45	1.40	1.15
CH_2=C-	1.75	1.85	1.60	1.20
CHBr=CH-	1.80	1.85	1.60	1.20
CH_2OHCH_2-	0.90	1.00	0.90	1.05
CF_3CH_2	1.95	2.10	1.60	1.20
$-SO_3CH_2$-	0.25	0.30	0.25	0.55
$HOOCCH_2$-	1.10	1.15	0.65[d]	1.10
CH_3-	1.80	1.80	1.50	1.20
CH_3CH_2-	1.90	1.80	1.60	1.20
$CH_3CH_2CH_2$-	2.05	2.15	1.70	1.20
C5'-deoxy-adenosyl	0.65	0.70	0.65	0.95
		Cobinamides		
H_2O	0.20-0.50[b]	0.35-0.50[b]	0.20[b]	0.80-1.00[b]
N≡C-	0.60,1.10[c]	0.80,1.20[c]	0.50, 0.80[c]	0.95, 1.05[c]
CH_2≡CH-	1.45	1.65	1.35	1.15
CH_3-	1.35	1.55	1.25	1.15
CH_3CH_2-	1.40	1.55	1.30	1.15
$CH_3CH_2CH_2$	1.60	1.75	1.50	1.20
SO_3^{2-}	0.75	0.74	0.59	0.83

[a]Solvent I, *sec*-butanol-water (9.5:4); solvent II, *sec*-butanol-glacial acetic acid-water (100:1:50); solvent III, *sec*-butanol-0.88 ammonia-water (9.5:0.675:4); and solvent IV, *n*-butanol-glacial acetic acid-water (4:1:5). Retention values relative to the R_f of cyanocobalamin.

[b]Elongated spots due to equilibration between aquo- and hydroxo-forms on the plate.

[c]Double spots due to the existence of stereoisomers involving the axial ligands.

[d]Ionized carboxylic acid.

Source: Ref. 89.

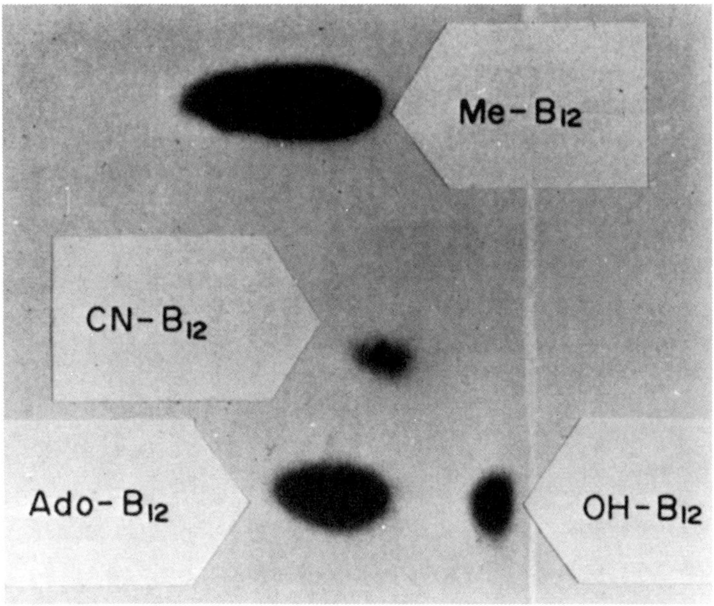

Figure 12.9 Two-dimensional chromatography and bioautography of plasma cobalamins from a normal subject, the origin marked at the left-hand corner. (From Ref. 54.)

TABLE 12.4 Precision of the Relative Amounts of the Cobalamins of a Plasma Pool Analyzed Five Times in Each of 4 Consecutive Weeks

		Standard deviation (SD)	
Cobalamin	Mean	Total	Intra-assay
SO_3-Cbl	0.16	0.10	0.07
MeCbl	0.65	0.09	0.06
CN-Cbl	0.02	0.02	0.02
AdoCbl	0.15	0.04	0.03
OH-Cbl	0.04	0.03	0.02
OH-Cbl + SO_3-Cbl	0.20	0.09	0.08

Source: From Ref. 57.

TABLE 12.5 R_f Values for Cobalamins on Reversed- and Normal-Phase TLC

	R_f				
TLC	OH-Cbl	MeCbl	AdoCbl	So_3-Cbl	CN-Cbl
Reversed[a]					
1:60:140	0.10	0.16	0.40	0.71	0.42
1:80:120	0.11	0.40	0.63	0.83	0.66
1:100:100	0.25-0.45	0.64	0.80	0.89	0.80
5:30:65	0.37	0.21	0.19	0.65	0.46
10:30:60	0.46	0.35	0.40	0.63	0.56
Normal phase[b]	0.02	0.52	0.27	0.61	0.45

[a]Prepoured silanized silica gel plates developed for 1-2 hr with acetic acid-methanol-water in the indicated proportions.
[b]Prepoured silica gel plates (IB 2) developed for 10 hr with ammonium hydroxide-2-propanol-2-butanol-water (1:50:50:50).
Source: From Ref. 55.

the separation on reversed-phase plates is based on partition chromatography and normal-phase on adsorption chromatography, the technique may be of value for isolating and identifying unusual cobalamins. An additional advantage of the method is the short developing time of 1-2 hr.

4.5 High-Pressure Liquid Chromatography

The latest development in the field of cobalamin separation is high-pressure liquid chromatography. An overview of methods and applications is presented in Table 12.6. For the separation of vitamin B_{12} from the vitamins B_2, K_3, C_1, D_2, E, and A in a multivitamin preparation, Schmit et al. [93] proposed a gradient elution on a Zibax ODS "Permaphase" column with a gradient changing from 100% water to 100% methanol at 3% per minute and a flow rate of 0.9 ml/min. The separation of vitamin B_{12} from vitamin B_2 on the one side and from vitamin K_3 on the other was not impressive, but as the authors stated, chromatographic conditions can be adjusted to suit the special requirements of a particular vitamin. Optimal conditions for the separation of a mixture of seven water-soluble vitamins on μBondapak C18 and μBondapak NH_2 have been more thoroughly investigated by Wills et al. [94], as demonstrated in the Figs. 12.10 and 12.11. The best combination for

TABLE 12.6 HPLC Separation of Cobalamins and Related Corrinoids

Column	Column dimensions (mm)	Mobile phase	Flow rate (ml/min)	Separated compounds	Reference
Partisil ODS 10μ	250 × 4.5	Methanol-0.2% ammonia (90:10, v/v)	2.6	OH-Cbl; methyl-/Cn-Cbl	94
Lichrosorb RP8 10μ	250 × 4.6	Acetonitrile-water (30:70, v/v), or 10 min linear gradient 10-25% acetonitrile in 0.083 M phosphoric acid-triethanolamine, pH 3.3	2.0 2.0	OH-, CN-, Ado-, MeCbl OH-, CN-, Ado-, MeCbl	98
Lichrosorb RP8 10μ	150 × 4.7	6 min linear gradient 15-45% methanol in 0.05% sulfuric acid	1.33	OH-, CN, Ado-, MeCbl; CN-Cbl from other water-soluble vitamins	95
Lichrosorb RP8 10μ	250 × 4.6	Acetonitrile-1 mM ammonium acetate, pH 4.4 (30:70, v/v), or 10 min linear gradient 5-30% acetonitrile in 0.05 M sodium phosphate pH 3.0	2.0 2.0	OH-, CN-, Ado-, MeCbl Cbl-, Cbi-analogs	58
Lichrosorb RP8 10μ	250 × 4.6	17 min linear gradient 0-25% acetonitrile in 0.085 M phosphoric acid-triethanolamine, pH 3.3	2.0	OH-, CN-, Ado-, MeCbl from human plasma	70
Lichrosorb Si-60 5μ	250 × 4.6	Methanol-0.01% sodium cyanide (40:60, v/v)	0.8	diCN-Cbl in human plasma	100
Zipax ODS "permaphase"	100 × 2.1	33 min linear gradient 100% water to 100% methanol (50°C)	0.9	Multivitamin preparation (B_2, B_{12}, K_3, C, D_2, E, A)	93
μBondapak C18	300 × 3.9	Methanol-1% KCl (20:80, v/v); methanol-1.5% tetrabutylammonium-phosphate (30:70, v/v)	1.0 1.0	Multivitamin preparation (B_1, B_2, B_6, B_{12}, C, folic acid, niacine, niacinamide)	96
μBondapak NH_2	300 × 3.9	Methanol-0.125% citrate, pH 2.4 (80:20, v/v)	1.0	Multivitamin preparation (B_1, B_2, B_6, B_{12}, C, folic acid, niacine, niacinamide)	96
μBondapak C18 Zorbax ODS 6μ	300 × 3.9 250 × 4.6	10 min concave gradient 23-70% methanol in 0.05 M NaH_2PO_4 (40°C)	1.8	OH-, CN-, Ado-, MeCbl, cobalamin analogs: SO_3-Cbl, monobasic acids of CNCbl, alkanolamine analogs, anilide derivative of monobasic acids	97
μBondapak C18	300 × 3.9	H_2O-acetic acid-isopropanol (90:1:9, v/v)	0.3	CN-Cbl; analogs of CN-Cbl, CN-Cba	99
		10 min linear gradient 3-21% isopropanol in 0.005 M 1-heptane sulfonic acid	1.0	SO_3^-, CN-, OH-, Ado-, MeCbl	99
μBondapak NH_2	300 × 3.9	58 mM pyridine acetate, pH 4.4-tetrahydrofuran (96:4, v/v)	0.3	Cbl, Cbl-monocarboxylic acids	99

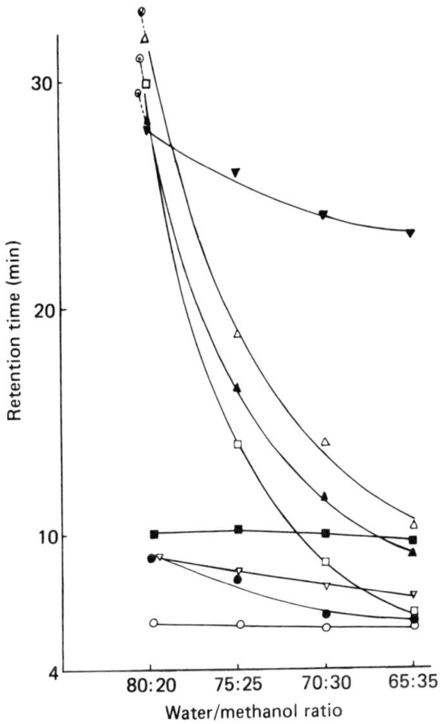

Figure 12.10 Effect of water-methanol ratio with tetrabutylammonium phosphate reagent on retention times of water-soluble vitamins on μBondapak C18: ascorbic acid (○), niacin (●), folic acid (□), pyridoxine (■), riboflavine (△), vitamin B_{12} (▲), niacinamide (▽), thiamine (▼), and ⊠ not eluted from column. (Reproduced from Ref. 94 by permission of Preston Publications, Inc.)

each of the individual vitamins in the preparation can be chosen from these data. A first attempt to separate different cobalamin forms was presented by Pellerin et al. [95], who described the separation of OH-Cbl from CN- and MeCbl on a Partisil ODS 10 μm column and methanol-0.2% ammonia (90:10) as a solvent at a flow rate of 2.6 ml/min. Complete separation of OH-, CN-, Ado-, and MeCbl was obtained by Mourot et al. [96] by the introduction of gradient elution from 15 to 45% methanol in 0.05% sulfuric acid with an increase of 5% per minute at 80 ml/hr on a Lichrosorb RP8 10 μm column (150 × 4.7 mm). For the quantitative analysis of cyanocobalamin in veterinary medicaments, they proposed isocratic elution with 0.05% sulfuric acid-acetonitrile (85:15) at 1 ml/min on a 10 cm Lichrosorb RP8 column. By UV monitoring at 350 nm, a

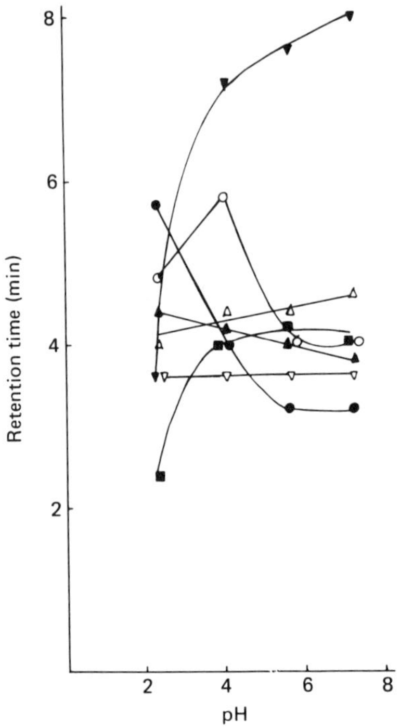

Figure 12.11 Effect of pH of 1% citrate in methanol (20:80) on retention times of water-soluble vitamins on μBondapak NH_2: ascorbic acid (○), niacin (●), pyridoxine (■), riboflavine (△), vitamin B_{12} (▲), niacinamide (▽), and thiamine (▼). (Reproduced from Ref. 94 by permission of Preston Publications, Inc.)

lower detection limit for specific analysis of cyanocobalamin of about 100 ng became possible.

A similar gradient elution for the separation of different cobalamin forms and various synthetic cobalamin analogs was described by Frenkel et al. [56,97]. A linear quantitative relationship between 100 ng and 50 μg of injected cobalamin was reported by monitoring UV adsorption at 254 nm.

Jacobsen and Green [98] achieved complete separation of the four cobalamin forms by isocratic elution on a C8-reversed-phase column using acetonitrile-water (30:70) at 2 ml/min, but separation was improved by elution with a gradient from 10 to 25% acetonitrile in 0.083 mol/liter phosphoric acid, titrated to pH 3.3 with triethanolamine, in 10 min at 2 ml/min. In a later report, Jacobsen et al. [58] have compared the separation by gradient elution and isocratic elution of the naturally occurring cobalamins, coenzyme analogs, aminoalkylcobalamins, and of various cobinamide forms (Table 12.7). In both elution systems a Lichrosorb C8 10 μm column was used. Isocratic elution was carried out with 30% acetonitrile in 1 mM ammonium acetate pH 4.4; the gradient system consisted of 0.05 M phosphoric acid titrated to pH 3.0 with concentrated ammonia and acetonitrile from 5 to 30% or 50% (as required) at a rate of 2.5% per minute and a flow rate of 2.0 ml/min (Fig. 12.12). In general separation in the isocratic system was satisfactory despite minor disadvantages of small pH-dependent mobility changes of some cobalamin forms (Ado- and H_2O-Cbl) and prolonged retention of some analogs. The elution methods also proved excellently suitable for the analysis of the conversion of radioactive cyanocobalamin into other cobalamin forms after uptake by either *Lactobacillus leichmanii* or L1210 lymphoblasts.

An even larger number of cobalamin analogs was investigated for their retention times on either μBondapak C18 columns or μBondapak NH_2 columns by Binder et al. [99] (Table 12.8). The elution solvent on C18 columns was H_2O-acetic acid-isopropanol (90:1:9), and the flow rate was 0.3 ml/min. NH_2 columns were eluted with 58 mM pyridine acetate, pH 4.4, with tetrahydrofuran (96:4) at a flow rate of 0.3 ml/min. The latter column was used to separate different monocarboxylic derivatives of cyanocobalamin (Fig. 12.13). For the separation of the five naturally occurring cobalamins the authors used elution of a C18 column with a 10 min linear gradient from 3 to 21% isopropanol in 5 mM 1-heptane sulfonic acid. Cobalamins were monitored at 254 nm (Fig. 12.14). Elution of only 25 pg CN[^{57}Co]Cbl on a Lichrosorb RP8 column according to Jacobsen and Green [96] and measuring the radioactivity in 0.5 ml column eluate fractions demonstrated also that this small amount could be separated in a sharp peak with more than 99% recovery [70]. This observation made it possible to separate cobalamins in extracts from human plasma by HPLC, coupled with a sensitive radioisotope dilution assay for the detection of cobalamins in the column fractions [70]. Because of the specific problems of cobalamin extraction, Van Kapel et al. [70] investigated the effect of various agents on the separation of a mixture of reference cobalamins (Fig. 12.15 and 12.16). The effects of NaN_3 on the separation profile are not clear; $Na_2S_2O_5$ causes the conversion of both OH- and CN-Cbl to SO_3-Cbl, and $NaNo_2$ leads to the partial formation of nitritocobalamin, eluting very close to Ado-Cbl. From the SH-blocking agents tested, the separation profile of the reference cobalamins was disturbed by chloroacetophenone, *p*-hydroxomercuribenzoate, and iodoacetamide, giving rise to the formation of unknown derivatives of one or more of the reference cobalamins. A typical distribution pattern of cobalamin and cobalamin analogs extracted from human plasma in the presence of SH-blocking *N*-ethylmaleimide is presented in Fig. 12.17. Recovery of injected cobalamin (about 1 pmol) in the eluate fraction varied from 97 to 100%.

TABLE 12.7 Retention Times of Cobalamin Coenzymes and Related Corrinoid Analogs on a LiChrosorb C8 Column

Corrinoid	Gradient elution (min)	Isocratic elution (min)
Naturally occurring		
Aq-Cbl	9.0	4.6-4.8[a]
HSO_3-Cbl	9.7	1.1
CN-Cbl	10.2	1.6
Ado-Cbl	11.4	2.8-3.0[a]
Me-Cbl	12.8	2.2
Coenzyme analogs		
CN-Cbl	10.9	5.2
Tub-Cbl	11.1	9.6
AraA-Cbl	11.2	3.3
For-Cbl	11.1	4.3
ε-Ado-Cbl	11.7	3.5
Pr-Cbl	14.5	9.1[b]
Dap-Cbl	16.8	27.1[b]
Aminoalkylcobalamins		
AC_2-Cbl	9.8	5.9
AC_3-Cbl	10.5	10.4[b]
AC_5-Cbl	11.2	nd
AC_8-Cbl	12.8	nd
AC_{11}-Cbl	14.9	nd
Cobinamides		
$(Aq)_2$-Cbi	7.8	33.8
$(CN)_2$-Cbi	—[c]	1.6
CN-Cbi	9.0, 10.5[d]	4.9, 5.7[d]
AC_3-Cbi	9.2	nd
Ado-Cbi	10.2	6.1
Me-Cbi	12.7	9.6
Dap-Cbi	17.0	23.9

[a]Sharp peaks but positions somewhat variable.
[b]Spreading.
[c]Compound unstable at pH 3.0; monocyano isomers observed
[d]Isomeric forms.
Source: From Ref. 58.

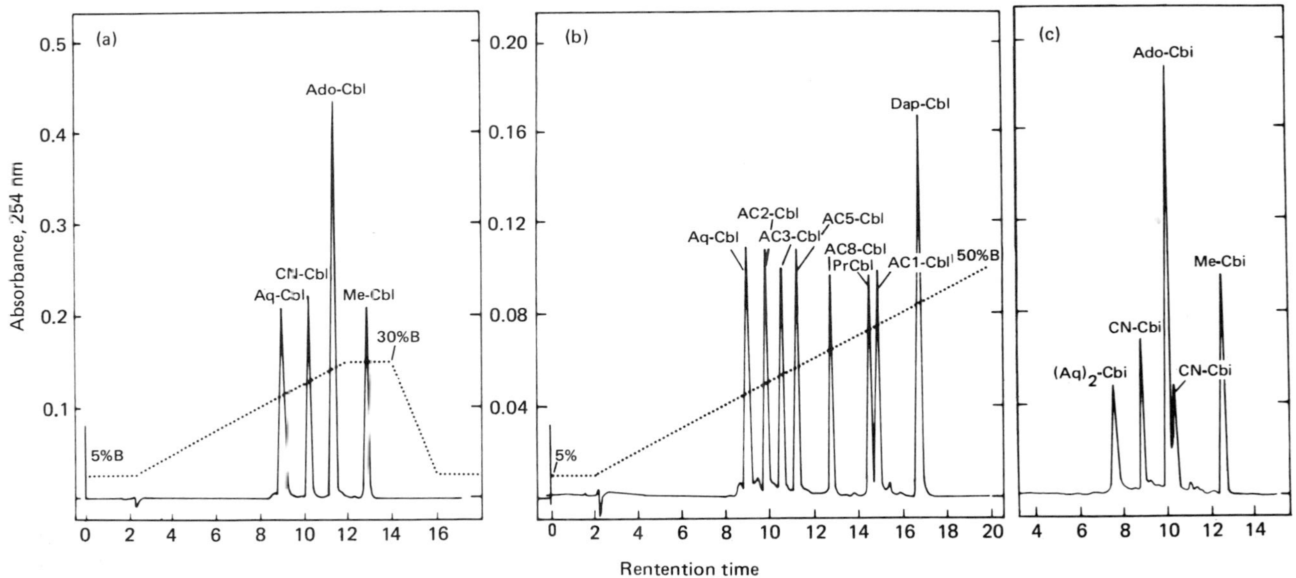

Figure 12.12 Gradient separation of cobalamin and cobinamide standards on a Lichrosorb C8 column. The gradient profile is indicated in (a) and (b) by the dotted line. (a) Separation of the naturally occurring cobalamins, 5 nmol each. (b) Separation of a mixture of cobalamin containing different upper-axiomal ligands. (c) Separation of cobinamide analogs, 5 nmol each, with a gradient identical to that in (a). (From Ref. 58.)

TABLE 12.8 Separation of Cobalamin Analogs by HPLC[a]

	Retention time	
Cobalamin	Min 30.0	Relative to CN-Cbl 1.00
Alterations adjacent to corrin ring		
CN-Cbl[C_8-OH,c-OH]	22.0	0.73
CN-Cbl[c-lactam]	24.0	0.80
CN-Cbl[c-lactone]	35.0	1.17
CN-Cbl[10-Cl,c-lactone]	68.3	2.28
CN-Cbl[13-epi]	21.1	0.72
CN-Cbl[b-OH]	38.5	1.30
CN-Cbl[d-OH]	41.8	1.38
CN-Cbl[e-OH]	45.8	1.53
CN-Cbl[C_8-OH,dc-OH]	23.7	0.79
CN-Cbl[C_8-OH,ec-OH]	27.7	0.92
CN-Cbl[C_8-OH,bc-OH]	26.7	0.89
CN-Cbl[de-OH]	61.4	2.04
CN-Cbl[bd-OH]	52.3	1.74
CN-CBl[be-OH]	53.2	1.77
CN-Cbl[bde-OH]	90.2	3.01
CN-Cbl[13-epi,d-OH]	29.4	0.97
CN-Cbl[13-epi.e-OH]	42.3	1.38
CN-Cbl[13-epi.h-OH]	27.2	0.92
CN-Cbl[c-lactam.d-OH]	27.8	0.93
CN-Cbl[c-lactam.e-OH]	24.2	0.81
CN-Cbl[c-lactam.b-OH]	20.9	0.70
CN-Cbl[c-lactone.d-OH]	52.3	1.74
CN-Cbl[c-lactone.e-OH]	46.6	1.55
CN-Cbl[c-lactone.b-OH]	38.4	1.28
Alterations in nucleotide		
[5,6-Cl_2BZA]CN-Cba	43.1	1.44
[NZA]CN-Cba	34.8	1.16
[5,(6)-MeBZA]CN-Cba	25.7	0.86
[5,(6)-C BZA]CN-Cba	28.8	0.96
[5,(6)-NO_2BZA]CN-Cba	19.5	0.65
[5,(6)-OCH_3BZA]CN-Cba	23.4	0.78
[5,(6)-COOH BZA]CN-Cba	19.5	0.65
[5,(6)-OH BZA]CN-Cba	21.1	0.70
[BZA]CN-Cba	23.6	0.79
[3,5,6-Me_3BZA](CN,OH)Cba	36.4; 121.7[b]	1.21; 4.06
[CN,OH]Cbi	22.7; 14.0[b]	0.76; 0.47
[2-MeAde]CN Cba	20.4	0.68
[Ade]CN-Cba	21.0	0.70

[a]Separation was performed on a Waters μBondapak C18 reversed-phase column (10 μm particle size, 300 × 3.9 mm ID) using H_2O-acetic acid-isopropanol (90:1:9) at a flow rate of 0.3 ml/min. Samples of 0.1-1.0 nmol were injected in a volume of 10 μl and detected based on absorbance at 365 nm.

[b]Two peaks are seen with these analogs in which the base is absent or cannot coordinate as the lower axial ligand; thus, the CN can be present as either the upper or the lower axial ligand.

Source: Ref. 99.

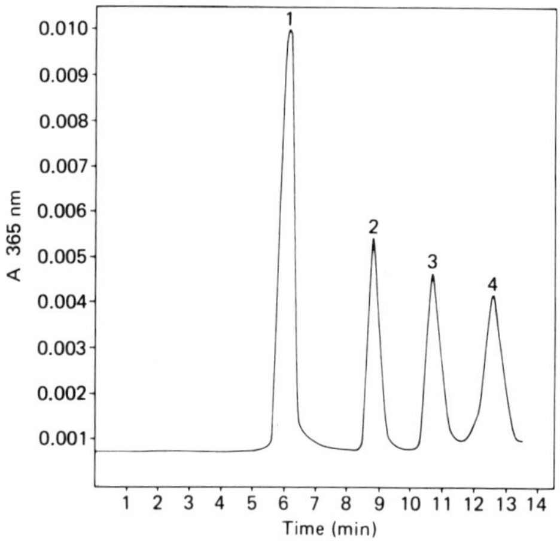

Figure 12.13 HPLC separation of Cbl (20 pmol) and Cbl monocarboxylic acids (10 pmol each) on μBondapak NH_2. CN-Cbl (1), CN-Cbl [d-OH] (2), CN-Cbl [b-OH] (3), and CN-Cbl [e-OH] (4). (From Ref. 99.)

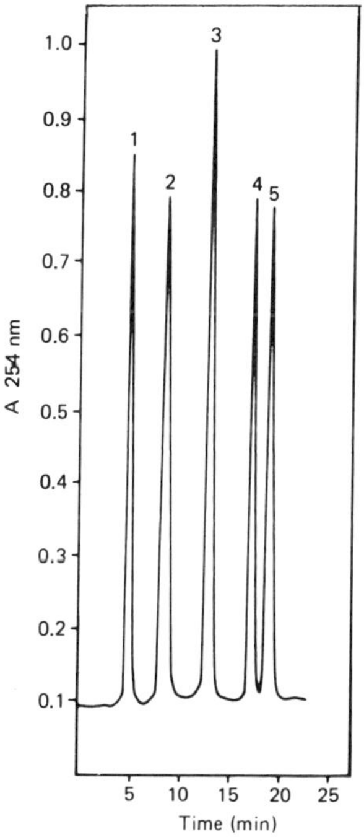

Figure 12.14 HPLC separation of the naturally occurring cobalamins on μBondapak C18 with an isopropanol gradient from 3 to 21% in 5 mmol/liter 1-heptane sulfonic acid: SO_3-Cbl (1), CN-Cbl (2), OH-Cbl (3), MeCbl (4), and AdoCbl (5). (From Ref. 99.)

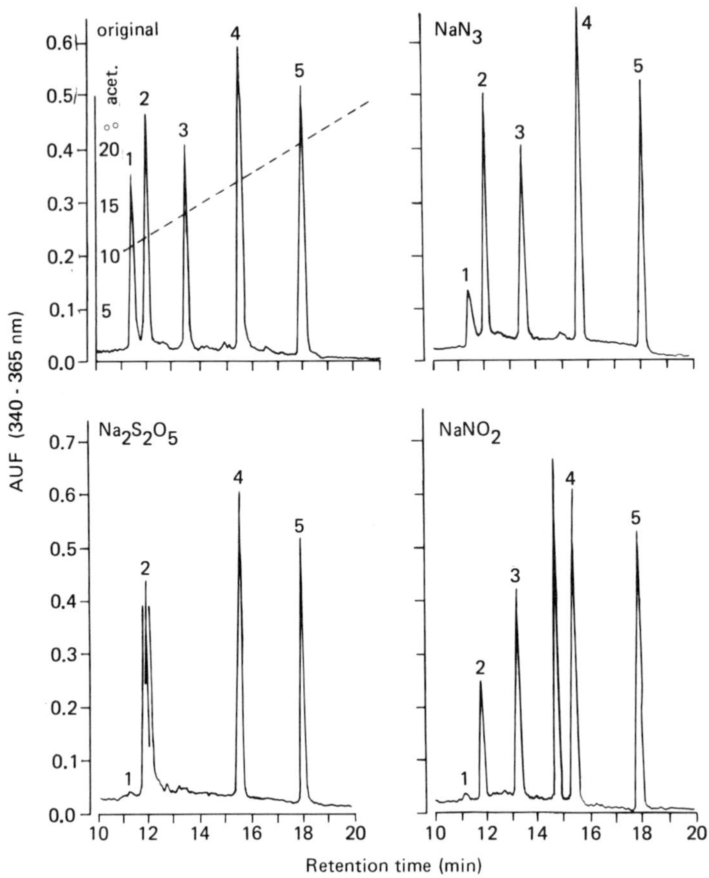

Figure 12.15 HPLC separation of a mixture of reference cobalamins (1.4 nmol each) on Lichrosorb RP8 with an 8-25% acetonitrile gradient in 85 mmol/liter phosphate-triethanolamine buffer, pH 3.3, after extraction from plasma without (original) or with additives, NaN3 (10 mmol/liter), $Na_2S_2O_5$ (1 mmol/liter), or $NaNO_2$ (5 mmol/liter): OH-Cbl (1), SO_3-Cbl (2), CN-Cbl (3), AdoCbl (4), and MeCbl (5). (From Ref. 70.)

Day-to-day variations of retention time for individual cobalamins were less than 10 sec.

A special application of liquid chromatographic separation of cobalamins was presented by Butte et al. [100], implying the quantitative separation of CN-Cbl as a tool in the estimation of the glomerular filtration rate. Cyanocobalamin (2 ng) is injected intravenously to saturate

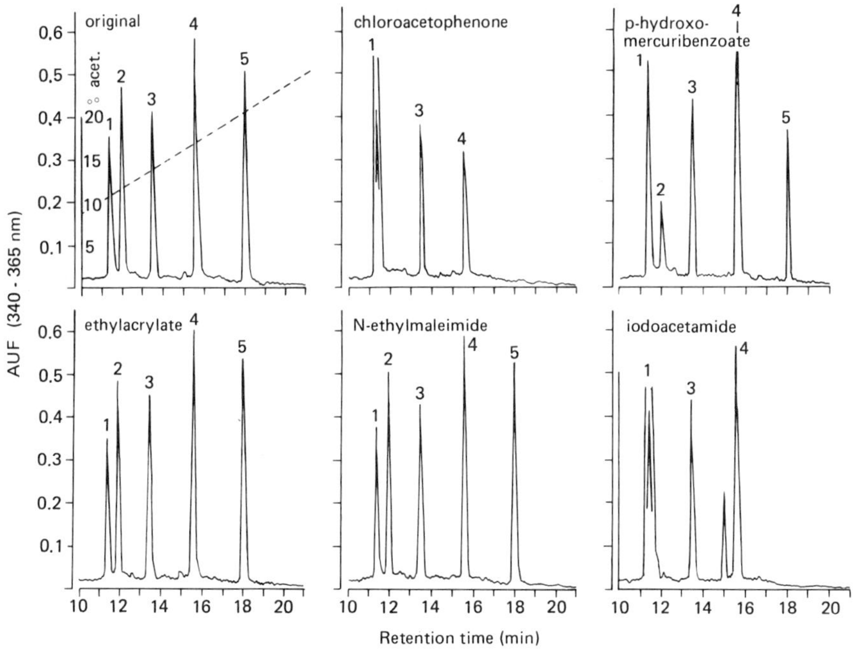

Figure 12.16 HPLC separation of a mixture of reference cobalamins (1.4 nmol each) on Lichrosorb RP8 with an 8-25% acetonitrile gradient in 85 mmol/liter phosphate-triethanolamine buffer, pH 3.3, after extraction from plasma without (original) or with thiol-blocking agents (5 mmol/liter each): OH-Cbl (1), SO_3-Cbl (2), CN-Cbl (3), AdoCbl (4), and MeCbl (5). (From Ref. 70.)

binding sites in blood and cells. Cyanocobalamin (5 ng) is administered intravenously 8-14 hr later, and blood samples are taken at time intervals of up to 90 min. The plasma is extracted with TCA, and bromecresyl green is added as an internal standard. Desalting is carried out on Sep-pak C18. After elution from the Sep-pak with methanol, a sample is injected on a Lichrosorb Si60 column and eluted with methanol-sodium cyanide, 0.1 g/liter (40:60) at a flow rate of 0.8 ml/min. The relative peak area is taken as a measure for the amount of cobalamin in the sample in comparison with a series of standard CN-Cbl concentrations

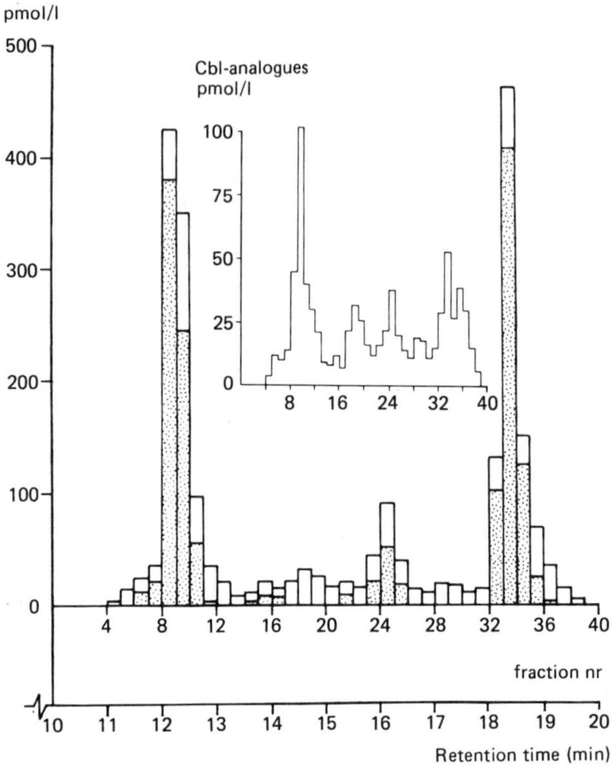

Figure 12.17 Distribution pattern of cobalamins and cobalamin analogs (insert) in normal human plasma, measured in the HPLC eluate by radioisotope dilution assay using R-binding (▭) or "R-binder blocked" intrinsic factor (▨): OH/SO_3-Cbl (fractions 7-12), CN-Cbl (fractions 15-17), AdoCbl (fractions 22-26), and MeCbl (fractions 33-36); fraction volume, 0.5 ml. (From Ref. 70.)

in plasma. The analytical recovery of the procedure is about 65%. Precision varies from 4% at 500 μg/liter to 7% at 200 μg/liter cobalamin in plasma.

5. DISTRIBUTION OF COBALAMINS

5.1 Food

The distribution of cobalamins in food has been extensively investigated by Farquharson and Adams [59], who presented a list of semiquantitative data on the relative amounts of separate cobalamin forms in 24 items of animal food products prepared in various ways. All five cobalamins, OH-, CN-, SO_3-, Ado-, and MeCbl, were present, but AdoCbl and OH-Cbl were the predominant forms. The presence of SO_3-Cbl, almost exclusively in canned foods, was regarded as an artefact. It is remarkable that, whereas all foods had been exposed to light, considerable amounts of Ado- and MeCbl were still recovered.

5.2 Tissues

The distribution of cobalamins in the tissues of normal individuals, pernicious anemia patients, and patients with various congenital disorders of cobalamin metabolism have been investigated by Linnell et al. [54] and Dillon et al. [101]. The distribution in human biopsy and autopsy specimens of various organs is shown in Table 12.9. Although Ado-Cbl is the most abundant form in all tissues, the proportion of MeCbl shows the largest variation. Cyanocobalamin was barely detectable in solid tissues, but higher values were found in blood cells. In pernicious anemia the tissue cobalamin distribution remains normal, but the cobalamin pattern in leukocytes and bone marrow is characterized by a raised CN-Cbl fraction and a decrease of Ado- and MeCbl.

The tissue distribution of cobalamins in rats, cats, and guinea pigs has been investigated by Quadros et al. [102] using the chromatobioautographic technique. In the rat the total cobalamin levels in most tissues are similar to those in the human, but the relative amount of MeCbl is much lower; AdoCbl predominates in most tissues. Moreover, unidentified corrinoids were detected in 50% of the rat tissues and 60% of guinea pig tissues. In erythrocytes of all three animal species the

TABLE 12.9 Tissue Cobalamins in Normal Human Subjects

Tissue	Total cobalamin (nmol/g)	OH-Cbl (%)	CN-Cbl (%)	AdoCbl (%)	MeCbl (%)
Liver	615 ± 122	44	2.0	45	9.0
Kidney	169 ± 30	28	0.6	53	18
Spleen	23.6 ± 2.8	24	0.8	50	26
Brain	38.3 ± 7.4	24	0.5	63	13
Erythrocytes	0.15 ± 0.02	26	6	53	15
Leukocytes	2.90 ± 0.32	28	4	48	20
Bone marrow	9.56 ± 1.06	30	2	55	13

Source: Data from Ref. 54 and 101.

levels of MeCbl and OH-Cbl were higher than in plasma, whereas in humans the reverse situation is found.

Of special interest is the effect of specific dietary measures on the cobalamin distribution in plasma and tissues. Gimsing et al. [64] demonstrated that, irrespective of the cobalamin form, OH- or CN-Cbl, administered parenterally to patients with vitamin B_{12} deficiency, the distribution of cobalamin in the tissues remains unchanged. Recently, Linnell et al. [103] investigated the effects of dietary deficiencies of methionine, choline, and/or cyanocobalamin on the tissue levels of total cobalamins and methylcobalamin in the rat. Cobalamin deprivation always led to lower tissue cobalamin levels, without affecting the cobalamin distribution. Choline deprivation caused a relative decrease of methylcobalamin in the liver; methionine deprivation exerted the opposite effect.

5.3 Plasma

Differences in extraction, separation, and detection techniques for the analysis of plasma cobalamins have led to different figures for the substance concentrations of individual cobalamins. The data available at present are summarized in Table 12.10. It is remarkable that, whereas Gimsing et al. [57] claim to have improved the extraction efficiency of OH-Cbl by a factor of 2-3, due to the effect of CdAc, their data on the concentration of OH-Cbl do not differ significantly from those of Linnell et al. [76] and of Van Kapel et al. [70] obtained without *N*-ethylmaleimide. The use of the thiol-blocking agent *N*-ethylmaleimide during extraction appeared to cause an increase in the concentration of OH-Cbl at the expense of MeCbl in particular, and recovery studies with plasma enriched with the individual cobalamin forms are indicative that the figures obtained by Van Kapel et al. [70] are correct. A representative distribution of plasma cobalamins thus analyzed by HPLC and quantitated with a radioisotope dilution assay is shown in Fig. 12.17. By utilizing the special binding affinities of intrinsic factor and R binder in the radioassay we were able to determine the distribution of plasma cobalamin analogs over the chromatogram. Obviously, the analogs represent a collection of derivatives scattered over the peaks of the established cobalamin forms.

As a consequence of the findings of Van Kapel et al. [70], the data on cobalamin distributions in plasma, and presumably in tissues as well, published previously [76], will need correction with respect to the contribution of OH-Cbl. This applies also to the studies of Nexø et al. [92,104], which indicate that in human plasma MeCbl accounts for most of the cobalamins attached to transcobalamin I, whereas the predominant form on transcobalamin II is AdoCbl. This observation is of interest in the comparison between humans and animals with respect to plasma transcobalamin and cobalamin distribution: only in humans is a considerable

TABLE 12.10 Distribution of Cobalamins in Human Plasma (Range in Parentheses)

Cobalamin	Linnell et al. [76][a]	Gimsing et al. [57,63][b]		Van Kapel et al. [70][c]	
		-CdAc	+CdAc	-NEM	+NEM
OH-Cbl (+SO_3-Cbl) (pmol/l)	47 (35-145)	(0-30)	38 (10-104)	47 (0-95)	161 (45-280)
CN-Cbl (pmol/l)	2.4 (0-40)	(0-70)	13 (2-48)	35 (0-90)	15 (0-30)
AdoCbl (pmol/l)	60 (22-185)	(6-90)	20 (2-77)	55 (25-85)	32 (10-52)
MeCbl (pmol/l)	276 (117-458)	(212-512)	250 (135-427)	252 (80-428)	182 (80-282)

[a]Extraction with hot ethanol and phenol, TLC, and bioautography.
[b]Extraction with hot ethanol and XAD-2, TLC, and bioautography.
[c]Extraction with 1% acetic acid and silanized silica 60 H, HPLC, and radioisotope dilution assay.

amount of MeCbl present in plasma, and only in humans is the major part of plasma cobalamins bound to transcobalamin I. In most animals transcobalamin I does not occur in plasma and the MeCbl is present in less than 10% of the total cobalamin [105].

6. CONCLUDING REMARKS

In the last decade the chromatography of cobalamins has been following the trend toward more sophisticated technology. The use of high-pressure liquid chromatography now allows for a fast separation on the same columns of a few femtomoles to several micromoles of cobalamins with a high degree of resolution. There seems to be no area of analytical corrinoid chromatography in which HPLC is not becoming the method of choice. However, for the laboratories that do not have access to HPLC equipment, other forms of chromatography, summarized in this chapter, are still valuable.

The improvement in chromatographic separation methods is putting higher demands upon sample preparation and detection methods as well. It has become evident that the plasma extraction precedures used until recently have been unsatisfactory with respect to the recoveries of OH-Cbl and SO_3-Cbl by causing overestimation of the other cobalamin forms, MeCbl in particular. It remains to be established whether the application of an improved HPLC separation method from the authors' laboratory will also lead to adjustments of the data available at present on cobalamin distribution in food, tissue, and cell extracts.

The identification and quantification of cobalamin fractions depends on the selectivity and the sensitivity of the detection methods. For example, in bioautography, in which the use of *E. coli* is universal, it may not be possible to discern bacterial growth stimulated by certain cobalamin analogs from growth stimulated by known cobalamins. This differentiation has recently become possible in a radioisotope dilution assay that takes advantage of the affinity of intrinsic factor for only biologically active cobalamins and also of R binders for inactive analogs. Equal affinity of all cobalamin forms for the binding protein has then to be induced by prior conversion to the CN form. The actual presence and identity of cobalamin analogs and their significance in biological systems are still largely unknown, and this will be a stimulus for future chromatographic analysis.

With the various chromatographic techniques, biosynthetic pathways of the corrinoids, isolation of synthetic and natural analogs, conversion of oxidized cobalamin into reduced coenzyme forms in biological systems, and distribution of cobalamins in tissues and body fluids have been studied with remarkable success. In the medical area the advent of better equipment and improved methods have opened the way to more thorough evaluation of the distribution of cobalamins in plasma and tissues in various disorders, e.g., neoplasms, and the effects of chemotherapy on cobalamin metabolism. In this regard the cytostatic effect of nitrous oxide through its direct influence on cellular cobalamin metabolism seems to be a promising area of research.

REFERENCES

1. G. H. Whipple, F. S. Robscheit, and C. N. Hooper, *Amer. J. Physiol.*, *53*, 236 (1920).
2. G. R. Minot and W. P. Murphy, *J. Amer. Med. Ass.*, *87*, 470 (1926).
3. E. L. Rickes, N. G. Brink, F. R. Koiuszy, T. R. Wood, and F. Folkers, *Science*, *107*, 396 (1948).
4. E. L. Smith, *Nature 161*, 638 (1948).
5. R. Bonnett, J. R. Cannon, V. M. Clark, A. W. Johnson, L. F. J. Parker, E. L. Smith, and A. R. Todd, *J. Chem. Soc.*, *1*, 1158 (1957).
6. D. C. Hodgkin, J. Kamper, M. Mackay, J. W. Pickworth, K. N. Trueblood, and J. G. White, *Nature*, *178*, 64 (1956).
7. H. A. Barker, H. Weissbach, and R. D. Smyth, *Proc. Nat. Acad. Sci. U.S.*, *44*, 1093 (1958).
8. H. Weissbach, J. I. Toohey, and H. A. Barker, *Proc. Nat. Acad. Sci. U.S.*, *45*, 521 (1959).
9. J. I. Toohey and H. A. Barker, *J. Biol. Chem.*, *236*, 560 (1961).
10. K. Lindstrand, and K.-G. Ståhlberg, *Acta Med. Scand.*, *174*, 665 (1963).
11. K.-G. Ståhlberg, *Scand. J. Haematol.*, *1*, 220 (1964).
12. IUPAC-IUB Commission on Biochemical Nomenclature, *Biochemistry*, *13*, 1555 (1974).
13. IUPAC-IUB Commission on Biochemical Nomenclature, *Biochem. J.*, *147*, 1 (1975).
14. R. Bonnett, J. R. Cannon, A. W. Johnson, and A. R. Todd, *J. Chem. Soc.*, 1148 (1957).
15. M. J. Mahoney,and L. E. Rosenberg, *J. Lab. Clin. Med.*, *78*, 302. (1971).
16. R. A. Firth, H. A. O. Hill, J. M. Pratt, R. G. Thorp, and R. J. P. Williams, *J. Chem. Soc.*, *A*, 381 (1969).
17. J. B. Armitage, J. R. Cannon, A. W. Johnson, L. F. J. Parker, E. L. Smith, W. H. Stafford and A. R. Todd, *J. Chem. Soc.*, 3849 (1953).
18. W. Friedrich and K. Bernhauer, *Z. Naturforsch.*, *9b*, 685 (1954).
19. D. Dolphin, *Methods Enzymol.*, *18*, 34 (1971).
20. J. L. Peel, *Biochem. J.*, *88*, 296 (1963).
21. G. N. Schrauzer and J. W. Sibert, *Arch. Biochem. Biophys.*, *130*, 257 (1969).
22. R. O. Brady and H. A. Barker, *Biochem. Biophys. Res. Commun.*, *4*, 373 (1961).
23. H. P. C. Hogenkamp, *J. Biol. Chem.*, *238*, 477 (1963).

24. H. P. C. Hogenkamp, J. N. Ladd, and H. A. Barker, *J. Biol. Chem.*, *237*, 1950 (1962).
25. J. N. Pratt, *J. Chem. Soc.*, 5154 (1964).
26. H. P. C. Hogenkamp, *Biochemistry*, 5, 417 (1966).
27. A. W. Johnson, N. Shaw, and F. Wagner, *Biochim. Biophys. Acta*, *72*, 107 (1963).
28. H. P. C. Hogenkamp, in *Cobalamin* (B. M. Babior, ed.), Wiley Interscience, New York, 1975, p. 21.
29. J. M. Wood, F. S. Kennedy, and C. G. Rosen, *Nature*, *220*, 173 (1968).
30. K. Bernhauer, P. Renz, and F. Wagner, *Biochem. Z.*, *335*, 443 (1962).
31. D. Dolphin, A. W. Johnson, and N. Shaw, *Nature*, *199*, 170 (1963).
32. J. A. Hill, J. M. Pratt, and R. J. P. Williams, *J. Chem. Soc.*, 5149 (1964).
33. H. C. Friedman, in *Cobalamin* (B. M. Babior, ed.), Wiley Interscience, New York, 1975, p. 75.
34. H. C. Friedman, in *Vitamin* B_{12} (B. Zagalak and W. Friedrich, eds.), Walter De Gruyter, Berlin, 1979, p. 331.
35. A. R. Battersby and E. McDonald, in B_{12} Vol. 1 (D. Dolphin, ed.), Wiley Interscience, New York, 1982, p. 107.
36. R. O. Brady, E. G. Castanera, and H. A. Barker, *J. Biol. Chem.*, *237*, 2325 (1962).
37. G. A. Walker, S. Murphy, and F. M. Huennekens, *Arch. Biochem. Biophys.*, *134*, 95 (1969).
38. J. Pawelkiewicz, M. Gorna, W. Fenrych, and S. Magar, *Ann. N.Y. Acad. Sci.*, *112*, 641 (1964).
39. G. S. Kerwar, C. Spears, B. McAnslan, and H. Weissbach, *Arch. Biochem. Biophys.*, *142*, 231 (1971).
40. R. T. Taylor in B_{12} Vol. 2 (D. Dolphin, ed.) Wiley Interscience, New York, 1982, p. 307.
41. B. M. Babior, in *Cobalamin* (B. M. Babior, ed.), Wiley Interscience, New York, 1975, p. 145.
42. G. Morrow, L. A. Barness, G. J. Cardinale, R. H. Abeles, and J. G. Flaks, *Proc. Natl. Acad. Sci. U.S.*, *63*, 191 (1969).
43. J. M. Poston and T. C. Stadtman, in *Cobalamin* (B. M. Babior, ed.), Wiley Interscience, New York, 1975, p. 111.
44. L. G. J. Ljungdahl and H. G. Wood, in B_{12}, Vol. 2 (D. Dolphin, ed.), Wiley Interscience, New York, 1982, p. 165.
45. J. Pawelkiewicz, in *Vitamin B12 und Intrinsic Factor* (H. C. Heinrich, ed.), F. Enke Verlag, Stuttgart, 1962, p. 280.
46. W. Friedrich, *Vitamin B12 und verwandte corrinoide*, Georg Thieme Verlag, Stuttgart, 1975.
47. G. Matsuda, *J. Vitaminol.*, *1*, 221 (1955).
48. H. Kubo, S. Fujita, and H. Ogino, *Chem. Abstr.*, *54*, 13153 (1960).

49. W. Friedrich and K. Bernhauer, *Z. Naturforsch.*, *9b*, 755 (1954).
50. M. W. Penley, D. G. Brown, and J. M. Wood, *Biochemistry*, *22*, 4302 (1970).
51. V. B. Koppenhagen, B. Elsenhaus, F. Wagner, and J. J. Pfiffner, *J. Biol. Chem.*, *249*, 6532 (1974).
52. E. V. Quadros, A. Hamilton, D. M. Matthews, and J. C. Linnell, *J. Chromatogr.*, *160*, 101 (1978).
53. S. H. Ford and H. C. Friedman, *Arch. Biochem. Biophys.*, *175*, 121 (1976).
54. J. C. Linnell, A. V. Hoffbrand, H. A.-A. Hussein, I. J. Wise, and D. M. Matthews, *Clin. Sci. Mol. Med.*, *46*, 163 (1974).
55. W. A. Fenton and L. E. Rosenberg, *Anal. Biochem.*, *90*, 119 (1978).
56. E. P. Frenkel, R. Prough, and R. L. Kitchens, *Methods Enzymol.*, *67*, 31 (1980).
57. P. Gimsing, E. Nexø, and E. Hippe, *Anal. Biochem.*, *129*, 296 (1983).
58. D. W. Jacobsen, R. Green, E. V. Quadros, and Y. D. Montejano, *Anal. Biochem.*, *120*, 394 (1982).
59. J. Farquharson and J. F. Adams, *Brit. J. Nutr.*, *36*, 127 (1976).
60. H. Kondo, J. F. Kolhouse, and R. H. Allen, *Proc. Natl. Acad. Sci. U.S.*, *77*, 817 (1980).
61. H. Vogelmann and F. Wagner, *J. Chromatog.*, *76*, 359 (1973).
62. J. C. Linnell, H. M. Mackenzie, J. Wilson and D. M. Matthews, *J. Clin. Pathol.*, *22*, 545 (1969).
63. P. Gimsing, E. Hippe, and E. Nexø, in *Vitamin B12* (W. Friedrich and B. Zagalak, eds.), Walter de Gruyter, Berlin, 1979, p. 665.
64. P. Gimsing, E. Hippe, I. Helleberg-Rasmusen, M. Moergaard, J. Lanng Nielson, P. Bastrup-Madsen, R. Berlin, and T. Hansen, *Scand. J. Haematol.*, *29*, 311 (1982).
65. M. C. Mahoney and L. E. Rosenberg, *Am. J. Med.*, *48*, 584 (1970).
66. P. Gimsing, E. Nexø, and E. Hippe, *Anal. Biochem.*, *129*, 288 (1983).
67. W. R. Bauriedel, J. C. Picken, Jr., and L. A. Underkofler, *Proc. Soc. Exp. Biol. Med.*, *91*, 377 (1956).
68. R. T. Taylor and M. L. Hanna, *Arch. Biochem. Biophys.*, *141*, 247 (1970).
69. E. L. Lien and J. M. Wood, *Biochim. Biophys. Acta*, *264*, 530 (1972).
70. J. Van Kapel, L. J. M. Spijkers, J. Lindemans, and J. Abels, *Clin. Chim. Acta*, *131*, 211 (1983).
71. G. Markus and F. Karush, *J. Amer. Chem. Soc.*, *79*, 134 (1957).
72. A. Lorber, C. C. Chang, D. Masuoka, and I. Meacham, *Biochem. Pharmacol.*, *19*, 1551 (1970).
73. R. B. Martin and J. T. Edsall, *J. Amer. Chem. Soc.*, *81*, 4044 (1959).

74. J. Sedlak and R. H. Lindsay, *Anal. Biochem.*, *25*, 192 (1968).
75. C. F. Brewer and J. P. Riehm, *Anal. Biochem.*, *18*, 248 (1967).
76. J. C. Linnell, A. V. Hoffbrand, T. J. Peters, and D. M. Matthews, *Clin. Sci.*, *40*, 1 (1971).
77. J. F. Kolhouse, H. Kondo, N. C. Allen, E. Podell, and R. H. Allen, *N. Engl. J. Med.*, *299*, 785 (1978).
78. G. Tortolani, P. Bianchini, and V. Mantovani, *Farm. Ed. Prat.*, *25*, 772 (1970).
79. G. Tortolani, P. Bianchini, and V. Mantovani, *J. Chromatogr.*, *53*, 577 (1970).
80. R. A. Gams, E. M. Ryel, and L. M. Meyer, *Proc. Soc. Exp. Biol. Med.*, *149*, 384 (1975).
81. R. A. Gams, E. M. Ryel, and F. Ostroy, *Blood*, *47*, 923 (1976).
82. P. Gimsing and E. Hippe, *Scand. J. Haematol.*, *24*, 243 (1978).
83. J. A. Begley and C. A. Hall, *J. Chromatogr.*, *177*, 360 (1979).
84. S. L. Lee, *Methods Enzymol.*, *67*, 3 (1980).
85. R. H. Allen and P. W. Majerus, *J. Biol. Chem.*, *247*, 7695 (1972).
86. M. Katada, S. Tyagi, A. Nath, R. L. Petersen, and R. K. Gupta, *Biochim. Biophys. Acta.*, *584*, 149 (1979).
87. V. B. Koppenhagen, F. Wagner, and J. J. Pfiffner, *J. Biol. Chem.*, *248*, 7999 (1973).
88. J. F. Kolhouse and R. H. Allen, *J. Clin. Invest.*, *60*, 1381 (1977).
89. R. A. Firth, H. A. O. Hill, J. M. Pratt, and R. G. Thorp., *Anal. Biochem.*, *23*, 429 (1968).
90. T. M. Vickrey, R. N. Katz, and G. N. Schrauzer, *J. Amer. Chem. Soc.*, *97*, 7248 (1975).
91. J. C. Linnell, H. A.-A. Hussein, and D. M. Matthews, *J. Clin. Pathol.*, *23*, 820 (1970).
92. E. Nexø and J. Andersen, *Scand. J. Clin. Lab. Invest.*, *37*, 723 (1977).
93. J. A. Schmit, R. A. Henry, R. C. Williams, and J. F. Dieckman, *J. Chromatogr. Sci.*, *9*, 645 (1971).
94. R. B. H. Wills, C. G. Shaw, and W. R. Day, *J. Chromatogr. Sci.*, *15*, 262 (1977).
95. F. Pellerin, J.-F. Letavernier, and N. Chanon, *Ann. Pharm. Fr.*, *9-10*, 413 (1977).
96. D. Mourot, B. Delépine, J. Boisseau, and G. Gayot, *Ann. Pharm. Fr.*, *5-6*, 235 (1979).
97. E. P. Frenkel, R. L. Kitchens, and R. Prough, *J. Chromatogr.*, *174*, 393 (1979).
98. D. W. Jacobsen and R. Green in *Vitamin B12* (W. Friedrich and B. Zagalak, eds.), Walter de Gruyter, Berlin, 1979, p. 663.
99. M. Binder, J. F. Kolhouse, K. C. Van Horne, and R. H. Allen, *Anal. Biochem.*, *125*, 253 (1982).

100. W. Butte, H.-H. Riemann, and A. J. Walle, *Clin. Chem.*, *28*, 1778 (1982).
101. M. J. Dillon, J. M. England, D. Gompertz, P. A. Goodey, D. B. Grant, H. A.-A. Hussein, J. C. Linnell, D. M. Matthews, S. H. Mudd, G. H. Newns, J. W. T. Seakins, B. W. Uhlendorf, and I. J. Wisse, *Clin. Sci. Mol. Med.*, *47*, 43 (1974).
102. E. V. Quadros, D. M. Matthews, I. J. Wisse, and J. C. Linnell, *Biochim. Biophys. Acta.*, *421*, 141 (1976).
103. J. C. Linnell, M. J. Wilson, Y. B. Mikol, and L. A. Poirier, *J. Nutr.*, *113*, 124 (1983).
104. E. Nexø, *Scand. J. Haematol.*, *18*, 358 (1977).
105. J. C. Linnell, L. Collings, M. C. Down, and J. M. England, *Clin. Sci.*, *57*, 139 (1979).

INDEX

C

D

G

H

I

Q

R

U

V